MW00835430

HPLC Made to Measure

Edited by
Stavros Kromidas

Related Titles

S. Kromidas

Practical Problem Solving in HPLC

2000
ISBN 3-527-29842-8

S. Kromidas

More Practical Problem Solving in HPLC

2004
ISBN 3-527-31113-0

V. R. Meyer

Practical High-Performance Liquid Chromatography

2004
ISBN 0-470-09378-1

P. C. Sadek

Troubleshooting HPLC Systems

A Bench Manual

2000
ISBN 0-471-17834-9

U. D. Neue

HPLC Columns

Theory, Technology, and Practice

1997
ISBN 0-471-19037-3

L. R. Snyder, J. J. Kirkland, J. L. Glajch

Practical HPLC Method Development

1997
ISBN 0-471-00703-X

HPLC Made to Measure

A Practical Handbook for Optimization

Edited by
Stavros Kromidas

WILEY-VCH

WILEY-VCH Verlag GmbH & Co. KGaA

The Editor

Dr. Stavros Kromidas
Rosenstrasse 16
66125 Saarbrücken
Germany

Library of Congress Card No.: applied for

British Library Cataloguing-in-Publication Data
A catalogue record for this book is available from the British Library

Bibliographic information published by Die Deutsche Bibliothek
Die Deutsche Bibliothek lists this publication in the Deutsche Nationalbibliografie; detailed bibliographic data is available in the Internet at <http://dnb.ddb.de>

Printed in the Federal Republic of Germany
Printed on acid-free paper

Cover Design   SCHULZ Grafik-Design, Fußgönheim
Typesetting   Manuela Treindl, Laaber
Printing   betz-druck GmbH, Darmstadt
Binding   J. Schäffer GmbH, Grünstadt

ISBN-13: 978-3-527-31377-8
ISBN-10: 3-527-31377-X

Foreword

HPLC has become the analytical method against which all others are measured and compared. It is perhaps the most widely employed method of analysis of all those instrumental approaches that have ever been or are now in vogue. Having been involved with HPLC for perhaps the past 35 years, since the early 1970s, I have seen the technique and field grow and prosper, academically and commercially. It has become an incredible commercial success, and the cornerstone of many academic careers in analytical and other fields of chemistry. Annual, dedicated meetings, as well as major parts of ACS, ASMS, AAPS and AAAS meetings, are routinely devoted to talks and discussions on or involving HPLC. Though it has not quite displaced GC or flat-bed electrophoresis, it has surely been highly competitive for volatiles and biological macromolecules, respectively. Indeed, one could argue that it is the very first technique that most analysts, biologists or biochemists would consider investigating and applying for virtually any class of analytes, regardless of molecular weight, size, volatility, ionic charges, polarity, hydrophobicity, or other physical or chemical properties. HPLC has become a technique that can be applied to virtually any analyte or class of analytes, almost without regard to the properties thereof. There are very few other analytical methods for which this can be claimed. HPLC has truly become the "800 pound gorilla", and it may be virtually impossible for any other technique to displace it from this niche in the analytical world, not even CEC or 2DE or multidimensional CE.

Why then another book dealing with this same topic? I have read some other texts by *Stavros Kromidas,* and was thus eager to preview this current one. This text is really an edited book, though *Stavros Kromidas* has contributed several excellent chapters of his own. The other contributions come from an international group of invited authors, mainly from the US, Canada, and Western Europe. Virtually all of these individuals are well known in the HPLC community, such as *Uwe Neue, Michael McBrien, Lloyd Snyder, John Dolan, Klaus Unger,* and so forth. Most, if not all, have been heavily involved in HPLC matters for decades, and have, in their own right, become well-regarded and recognized experts in their various fields. The book is heavily practice and practicality oriented, in that it aims to help the readers become more knowledgeable and better adept at using various forms of and approaches in HPLC. However, it is not a "Methods"-type text, such as those published by Humana Press; it is not just a compilation of practice-oriented HPLC methods for various analytes.

HPLC Made to Measure: A Practical Handbook for Optimization. Edited by Stavros Kromidas
Copyright © 2006 WILEY-VCH Verlag GmbH & Co. KGaA, Weinheim
ISBN: 3-527-31377-X

Rather, the text is quite detailed, with scientific discussions and theory, lots of equations and principles, reference to a variety of practical software, and with an emphasis on understanding the fundamentals in each and every chapter. This is not an introductory text; it is not meant as a text for a graduate Analytical Separations type course. Rather, it is quite an advanced text, dealing with many recent and contemporary aspects of HPLC. It deals with approaches for method optimization, currently available software and practices, chemometrics, principal component analysis, the selection of ideal stationary phases, and tools for column characterization and method optimization. Of course, it deals extensively with reversed-phase HPLC, but it also covers many other areas, including GPC/SEC, affinity chromatography, chiral separations, microLC, nanoLC, and even micro-chip-based LC instrumentation and techniques. It also deals with immuno-chromatographic methods, two-dimensional HPLC (MDLC), LC-MS, LC-NMR, and even how magic-angle spinning NMR spectroscopy can be used to better understand the selectivity of stationary phases in HPLC. All in all, there are over two-dozen individual chapters, some authored by the same author(s), but most not. It is to the Editor's credit that he has not written most or even close to 50% of the total chapters, but rather that he has invited the most highly regarded and best-known authors, young and old, to contribute in areas of their unique expertise. He has made an exceptionally good selection of such authors, each of whom has done an admirable job in their final writings and efforts.

This is not a book that you will pick up and read in a single sitting; that would appear impossible, even for those of us who have already devoted a major portion of our careers to researching and developing HPLC areas. It is not an easy read; it is not a trivial text. Rather, it is clearly an advanced, involved, and detailed text. It is a book to be read slowly and carefully, because it contains an incredible amount of useful and important practical knowledge. It also covers the very latest developments in HPLC, not just the fundamentals, but where the field stands today, and where it is going tomorrow. It is a practical handbook for the optimization of HPLC and its ultimate application, but it is far from being just a handbook or "how to do it" text. It is really more of a summary of where HPLC stands today, what can be done with its various techniques and instrumentation, and what is important to know about its future developments and applications. It is an incredibly useful and practical tome, collated by experts pooling their expertise, and it will make better chromatographers out of those of us who take up the book, study it carefully, and then apply its lessons to our own future needs. It is not a simplistic methods development type book, though it does aim to help us optimize and improve our methods development approaches. It is far more than "just" a Practical Handbook for Optimization, though the subtitle might make that suggestion.

It is my hope that those of you thinking of purchasing this particular, newer text on HPLC, and those who have already made this wise decision and are about to pursue the text itself, will benefit from these choices. They were and are wise choices; now it is up to you to make the most of the book, which means not just reading the text and studying the figures and tables, but making every effort possible to really understand what the authors are trying to impart to the readers. This may require re-reading of the same chapter more than once; I did – actually several times, as these are not easy chapters or contributions. However, in the long run, the time will be well spent and such efforts will be rewarded, for the book is truly a wealth of useful and practical information. Obviously, I highly recommend the book to those contemplating purchase and study, for it is really one of the better texts to have come along in many years dealing with this, one of our very favorite subjects, HPLC.

January 2006

Ira S. Krull
Associate Professor
Department of Chemistry and Chemical Biology
Northeastern University
Boston, MA, USA

Preface

The optimizing of practices and processes constitutes an essential prerequisite for long-term success. The objective and the motives may be very different: self-preservation among living things, "saving lives" among volunteers in Africa, maximizing profits among marketing strategists, new discoveries among scientists. This principle is of course also valid in chemistry and in analytics.

This book deals exclusively with the subject of optimization in HPLC. The aim is to examine this important aspect of HPLC from diverse perspectives. First, we have set out the fundamental aspects, encompassing the principal considerations and background information. At the same time, we have endeavored to present and discuss as many practical examples, ideas, and suggestions as possible for the everyday application of HPLC. The implementation of concepts for rapid optimization should equally aid and support the planning of effective method development strategies as in daily practice at the laboratory bench. The aim of the book is to contribute to purposeful, affordable, forward-looking method development and optimization in HPLC.

To this end, internationally renowned experts have offered their knowledge and experience. I extend my sincere thanks to these colleagues. I also thank Wiley-VCH, in particular Steffen Pauly, for their valued collaboration and good cooperation.

Saarbrücken, January 2006   *Stavros Kromidas*

HPLC Made to Measure: A Practical Handbook for Optimization. Edited by Stavros Kromidas
Copyright © 2006 WILEY-VCH Verlag GmbH & Co. KGaA, Weinheim
ISBN: 3-527-31377-X

Contents

HPLC Made to Measure: A Practical Handbook for Optimization. Edited by Stavros Kromidas
Copyright © 2006 WILEY-VCH Verlag GmbH & Co. KGaA, Weinheim
ISBN: 3-527-31377-X

List of Contributors

Klaus Albert
Institute of Organic Chemistry
University of Tübingen
Auf der Morgenstelle 18
72076 Tübingen
Germany

Bonnie A. Alden
Waters Corporation, CRD
34 Maple Street
Milford, MA 01757
USA

Mario Arangio
CarboGen AG
Schachenallee 29
5001 Aarau
Switzerland

Wolf-Dieter Beinert
VWR International GmbH
Scientific Instruments
Hilpertstrasse 20A
64295 Darmstadt
Germany

Roberto Biancardi
Solvay Solexis SpA
Viale Lombardia, 20
20021 Bollate (MI)
Italy

Hans Bilke
Sandoz GmbH
Biochemiestrasse 10
6250 Kundl
Austria

Yung-Fong Cheng
Cubist Pharmaceuticals
65 Hayden Ave.
Lexington, MA 02421
USA

Maristella Colombo
Oncology – Analytical Chemistry
Nerviano Medical Sciences
Via le Pasteur, 10
20014 Nerviano (MI)
Italy

Diane M. Diehl
Waters Corporation, CAT
34 Maple Street
Milford, MA 01757
USA

John W. Dolan
BASi Northwest Laboratory
3138 NE Rivergate
Building 301C
McMinnville, OR 97128
USA

HPLC Made to Measure: A Practical Handbook for Optimization. Edited by Stavros Kromidas
Copyright © 2006 WILEY-VCH Verlag GmbH & Co. KGaA, Weinheim
ISBN: 3-527-31377-X

Melvin R. Euerby
AstraZeneca R&D Charnwood
Analytical Development
Pharmaceutical and Analytical R&D
Charnwood/Lund
Bakewell Road
Loughborough
Leicestershire LE11 5RH
United Kingdom

Sergey Galushko
Dr. S. Galushko Software
Development
Im Wiesengrund 49b
64367 Mühltal
Germany

Eric S. Grumbach
Waters Corporation, CAT
34 Maple Street
Milford, MA 01757
USA

Marc D. Grynbaum
Institute of Organic Chemistry
University of Tübingen
Auf der Morgenstelle 18
72076 Tübingen
Germany

Heidi Händel
Institute of Organic Chemistry
University of Tübingen
Auf der Morgenstelle 18
72076 Tübingen
Germany

Tom Hennessy
Biopolis
Biomedical Science Group
20 Biopolis Way
Singapore 1 38668
Singapore

Pamela C. Iraneta
Waters Corporation, CRD
34 Maple Street
Milford, MA 01757
USA

Markus Juza
Siegfried Ltd.
Untere Brühlstrasse 4
4800 Zofingen
Switzerland

Marianna Kele
Waters Corporation, CRD
34 Maple Street
Milford, MA 01757
USA

Peter Kilz
PSS Polymer Standard Service GmbH
POB 3368
55023 Mainz
Germany

Stavros Kromidas
Rosenstrasse 16
66125 Saarbrücken
Germany

Manfred Krucker
Institute of Organic Chemistry
University of Tübingen
Auf der Morgenstelle 18
72076 Tübingen
Germany

Hans-Joachim Kuss
Psychiatric Department (LMU)
University of Munich
Nussbaumstrasse 7
80336 Munich
Germany

Jörg P. Kutter
MIC – Department of Micro and
Nanotechnology
Technical University of Denmark
2800 Lyngby
Denmark

Christiane Lohaus
Medical Proteom-Center
Center for Clinical Research
Ruhr-University of Bochum
Universitätsstrasse 150
44780 Bochum
Germany

Ziling Lu
Waters Corporation, CAT
34 Maple Street
Milford, MA 01757
USA

Egidijus Machtejevas
Institute for Anorganic Chemistry
and Analytical Chemistry
Johannes-Gutenberg-University
Duesbergweg 10–14
55099 Mainz
Germany

Jürgen Maier-Rosenkranz
GRACE Davison – Alltech Grom
GmbH
Discovery Sciences
Etzwiesenstrasse 37
72108 Rottenburg-Hailfingen
Germany

Friedrich Mandel
Agilent Technologies
Hewlett-Packard-Strasse 8
76337 Waldbronn
Germany

Katia Marcucci
Sienabiotech SpA
Via Fiorentina, 1
53100 Siena
Italy

Katrin Marcus
Medical Proteom-Center
Center for Clinical Research
Ruhr-University of Bochum
Universitätsstrasse 150
44780 Bochum
Germany

Jeffrey R. Mazzeo
Waters Corporation, CAT
34 Maple Street
Milford, MA 01757
USA

Michael McBrien
Advanced Chemistry Development
Inc.
110 Yonge Street
Toronto, Ontario M5C 1T4
Canada

Alberto Méndez
Waters Cromatografia S.A.
Parc Tecnològic del Vallès
08290 Cerdanyola del Vallès
Barcelona
Spain

Helmut E. Meyer
Medical Proteom-Center
Center for Clinical Research
Ruhr-University of Bochum
Universitätsstrasse 150
44780 Bochum
Germany

Veronika R. Meyer
EMPA St. Gallen
Materials Science and Technology
Lerchenfeldstrasse 5
9014 St. Gallen
Switzerland

Egbert Müller
Tosoh Bioscience GmbH
Zettachring 6
70567 Stuttgart
Germany

Uwe D. Neue
Waters Corporation, CRD
34 Maple Street
Milford, MA 01757
USA

Patrik Petersson
AstraZeneca R&D Lund
Analytical Development
Pharmaceutical and Analytical R&D
Charnwood/Lund
22187 Lund
Sweden

Michael Pfeffer
Schering AG
In-Process-Control
13342 Berlin
Germany

Karsten Putzbach
Institute of Organic Chemistry
University of Tübingen
Auf der Morgenstelle 18
72076 Tübingen
Germany

Oleg Pylypchenko
Institute of Bioorganic Chemistry
of Ukrainian National Academy
of Sciences
Murmanskaja str., 1
02660 Kiev-94, MCP-600
Ukraine

Milena Quaglia
LGC
Analytical Technology
Queens Road
Teddington, Middlesex, TW11 OLY
United Kingdom

Giuseppe Razzano
Via D. Manin, 18
Magenta (Cap. 20013)
Milano
Italy

Vincenzo Rizzo
CISI – University of Milan
Via Fantoli, 16/15
20138 Milano
Italy

Heike Schäfer
Medical Proteom-Center
Center for Clinical Research
Ruhr-University of Bochum
Universitätsstrasse 150
44780 Bochum
Germany

Stefan Schömer
pro-isomehr
Altenkesseler Strasse 17
66115 Saarbrücken
Germany

Irina Shishkina
Institute of Bioorganic Chemistry of
Ukrainian National Academy of
Sciences
Murmanskaja str., 1
02660 Kiev-94, MCP-600
Ukraine

Dirk Sievers
Waters GmbH
Hauptstrasse 87
65760 Eschborn
Germany

Federico R. Sirtori
Nerviano Medical Sciences
Oncology – Analytical Chemistry
Viale Pasteur 10
20014 Nerviano (MI)
Italy

Urban Skogsberg
Cambrex Karlskoga AB
R&D Analysis
69185 Karlskoga
Sweden

Lloyd R. Snyder
LC Resources Inc.
26 Silverwood Ct.
Orinda, CA 94563
USA

Frank Steiner
Dionex Softron GmbH
Dornierstrasse 4
82110 Germering
Germany

Cinzia Stella
Imperial College
Biological Chemistry Department
Biomedical Sciences Division
Sir Alexander Fleming Building
London, SW7 2AZ
United Kingdom

Vsevolod Tanchuk
Institute of Bioorganic Chemistry of
Ukrainian National Academy of
Sciences
Murmanskaja str., 1
02660 Kiev-94, MCP-600
Ukraine

KimVan Tran
Waters Corporation, CAT
34 Maple Street
Milford, MA 01757
USA

Klaus K. Unger
Institute for Anorganic Chemistry
and Analytical Chemistry
Johannes-Gutenberg-University
Duesbergweg 10–14
55099 Mainz
Germany

Jean-Luc Veuthey
Faculty of Sciences
School of Pharmaceutical Sciences
University of Geneva
20, Bd d'Yvoy
1211 Genève 4
Switzerland

Knut Wagner
Pharma Analytical Laboratory
Merck KGaA
Frankfurter Strasse 250
64293 Darmstadt
Germany

Michael G. Weller
Bundesanstalt für Materialforschung
und -prüfung (BAM)
Unter den Eichen 37
12205 Berlin
Germany

Norbert Welsch
Institute of Organic Chemistry
University of Tübingen
Auf der Morgenstelle 18
72076 Tübingen
Germany

Loren Wrisley
Analytical and Quality Sciences
Wyeth Research
401 N. Middletown Road
Pearl River, NY 10965
USA

Structure of the Book

The book consists of five parts:

Part 1. Fundamentals of Optimization

The aim of Part 1 is to provide an overview of important aspects of optimization in HPLC from various viewpoints. In Chapter 1.1 (*Stavros Kromidas*), the principles of optimization are illustrated using RP-HPLC as an example, and recommendations for method development are made. Fast gradients on short columns lead more often than one might think to sufficient resolution in the shortest of analysis times, and this topic is discussed in Chapter 1.2 (*Uwe D. Neue*). For the separation of polar/ionic substances, pH is by far the most important factor in optimization procedures. The next two chapters (1.3 *Uwe D. Neue*, 1.4 *Michael McBrien*) are devoted to this aspect. Optimization means more than merely the "correct" choice of method parameters. Efforts to obtain as much information as possible, or at least the necessary information, are also a part of optimization. In this context, the evaluation of chromatographic data and calibration take on a special significance. These topics are dealt with in Chapter 1.5 (*Hans-Joachim Kuss*) and Chapter 1.6 (*Stefan Schömer*).

Part 2. Characteristics of Optimization in Individual HPLC Modes

In Part 2, specific aspects of optimization in individual techniques are considered. In RP chromatography (Section 2.1), besides the choice of eluents (for this, see also Chapters 1.1 to 1.4), above all the choice of column represents a difficult and time-consuming task. The subject of RP columns is covered by a total of six authors: two authors (2.1.1 *Stavros Kromidas*, 2.1.2 *Uwe D. Neue*) focus on the more practical aspects of this issue, while *Frank Steiner* (Chapter 2.1.5) and *Lloyd R. Snyder* (Chapter 2.1.6) present more fundamental, theoretical considerations, with nevertheless real practical relevance, on the questions of column characterization and column selection. Naturally, the meaningfulness of results increases with the number of experimental data, and so the handling of figures, and above all the identification and interpretation of correlations, is only possible with the aid of mathematical tools. Chemometrics is a suitable tool, for example, for establishing the similarity of columns on the basis of chromatographic data. The application of chemometrics from a practical viewpoint is briefly described in Chapter 2.1.1 (*Stavros Kromidas*) and extensively detailed in Chapters 2.1.3 (*Melvin R. Euerby*)

HPLC Made to Measure: A Practical Handbook for Optimization. Edited by Stavros Kromidas
Copyright © 2006 WILEY-VCH Verlag GmbH & Co. KGaA, Weinheim
ISBN: 3-527-31377-X

and 2.1.4 (*Cinzia Stella*). To conclude Section 2.1 on RP-HPLC, *Urban Skogsberg* (Chapter 2.1.7) shows how magic-angle spinning NMR spectroscopy can be used to obtain precise information on the interactions and arrangements of functional groups on RP surfaces. Thereafter, optimization, as well as troubleshooting and avoiding errors, are considered for the following fields of HPLC: normal phase (2.2 *Veronika R. Meyer*), GPC (2.3 *Peter Kilz*), gel filtration (2.4 *Klaus K. Unger*), affinity chromatography (2.5 *Egbert Müller*), and enantiomer separations (2.6 *Markus Juza*). Three different approaches have been selected to illustrate the topic of miniaturization: in Chapter 2.7.1 *Jürgen Maier-Rosenkranz* deals with micro/nano-LC, in Chapter 2.7.2 *Jörg P. Kutter* presents flow-chromatographic separations on chips, and in Chapter 2.7.3 *Uwe D. Neue* describes the possibilities and frontiers of a new variant of classical HPLC, namely UPLC. All three techniques offer remarkable time savings, although their limitations and difficulties are also mentioned.

Part 3. Coupling Techniques

Part 3 is exclusively devoted to coupling techniques. The more demanding the analytical problem to be addressed (complexity and number of sample components, greater chemical similarity of the analytes to be separated, etc.), the more necessary coupling techniques become. A coupling of different separation techniques can lead to an improved chromatographic resolution, as, for example, in immunochromatography (3.1 *Michael G. Weller*) and LC-GPC coupling (3.2 *Peter Kilz*). In other cases, at a given resolution, LC-spectroscopy couplings can yield specific information. The most popular coupling techniques are LC-MS (3.3 *Friedrich Mandel*) and LC-NMR (3.4 *Klaus Albert*).

Part 4. Computer-Aided Optimization

Automation can generally lead to the elimination of errors and to time saving. Meanwhile, fully automated computer-aided method development and semi-automated optimization in HPLC have reached a remarkable level of maturity and sophistication. Through several real examples, *Lloyd R. Snyder* (Chapter 4.1) and *Sergey Galushko* (Chapter 4.2) describe the possibilities offered by the software packages DryLab® and ChromSword®, respectively. *Michael Pfeffer* (Chapter 4.3) compares the two software concepts from the point of view of the user, and presents a new software tool that also incorporates automatic column selection.

Part 5. User Reports

In the final part, users are able to express their opinions. In four different cases, rather sophisticated and/or new techniques/concepts are presented for the solution of particular chromatographic problems, although equally from the point of view of the user and with practical relevance. One or other of the presented solutions to these problems may be of interest to some readers. *Katrin Marcus* (Chapter 5.1) presents the application of LC-MS/MS coupling in proteomics, *Hans Bilke* (Chapter 5.2) demonstrates ways of testing for robustness in RP-HPLC, *Knut Wagner* (Chapter 5.3) describes a hardware solution for the separation of complex mixtures, and *Mario Arangio* (Chapter 5.4) considers the potential of multiple detection (UV, MS, CLND) in the characterization of libraries of newly synthesized substances.

The five parts describe discrete units of subject matter, but nevertheless the book does not necessarily have to be read in a linear fashion from beginning to end. The individual chapters have been written so that they constitute self-contained modules, and so one can always be skipped. In this way, we have tried to make the character of the book meet the criteria of a reference work. Different interpretations of a topic by different authors have been accepted, as has some repetition, so as not to disrupt the flow of the writing. Finally, some important areas have been covered by several authors, who have naturally placed more emphasis on certain aspects. This applies, for example, to pH (*Uwe D. Neue, Michael McBrien*), weighted regression (*Hans-Joachim Kuss, Stefan Schömer*), the selectivity of stationary RP phases (*Stavros Kromidas, Uwe D. Neue, Melvin R. Euerby, Cinzia Stella, Lloyd R. Snyder*), chemometrics (*Stavros Kromidas, Melvin R. Euerby, Cinzia Stella*), and LC-MS (*Friedrich Mandel, Katrin Markus*). The reader may benefit from the different descriptions of the topics and from the individual evaluations of the authors.

1
Fundamentals of Optimization

HPLC Made to Measure: A Practical Handbook for Optimization. Edited by Stavros Kromidas
Copyright © 2006 WILEY-VCH Verlag GmbH & Co. KGaA, Weinheim
ISBN: 3-527-31377-X

1.1
Principles of the Optimization of HPLC Illustrated by RP-Chromatography

Stavros Kromidas

First of all, some questions will be discussed, which should reasonably be answered before beginning method development. Subsequently, we will treat the principal possibilities for improving the resolution in HPLC. It follows a discussion about efficiency and the "right" sequence of such measures for the isocratic and the gradient mode. There is a particular focus on strategies and concepts for developing a method and checking peak homogeneity.

The last section will show ways to achieve other aims than "better separation": "make it faster", "raise sensitivity" or "save money". The chapter ends with a conclusion and an outlook.

1.1.1
Before the First Steps of Optimization

For economic reasons, one really ought to address the following questions prior to commencing the development of a method or the optimization of a given separation.

- What do I want? In other words, what is the true intention of the separation?
- What do I have? That is to say, what relevant information about the analytical purpose and the samples is available?
- How should I do it? Do I have all what I need, and is what I want to do really possible?

At first glance, these questions might appear too theoretical or even over-critical. Nevertheless, careful consideration of the actual aims and realistic possibilities for solving an analytical problem would seem to be important at the outset. An early discussion with my boss, a colleague or my client – if you are short, even with yourself – can later prevent a good deal of trouble, time expenditure, and last but not least costs. This time-saving can be considered a good investment.

As regards the first question: "What do I want?"

If it is at all possible, at the outset the following or similar questions should be answered:

- Do I need a method for the accurate quantification of *this* toxic metabolite, or is the aim that the authorities just accept my method?
- What is most important in *this* case: short analysis times, durable columns, robust conditions, or simply optimal specificity?
- Must the relative standard deviation S_{rel} be no higher than 2%? What loss of quality would be incurred if S_{rel} were to be 2.5%? Is there actually a correlation between the cost of the analysis and real improvement in the quality of the product?

HPLC Made to Measure: A Practical Handbook for Optimization. Edited by Stavros Kromidas
Copyright © 2006 WILEY-VCH Verlag GmbH & Co. KGaA, Weinheim
ISBN: 3-527-31377-X

In other words, is the aim just to meet the requirements in this specific case or is the real "truth" at stake, i.e., are formal aspects or analytical questions in the foreground? This question should be consciously and truthfully answered because of the possible consequences.

How difficult it can be to stand by meaningful and well-considered decisions without being regarded as outlandish or as a troublemaker has been documented elsewhere [1]. Where possible, one should question all aspects. Unconventional questions frequently result in simple solutions.

As regards the second question: "What do I have?"

Information on the sample makes the design of a suitable method easier.
  Some examples:

- What is written in the report of colleagues from the chemical development department on the light sensitivity and the sorption properties on glass surfaces of a new drug?
- Can I contact these colleagues quickly? That is to say, can I get relevant information with a minimum of effort?
- There may be information about similar separations in the past, which were not pursued further, in an internal database (which is perhaps rarely updated and even more rarely accessed).
- May I quickly calculate the pK_a value of the known main component in the sample with appropriate software (see Chapter 1.4)?
- Has a colleague in a neighboring department worked in the past with similar compounds and might therefore be able to provide valuable insights?

As far as possible, all means of communication with colleagues should be pursued to gather information. At times it may be helpful not to make this public.

As regards the third question: "How should I do it?"

One should assess the feasibility of the proposed work absolutely unconditionally.
  Some examples:

- Can I convince my boss that it is useful from the overall company point of view to discuss in advance with the later routine users the design of the method and additional details? If fear of loss of know-how or questions of budget or other psychological and social barriers make impossible de facto a discussion with "the others", it is a bitter reality that one must accept.
- On the other hand, is it worth fighting for a change of the following well-known and accepted situation? A deadline is fixed and therefore a validation must be finished in two weeks. Later, the burden of subsequent, substantial costs for a repeat of the measurements, complaints, out-of-spec situations, etc., which inevitably result because an analytical method can hardly be validated within two weeks under real conditions, is not placed on "us" but on quality control, and as testing costs they have been accepted since decades in the absence of overall considerations. The reader may imagine the consequences, or viewed more positively, the possibilities for improvement.

- Is it really worthwhile in the case of the development of a routine method, which shall be applied all over the world, to opt for a polar RP-phase because of the frequently observed higher selectivity, even if one has to expect problems with the charge-to-charge reproducibility? Might a hydrophobic, more rugged column with a lower but still sufficient selectivity the better choice?
- Is it useful to demonstrate my analytical "knowledge" by further trimming the relative standard deviation of a method being used in diverse plant laboratories to a value of 0.7%?

Realities – and opinions are also realities – which determine the success or failure of an analytical activity should, wherever possible, influence the design of the method. It is useful if the number of meetings can be reduced to "a cup of coffee" or "lunch often together". The point is to improve the communication and this can turn out to be easier in a less formal situation.

In conclusion, two basic preconditions for successful method development may be noted:

1. Expert knowledge exists or can be loaned or sold.
2. The analytical possibilities correspond with the requirements, and it is possible to talk about them.

In the author's opinion, a clear definition of requirements, unequivocally formulated, understandable goals for all involved persons, shortcuts to information, and a critical estimation of possibilities/risks are more important, not only in analytics, than obtaining exemplary results such as low detection limit, correlation factors around 0.999, S_{rel} smaller than 1%, or 30% less expensive equipment.

1.1.2
What Exactly Do We Mean By "Optimization"?

Optimization of a separation is principally directed by the following goals:

- to separate better (higher resolution),
- to separate faster (shorter retention time),
- to see more (lower detection limit),
- to separate at lower cost (economic effort),
- to separate more (higher throughput).

The three first-named goals may be most important, and of these the improvement of the resolution is the prime concern. Therefore, we will treat this topic before we start to deal with the other aspects. Preparative HPLC is not the subject of this book.

Preliminary Remarks

The theory of chromatography is fundamentally valid for all chromatogaphic techniques. Therefore, basically the same principles are pursued. However, it is evident that the priorities and the weighting of the specific steps are very different, for example in GPC and in μ-LC-MS(MS). In the following, the possibilities for optimization are presented and proposals for the most popular liquid-chromatographic technique, RP-HPLC, are given in short form.

The characteristics of the different modes are treated in Chapters 2.1 to 2.7. It is assumed that the reader is familiar with the chromatographic rules and the theory of HPLC, so these are not treated in detail, but where necessary some technical terms are briefly explained.

The immediately following remarks relate to isocratic separations.

1.1.3
Improvement of Resolution ("Separate Better")

Resolution (R), in simplest terms, is the distance between two neighboring peaks at the base of the peaks. An increase in this distance is what every chromatographer routinely strives for.

The corresponding equation is:

$$R = \frac{1}{4} \sqrt{N} \cdot \frac{\alpha - 1}{\alpha} \cdot \frac{k_2}{k_2 + 1}$$

where N is the plate number, a measure of the performance or the efficiency of the column.

The number of plates is in effect a measure of the widening of the substance band because of diffusion effects. The basic question is whether the molecules of the analyte that reach the detector are contained in a small or a large (peak) volume, i.e., will one get sharp or broad peaks?

Strictly speaking, one should distinguish between the theoretical and the effective plate number. The theoretical plate number is the number of plates of an inert component (see below) and therefore a characteristic and constant value for a column under defined conditions. The effective plate number is the number of plates of a specified retained component, and the retention factor (see below) enters into the calculation. Today, however, this distinction is not made everytime; one speaks only about plate number. In the most cases, the theoretical plate number is calculated, but of retained substances. In this context, it should be made clear that the plate number depends on a lot of factors, e.g. the injection volume, the temperature, the composition of the eluent, the flow rate, the retention time, the analyte, and last but not least the equation used for the calculation, i.e. peak width at the peak base, at 10% or at 50% peak height. Therefore, the comparison of literature values of plate numbers is inherently difficult.

α: *Separation factor*, formerly *selectivity factor*.
α is a measure of the capability of a chromatographic system (chromatographic system: the actual combination of the stationary phase, the mobile phase, and the temperature) to distinguish two given compounds.

The α value is the quotient of two net retention times, i.e. the quotient of the dwell times of the two components within the stationary phase.

The point is that if *this* particular chromatographic system is selective for these two compounds, then in principle they are separable. Selectivity, in simplest terms, is the distance between two peaks, from the top of one peak to the top of the other. This is different from the resolution in that for the determination of the selectivity the form of the peak (plate number) is not considered, because α is only the quotient of two (retention) times. The separation factor depends only on the chemistry; see below on the subject of retention factor.

k: *Retention factor,* formerly *capacity factor k'.*
k is a measure of the strength of the interaction of a given compound in a given chromatographic system. It expresses for how much longer a given compound remains on the stationary phase compared to the mobile phase.

The *k* value is an index like the α value, and is thus independent of instrumental conditions such as the dimensions of the column or the flow rate. The *k* value changes only if parameters that have something to do with the interaction are changed, i.e., the chemistry: stationary phase, mobile phase, temperature. As long as these parameters are kept constant, the *k* value also stays constant, irrespective, e.g., of the flow rate or whether a 10 or 15 cm column is used.

Although the dead time does not appear explicitly in the equation for the resolution, it is useful for the following explanation to briefly deal with this term.

t_m: *dead time, breakthrough time, front, "air peak":* This is the dwell duration of an inert component in the HPLC equipment. A component is designated as inert if it is able to penetrate everywhere without steric hindrance, including, of course, in the pores of the stationary phase, but is not retained there. In other words, the dead time is the time of the presence of any not excluded component in the mobile phase – also in the standing mobile phase (i.e., within the pores), but again there is "no" interaction with the stationary phase. Therefore, the dead time only changes if something "physically" or "mechanically" is altered, e.g. a change in the length or the inner diameter of the column, the packing density, or the flow rate. The dead time is a time, which is independent from the particular compound just analysed. As all components move equally quickly in the eluent, the time that the compounds spend in the eluent makes no contribution to the separation. A separation is only possible if the substances stay in the stationary phase for different lengths of time.

The resolution R – the distance from peak base to peak base – depends only on the following three factors:

- the strength of the interaction between the compound and the stationary phase (if the peak comes soon or late), i.e. on the *k* value,
- the ability of the chromatographic system to distinguish between the two components of interest, i.e. the α value,
- if the relevant peaks are sharp or wide, i.e., the plate number.

Consequently, to improve resolution, there are in principle only three possibilities,

- namely a general increase in the interaction (k value increases),
- an analyte-specific change in the interaction (α value increases), or
- an increase in the efficiency of the separation (N value increases).

1.1.3.1 Principal Possibilities for Improving Resolution

The aforementioned topic is illustrated by way of a hypothetical example; see Fig. 1. Starting with a poor resolution (see Fig. 1, upper chromatogram), what are the principle possibilities for improving the resolution?

Remark: The dead time only changes in case 1 ("↑t_m").

Possibility 1: One ensures that all components – including one possible inert component (increase of dead time, see above) elute later. Because now the dead time increases too a physical process must be responsible. This could be a longer column, a larger inner diameter of the column, decrease of the flow rate. (A larger inner diameter leads to peak broadening which rules it out for all practical purposes).

Possibility 2: The retention time stays largely constant, one seeks only a better peak form. Here, there are somewhat more possibilities: reduction of the dead volume (e.g., thinner capillaries, smaller detector cell), reduction of the injection volume (remark: local overload of the column happens more frequently than one

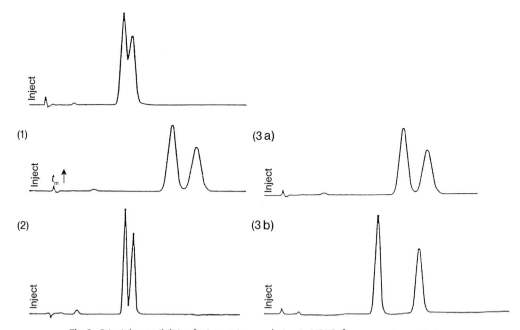

Fig. 1. Principle possibilities for improving resolution in HPLC; for comments, see text.

might imagine! Peak broadening caused by the injection is inversely proportional to the injection volume.) at an elution composition with equal solvent strength parameter, replacement of methanol with acetonitrile owing to the lower viscosity of the latter (approximately constant retention time can be expected), using smaller particles, or use of a newer, better packed column. In this context, one should also consider an optimization of the injection step as this also improves the peak form and consequently increases the plate number. The solvent of the sample should be weaker than the eluent; to this end, one uses a little bit more water in comparison to the eluent composition when preparing the sample solution.

In this way, it is possible to increase the concentration of the substance band at the head of the column, and the result is a better peak form. Finally, one should also consider various settings, such as sample rate (sampling time, sampling period), bunching factor, peak width, slit width in the case of a diode-array detector, etc. In this way, the peak form can also be improved measurably, without changing the "real" method parameters such as column or eluent.

Possibility 3a: This involves the increase of the interaction between the sample and the stationary phase, and for this there are only three "chemical" possibilities, as mentioned above: change of the eluent (e.g., increase of the water content), decrease of the temperature, and change of the stationary phase (e.g., use of a more hydrophobic phase).

The interactions increase for both/all components to be separated to the same extent, and so the retention times will also increase equally.

Possibility 3b: The same procedure as in Possibility 3a, but here it is possible to change the interactions of the two components to different extents, i.e. one component responds more strongly to a change, e.g., a change in the pH, than the other one.

Other possibilities do not exist in principle, because $R = f(N,a,k)$. This means that when trying to improve the resolution, one can only change consciously or intuitively these three factors.

1. One may successfully increase the interaction of the components of interest with the stationary phase per se, i.e. "the whole" comes later (case 3a, increase of k, e.g., by increasing the water content in the eluent) or one successfully increases the interaction of the components with the stationary phase individually, i.e. one component responds more strongly to the change than the others (case 3b, increase of α, e.g., change of the pH in the case of polar/ionic components). Both cases relate to the "chemistry": change of the temperature or change of the eluent (this includes the pH and other additives or modifiers, of course) or change of the stationary phase.
2. One may increase the plate number, either at (theoretically) constant retention time, case 2, or can concomitantly increase the retention time, case 1.

Other possibilities do not exist in principle.

Remark: In the case of a change of α and/or k, N varies simultaneously too, of course.

Having established how the resolution can be improved in principle, two questions now arise:

1. Which of the three possibilities has the greatest effect?
2. In which sequence should one try to optimize these parameters, i.e., which procedure is most economical?

1.1.3.2 What has the Greatest Effect on Resolution?

Consider once more the equation for the resolution:

$$R = \frac{1}{4}\sqrt{N} \cdot \frac{\alpha - 1}{\alpha} \cdot \frac{k_2}{k_2 + 1}$$

As one can see from this equation, R responds most sensitively to a change of α. Therefore, a variation of the selectivity is the most effective, but frequently also the most difficult procedure for improving the resolution. For better illustration, consider two numerical examples:

1. If the α value is 1.01 for two neighboring peaks, some 160,000 plates would be needed to separate them perfectly (baseline separation). If the α value could be increased to 1.05, only some 2000 plates would be needed.

   In other words, if the selectivity can be slightly improved, far fewer plates will be needed for a given resolution, i.e., the flow rate can be increased or a shorter column can be used, and both will result in a shorter separation time. The "few" plates that are lost due to the increased flow rate or the shorter column, respectively, do not matter as the selectivity is improved.

2. Let us assume that when using a column of 9000 plates the second of two components to be separated elutes with a k value of 2 and an α value of 1.05. The resolution resulting from these values is 0.75.

$$R = \frac{1}{4}\sqrt{9000} \cdot \frac{1,05 - 1}{1,05} \cdot \frac{2}{2 + 1} = 0,75$$

   The demand is: "a resolution of at least $R = 1$".

What are the possibilities now?

If it is possible to increase the α value to 1.10 at approximately constant interaction strength, then the resolution improves by a factor of almost 2.

$$R = \frac{1}{4}\sqrt{9000} \cdot \frac{1,10 - 1}{1,10} \cdot \frac{2}{2 + 1} = 1,44$$

If one considers the sequence of the most efficient steps for improving the resolution, then the efficiency comes directly after the selectivity. An increase in the plate number can be achieved either by the classical approach of increasing the length of the column – in this case, a doubling of the column length and consequently a doubling of the retention time and the plate number increases the resolution by $\sqrt{2}$, that is, by a factor of just 1.4 – or by reduction of the particle size at constant dimensions of the column and therefore at constant retention time, which is the more frequently applied method today. Other possibilities for enhancing the plate number, such as better packing technique or reduction of the dead volume of the equipment (see below), should be common knowledge and are only mentioned here in passing.

The third, most simple method, although quite inefficient for k values > ca. 4, is the enhancement of the k value. This should also be shown by our numerical example. First, one might increase the interaction, in the first step to a k value of 4 and in a second step to 6. In this case, the resolution would improve to 0.90 and 0.97, respectively.

$$R = \frac{1}{4}\sqrt{9000} \cdot \frac{1,05 - 1}{1,05} \cdot \frac{4}{4 + 1} = 0,90$$

$$R = \frac{1}{4}\sqrt{9000} \cdot \frac{1,05 - 1}{1,05} \cdot \frac{6}{6 + 1} = 0,97$$

One would obtain a quite small improvement in the resolution as a result of a reasonable extension of the separation time (k value of 6). Therefore, the enhancement of the interaction (e.g., the increase of the water content of the eluent) is a common, easily implemented, but mostly quite inefficient procedure for routinely improving resolution. If, however, the plate number could be improved to 15,000, for example (smaller particles, injection tricks, etc.), in the case of a k value of 4, then the resolution would improve to 1.17.

$$R = \frac{1}{4}\sqrt{15\,000} \cdot \frac{1,05 - 1}{1,05} \cdot \frac{4}{4 + 1} = 1,17$$

1.1.3.3 Which Sequence of Steps is Most Logical When Attempting an Optimization?

In Fig. 2, an economical strategy is presented.

Remark: If the possibility of LC/MS coupling is available, it should be applied at the outset or at least at an early stage of the separation attempts; see below.

Explanations Relating to Figure 2

Question 1: "Reasonable" interactions, acceptable analytical run time?

After a first run with an unknown sample, or if one has to optimize an existing method, the first question ahead of any optimization steps should be: "Are there reasonable interactions?". In an isocratic separation, the components of interest

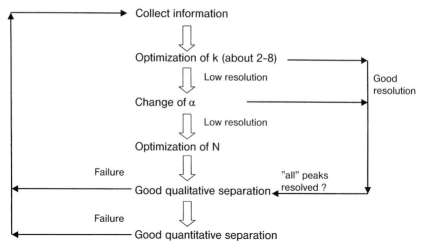

Fig. 2. Strategy for method development in HPLC.

should elute in the range of ca. $k = 2$–8. In the case of a gradient separation the $\bar{k}$ value should be ca. 5, in the middle of the chromatogram (see Section 1.1.3.4.2). These ranges are a good compromise between run time, robustness, and resolution. If the interactions are adequate (check k values) but the run time is not acceptable, then the flow should be increased or, alternatively, a shorter column should be used.

Question 2: Is it possible with the help of appropriate settings to get information on the peak homogeneity?

If one is satisfied with the run time but not with the resolution, then one has to consider whether or not the settings such as sampling time, sampling period, peak width or wavelength are optimal for this particular separation (early/late, narrow/broad peaks). One should then deal with the most effective of the three separation parameters, i.e. the selectivity.

Question 3: Is the selectivity sufficient?

If one obtains a short run time and a satisfactory selectivity (separation factor of the critical pair ca. 1.05–1.1) but the resolution is not yet satisfactory, then one should enhance the efficiency – to put it simply "make the peak sharper!".

Question 4: Is it now possible to enhance efficiency or must the selectivity be further improved in order to get a better resolution?

As a rule, this is a more economical way to improve peak shape than to continue to improve the selectivity (change of chemistry, which means a change of columns and eluents).

Two real examples, taken from Ref. [1], illustrate these statements:

Example 1

In Fig. 3a, the chromatogram of a validated method from a pharmaceutical company is shown.

The parameters of the method are as follows: linear gradient from 10% to 90% methanol, common commercial C_{18} phase, 5 µm, column 125×4 mm, flow rate 1 mL min^{-1}, ambient temperature, injection of 30 µL sample dissolved in THF/ MeCN. The large peak at the dead time results from matrix components and is not a problem. The method, which is not really optimal, should be optimized in a quick and easy way.

Question 1: Is the run time OK?

No. The first troubling thing one notes is the really long run time: 16 min for two peaks is not acceptable for a routine method today. First, the flow was increased to 2.6 mL min^{-1}. As expected, a shorter run time was obtained without a loss of resolution: the gradient was changed in such a manner that the gradient volume was the same as at the start (see Chapter 1.2), Fig. 3b.

However, a retention time of 10 min for two peaks is still too long. While the pressure was ca. 345 bar at the flow rate of 2.6 mL min^{-1}, the starting conditions of the gradient were changed in the next step. It was not started at 10% but at 40% methanol; see Fig. 3c. The retention time of 3–4 min for two peaks was then satisfactory.

Questions 2 and 3: Are the settings and selectivity OK?

Yes, see the distance between the tops of the peaks in Fig. 3c (that is simply to say "selectivity"). Although the selectivity is obviously not bad, the resolution cannot be described as satisfactory because of the fronting.

Question 4: Can I improve the peak shape and hence the plate number?

A simple test was then performed: the sample solution was diluted twofold with the eluent (MeOH/H$_2$O, 40 : 60) and the resulting 120 µL was injected. Finally, the result thus obtained was satisfactory; see Fig. 3d.

Note: It is better to inject 100 or 150 µL of sample solution with an eluent-like solvent than to inject 20 or 30 µL of sample solution with a stronger solvent than the eluent (e.g., containing more MeCN than the eluent).

To sum up, in this example first the run time was reduced and then, due to the apparently adequate selectivity, resolution was improved merely through increasing efficiency (better peak shape).

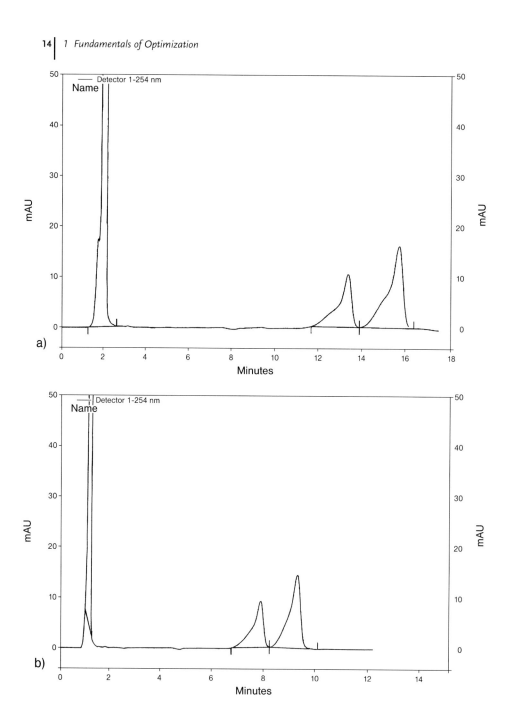

Fig. 3. Simple ways to improve the resolution of an existing RP method.
For details, see text.

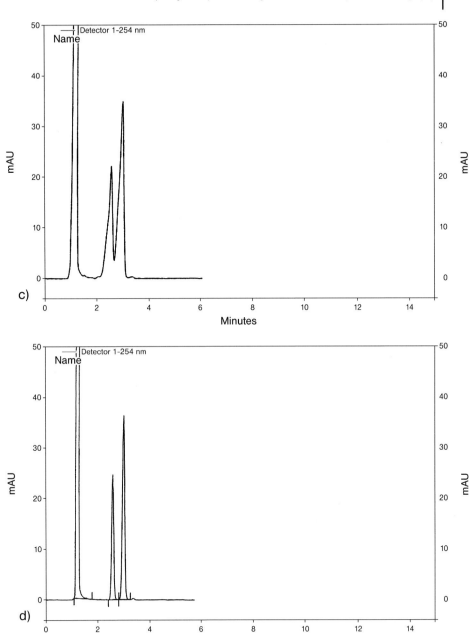

Fig. 3. (continued)

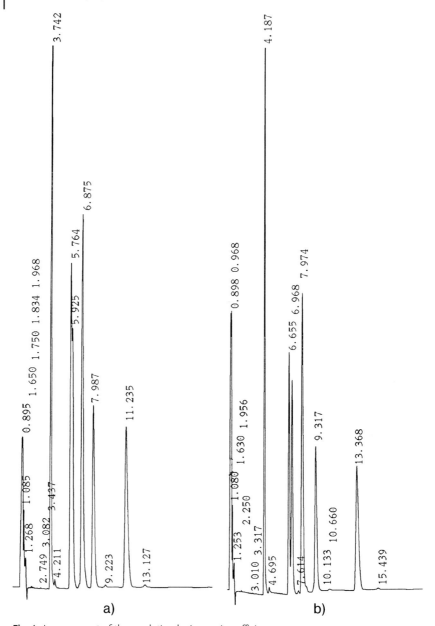

Fig. 4. Improvement of the resolution by increasing efficiency.

Example 2

Figure 4 shows the isocratic separation of metabolites of some tricyclic anti-depressants. In (a), separation on a 5 µm Luna 2 C_{18} column is shown. The critical pair of peaks 2 and 3 at ca. 5.8 min (α value 1.05) cannot be separated.

Here, the resolution is not sufficient and hence there is a need for action. In this case, the answer to the first question, if the run time is okay, is "yes". As regards the selectivity, the decision is less clear-cut. If one were to try to improve the separation in this case by changing the column or eluent (selectivity, therefore "chemistry"), this would be unlikely to be a quick solution.

In (b), the separation obtained on a 3 µm column is depicted, under otherwise completely identical conditions. One nearly obtains baseline separation at similar selectivity (α value = 1.04).

The use of a column with smaller particle size is a quick and usually also an economical way to proceed. Even if the obtained selectivity is "only" 1.05, one should first consider an improvement of the resolution by efficiency (N), instead of continuing to try the more large-scale "chemical" possibilities: column, eluent, temperature.

As simplified conclusion and taking into consideration Examples 1 and 2, one can state the following ($kαN$ principle):

1. First look for "reasonable" interactions (k) and acceptable retention times (e.g., by variation of the eluent composition and if necessary the flow).
2. Try to make the selectivity as good as possible through an acceptable degree of effort. This means an α value of around 1.05–1.1 (e.g., replace acetonitrile with methanol, add a modifier, change the pH).
3. If the resolution is still not satisfactory, improve the peak form (N). On the one hand, this can be accomplished by "scientific" procedures, e.g. using smaller particles, most effectively in the following combination: decrease column length and particle size, increase flow rate and possibly temperature. On the other hand, one should also consider simple but no less effective tricks that also result in a better peak form, i.e. smaller injection volume, weaker solvent to dilute the sample, thin/short connection capillaries between autosampler and detector.

1.1.3.4 How to Change *k*, α, and *N*

1.1.3.4.1 Isocratic Mode
Before we start to consider strategies for developing methods for unknown samples and for checking the peak homogeneity, let us consider the established procedures that are available to change or increase k, α, and N (see Fig. 5).

Remark: In the given sequence of the different possibilities in Fig. 5, the efficiencies (time factor and importance) generally decrease. That is not to say that the stationary phase plays a subordinate role with regard to selectivity. Not at all! However, one should test the other faster things before trying out a completely new column; see below.

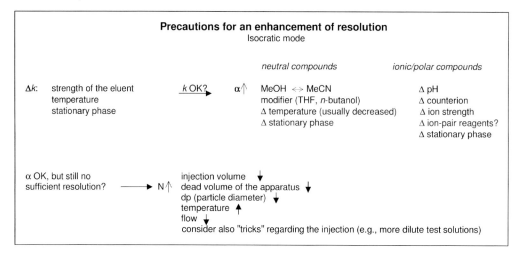

Fig. 5. Possibilities for improving resolution in the RP-HPLC (isocratic mode).

1.1.3.4.2 Gradient Mode

Preliminary Remarks

If in the case of a gradient separation one expects no more than 8–10 peaks and if the matrix under consideration is not a difficult one such as fermentation broth, urine, a cream, a plant extract, etc., then 100 or even 125 mm long columns are, as a rule, definitely too long; see Ref. [1] and Chapters 1.2 and 2.7.3. In many cases, we have been able to separate 4–6 peaks with a 10 mm/2 mm/2 μm C_{18} column and "everyday" equipment.

For gradient runs, the following equation applies:

$$\bar{k} = \left(\frac{t_G}{\Delta\%B}\right) \cdot \left(\frac{F}{V_m}\right) \cdot \left(\frac{100}{S}\right)$$

where
$\bar{k}$ = mean k value; the analyte is in the middle of the column (lengthwise)
t_G = duration of the gradient (min)
F = flow (mL min^{-1})
V_m = column dead volume
$\Delta\%B$ = changes in B from the beginning to the end of the gradient
S = slope of the $\%B/t_G$ curve; for smaller molecules, S is set at about 5

Bearing in mind that an optimal $\bar{k}$ value would be about 5, $\bar{k}$ and α can be changed in the following way:

- gradient volume (by the gradient time or more smartly by the flow rate),
- steepness of the gradient,

- % B, i.e. start and end conditions,
- profile of the gradient (linear, convex, concave); if a transfer of the method is intended, then only linear gradients should be used, because of their simple transferability. The inclusion of isocratic steps, on the contrary, should cause no great problems,
- temperature,
- stationary phase.

From a practical point of view, one should think in terms of high flow rates in the course of gradient runs. Increasing the flow while keeping the gradient time constant yields a better resolution because the gradient volume (flow × time) increases. Peak capacity (number of peaks per time unit) and thus resolution itself also increases with increasing gradient volume. Even if resolution is satisfactory, the flow should be increased. For example, if the flow is increased by factor 2 while simultaneously decreasing the gradient time also by a factor of 2 and adapting the gradient accordingly, the resolution will remain the same because the gradient volume remains constant – in only half the time! The disadvantages of using a higher flow rate are the higher pressure and the decrease in the peak area. In the case of gradient separations, the enhancement of the plate number is usually a minor goal because the peaks tend to be inherently sharp.

1.1.3.4.3 Acetonitrile or Methanol?

This question is treated in detail elsewhere [2, 3]; see also Chapter 2.1.4. Here, we consider only the results obtained through numerous experiments with various classes of substances under different conditions. It would seem that in the case of mixtures with similar solvent strengths, methanol yields the better selectivity in comparison with acetonitrile, possibly because of the preference for polar interactions (protic *vs.* aprotic solvent). This is most apparent in the case of small molecules such as primary amines.

At the same time, one generally observes a worse peak form because of the enhanced viscosity compared with acetonitrile. This is illustrated by two examples, see Figs. 6 and 7.

Figure 6 shows the injection of uracil, pyridine, benzylamine, and phenol in an acidic methanol/phosphate buffer (a) and in an acidic acetonitrile/phosphate buffer (b). These rather unsuitable eluents (strong bases in an acidic medium) have been deliberately chosen to test the selectivity of methanol and acetonitrile for polar analytes in difficult situations. In methanol, the bases were at least partly separated, which was not achieved in acetonitrile. Figure 7 shows the same separation under neutral conditions in methanol (a) and acetonitrile (b). Again, the better selectivity in methanol is striking. Not only are polar contaminants almost completely separated from uracil, whereas they are barely discernible in acetonitrile (see arrow), but in methanol phenol can also be separated from benzylamine, which is not achieved in acetonitrile. We can thus conclude that methanol gives rise to better selectivity while acetonitrile gives rise to better peak symmetry. This can be observed for many categories of substances.

Possible Explanation: An idea currently discussed is the existence of active or high energy sites on the surface of RP material. Although they take up only 0.4% of the total surface, they play a dominant part in selectivity, as these are the sites that are available to polar/ionic components at a methanol content in the eluent between 0 and 60%. Assuming an experimentally measured solvent layer of 2.5 Å (methanol) or 13 Å (acetonitrile) on the surface, in an acetonitrile scenario, the hydrophobic residue of polar/ionic molecules can only interact with the ends of alkyl chains anchored to the surface. With methanol in the eluent, the molecules can diffuse through the significantly thinner solvent layer, resulting in stronger/additional interactions. Furthermore, the formation of labile methanolates could facilitate polar interactions leading to good polar selectivity in methanol. Finally, we would like to point out three further observations made in our experiments:

1. Differences in selectivity are insignificant where the methanol/acetonitrile concentration in the eluent is low.
2. Selectivity differences between eluents containing methanol or acetonitrile are particularly noticeable in strongly hydrophobic stationary phases and least pronounced in polar stationary phases such as CN.
3. After the addition of methanol, the pH value of a solution shifts to the alkaline, see Fig. 7c (source: Dr. Bernhard Dreyer). See also comments in Chapter 1.3 and Table 1 in Chapter 1.4.

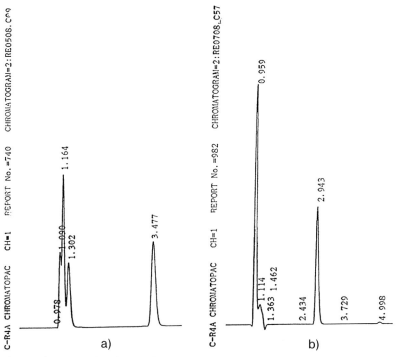

Fig. 6. Selectivity in phosphate buffer/methanol (a) and in phosphate buffer/acetonitrile (b); for comments, see text.

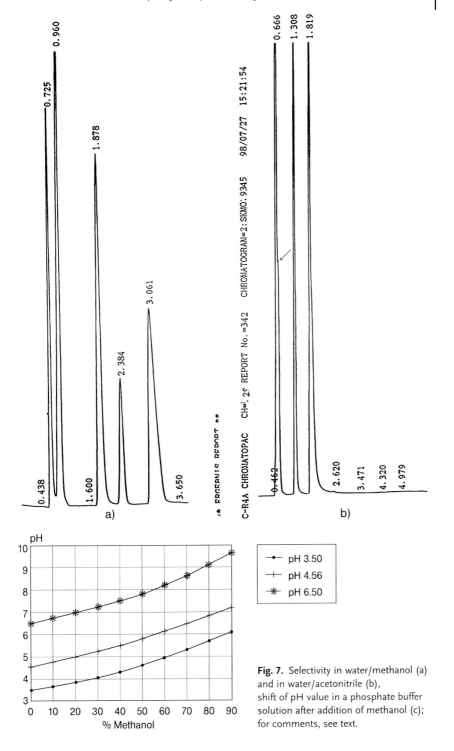

Fig. 7. Selectivity in water/methanol (a) and in water/acetonitrile (b), shift of pH value in a phosphate buffer solution after addition of methanol (c); for comments, see text.

Figure 7c shows the change in pH value of a 20 mMol Na buffer after the addition of methanol. This explains why a column that is run with an eluent at a nominal pH of 6 or 7 has a comparatively low lifespan. After the addition of methanol or acetonitrile, the pH value shifts towards the alkaline. Most silica gels detach and partially dissolve above a pH of 8.

1.1.4
Testing of the Peak Homogeneity

If a possible "rider" is seen on the flank of a peak, how can one improve the resolution? Testing the peak homogeneity and improvement of the resolution are related questions. For both, a three-step concept can be applied.

Step 1: A quick and cheap action for testing the peak homogeneity – the "1/2 hour method"

Supposing that time is limited and/or one has to work within given terms or is bound by requirements of a customer; in other words, the eluent, column, and temperature cannot be changed. What possibilities remain to better depict the chromatogram, or even to improve the separation? The use of contemporary standard equipment is assumed.

1. Make the most of the possibilities offered by the diode array/software, e.g., ratio plot, match factor, contour plot, 3D plot, first and second derivatives of the relevant peak.
2. One should ask whether all settings of the detector and of the software are optimal and whether the hardware can be further optimized.
   In the following, some possibilities are presented and some numerical values are recommended. For quite old equipment, reduction of the time constant to 0.1 s and use of the 10 or 100 mV output of the detector are recommended. With more modern equipment, a similar manipulation is possible by means of the settings "bunching factor" and "bunching rate"/"sampling period" of the software. Set the peak width during the integration to 0.01 min and the sample rate between 5–10 data points per second, i.e. a sampling period of ca. 200 ms. These settings are very important and necessary in the event of early, sharp peaks. If you use very short/narrow columns and obtain "really" fast peaks, 20–40 Hz sample rate are an absolute "must". Are wavelength, slit width, and reference wavelength optimally adjusted? Is the inside diameter of the capillary between the autosampler and detector 0.13 mm or at most 0.17 mm?
3. Remember the injection. To test the peak homogeneity, inject just 1 or 2 μL; there is frequently a risk of local overload of the stationary phase surface. Dilute the sample solution with water or with the eluent and inject again.

A change of settings or a repeated injection takes a matter of seconds or at most a few minutes. The following three examples show that it is profitable to remember such simple actions (see Figs. 8 to 10).

Fig. 8. Effect of sampling rate on the peak shape/resolution for early peaks; for comments, see text.
(a) Chromatogram taken at 0.01 min peak width.

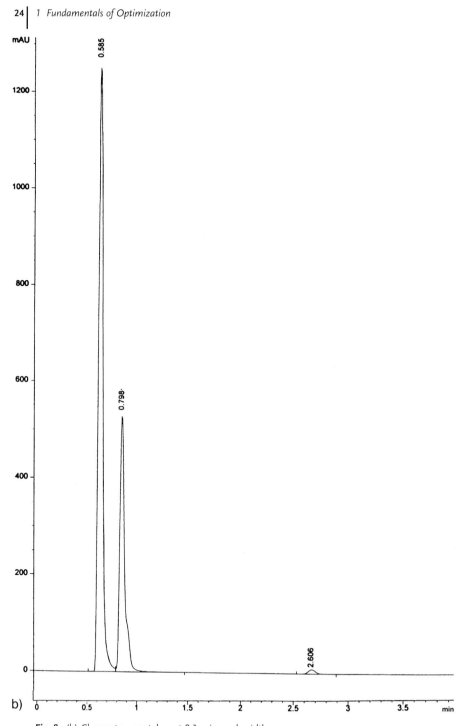

Fig. 8. (b) Chromatogram taken at 0.1 min peak width.

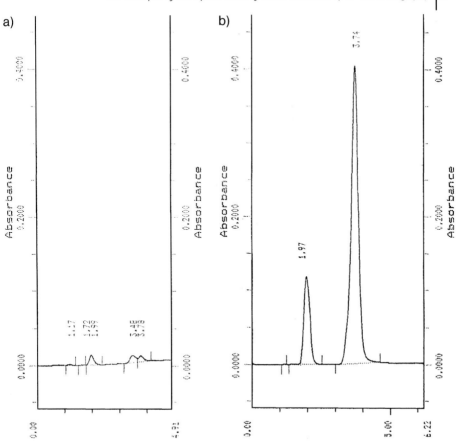

Fig. 9. Effect of injection volume on the resolution of neutral compounds.

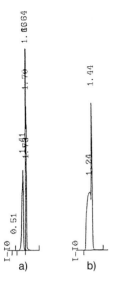

Fig. 10. Effect of injection volume on the resolution
of polar compounds.

Figure 8 shows the same separation with 10 data points per second (a) and 1 data point per second (b). Using a low sample rate, there is an unnecessary loss of resolution for the early peaks.

Figure 9 shows the injection of 20 µL of acetophenone in a standard RP system (b). One may assume that the first peak relates to an impurity, and that the second one is acetophenone. Only the injection of 1 µL – see chromatogram (a) – "unmasks" the homogeneity of the impurity and reveals a possible impurity in the "acetophenone" peak.

Figure 10 shows the injection of 20 µL of benzoic acid (b). As can be easily identified, there is an unknown compound for which the peak is poorly resolved. In (a), one can see a much better resolution if 5 µL is injected. The risk of local overloading is increased in the case of polar compounds due to possible dual-interaction mechanisms.

Step 2: Change of chromatographic parameters

This step encompasses procedures that are usually performed within the scope of an optimization. In doing so, mostly the interactions between the sample and the stationary phase are changed. The intention is a change of the retention factor k (mostly enhancement), but ideally also of the separation factor α. Otherwise, at constant interaction strength ("chemistry" constant and therefore k and α as well), one attempts to enhance the plate number or in the case of a miniaturization to prevent dilution or to enhance the relative mass sensitivity. Using a trial-and-error procedure, one needs 1–2 weeks. Using a systematic procedure, aided perhaps by an optimization program, the time can be reduced measurably; see Part 4 for chapters on computer-aided optimization.

Some possibilities are listed below:

1. Eluent: Change of polarity, replacement of acetonitrile by methanol (or reverse), addition of ca. 5–10% of a modifier such as THF, isopropanol or *n*-butanol in the case of neutral components, or addition of amines or acid in the case of polar/ionic components; change of pH, buffer type, and/or molarity.

2. Stationary phase: The broad spectrum of commercially available columns is a great blessing and also a plague. For some rules and some theoretical background for the selection of RP columns, see Chapter 2.1. In this context, the double-column technique should be mentioned. An example of this is as follows: the separation of five quite polar components on a Nucleosil C_{18} column is not especially good; see Fig. 11. On a CN column, the separation is even worse (see Fig. 12). However, when the two columns are connected in series, a very nice separation under identical, constant conditions is obtained (see Fig. 13).

Better resolution is achieved, especially in the first part of the chromatogram. The last apolar peak is slightly delayed by the polar CN material, but this small increase in run time is a price worth paying. Moreover, we have been able to separate tricyclic antidepressants and their metabolites (a total of 12 peaks) in one isocratic run using Zorbax Bonus/Chromolith Performance and AQUA/Zorbax Extend coupled in series.

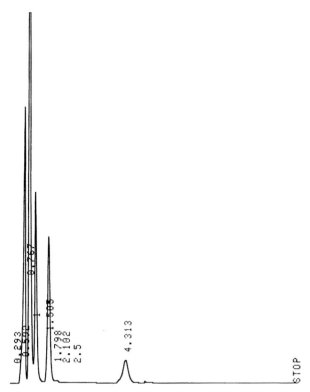

Fig. 11. Isocratic separation of five peaks on a Nucleosil C$_{18}$ column.

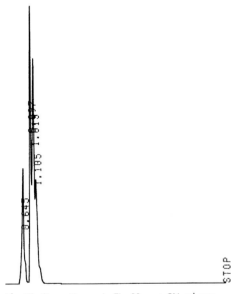

Fig. 12. Separation as in Fig. 11 on a CN column.

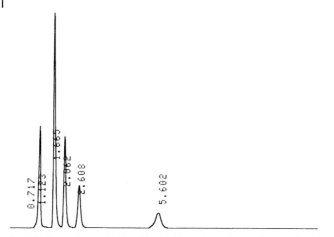

Fig. 13. Separation as in Figs. 11 and 12 on CN and C_{18} columns placed in series.

3. Temperature: At low temperatures, the influence of the mobile phase recedes into the background and the individual properties of the stationary phase come to the fore. The enthalpy differences in the interactions between the individual components and the stationary phase are larger than at higher temperatures and therefore a differentiation (= selectivity) is often easier.

If the temperature is reduced, then the kinetics decreases and therefore also the number of plates. The peaks become wider. On the other hand, the selectivity generally increases. The last noted advantage predominates mostly in the case of RP mode if isomers are to be separated, or with enantiomers (see Chapter 2.6), and therefore the best resolution for difficult separations is frequently obtained at low temperatures. In cases where very slow kinetics is seen as a result of the mechanism, one should work at higher temperatures. At temperatures well above 100 °C one enters the realm of high-speed kinetics, and this results in very good resolution.

The following can be established: Lowering the temperature seems to be an advantage where the steric aspect is crucial in otherwise similar components undergoing separation, e.g. enantiomers, spirally twisted structures, double-bonding isomers etc. In conventional RP separations, however, raising the temperature is usually the preferable option, as this decreases retention time as well as back pressure. The latter effect makes it possible to use 3 or even 1.7–2 µm particles without any trouble. Improved efficiency (higher number of theoretical plates) is achieved not only by using small particles, but additionally lowering the viscosity of the eluent, thus increasing kinetics. Accordingly, at a flow rate of 2 mL/min at 80 °C in four 10 cm columns in serial arrangement, the retention time is equal to that of a 25 cm column at 30 °C and a flow rate of 1 mL/min, but efficiency has been increased by a factor of 4. As the increase in temperature leads to a decrease of polarity in the eluent, fewer organic components are needed in the eluent ("green" HPLC). Finally, the use of acidic buffers suppresses the ionization of bases. This, in turn, increases hydrophobic interaction, and the resulting faster kinetics yield significantly improved peak shapes in ionic/ionizable species.

Step 3: Coupling orthogonal separation techniques

If the sample of a particular separation problem is very important and its identity is truly unknown, one should consider more reliable possibilities for testing the peak homogeneity. Because some of these possibilities are discussed in detail in subsequent chapters, they are mentioned here only in brief. Thus, one may couple HPLC with spectroscopy, e.g. LC-MS(MS) (Chapters 3.3, 5.1, and 5.4) or LC-NMR (Chapter 3.4).

Upon *coupling HPLC with a spectroscopic method,* it is the specificity, not the resolution, that is improved. Therefore, the reliability of a qualitative analysis strongly increases.

Orthogonal separation techniques (= combination of different separation principles/ different mechanisms), 2D or multi-D chromatography. By the coupling of two chromatographic techniques, e.g. LC-GC, gel filtration-ion exchange or LC-DC (2D separation, see Chapter 3.2), or the coupling of a chromatographic and another technique, e.g., gel electrophoresis-HPLC, HPLC-ELISA (see Chapter 3.1) or HPLC-CE, the (chromatographic) resolution increases. Possible subsequent coupling with spectroscopy, e.g. LC-GC-MS or LC-CE-MS, results in an additional increase in specificity.

The possibilities of the coupling technique are illustrated here through the example of LC-DC coupling.

Figure 14 shows the RP gradient separation of an emulsifier on a 2 mm C_{18} column (source: G. Burger, Bayer Dormagen, Germany). A lot of peaks are obtained, some of which are well-separated while others are poorly separated. Consider the last peak marked "D". This peak is quite broad, but there is no strong indication of peak inhomogeneity. If one applies this peak "D" on-line to a DC plate and performs a thin-layer separation with this fraction (two-step gradient; rotation by 90 degrees), one obtains the chromatogram illustrated in Fig. 15. The benefit should be obvious; on coupling two chromatographic techniques, the peak capacities are multiplied.

A "light" version of the orthogonal principle is as follows: one performs a second separation on a completely different column, using another eluent, a reduced injection volume, and/or at another wavelength. The possibility that two or more compounds will show the same behavior under different conditions, i.e. that the same chromatogram results again, is quite low. Guidelines for the selection of RP columns appropriate for orthogonal experiments can be found in Chapters 2.1.1, 2.1.3, and 2.1.6.

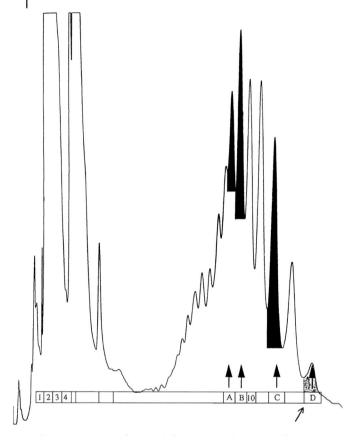

Fig. 14. Separation of an emulsifier in RP gradient mode; for comments, see text.

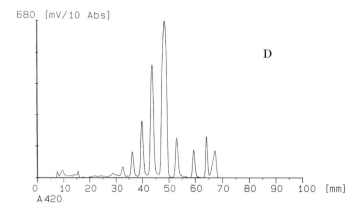

Fig. 15. Further separation of fraction "D" (see Fig. 14) on a thin-layer RP plate in two-step gradient mode; for comments, see text.

Example: An optimization has reached its final step. By injection of 20 µL of the sample on a hydrophobic phase, e.g. Symmetry, one obtains a nice chromatogram using an MeCN/water eluent at pH 3. For confirmation of the peak homogeneity/testing the selectivity, one now uses a polar column, e.g. LiChrospher, and an MeOH/water eluent, and injects 2 or 5 µL. If one obtains the same number of peaks, probably at different retention times and with different peak forms, one may assume that no information has been lost/obscured. In the simplest case, one can use a column with merely another surface, i.e. chemistry, under the same conditions as previously.

The efficiency of such simple testing is illustrated by three real examples:

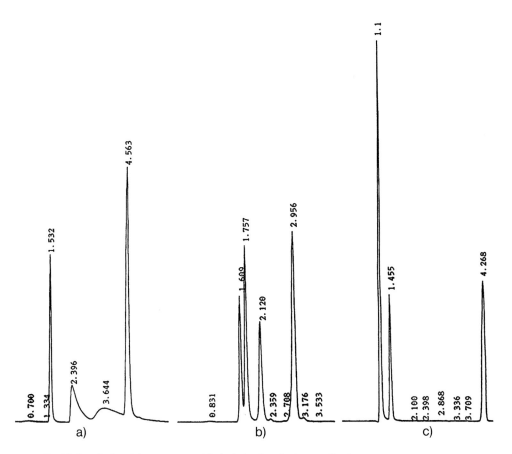

Fig. 16. Proof of peak homogeneity with the help of an "orthogonal" column. Pyridine/benzylamine/phenol test on three phase types in acidic methanol phosphate buffer. For comments, see text.

Example 1

In Fig. 16c , one can see the separation when a sample containing uracil, pyridine, benzylamine, and phenol is injected onto a modern, hydrophobic, endcapped column using an acetonitrile/phosphate buffer at pH 2.7.

If the fact that four peaks were to be expected was not known, no one would suspect the first, sharp, quite symmetrical peak at 1.1 min corresponds to more than one compound. In fact, this "single" peak corresponds to the two bases pyridine and benzylamine, which co-elute even before the dead time, i.e. they are excluded. Indeed, we can hardly expect such a hydrophobic stationary phase to show sufficient polar selectivity. Such a phase is unable to discriminate strong polar analytes in any case. In pharmaceutical laboratories one often sees the following situation: one uses modern, hydrophobic stationary phases because of their good peak symmetry when separating basic drugs. The chromatogram looks beautiful, the match factor of the PDA is around 990, and everything seems perfect, so peak homogeneity is claimed. However, the good peak symmetry also suggests good selectivity. Only by applying two polar stationary phases, Fluofix (Fig. 16 b), Hypersil ODS (Fig. 16 a), under constant conditions is the true situation revealed. Poor peak symmetry (Fig. 16 a) is not of concern here as the aim is just to ascertain the number of peaks.

Example 2

In Fig. 17, chromatograms obtained by the injection of metabolites of tricyclic antidepressants onto three columns are shown. The analytes are small, polar molecules, so two polar stationary phases were employed, an "old" packing (LiChrosorb (a)) and a new one (Reprosil AQ (b)). The apparent result is "five peaks", and so there is no reason for any suspicion. However, using a stationary phase with a pore diameter of 300 Å, one can see that the second, quite sharp peak is not homogeneous (six peaks in chromatogram (c)). Even for such small molecules, steric aspects may be relevant.

Another version of the orthogonal principle with only one column is as follows: one needs a stationary phase with a surface bearing two widely differing functional groups. This might be, e.g., an EPG phase (embedded polar group) with a hydrophobic alkyl chain along with a polar group in ionic form. Another possibility would bear usual C_{18}/C_8 alkyl chains together with very polar groups, e.g. "aggressive" (acid) silanol groups, terminal amino groups, etc. These different groups are responsible for different interactions, of course. Depending on the eluent composition, one or other of the mechanisms can dominate, i.e., through the choice of eluent one can determine whether hydrophobic or polar/ionic interactions prevail. As in the case of the version outlined above, it is very improbable that the analytes will show the same behavior in both cases, in other words that the resulting chromatogram will be almost the same. During these measurements, the wavelength can also be changed, of course. This version is also illustrated by an example (source: SIELC Technologies).

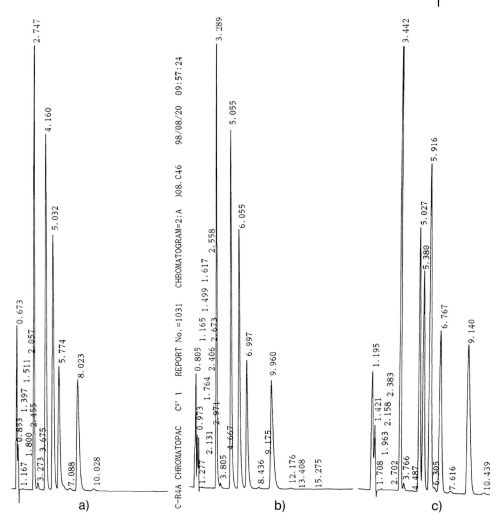

Fig. 17. On steric selectivity ("shape selectivity") even for small molecules; for comments, see text.

Example 3

In Figs. 18 and 19, the separation of six compounds on Primesep 100 with two eluents is illustrated. The surface of this "mixed-mode" stationary phase bears hydrophobic alkyl chains and embedded hydrophilic charged groups. Depending on the eluent, these two functionalities facilitate different interaction mechanisms. In other words, by choosing the appropriate composition of the eluent, one can dictate the mode of interaction.

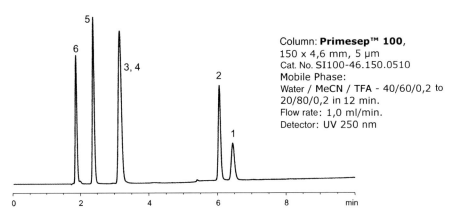

Column: **Primesep™ 100**,
150 x 4,6 mm, 5 µm
Cat. No. SI100-46.150.0510
Mobile Phase:
Water / MeCN / TFA - 40/60/0,2 to
20/80/0,2 in 12 min.
Flow rate: 1,0 ml/min.
Detector: UV 250 nm

Fig. 18. Reversed-Phase Separation: Separation on a "mixed-mode" stationary phase with an RP eluent; for comments, see text.

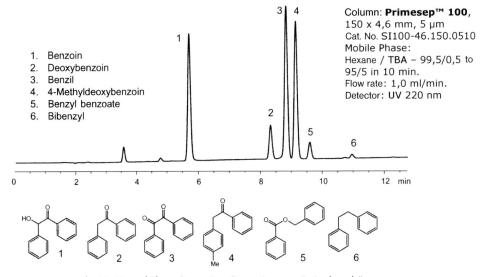

1. Benzoin
2. Deoxybenzoin
3. Benzil
4. 4-Methyldeoxybenzoin
5. Benzyl benzoate
6. Bibenzyl

Column: **Primesep™ 100**,
150 x 4,6 mm, 5 µm
Cat. No. SI100-46.150.0510
Mobile Phase:
Hexane / TBA – 99,5/0,5 to
95/5 in 10 min.
Flow rate: 1,0 ml/min.
Detector: UV 220 nm

Fig. 19. Normal-Phase Separation: Separation on a "mixed-mode" stationary phase (same column and sample as in Fig. 18) with an NP eluent; for comments, see text.

Figure 18 shows the separation with a standard RP eluent; one observes five peaks. In Fig. 19, the separation is shown using the same column but with an NP eluent and at a different wavelength. Firstly, because of the different mechanism there is an inverse elution order, and secondly compounds 3 and 4 are now baseline-separated.

Summary of the "light" version of orthogonal separations

1. Employ your actual eluent on two as different as possible RP phases.
   Variation: Replace e.g. 50% MeCN by methanol.
2. Use a column with different functionalities on the surface of the stationary phase with two as different as possible eluents.

The probability of achieving a successful separation is further increased if one decreases the injection volume and/or changes the wavelength.

1.1.5
Unknown Samples: "How Can I Start?"; Strategies and Concepts

Obviously, there are several possible protocols for addressing this problem. If a procedure is established in your laboratory, which is adjusted to your samples and your environment, and which is effective according to your criteria, then there is no reason to change anything. However, if the development of the crude method for essentially similar samples needs more than about two weeks, then there is probably room for improvement.

In the following, some proposals for straightforward method development are presented, preceded by two preliminary remarks.

1. If it is possible and if it is relevant to the question, the pH of the sample solution should be measured. Sometimes, the conductivity can also be useful. A rough indication of the acidic or alkaline reaction of the sample solution is a welcome aid for the pre-selection of the mobile and stationary phases. Even more effective is the calculation of pK_a values in the case of known or expected substances on the basis of their structures. The most interesting pH range with regard to selectivity as well as robustness experiments is ca. ±0.5 pH units around the pK_a value. One should begin with experiments in this range (see also Chapters 1.3 and 1.4).

2. If the possibility exists, an LC-MS(MS) or even an LC-NMR run at the beginning of the development of a method can quickly yield extensive information. In general, the coupling LC-spectroscopy is suitable either for the last step in an analytical sequence to test the peak homogeneity or for the first step after a successful chromatographic separation (HPLC effectively serving as a better sample preparation technique) to get an initial impression of the sample. Considering that a given mass spectrum can rapidly be compared with ca. 250,000 spectra in spectral libraries (e.g., NIST) via the Internet or via one's

own intranet with spectra of similar company-specific compounds, one should always try to use this possibility. Incidentally, the costs of 1000–2000 Euro, typical for the measurement of an LC-MS(MS) or (LC)NMR spectrum by an external service, are small in relation to the overall analytical costs associated with an important sample.

1.1.5.1 The "Two Days Method"

Applying a systematic procedure and using adequate equipment, it is realistic to be able to develop a method in roughly two or three days.

Consider a hypothetical case, in which no information is known about the sample, only that it can be dissolved in MeOH/water, MeCN/water or THF/water mixtures and therefore an RP separation is possible. In the following, a possible strategy is outlined; see also Ref. [1].

Step 1: Development of a "useful" gradient run

One could start this procedure after lunch, let's say at 13.00–13.30. One applies a linear gradient, e.g. 10 to 90% MeCN with an acidic pH value (pH ≈ 2.5–3.5) at ca. 30 °C. The pH is usually adjusted with phosphoric acid, but perchloric acid, hydrochloric acid or trifluoroacetic acid are interesting alternatives. One will always like to start with one's favorite column, which is okay. However, it should be borne in mind that in this hypothetical case, one does not know how many peaks to expect, so the column should have ca. 8000–10,000 plates, which is obtained by filling a 125 mm column with 5 μm particles or filling a 100 mm column with 3 μm particles. If no more than five to ten peaks are expected, a shorter column can/should be used. The flow rate should be ca. 2 mL min^{-1} and the gradient time ca. 20 min. With these values, a gradient volume of ca. 40 mL results, which is sufficient in the first instance. Theoretically, at least 30–35 components would be separable under these conditions. In any case, one should inject a quite dilute solution and the volume of the injection should be no more than 20–30 μL. Assuming that one is equipped with a DAD, one should exploit its full potential, and this means more than just working at different wavelengths. If time permits, it is very desirable to also test methanol as organic solvent at a pH of ca. 7–8. It needs ca. 2.5–3 hours to get "any" chromatogram after these experiments. If nothing intervenes, the time is now ca. 16.30.

Step 2: Choice of a selective column I

A column selecting valve, which should be an indispensable tool in any development laboratory, can now be charged with 6 or 12 columns. In the case of a completely unknown sample – a situation that rarely arises in real life – one should use "a little bit of everything", e.g., two polar phases, two hydrophobic phases, as well as a 60 Å and a 300 Å material of both. For a detailed discussion of the choice of an RP column, see Chapter 2.1.1.

Under the conditions that result in the "best gradient" in the preliminary experiments (step 1), one should perform a run overnight using the six columns.

To get enough "buffer" in the run time, the analytical time should be set at 45 min. Six interesting chromatograms should be obtained in the morning.

Step 3: Choice of a selective column II

At the end of the working day, a further six columns should be inserted into the column selecting valve, preferably columns with phases that are similar to the best column of the previous night's experiment. This is the column that yielded the most peaks.

Step 4: Fine optimization

Using the most selective of the 12 columns from the two overnight experiments, one should perform additional experiments to achieve optimal resolution. This includes a change of the gradient, including pH, as well as a variation of the temperature. If method development software is available (see the chapters of Part 4), then finding the optimal conditions can be a smart and efficient procedure.

By treating the last step systematically, with or without method development software, there is a good chance of devising a more or less useful method at the end of the day. Clearly, the resulting separation will have to be examined and probably optimized as well.

Because the choice of the column is made using the column selecting valve overnight, the actual working time that has to be invested for this procedure is only a day and a half.

Recently, the major manufacturers have started to offer HPLC equipment with the possibility of automatic operation of the column selecting valve and optimization of the separation according to the requirements of the user by means of suitable software. For more details, see the chapters of Part 4.

As one might imagine, there are different versions of the aforementioned procedure. In the following, some of these are briefly outlined. The reader should decide which concept corresponds most closely with his or her requirements.

(A) The simplified principle of the concept is as follows:
"We want to use at most three different columns and to reach the optimization mostly using variations in the eluent. In this way, it is not necessary to buy a lot of columns." In such a case, one should use, for example, a hydrophobic phase, which one can designate as a "universal column" (Symmetry, Luna C_{18} 2, YMC Pro C_{18}, Nucleodur Gravity, Purospher, etc.), a polar "universal column" (LiChrospher, Zorbax SB C_8, Atlantis d C_{18}, SynergiPOLAR RP, Polaris, etc.), and a column that proved useful in similar cases in the past. If there is no experience of this type in the laboratory, for the third column one could use a completely different type, e.g., Nucleosil 50 or Jupiter (both of which give rise to a steric effect because of the small or large diameter of the pores), SMT OD C_{18} (polymer layer), Fluofix (fluorinated alkyl chain), Hypercarb, ZircChrom (different matrix and therefore

different chemistry). In particularly complicated situation where selectivity needs to be tested in extreme conditions, it would be advisable to use a column that retains its stability in the acidic as well as in the alkaline or at very high temperatures, such as Pathfinder MS/PS (Shimadzu) or Blaze$_{200}$C$_{18}$ (Selerity Technologies). In this case, one would also start with a linear MeCN gradient of the eluent at three different pH values, e.g. pH 2.5, 4.5, and 7.5. Alternatively, and this is recommendable, methanol can be used. If the pK_a values of the expected components are known or can be calculated, the pH can, of course, be carefully adjusted. If one expects neutral and/or weakly polar analytes, then one should use in any case a neutral gradient containing 3–10% modifier. THF, isopropanol, and *n*-butanol are suitable modifiers.

(B) Another concept is as follows:
"We vary the stationary phase and keep two quite different eluents constant in order to ascertain which phase type and therefore which separation mechanism yields the best selectivity. Thereafter, we make a fine adjustment, first by variation of the eluent and the temperature."

This procedure is the same as that described above. Versions of this are as follows:

Version 1

- Overnight, six polar columns should be inserted into the column selecting valve (e.g., from very polar, like Platinum EPS or Hypersil ADVANCE, to moderately polar, like LiChrospher or XTerra).

- Over the next night, one should test only hydrophobic phases (e.g., from quite hydrophobic like Purospher and Discovery C$_{18}$ to very hydrophobic like Ascentis C$_{18}$ or SynergiMAX RP).

- A further night is spent on "specialities" of selectivity. The following possibilities may be considered:

    – One should consider the steric effect (e.g., Novapak, Nucleosil 50, and Spherisorb ODS 1 up to Zorbax SB 300, ProntoSil 300, and Symmetry 300).
    – One uses six completely different columns, e.g., one of moderate hydrophobicity but elevated polarity (e.g., SynergiFUSION RP), a monolith (e.g., Chromolith Performance), one with a long alkyl chain and hydrophilic endcapping (e.g., Develosil), one with incorporated ionic groups (Primesep A), and two "high-speed" columns (20–30 mm, 2–3 mm, 1.5–2 µm).
    – One works with an alkaline eluent for this and therefore has to use an alkaliresistant phase, e.g., Gemini, Zorbax Extend, XBridge, Asahipak, ZircChrom, Hypercarb.

In this way, one can test overnight some quite different phases for selectivity for the given separation problem, without the necessity of working actively with the equipment.

Version 2

If a quaternary low-pressure gradient system and a six-column selector are available, a two-eluent-six-column combination can be tested overnight. For the four mixer chamber inlets use, e.g., the following: MeCN, "acidic" water, "basic" water, rinsing liquid, e.g.:

- First night:
  Eluent 1: acidic MeCN/water gradient
  Eluent 2: alkaline MeCN/water gradient ... on six columns

- Second night:
  Eluent 3: acidic MeOH/water gradient
  Eluent 4: alkaline MeOH/water gradient ... on six columns

Where necessary, additional eluent/column combinations and pH profiles should be tested.

Summing up the method development concept, the proof of peak homogeneity as well as the robustness of the method may be performed as follows:

1.1.5.2 "The 5-Step Model"

This approach draws on experiences from various method development and optimization projects. Variations in the description of the first steps reflect individual preferences of the laboratories involved as well as differences in the hardware used.

Preliminary Remarks

1. Supposing you have access to a powerful automatic gradient mixer with very little dead volume, featuring 2–4 solvent inlets with corresponding eluent switching valve, a coolable column oven, a column-switching valve and a PDA detector, LC-DAD-MS-coupling would be the method of choice. Although using a 12-column selector would provide a wider range of possible variation, we suggest using the more widely available 6-column selector.

2. Furthermore, for economical reasons, we urge you to use only short columns (20–30 mm, 1.8–2 μm) – at least in the initial stages when you are still trying to find your way. Thus, through standards and fast runs, even fairly conventional LC machines could detect trends and roughly identify important parameters such as pH value, solvent and gradient very quickly. Alternatively, if the opportunity arises, a UPLC (Waters), Ultra-Fast-LC (Agilent), X-LC (Jasco), UltiMate (Dionex) etc could be used. In a second step, Fast LC or UPLC results can be transferred to HPLC or those from a conventional LC machine adapted to the expected routine situation if necessary. Where complex samples can be expected in routine separations (mother liquor, stress solutions, contaminated samples, complex matrix etc) the acquired knowledge from testing stationary phases could be used to find appropriate longer and therefore more robust columns.

3. The model shown below is based on a worst case scenario, i.e. no additional information about the sample is available. If the components of a sample are known or within a certain range of expectation, it is worth making a reasonable effort to collect preliminary information about their chromatographic behavior from various sources (internal/external databases, the Internet, software tools such as acdlabs, ChromSword etc) or simply about compound data, e.g. pK_S value (see Chapter 1.4) and degree of dissociation. This could be helpful when deciding on what type of column and what eluent to use. Finally, no further persuasion should be needed in favor of using small columns. Otherwise the amount of time needed for the experiments will go up accordingly.

Step 1: Orientation (day experiment)

Start off running three quick linear gradients (e.g. 5 to 95% acetonitrile) at three different pH-settings (e.g. 3, 7, 9) on six different columns (e.g. C_{18} manufacturer 1, C_{18} manufacturer 2, phenyl, EPG 1 (EPG: phase with **E**mbedded **P**olar **G**roup), EPG 2, "AQ" (AQ: hydrophilic endcapped phase) at an optimum wavelength. Assuming that the gradient duration is 3–6 min, the timescale will be as follows: 3 pH values × 6 columns × approx. 14 min (10 min maximum separation time plus 4 min equilibration time) = approx. 250 min, which is roughly 4 hours. The optimum column-pH value combination (criteria: primarily number, secondarily shape of peaks) is used with the following changes:

1. The pH value is set using a second acid/base (e.g. formic or perchloric acid instead of phosphoric acid; ammonia or triethylamine instead of sodium hydroxide).
2. 50% of the acetonitrile component in the eluent is replaced by 50% methanol (e.g. instead of a 5 to 95% acetonitrile gradient 2.5% MeCN plus 2.5% MeOH to 47.5% MeCN plus 47.5% MeOH) or by 10% THF. The time required for these three runs is approx. 30 min.

This first step which takes approx. 5–5.5 hours should answer the following question: Which stationary phase type, which modifier and which organic solvent in the eluent yields the highest number of peaks? If it emerges that the highest selectivity can be expected in the alkaline, six alkaline-stable columns should be tested at three higher pH levels (e.g. 10, 11, 12) by using two modifiers (e.g. ammonia, ammonium carbonate or borate buffer). Columns to be considered: Gemini, Pathfinder, Zorbax Extend, XBridge C_{18}/Shield, Kromasil, Hamilton PRP/Asahipak.

Note: In certain applications (ion exchange, RI detection, gel filtration), an eluent switching/low pressure valve is used to send naturally isocratic solvent mixtures along the six columns, modifying the solvent, pH value or salt content.

Variation

1. Three columns can be tested (e.g. C_{18}, phenyl, diol) in fast gradients using different solvents (usually acetonitrile and methanol or, rather than pure

methanol, 50% acetonitrile/50% methanol) and three different pH levels. Otherwise, proceed as described above. Number of runs: 3 columns × 2 solvents × 3 pH values = 18 runs. Required time: 18 runs × 14 min (10 min separation time plus 4 min equilibration time) = approx. 4 hours.

2. On the basis of the chromatograms obtained by these runs, it is then decided which parameter yields greater variance – the solvent or the pH value. Then a further test is run on three columns, with either two pH levels and the "better" solvent or with the "best" pH value and two solvent compositions. Number of runs: 3 columns × 2 pH levels or 2 solvent compositions = 6 runs. Required time: 6 runs × 14 min (10 min separation time plus 4 min equilibration time) approx. 1.5 hours.

3. Finally, the best column with the optimum pH level and the better solvent undergo a last small optimization process. Combine each of two gradients with two temperatures, resulting in 4 runs. Required time 4 runs × 14 min (see above), approx. 1 hour.

Total time required for the three steps: 4 + 1.5 + 1 = 6.5 hours. Allowing a little extra time for setting temperatures and other preparations, these tests would be feasible in a day. They also help recognize trends and yield similar information as the previously described runs – which stationary phase type, organic solvent, pH level, gradient and temperature are the most suitable to obtain the highest number of peaks.

Step 2: Choice of Column (Night Experiment)

The column-switching valve is equipped with the best of the six tested columns and five further columns (details about the choice of columns see Chapter 2.1). Two gradients are run in the six columns over night at a pH level ± 0.5 of the pH level previously considered as optimum. 2 pH values × 6 columns = 12 runs.

Step 3: Fine optimization, Method Robustness (Day Experiment)

The heretofore best combination of eluent, pH value and column now undergoes further fine optimization/adjustment – variation of gradient (initial and final conditions, slope, volume and profile of gradient), temperature and, where applicable, particle size and column length. This is also the time when for economic reasons, while chromatographic parameters undergo systematic variation, a first robustness check of the method used should take place (e.g. impact of small variations in pH level etc).Let me mention in this context that optimizing procedures and experiments to ensure method robustness are greatly helped by powerful optimizing programs that are commercially available (see Chapters 4 and 5.2). But even without such aids, it is realistic to set aside two days and one night to roughly work out the method. At this stage at the latest, critical input from an expert not involved in the project is needed.

Step 4: Checking Peak Homogeneity, Cross-Experiments (Night Experiment)

In principle, peak homogeneity can be checked using spectroscopy and ortho-gonal techniques, see above. Alongside PDA we recommend MS- (ESI-MS, TOF-MS, MALDI) and NMR offline/online coupling. 2D or Multi-D-chromatographic separation – often in connection with spectroscopic techniques – have become the tools of choice for a definitive check on peak homogeneity or for the separation of highly complex samples and/or matrices. For more details, see Chapter 5.3. How much work should go into these checks must be decided on a case-to-case basis. Evidence seems to suggest that orthogonal experiments (cross-experiments) are fairly reliable without the need for laborious coupling techniques. The peaks need not be identified, as long as their number is confirmed. The following schema gives an overview of simple and complex ways of checking peak homogeneity.

Note: It is understood that optimum hardware (capillaries, detector cell) and optimum settings have been chosen, e.g. wave length, reference wave length, data rate setting, bandwidth, time constant etc.

Checking Peak Homogeneity in RP HPLC

1. Make the most of what your machine has to offer without changing methodo-logical parameters: PDA, perhaps LC-MS.

2. Here is what you can check in isocratic runs: Are the quotients peak width/retention time on a straight line for all peaks? Could an outlier be an indication of the inhomogeneity of the peak in question?

3. Easy and fast checks: decrease injection volume, dilute the sample with water/eluent and reinject, use the same stationary phase with smaller particles.

4. Orthogonal tests:
   • Same column, different eluent (e.g. methanol instead of acetonitrile, or pH level Y instead of X).
   • Same eluent, different column (e.g. nonpolar instead of polar stationary phase).

5. Fractionate main peak (front flank, tip, back flank), concentration of the fractions is hardly ever needed.
   a) Reinject the fractions one by one.
   b) After a second column has been switched in series, the fractions are again injected one by one.
   c) The fractions are examined using GC, DC or CE.
   d) The fractions are examined using IR, MS or NMR spectroscopy.

Tests 1 to 4 and/or 5a or 5b should always be carried out on important samples. Spending a day on checking points 1 to 4 or, respectively, about two days in addition for points 5a and b is worth the effort, while decision to invest in two additional days for point 5a and b depends on the particular circumstances.

Step 5: Robustness of Method (Day Experiment),
Column Stability (Weekend Experiment)

The last step of a method development project should be a check of the ruggedness, if the method under consideration is not to be used for a single application. This means the capability of a method for routine work. The effort and the tests required depend of course on the actual question. Although robustness/ruggedness are very important issues, they are not topics of this book, and therefore the possible tests are only mentioned in brief.

Important Notice: It is vital that robustness is checked using real samples. If these are unavailable, samples should be prepared that resemble as closely as possible those that will be used in routine separations later: Matrix/placebos/solvents/excipients/stress solutions plus analyte. There is little point in testing the robustness of a method on standard solutions – or else there may be a few nasty surprises in store!

(A) Compatibility Sample-Stationary Phase

1. Catalytic Effect of Silica Gel?

Silica gel as a stationary phase or matrix in RP phase is a very good catalyst for solids. It is often found that many small peaks in the chromatogram stem from compounds that developed "in situ", i.e. through the catalytic impact of the column and could not be found in the original sample.

Double-check: Inject the sample as usual and switch off the pump while the sample is in the column. Leave the sample in the column for 20–30 min and then switch on the pump again. Do peak numbers and areas remain constant?

2. Irreversible Sorption in the Stationary Phase?

- Once the column has been equilibrated, inject the sample several times. Does the peak area remain constant?
- Inject a small sample volume, circumventing the column by connecting the sampler to the detector. Without the column, only one peak will appear. Make a note of its area. Now connect the column and inject the same sample volume. Depending on the column selectivity, 1, 2, 3 or more peaks will appear. Add all the peak areas and compare the sum to the peak area obtained from the run without column. If the difference is less than 5–8%, irreversible sorption in the stationary phase is negligible.

(B) Impact of the Sample Solvent

The sample solvent and/or the sample medium can have an impact on retention time, peak shape and peak area. Change one of the following parameters in the sample solution as appropriate: pH value, organic component, matrix, constitution of the sample, air content in sample solvent. When dealing with samples from a production plant, it is a good idea to be in close contact with the colleagues working in the plant concerned who could provide current updates. It is the small changes in sampling or the actual medium of the sample, often regarded as negligible in day-to-day work, which can cause major problems in analysis.

(C) Compatibility of Eluent and Stationary Phase (Column Stability)

- An intentionally aggressive eluent is applied to a column over a whole weekend if a high durability of the column is important for routine work. In this way, ageing of the column can be simulated, and after just two days one will have a good idea of whether one is dealing with a selective and not a robust method suited for routine work, which is better than acquiring this knowledge later in the course of routine work. "Aggressive" conditions have to be defined for individual cases and result from the preceeding optimization experiments, e.g.:

  - high water content (e.g., 95–100%)
  - extreme pH values (e.g., pH 1.5/9.5)
  - high salt content (e.g., 50–100 mmol)
  - increased flow (e.g., 2.5–3.5 mL min^{-1})
  - high temperature (e.g., ca. 50 °C)

  or in the extreme case all of these conditions simultaneously.

With polar RP-phases, two or three charges should be tested in any case.

(The laboratory precision would probably be checked in the framework of an official validation procedure, along with the reproducibility, if necessary.)

To sum up the procedures described above – here are the key figures:

Time required: four days, two nights and a weekend
Labor input: approx. 50–55 runs
List of what could be tested:

- 5 pH levels including fine-tuning;
- 15 columns including columns for cross-experiments;
- 3 organic solvents, 2 modifiers, and, where strong bases are involved, three further pH levels as well as two modifiers and further special columns;
- optimization of gradient and temperature as well as column dimensions and particle size where applicable;
- finally, data concerning peak homogeneity, method robustness and column stability.

Of course, this list is open to variations on a theme. Let me just mention two of them:

Variation 1

Hardware required: Low pressure gradient with eluent-switching valve, i.e. six solvent inlets or, if a high pressure gradient is used, two pumps with three solvent inlets each.

Alternative to steps 1–2, as follows:

1. *Night experiment:* In order to get an overview, run a gradient at five different pH levels (plus a rinsing solution) on six columns (5 gradients × 6 columns = 30 runs)
2. *Day experiment:* The optimum column is run under optimum conditions (acid/base, modifier, see above) to serve as reference in a comparison to five further columns (6 runs).
3. *Night experiment:* Six further columns are tested at two pH levels (±0.5 pH units off the optimum pH level) 2 pH levels × 6 columns = 12 runs.

Step 3 to 5 (fine optimization and column stability) as above.

Time required: four days, three nights, one weekend
Labor input: approx. 54–58 runs
Comments: In this variation, two more pH levels would be tested on a higher number of columns – 17 rather than 11. This may be helpful in difficult samples containing a wide range of polar/ionic species.

Variation 2

Hardware required: Two quaternary pumps
Alternative to steps 1–2, as follows:

1. *Night experiment:* Three pH levels, three eluents (each of the two pumps has one solvent inlet reserved for the eluent), six columns, 6 eluents × 6 columns = 36 runs.
2. *Night experiment:* The best column so far is used as reference column for two pH levels (±0,5 pH unit of the optimum pH level and the optimum eluent to be compared with further five columns (6 runs) 2 pH levels × 6 columns = 12 runs. Step 3 to 5 (fine optimization and column stability) as above.

Time required: three days, three nights, one weekend
Labor input: approx. 50–53 runs
Comments: In this variation, variants of organic solvent could be tested on more columns. This could be helpful in laboratories who are short of staff (three nights and three days compared to two nights and four days as in variation 1). Also useful for samples containing many components tending to neutral.

Final Remarks

The scope of this five-step program for effective method development/optimization is based on a worst case scenario, and depending on the individual situation, the five steps could be modified and shortcuts taken. Usually, some information about a sample is available, perhaps also about its area of use. Such information must be used in order to shape the testing conditions, e.g.

• Known pK_S value – experiments regarding the pH level can be targeted and thus reduced in their number.

- Decreased pH stability at pH = X or in solvent Y? Avoid pH level X or solvent Y.
- Expecting many low-concentration impurities? Choose 3 µm material.
- Complex matrix? Take great care in preparing the sample and use 5 rather than 3 µm material.
- Expecting many components in a complex matrix? Go for 2D chromatography from the start.

The following alternatives to the approach described could be considered in certain situations or service laboratories:

- Fully automatic method development (see Chapter 4).
- Super fast separations (see Chapter 2.7).
- Developing a fast LC-MS method (see Chapter 5.1).
- Barely sufficient chromatographic selectivity (using very short columns) combined with spectroscopic specificity (NMR, MALDI-TOF, FTIR, X-ray fluorescence).

Successful combinations could be listed as in Table 1 for quick reference. Various optimizing parameters (pH level, solvent, column etc.) in the table columns and the relevant chromatographic key data or results after an optimization step in the rows could be summarized in such a matrix. Rows A to D are individual criteria that can be defined and amended as needed. For example, the higher the numeric value of row B, the more effective the separation (peaks per time unit, peak capacity).

It goes without saying that only those optimization parameters should be recorded that led to useful results, i.e. the matrix should be filled in after each experiment. If solvent THF does not yield a reasonable situation, column 6 in Table 1 should be left out altogether or a different optimization parameter should take its place. If desired, A to D can be arranged in order of relevance. But even

Table 1 Overview of results following an optimization step.

Criteria	Parameter					
	pH = 2	pH = 2	pH = 5	pH = 5	pH = 8	pH = 8
	MeCN	MeOH	MeCN	MeOH	MeCN	THF
A	4	5	3	5	etc.	etc.
B	0,4	0,5	0,3	0,6	etc.	etc.
C	1,3	1,5	1	1,1	etc.	etc.
D	Good peak shape	Drift, high pressure	Tailing at Peak X	broad Peaks	etc.	etc.

Examples for individual criteria regarding the quality of an RP separation
A: Absolute number of peaks
B: Number of peaks per time unit, e.g. 4 peaks in 10 min, 4/10 = 0.4
C: Critical pair resolution
D: Comments, e.g., narrow peaks, base-line drift, high pressure, tailing peaks etc.

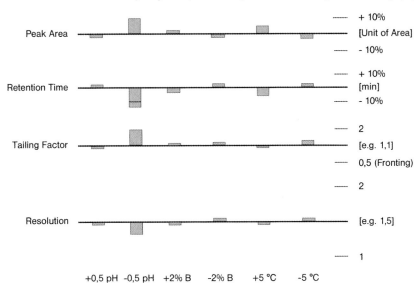

Fig. 20. Graphical representation of how specific parameters of a method influence chromatographic key data in a robustness experiment; for details, see text.

without such an order, one glance suffices to see which column-eluent combination is right for the specific criteria required.

A similar matrix gives a quick overview of the effect of various factors when testing rugedness (see Fig. 20). Chromatographic key data are entered on the left-hand side of the matrix. The changes they undergo due to variation in methodic parameters are to be analyzed. The changes carried out are entered into the line at the bottom of the matrix, which should also be specific – not only regarding the parameters to be analyzed (pH level, solvent ...) but also the scope of change (±05 pH units, X% B ...) When checking laboratory precision (intermediate precision), for example, "user 2" or "machine 2" etc. could be entered. The horizontal lines reflect the resulting numeric value (area, retention time ...) under the usual separation conditions – those that have been agreed upon when developing the method. The individual changes are visualized by hatched areas. Measurements can be given in % or as a numeric value. Based on the degree of change brought about by specific parameters, a decision can be taken which parameters should be considered as system suitability parameters for a later system suitability test and what criteria (accepted bandwidth of the numeric value in question) should be given. Figure 20 is an excellent example and should be interpreted as follows: The component is highly sensitive to a decrease in pH, apparently slipping near the pK_S region. This cuts the retention time by more than 10% of the time expected for similar separations. Similar findings apply to resolution – a neighboring peak is insufficiently separated. Furthermore, UV absorption seems to be pH level-dependent, as changes in the peak area suggest. It also seems that in this method, an increase in temperature can be critical (change in pH level – change through change in temperature?) As far as other parameters are concerned, the method seems to be rather robust. This kind of documentation could be enclosed to a validation report, as it facilitates decision-making further down the line if the method needs to be changed and the question is whether such a change would be only an adjustment (does it meet criterion X?) or revalidation is called for.

1.1.6
Shortening of the Run Time ("Faster Separation")

A decrease of the resolution, except in the case of gradient runs with a constant gradient volume, is associated with a shortening of the retention time. This can be very small, e.g. enhancement of the flow using 3 µm material, or it can be considerable, e.g. a change in the pH. One has to decide in each specific case whether the advantages exceed the disadvantages. Experience shows that one frequently works too cautiously and accepts needlessly long retention times, e.g. at 1 mL min^{-1}. A lot of possibilities are discussed above and outlined in brief below.

- enhancement of the flow,
- reduction of the column volume (shortening of the column length and/or diameter),
- change of the eluent combination (e.g., water content, modifier, pH, ionic strength),
- increasing the temperature,
- change of the stationary phase (e.g., use of a more polar phase or choice of another matrix, e.g. non-porous material, monolith),
- use of a gradient technique (do not overlook flow gradients),
- change of the gradient (e.g., starting conditions, steepness, gradient volume).

The potential of modern columns should be emphasized once more; with a column of dimensions ca. 20–50 mm × 1.5–2.1 mm, 1.7–2.0 µm, 10–15 peaks can be separated within 2–4 min; see also Chapter 2.7.

1.1.7
Improvement of the Sensitivity ("To See More", i.e. Lowering of the Detection Limit)

Aspects of this possibility have already been discussed above. Some of them are collected in Table 2.

1.1.8
Economics in HPLC ("Cheaper Separation")

Remark: It should be obvious that "economical" means more than just "cheap". Thus, aspects at the periphery of HPLC analytics play a more important role than the actual analytical aspects. Some of these questions were addressed at the start of this chapter.

Table 2. Possibilities for enhancing sensitivity/lowering the detection limit in RP-HPLC.

(a) Aspects due to the HPLC equipment
   (e.g., use of the appropriate modules, optimal settings, avoid noise)

- Increase the sensitivity (range/attenuation) – the peak-to-noise ratio improves.
- Use a small time constant (< 0.5 s, or better 0.1 s).
- Are the settings for *this* separation optimal? That is, sampling rate, sampling period, rise time (bunching factor, response time), peak width, wavelength. Is the detector the right one? Is perhaps a derivatization necessary?
- Consider electronic damping; are protective covers on cables/interfaces O.K.?
- Miniaturization successfully realized? ($L \downarrow$, $ID \downarrow$, $dp \downarrow$) Are cell volume and design appropriate for *this* separation?
- Are the cell and the lens clean?

(b) Chromatographic conditions

- Increase the plate number.
- Apply steeper gradient.
- Increase the injection volume/mass (overloading?).
- Improve the peak shape (pH, modifier); remark: a tailing factor of 1.5 leads to a decrease of one-third in the peak height and thus to a decrease of one-third in the sensitivity.
- Avoid band spreading – aim for "optimal" injection techniques; on-column concentration possible/necessary?
- Are the membranes of the degasser still OK?
- Avoid baseline drift in gradient runs; is *this* eluent the right one for *this* wavelength? Might the use of a UV absorber in eluent A decrease the drift?

In the following, points are listed without any ranking, which one should remember when aiming for efficiency enhancement or cost saving. *Here* lies the big potential, not in the 20 or 30% discount when buying columns or equipment.

- In the event of method transfers or a co-operation with contract partners, one should aim for a standardization of the equipment.
- The validation should be performed under real and not under optimal conditions. From the information point of view, what can be gained if a skilled expert using optimal equipment and clean standard solutions measures several times during a week and thereby obtains a relative standard deviation of 1% and a correlation factor of 0.999?
- The person responsible for the project should attend to the whole project, e.g. from the development of the new substances up to the registration and should use his or her knowledge and up-to-date information in discussion with and between the various departments. Such a person can be a very important mediator, but it is important that this person has the ability and the permission.
- The sequence of operations should be adapted to the requirements and the realities. Single steps in the daily life of the laboratory should be intelligently combined.

- One should efficiently organize recurrent activities such as purchasing, sample management, equipment testing.
  One should endeavor to find reasonable guidelines for quality management.
- There should be a well-functioning interchange of information – think communication!

These points are not topics of this book and therefore are only mentioned here as a brief indication on effectiveness and efficiency in the HPLC environment. If the "company philosophy" permits, one should always reflect on the effectiveness and efficiency of one's own activities, almost instinctively and unconsciously.

In the case of effectiveness, the question concerns the correct choice of the means ("am I doing the right thing?"), while in the case of efficiency the issue is the correct use of this means ("am I doing it in the right way?").

Examples involving effectivity

It should be considered, for example, whether it is sensible to use HPLC for the entry control of raw materials, for identity checks, for the measurement of content, or for dissolution tests. In the case of the first two of these problems, NIRS might be a more sensible option, while for the latter two a titration or an on-line UV measurement would be more economical. It is not necessary to obtain the information through a "separation" in all cases.

Dissolution tests often result in a single peak, in some cases, can only be separated using the solvent. This begs the following questions:

- Do I really need a column, as this may be a case of identification rather than separation? Could I perhaps just use a capillary instead, as in a FIA (*Flow Injection Analysis*)?
- If it has to be a column, would perhaps a 10 mm × 2.1 mm precolumn do as a separation medium?
- Do I need to use the expensive and toxic solvents acetonitrile or methanol, or would ethanol be a possible alternative?

Examples involving efficiency

Is the use of the available equipment optimal, for example the DAD or the elusaver? Is it perhaps possible to work with methods in such a way that cleaning time and reflushing of the equipment can be reduced?

Are there "clever sentences" in the validation report that make it possible to denote small changes in the method as an adjustment by observation of the given requirements?

In this way, expensive, large-scale revalidation can be reduced to the absolute minimum. Experience shows that the margins given by organizations such as the FDA, EPA, Pharmeura, etc., are not always adhered to. On the contrary, fear and widespread alibi mentality has the result that one is frequently overcautious.

Consider now the purely chromatographic possibilities for saving time and reducing costs.

Is a "good" separation also an "optimal" separation?

If one is dealing with an isocratic method with an MeCN content in the eluent of 60–80%, then one should take a good look at the overall separation. Such a high MeCN content may be an indication of the following: the interaction of the compound(s) with the stationary phase is perhaps so strong that this high organic content of the eluent is necessary to obtain a short, acceptable run time. It is more economical in this case to use a shorter or thinner column. This could, for example, be a hardware change by a factor of two. In the first case, this would result in a saving of eluent used by a factor of two, in the second case by a factor of four. Even a change from a 4 mm to a 3 mm column would lead to a saving of solvent of ca. 45%. In this context, one should think not only of the purchase price of the solvent, but more of the costs of removal and disposal and of environmental concerns.

As already mentioned, several aspects fall under the umbrella of "economics", including time. Provided that the packing quality and the properties of the stationary phase are the same, one obtains the same resolution using a 150 mm, 5 μm column as using a 100 mm, 3 μm column, or a 50 mm, 1.7 μm column, respectively. In the second case, however, the time saving is ca. one-third, in the third case ca. two-thirds. For more time-saving procedures, see above ("faster separation").

Finally, the following remark: An optimization step, however it is achieved, only really yields results if one is willing and ready to draw conclusions.

To illustrate this, just think of an everyday example: one enhances the flow of a simple separation to reduce the retention time. Possibly this advantage yields absolutely nothing, because, if no other action happens, it is really unimportant, if a series of measurements ends at midnight or at 3 am. However, if the laboratory routine and other working processes (integrated laboratory production/raw material delivery) can be adapted to the obtained time saving, then, for example, the autosampler can be reloaded in the evening and the production unit can get required information directly in the morning. A time saving of 10 hours in the laboratory can be important; a time saving of 10 hours in production is a lot. However, these are decisions for management to make and are beyond the scope of this book.

1.1.9
Final Remarks and Outlook

The following topics would seem to be worthy of note:

- Even in a strongly regulated environment, there should be scope for searching questions. The sincere explanation of requirements in relation to the actual goals makes the relations with others easier and also makes general economic sense.

- Some checks of robustness should be made at an early stage of the project, even during the development of the method. The later a weakness of the method becomes apparent, the more expensive it is.

- A small improvement of the communication – communication in this context means more than just the exchange of words, arguments, and opinions – can yield an remarkable effect. A 5% improvement of real communication can lead to an improvement of 30% in earnings [3]. Admittedly, it is difficult to prove such an improvement or the opposite (frustration of the employees) in numerical terms, but it is possible.

- In the case of an unknown sample, do not trust any one peak, even if it is very sharp.

- After a successful separation, one should use a completely different column-eluent combination and inject 1–2 µL, exploiting the possibilities of the DAD. Orthogonal separations yield more security than match factors, etc., of the DAD after a one-dimensional separation.

- If one is dealing with a really important sample and the truth is at stake, then 1000 or 2000 Euro for an LC-MS(MS) or LC-NMR/2D-NMR measurement or the consultation of an expert frequently represents a profitable investment.

- Is one aware that by injecting standard solutions one just measures the precision of the instrument and this test has nothing to do with *the* "system suitability test", even though everyone in the laboratory speaks as if this were the case? Unless, that is, I could prove in the validation process that matrix, all excipients, degradation products etc have no impact on resolution and sensitivity. If, in addition, the standard deviation is not exceeded more than 1–2 times a year, I could conclude that my system is robust and consider increasing the time span between system checks. This is often accepted by national authorities as well as by e.g. the FDA, as long as it can be documented through historic data and control cards that the control process is ISC (in statistical control).

- Is "top value" the same as "good analytics"? A lot of us are probably of the opinion that the proportions are doubtful if someone drives a Ferrari for 500 m from home to work place everyday or if a private modern baseball stadium is only used twice a year only because of a lack of time. In the same way, a critical discussion of top values/top performance is advisable in analytics: is it meaningful to equip the instruments for quality control with DADs only for the reason of image or the like, if de facto these facilities will be used at only 20% capacity, if at all? Is the requirement "$R > 2$" meaningful? Is it necessary to strive for optimal resolution in every case, or can it sometimes be better just to aim for resolution that is adequate for the analytical problem at hand? How many "out-of-spec" situations do I produce for my colleagues in quality control, if as method developer I claim a relative standard deviation of 0.8% for a method with a biological matrix or a contaminated process sample?

Outlook

From today's point of view, the following trends in HPLC in the context of method development/optimization can be discerned:

Coupling techniques: For the separation of complex samples, more and more coupling techniques will be used, partially in miniaturized form, e.g. "RP-ion exchange" in µ-bore design and subsequent on-line coupling with MS. Through the further development of interfaces in mass spectrometry, applications of MS as the dominant partner of LC will become more diverse. The LC-NMR coupling will only prevail in pure research laboratories, because of the high prices of 600–900 MHz NMR equipment. For automation in relation to coupling techniques, see below.

2D/Multi-D Chromatography: 2D- and multi-D-chromatography will become increasingly important, and there are two reasons for this: (1) Such systems can easily be put together using commercial modules, and any easily available software can be used for control functions. The wide variety of available separation media (columns, capillaries) makes practically every imaginable combination possible. (2) The complexity of samples and thus the need for chromatographic resolution is ever-increasing. All sorts of combinations are possible, as long as the basic requisite is met – the second dimension must be much faster than the first. If, for example, a micro bore column in the first dimension has an injection volume of 1 µL, the negative effect of a possible incompatibility of eluents in an NP-RP combination will be barely noticeable. Assuming a peak capacity of $n = 9$, switching a second column in series would improve peak homogeneity "only" by a factor of 1.4 ($n \cdot \sqrt{2}$, thus $10 \cdot 1.4 = 14$ peaks), whereas orthogonal switching would yield a (theoretical) increase by a factor of 10 ($n = 10^2 = 100$ peaks). Here is a list of successfully used combinations:

NP-RP, HILIC-RP, ZrO_2-RP, SEC-LC, IC-LC, LC-GC, LC-GC-GC, LC-LC-APCI-MS.

Miniaturization: Columns of length 20–30 mm packed with < 2 µm particles can be expected to become increasingly popular, at least in method development, because of the short retention times and the low consumption of solvent. The long-expected breakthrough has been held back because of the well-known circumstances in analytics, and may start in the next 5–8 years. Only recently have the major HPLC providers seriously addressed the issue of miniaturization techniques, not only in the field of the columns on offer, but also with regard to the accessory equipment. UPLC (ultra-performance liquid chromatography), Ultra Fast LC, HSC (high-speed chromatography), RR- and RR-HT (rapid resolution high-throughput) columns are some examples. The handling of columns filled with particulate material, even in the case of 1.5–1.8 µm particles, is easier than that of other separation media.

Therefore, for the time being, CLC (with particulate material or monolith), chip-LC, and non-porous materials remain interesting but specialized niche products.

Conclusion: The future of HPLC lies without any doubt in fast separations. From today's perspective, the most promising routes towards this goal are UPLC, Xtreme LC (Specific hardware, high pressures, ca. 1.7 µm material), Ultra-Fast-LC (conventional HPLC machine, perhaps with minor hardware/software modifications and 1.8–2 μm material), chip HPLC and monoliths, preferably in capillaries. Some of these technologies are already quite sophisticated. When taking a decision about possible equipment changes, the user should think very carefully if the time is right and which of the methods on offer meet the actual needs. Here are some of the questions that need to be answered before coming to a decision:

- How important is a retention time of 30 or 60 sec if the overhead time (injection, needle-washing etc) takes up 2 min or the sample preparation time is even 5 min? How many samples do I have to deal with on a daily basis?
- Am I prepared to become long-term dependent on a specific – albeit excellent – technology provided by a single firm?
- What seems to be more acceptable to my colleagues/contract partners – a flow rate of 10–20 μl/min or 5 ml/min? Depending on the matrix used and the dimension of the column/capillary, the desired decrease in separation time can be achieved in any case.
- What about validation/method transfer? Will the real life samples in routine separations provide the same consistency? How experienced is my staff?
- Can I make use of the substantial time savings and the faster information or would the advantage just lie in the very short analysis time?

In a situation where faster data availability helps to optimize the production process – by enabling me to analyze more experiments (e.g. stress samples under a variety of conditions, intermediate products of a synthesis process etc.) and thus underpin my conclusions, if the time for method development can/must be cut by a factor of five, or if I do not have to deal with difficult matrix, then I would take the plunge sooner rather than later. The potential scientific and economic benefits are immense.

Automation: A high level of automation in method development and effective time-saving procedures look set to become increasingly indispensable.

Some examples for illustration:

1. Instruments for automatic overnight method development with computer-aided optimization and a column selecting valve are now associated with high robustness and will be applied more and more. On the basis of calculated substance data, a computer-aided informed choice of column and eluent is possible, at least the "assiduous" choice of column and eluent, until a given chromatographic resolution is reached. In the future, more will be calculated before the measurements start: e.g., pK_a and log P values, and other chromatographically relevant data. One can expect that more and more evolutionary and generic algorithms are implemented in optimization software. After this, the appropriate chromatographic conditions (column, eluent, temperature) can be selected manually or automatically by means of optimization software. The chromatographic software of the future will use its "memory" to optimize the separation according to given requirements through interaction with databases via the Internet or an intranet, using individual filters and with the

modules of the HPLC equipment including columns containing inserted microchips. Communication with the user will be possible at all stages and the software will be able to "ask" the user in difficult situations. The systems will be self-starting, self-testing, capable of independent calibration and measurements, and, through built-in diagnostic tools, capable of recognizing errors that arise and in simple cases of compensating for these. The sample may then proceed to a second, parallel connected, preconditioned column – in case the first no longer provides the required quality (selectivity, efficiency) for the separation problem at hand.

The technological challenges of the future lie in:
- handling micro/nano capillaries and nano connectors,
- reproducibility of microcolumns and capillaries,
- packing chiral micro bore columns and capillaries,
- chip technology.

2. After a rough separation on a 10–20 mm column, one obtains the necessary information by a subsequent step having a higher level of specificity. This requires a splitting system; the sample is transferred in parallel or optionally to DAD-MS(MS), NMR-MS, ICP/OES, ICP/MS, FT-IR, fluorescence/X-ray fluorescence and, if necessary, also to a second (chromatographic) separation system (e.g., IEC, GPC or CE). By using sophisticated LC-UV-NMR-MS coupling techniques, a dream of separation scientists seems at last to be becoming reality – at least in specific cases: "enter the sample and out comes the structure!". Such systems have been available for some years in different versions. In contrast to the systems listed under 1, such coupling techniques are very complex and usually only usable to their full potential by the user who designed them. In general, one can conclude the following: as far as is possible, "physics" should be preferred to "chemistry". From the user's point of view, the handling of "physics" is easier in daily work. The measurement/testing has fewer variables and therefore fewer sources of error and results are obtained immediately. Frequently, the relationship effort/specificity = information is better.

"Chemistry" in the column: As is explained in detail in Chapter 2.1.1, polar/ionic interactions – being highly specific – play a rather more important role in RP-HPLC than hydrophobic interactions. For this reason, in the last decade more polar than "classical" hydrophobic RP phases have been developed. This trend looks set to continue. Any type of alkyl chain that enables the necessary retention of the components may be attached to the phase surface. On the other hand, polar groups reside on the surface of the stationary phase which are important for the differentiation of ionic to moderately polar analytes. Polar character of a phase can be imparted in different ways: polar and ionic groups in the alkyl chains ("embedded", EPG, zwitterion chemistry at the alkyl chain), polar terminal groups, polar groups on the matrix surface, low coverage, hydrophilic endcapping, sterically protected surfaces, etc. Because of the dual character of such phases, one or other mechanism can be selectively invoked by the selection of the eluent and the number of columns that need to be used decreases. Also, to decrease costs and for more environmentally friendly work, thermally stable columns will be developed (mostly based on zirconium oxide, and less frequently based on titanium oxide or polymer based) for use at 100–200 °C:

1. Supercritical water and/or ethanol can be used as cheap and ecologically benign "hydrophobic" eluents.
2. Instead of solvent gradients, temperature gradients can be used. Finally, types of stationary phases that undergo a change of a surface functional group from polar to hydrophobic depending on temperature are solvent-conserving. In this way, the selectivity can be obtained primarily by a temperature change. Such a column can be used with either a polar or an apolar eluent at either a lower or a higher temperature. The aim is to achieve a broad range of selectivities with only four runs and using a minimum number of different columns and different (cheap, non-toxic) eluents.

Final Comments

In analytics, there is a trend away from High Throughput Analysis towards High Content of Information. Producing simply large amounts of data of the same quality (e.g. chromatographic data following an RP separation) has not provided the desired results in difficult cases, as demonstrated by combinatorial chemistry. The density of information will enhance the quality of analysis if the data obtained from a sample are robust rather than numerous and can be correlated to each other (multiple statistical approaches, improving data analysis techniques). It is thus possible to achieve maximum resolution without unreasonable effort, and peak purity can be checked in 2D separations. Spectroscopic data (DAD, MS, ICP, MALDI-TOF) can be used to achieve very high specificity. Further information about the target component such as chirality, biorelevance is available at an early stage. Chemometrics, generic algorithms, optimizations tools etc. will be used to an ever larger extent. For economic reasons, there is a trend towards to preliminary calculations in order to predict results, discover interdependencies and approach method development more efficiently. Pro-active thinking and calculations have the edge over laborious and expensive measuring experiments. Rather than being achieved by testing at the end of the production chain, quality should be achieved through intelligent process design adapted to the problem at hand. Post factum quality control is going to be replaced by continuous optimization of the individual stages in the process, known as quality assurance. This allows early corrective steps to be taken where needed. It is a matter of establishing quality earlier in the process. The earlier critical factors can be discovered and checked, the more robust the method or process can become, which reduces quality control costs – in other words stop and think before you experiment. Not for the first time does such advice come from unexpected quarters, e.g. from authorities such as the FDA. Rather than insisting on compliance with formal requirements and strict reglementation (which was not very successful), continuous gentle pressure is supposed to help implement quality assurance in the early stages of development (Modern Quality Management Techniques, Real Time Quality Assurance). Another buzz word on these lines is PAT (Process Analytical Technology) – analytic facilities should be available all along the process, but high-tech is not always needed, as long as pH value, temperature, humidity, particle size etc. are checked at the critical points to identify potential flaws. Minimizing risks by preventative checkups using intelligent simple analytical tools is the order of the day.

The realization of perfected solutions clearly depends on a lot of factors, in analytics as in other fields. Therefore, a lot can be realized but not applied. Nevertheless, many of the trends outlined above are reality today. However, it is difficult to predict which of them will find their way into the laboratories most widely. Economic preconditions may on the one hand enhance the formalism, while on the one hand reducing the amount of regulatory requirements to a normal level. Frequently, both trends coexist. It is to be hoped that common sense can be given precedence over formal regulations.

References

1 S. Kromidas, "More Practical Problem Solving in HPLC", Wiley-VCH, Weinheim, 2005; ISBN 3-527-29842-8.

2 S. Kromidas, "Eigenschaften von kommerziellen C_{18}-Säulen im Vergleich", Pirrot Verlag, Saarbrücken, 2002; ISBN 3-930714-78-7.

3 Unpublished results from projects on cost reduction and improvement of efficiency in an analytical laboratory.

1.2
Fast Gradient Separations

Uwe D. Neue, Yung-Fong Cheng, and Ziling Lu

1.2.1
Introduction

The time involved in the execution of analytical procedures is a subject of major interest in today's pharmaceutical industry. It is also of increasing importance in other industries as well. The general goal is to reduce analysis time as much as possible in order to reduce the cost of the analysis and to increase sample throughput. At the same time, the quality of the analysis must not be compromised. In chromatography, the key parameters are the resolving power of a separation as well as the sensitivity of an assay. In this chapter, we show how the peak capacity of a gradient can be maximized through appropriate choice of the gradient operating conditions. We cover both the theoretical basis of this technique as well as solutions to practical difficulties.

The principles outlined here are especially relevant for the rapid analysis of plasma or urine samples by means of liquid chromatography coupled with mass spectrometry [1]. In the last few years, understanding of this analytical technique has continued to grow, and investigators have realized that ion suppression plays an important role in signal suppression and signal enhancement in HPLC/ESI-MS analyses [2]. This problem can often be avoided completely if the chromatographic resolution is maximized [3]. At the very least, the solution to this problem can be simplified by maximizing the separation power of the chromatographic system. The thought processes involved in this solution include the manipulation of the separation through pH change as well as the optimization of instrumental parameters such as gradient delay volume or detector sampling rate.

1.2.2
Main Part

1.2.2.1 Theory
The principles of fast gradient separations have already been outlined in the past [4, 5]. Here, we only reiterate the most important results. As a measure of the quality of a separation under gradient conditions we use the peak capacity P, which is defined as follows:

$$P = 1 + \frac{t_g}{w} \tag{1}$$

Here, t_g is the gradient duration, and w is the (average) peak width. The peak width itself is a function of the column plate count N, the retention factor at the column outlet k_e, as well as the column dead time t_0:

HPLC Made to Measure: A Practical Handbook for Optimization. Edited by Stavros Kromidas
Copyright © 2006 WILEY-VCH Verlag GmbH & Co. KGaA, Weinheim
ISBN: 3-527-31377-X

$$w = 4 \cdot \frac{t_0 \cdot (k_e + 1)}{\sqrt{N}} \tag{2}$$

The retention factor at the column exit depends on the gradient steepness G:

$$k_e = \frac{k_0}{G \cdot k_0 + 1} \tag{3}$$

where k_0 is the retention factor at the beginning of the gradient. The gradient steepness G is defined as follows:

$$G = B \cdot \Delta c \cdot \frac{t_0}{t_g} \tag{4}$$

Δc is the difference in the concentration of the organic modifier at the beginning and end of the gradient. B is the slope of the linear relationship between the logarithm of the retention factor and the solvent composition. The ratio of the gradient duration to the column dead time is called the gradient span.

In all cases of practical interest, the gradients extend over a large difference in solvent composition, and the analytes have a high retention at the start of the gradient. Under these circumstances, we can further simplify Eq. (3):

$$k_e = \frac{1}{G} \tag{5}$$

If we now combine all of these equations in order to express the peak capacity as a function of the gradient parameters, we obtain:

$$P = 1 + \frac{\sqrt{N}}{4} \cdot \frac{B \cdot \Delta c}{B \cdot \Delta \cdot \dfrac{t_0}{t_g} + 1} \tag{6}$$

With this equation, and the simplifications mentioned above, we can now estimate how the peak capacity changes as we vary the gradient parameters. Thus far, we have assumed that the plate count remains constant. However, what we would like to do is vary the flow rate at a constant run time. This means that the plate count will change as well. To introduce this variation, we use the van Deemter equation:

$$H = A \cdot d_p + \frac{B \cdot D_M}{u} + \frac{d_p^2}{C \cdot D_M} \cdot u = A \cdot dp + \frac{B \cdot D_M \cdot t_0}{L} + \frac{d_p^2}{C \cdot D_M} \cdot \frac{L}{t_0} \tag{7}$$

Here, d_p is the particle size, u the linear velocity, L the column length, and A, B, and C are the constants of the van Deemter equation. The diffusion coefficient of the analytes, D_M, depends on the sample and the composition of the mobile phase. In order to avoid excessive complications, we assume typical diffusion coefficients for the types of analytes that we are using. For the coefficients of the van Deemter equation, the following typical values can be found in the literature [6]: $A = 1.5$,

$B = 1$, $C = 6$. Using the known relationship between plate count and theoretical plate height, we obtain the peak capacity:

$$N = \frac{L}{H} \tag{8}$$

$$P = 1 + \frac{1}{4} \cdot \sqrt{\frac{L}{A \cdot dp + \dfrac{B \cdot D_M \cdot t_0}{L} + \dfrac{d_p^2}{C \cdot D_M} \cdot \dfrac{L}{t_0}}} \cdot \frac{B \cdot \Delta c}{B \cdot \Delta \cdot \dfrac{t_0}{t_g} + 1} \tag{9}$$

This form of the equation for the peak capacity can now be used to examine how the separation power under gradient conditions (= the peak capacity) changes as we vary the analysis time ($t_a = t_0 + t_g$) or the flow rate (which determines the column dead time t_0). We discuss these situations in the following paragraphs.

1.2.2.2 Results

1.2.2.2.1 General Relationships

Experience tells us that at constant flow rate the peak capacity improves when we increase the gradient duration. This is shown in Fig. 1. Here, we have plotted the measured peak capacity for a 3.5 µm column *versus* the gradient span. The strongly eluting solvent in the gradient was acetonitrile, and the sample was a mixture of typical pharmaceutical compounds with molecular weights of around 300. One can see that the peak capacity improves as the gradient span increases. It should be mentioned that a standard gradient as applied in a typical analytical laboratory typically extends over a gradient span of 20 or at most 40.

The relationship shown in Fig. 1 follows from Eq. (6). However, we need to realize that the gradient span can be expanded through two different approaches.

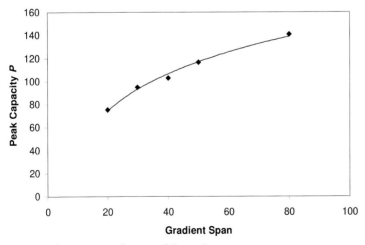

Fig. 1. Peak capacity as a function of the gradient span.

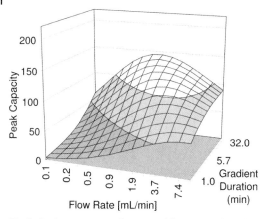

Fig. 2. Peak capacity as a function of flow rate and gradient duration.

On the one hand, we can increase the gradient duration while keeping the flow rate constant. On the other hand, we can decrease the column dead time by increasing the flow rate. However, a change in the flow rate changes the plate count. If we want to understand the influence of flow rate on the peak capacity at a constant analysis time, we need to examine Eq. (9) in its entire complexity. In order to do this, we can look at the dependence of peak capacity on flow rate and gradient duration in a three-dimensional graph. An example of such a graph is shown in Fig. 2 for a 4.6×50 mm 5 µm column. Both flow rate and gradient duration have been plotted using logarithmic scales. The flow rate covers the range from 0.1 to 10 mL min^{-1}, and the gradient duration extends from 1 min to 32 min. Let us examine the conditions for the best peak capacity for different gradient run times. For long gradients, the optimal flow rate is slightly lower than 1 mL min^{-1}. Under these conditions, we achieve a peak capacity of about 150. If we now run a fast, 1 min gradient at the same flow rate, the peak capacity is poor, only about 20. For such a rapid gradient, the optimum flow rate is around 5 mL min^{-1}. With this flow rate, we reach a peak capacity of around 65 – much better than at the slow flow rate. For somewhat less extreme gradients with a gradient duration between 2 and 4 min, one should use flow rates between 4 mL min^{-1} and 2.5 mL min^{-1}. This results in peak capacities ranging from 75 to more than 90. This means that the peak capacity for a 2 min gradient is only half of what can be achieved with a 30 min gradient, but with a 15-fold time saving this is surely not a disadvantage.

1.2.2.2.2 Short Columns, Small Particles
As we have seen, we need to use fast flow rates for fast gradients if we want to maximize the separation power. This also means that we should use short columns. In this section, we examine this proposition in more detail.

The plot in Fig. 2 stops at a flow rate of 10 mL min^{-1}. The reason for this limit is that at this flow rate we reach the pressure limit of the HPLC instrument. What will happen if we pack our 5 cm column with smaller particles? This is shown in

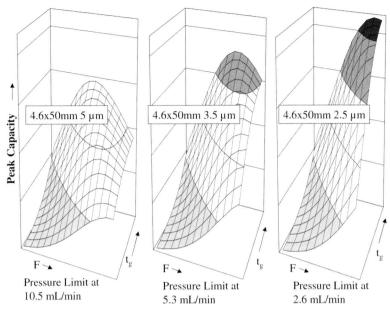

Fig. 3. Peak capacity as a function of flow rate and gradient duration for three columns of length 5 cm packed with 5 μm, 3.5 μm, and 2.5 μm particles.

Fig. 3. Here, we have created the same diagram as discussed above for a series of 5 cm columns packed with 5 μm, 3.5 μm, and 2.5 μm particles. As we can see, for slow analyses the separation power increases with decreasing particle size. With the 5 μm column, we reach a peak capacity of nearly 150 for a 30 min analysis, while with the 3.5 μm column a value of around 180 is reached. This performance is surpassed by the 2.5 μm column, with a peak capacity of roughly 220 for this half-hour analysis. However, for very rapid analyses, the column back-pressure limits the performance of the column. For a 1 min separation, the 3.5 μm column shows a somewhat better performance than the 5 μm column, but the peak capacity for the 2.5 μm column is lower than what was achievable with the 3.5 μm column. The reason for this is that the 2.5 μm column reaches the pressure limit imposed by the instrument. The main explanation for this is that, at a fixed column length, smaller particles reduce the flow rate that can be used. This can also be seen in Fig. 3. On the other hand, the 5 cm 3.5 μm column is ideal for analysis times in the 2–4 min range. Under these conditions, the 3.5 μm column exhibits a better separation power than the 5 μm column and still exceeds the performance of the 2.5 μm column.

There is still a way to increase the separation power for very rapid analyses, even within the pressure limits of older instrumentation. This is best accomplished by simultaneously varying the column length and the particle size [6, 7]. If the particle size is varied while the ratio of column length to particle size remains constant, the maximum achievable plate count remains the same. Thus, for example, a 10 cm 10 μm, a 5 cm 5 μm, and a 3 cm 3 μm column all have the

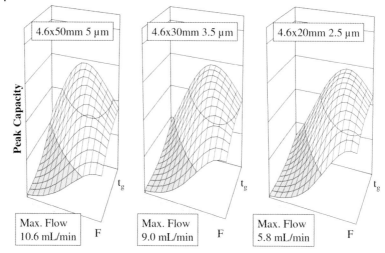

Fig. 4. Peak capacity as a function of flow rate for three columns of approximately equal ratio of column length to particle size; gradient duration from 1 min to 30 min.

same maximum plate count, and the same pressure drop for a particular column dead time. The difference between the three columns lies in the fact that columns packed with the smaller particles reach this plate count maximum at a higher analysis speed [6, 7]. The same thought process can be applied to gradient separations. In Fig. 4, we compare the performances of three columns of nearly equal column length to particle size ratio: a 5 cm 5 μm column, a 3 cm 3.5 μm column, and a 2 cm 2.5 μm column. For a 30 min separation, the peak capacity is practically identical for all three columns. However, for rapid, 1 min separations, better results are achieved with the smaller particles packed into shorter columns. The 2 cm 2.5 μm column reaches a peak capacity of nearly 90 under these conditions, while the 5 cm 5 μm column barely reaches a value of 65. The shorter column reaches the maximum at a flow rate that is twice as high as that for the 5 μm column, but this value is still far from the pressure limit of the HPLC instrument (see Chapter 2.7.3). For rapid separations, one should therefore use the shortest column with the smallest particles, unless extra-column effects, e.g. bandspreading and detector sampling rates, put a limit to such an endeavor.

1.2.2.2.3 **An Actual Example**
For the practitioner of HPLC, it is often best to view a problem in the form of an actual application example. For this purpose, we have selected the separation of five pharmaceuticals on a 3 cm 3.5 μm XTerra MS C_{18} column (Fig. 5). We used a high flow rate of 1.5 mL min^{-1} for the column with an internal diameter of 2.1 mm. We changed the gradient duration at this constant flow rate, which means that we moved parallel to the *y*-axis in Figs. 2 to 4. For a 4 min gradient the peak capacity was 140, for a 2 min gradient it was just above 100, and for a 1 min gradient we still achieved a peak capacity of 75. These results are in complete

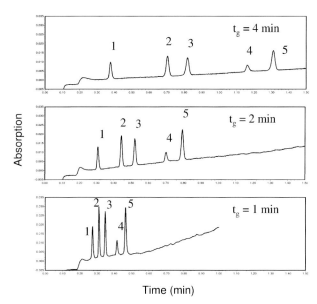

Fig. 5. Gradient separation of lidocaine, prednisolone, naproxen, amitriptyline, and ibuprofen. Column: 2.1 mm × 30 mm XTerra MS C_{18}, 3.5 μm. Temperature: 60 °C. Flow rate: 1.5 mL min^{-1}. Gradient from 8 to 95% acetonitrile. Top: 4 min; center: 2 min; bottom: 1 min.

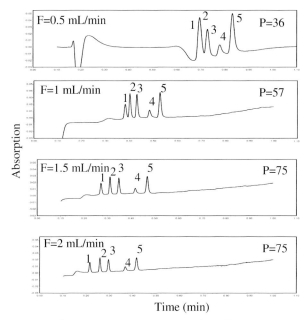

Fig. 6. Gradient separation as in Fig. 5. Gradient duration: 1 min. Flow rates (from top to bottom): 0.5 mL min^{-1}, 1.0 mL min^{-1}, 1.5 mL min^{-1}, 2.0 mL min^{-1}.

agreement with our expectations. As one can see, we were still able to achieve the separation even with the fastest gradient, despite the fact that it deteriorated on going from the slower to the faster gradients.

Figure 6 shows how the same separation changes as we change the flow rate at a constant gradient duration of 1 min. The flow rate was varied in steps of 0.5 mL min^{-1} from 0.5 mL min^{-1} to 2 mL min^{-1}. The peak capacity improved from 36 at 0.5 mL min^{-1} to 57 at 1 mL min^{-1} to 75 at both 1.5 mL min^{-1} and 2 mL min^{-1}. These values also agree with the theoretical expectations.

This real example shows that the thought processes outlined above are fundamentally correct. In our laboratory, we have demonstrated the same effects many times over, although not in the same detail as shown here. If one wants to achieve optimal results with rapid gradients, one needs to work at high flow rates. Often, these flow rates are at the limit of existing HPLC instrumentation. We discuss the limitations in the next paragraph.

1.2.2.3 Optimal Operating Conditions and Limits of Currently Available Technology

In Table 1, we have assembled a few guideline values as a rule of thumb for the selection of optimal gradient conditions, if the gradient is carried out over a large difference in solvent composition. We have selected several short columns with commonly used internal diameters of 4.6 mm and 2.1 mm and lengths of 50 mm, 30 mm, and 20 mm. The commercially available XTerra packing was used as the model for available particle size distributions: 5 µm, 3.5 µm, and 2.5 µm. The table shows the approximate values of the optimal flow rates for 1 min, 2 min, and 4 min gradients from 0% to 100% organic.

The table also shows that the optimal flow rates for these rapid gradients lie outside the pressure limit of current HPLC technology, if one wants to use columns of length 5 cm or more. It should be pointed out that the estimates assume that we want to separate typical pharmaceutical compounds with molecular weights between 250 and 600 with a water/acetonitrile gradient. The exact values are somewhat dependent on the solvent composition at which the compounds are

Table 1. Optimal flow rate as a function of the gradient duration and the column dimensions.

d_p	5 µm			3.5 µm			2.5 µm		
t_g	1 min	2 min	4 min	1 min	2 min	4 min	1 min	2 min	4 min
4.6 × 50	7.5	5.0	2.5	10.0 [a]	6.0 [a]	3.5	10.0 [a]	6.0 [a]	3.5 [a]
4.6 × 30	5.0	3.0	2.0	6.5	3.5	2.5	7.0 [a]	4.0	3.0
4.6 × 20	4.5	2.0	1.5	5.0	2.5	2.0	6.0	3.0	2.0
2.1 × 50	1.5	1.0	0.5	2.0 [a]	1.2 [a]	0.7	2.0 [a]	1.2 [a]	0.7
2.1 × 30	1.0	0.6	0.4	1.3	0.7	0.5	1.4 [a]	0.8	0.6
2.1 × 20	0.9	0.4	0.3	1.0	0.5	0.4	1.2	0.6	0.4

[a] Outside the pressure limits of the technology available at the time of writing.

eluted. Higher flow rates should be used if the elution starts at high acetonitrile content. This means that for practical use the values in Table 1 should be taken only as guidelines, and that the actual optimal flow rates should be determined experimentally. Nevertheless, these values clearly demonstrate that optimized rapid separations require high flow rates and higher pressures than are available today. In addition, we should also think about separations that are faster than 1 min (see Chapter 2.7.3).

1.2.2.4 Problems and Solutions

An optimal execution of rapid gradients also requires optimal settings of instrumental parameters. In the following, we discuss some important details.

1.2.2.4.1 Gradient Delay Volume

In order to execute rapid gradients, the gradients have to arrive at the column inlet without delay. However, most modern HPLC instruments operate with a single pump, and the gradients are generated in a gradient mixing chamber upstream of the pistons. This means that, before it starts to influence the separation in the column, the gradient needs to migrate through the volume of the mixing chamber, the connection tubing to the pump heads, the pump heads themselves, the tubing of the injector, and finally the tubing to the column. The sum of all these volumes is called the gradient delay volume, and the time that it takes to purge this volume at a particular flow rate is called the gradient delay time. Older instruments may have a gradient delay volume of several mL. In modern HPLC instruments, it is of the order of 1 mL. If a 4.6 mm i.d. column is used, a gradient delay volume of 1 mL has only a small effect on the separation. On the other hand, a short, 2 cm column with an internal diameter of 2 mm has a column dead volume of only 40 μL. This means that the gradient delay volume of the instrument has a value of over 20 times the column dead volume. This leads to a significant delay before the gradient reaches the column. The consequence is that the analysis takes much longer than it should. How can we avoid this time delay?

The solution to this problem is a delayed injection. The sample is injected shortly before the gradient reaches the column inlet. Now, the instrument executes the gradient as desired, and a prolonged analysis time is avoided. In addition, one can use the time saved under normal operating conditions and continuous gradient runs for column re-equilibration with the initial starting conditions. The way to do this is to program the end of the gradient as desired, perhaps with the inclusion of a column wash step, and then convert directly to the starting conditions of the gradient. The re-equilibration is then executed in exactly the same way as if the gradient delay volume did not exist. This is best shown by way of a diagram (Fig. 7). The upper trace shows the execution of the analysis without the delayed injection. The gradient only reaches the column after the gradient delay volume has been purged. The gradient is then executed, and the column is washed and re-equilibrated with the starting mobile phase. The next gradient is started only after completion of the entire procedure. The lower trace shows the same gradient scheme, but now we take advantage of the delayed injection feature. The first

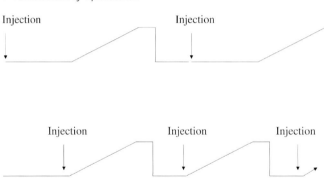

Fig. 7. Delayed sample injection to eliminate the gradient delay volume.

injection is executed just before the gradient reaches the column. In every subsequent gradient, the gradient run time, the purge cycle, and the column re-equilibration are all programmed as if the gradient delay volume does not exist. The point at which the gradient reaches the column only needs to be calculated once, and then one programs the injector to inject again at this time thereafter. With this trick, the influence of the gradient delay volume on the gradient run time can be largely eliminated.

In order to take full advantage of this feature, one needs to know the gradient delay volume. This is best measured by programming a linear gradient from water to water containing a small amount of a UV-absorbing compound, such as 1% acetone, without a column. The gradient delay time is then the difference between the time at which the start of the gradient was programmed and the time found by linear extrapolation of the UV trace recorded by the detector. The gradient delay volume is then calculated by multiplying this time by the flow rate.

1.2.2.4.2 Detector Sampling Rate and Time Constant

When running very rapid gradients, the peak width may be as little as 1 s. In the rapid gradient example shown above, the peak width was only 0.8 s. It is generally assumed that a peak can be represented well and is not unduly broadened if about 40 data points or more are measured over its width. This means that the sampling rate needs to be faster than 20 ms per data point. A similar argument can be put forward for the time constant, which is the measure of the speed of signal processing with older detectors. The time constant should be smaller than 50 ms. Today, the detector noise is often suppressed through an accumulation of data points obtained by the use of digital filters. Key characteristics of this process are the width of the filter, i.e. the number of data points used, as well as the filter algorithm, i.e. the way in which the data points are accumulated.

The experienced chromatographer will recognize two of the phenomena resulting from signal processing without difficulty. Edged peak shapes are a sign that the sampling rate is too low. A problem with the time constant results in tailing peaks. This tailing increases when the speed of the analysis is further increased. Difficulties arising from digital filtering are more difficult to recognize since the peaks are not visibly deformed. In order to check for this problem, it is best to run two identical chromatograms with different filter settings. If this results in a difference in the peak widths, less filtering should be used to maximize resolution. If these difficulties arise, it is highly advisable to carefully read the care and use manuals of the detector and the data system.

1.2.2.4.3 Ion Suppression in Mass Spectrometry

A completely different detector problem arises if one wants to perform a quantitative analysis of a complex sample, such as a plasma sample, by means of an LC/MS method. LC/MS coupling has been a major breakthrough for this type of analysis. Analysis times of as little as 1 min can be achieved nowadays. However, the simplicity and the speed of the analysis have resulted in a complete disregard for the value of the chromatographic separation. In due course, it was found that "invisible" sample constituents cause a falsification of the signal, for the most part a suppression of the MS signal [1, 2]. This problem can be minimized or even eliminated through the use of more efficient separation techniques or better sample clean-up. In any case, it is worthwhile to measure the magnitude of this effect. If one knows the problem, one can search for ways to eliminate it through improved sample preparation techniques or an improved chromatographic separation.

The method used for measuring signal suppression is rather straightforward (Fig. 8), and the relatively little effort is always justifiable. A solution of the analyte(s) of interest is mixed post-column with the column effluent via a T-connection. The concentration of the analytes should be in the middle of the detection range of interest. Then, a blank plasma sample (i.e., without analytes) is injected onto the column, and the analysis is executed as usual. The mass spectrometer monitors the signal for the analyte(s) during this process. If the substances that elute from the column cause signal suppression, the signal declines. If there is no suppression, the signal remains constant. From the chromatographic analysis, we know at which point in time the analyte(s) elute. If there is no ion suppression during this time, the method can be used for the analysis of the compound(s) of interest in plasma samples. If one encounters a significant amount of ion suppression, one needs to re-work some part of the analytical method. Either the separation of the analyte from the matrix interferences needs to be improved, or more effectively, the sample preparation procedure can be refined. Possible ways of improving the HPLC method are often an adjustment of the pH value or an expansion of the gradient. Using an effective solid-phase extraction procedure instead of a simple protein precipitation will reduce the matrix interferences carried over into the HPLC method. If an improvement needs to be made, it is often best to examine several alternative methods and to compare them with respect to ion suppression. The method described here for the determination of ion suppression is universally applicable. The only disadvantage is the fact that the sample is often in the form of a specific salt, whereas during elution in the HPLC method it is associated with the counterions of the HPLC buffer. However, we are not aware of any negative consequences of this fact.

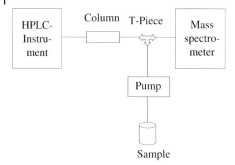

Fig. 8. Method for the determination of ion suppression.

References

1 P. R. Tiller, L. A. Romanyshyn, U. D. Neue, "Fast LC/MS in the analysis of small molecules", *Analytical and Bioanalytical Chemistry* 377 (2003), 788–802.

2 B. K. Matuszewski, "Strategies for the Assessment of Matrix Effect in Quantitative Bioanalytical Methods Based on HPLC-MS/MS", *Anal. Chem.* 75 (2003), 3019–3030.

3 Y.-F. Cheng, Z. Lu, U. D. Neue, "Ultra-Fast LC and LC/MS/MS Analysis", *Rapid Commun. Mass Spectrom.* 15 (2001), 141–151.

4 U. D. Neue, J. L. Carmody, Y.-F. Cheng, Z. Lu, C. H. Phoebe, T. E. Wheat, "Design of Rapid Gradient Methods for the Analysis of Combinatorial Chemistry Libraries and the Preparation of Pure Compounds", in *Advances in Chromatography* (Eds.: E. Grushka, P. Brown), Vol. 41, Marcel Dekker, New York, Basel, 2001, pp. 93–136.

5 U. D. Neue, J. R. Mazzeo, "A Theoretical Study of the Optimization of Gradients at Elevated Temperature", *J. Sep. Sci.* 24 (2001), 921–929.

6 U. D. Neue, "HPLC Columns – Theory, Technology, and Practice", Wiley-VCH, New York, 1997.

7 U. D. Neue, B. A. Alden, P. C. Iraneta, A. Méndez, E. S. Grumbach, K. Tran, D. M. Diehl, "HPLC Columns for Pharmaceutical Analysis", in Handbook of HPLC in Pharmaceutical Analysis (Ed.: M. Dong, S. Ahuja), Academic Press, Elsevier, Amsterdam, 2005, pp. 77–122.

1.3
pH and Selectivity in RP-Chromatography

Uwe D. Neue, Alberto Méndez, KimVan Tran, and Diane M. Diehl

1.3.1
Introduction

The majority of the sample compounds analyzed by reversed-phase (RP)-chromatography have ionizable functional groups, such as carboxylic acid, sulfonic acid or amino groups. The retention of a sample strongly depends on the ionization of the functional group [1]. Often, there is roughly a 30-fold difference in the retention time between the neutral and the ionized form of the same analyte. The degree of ionization is determined by the pH of the mobile phase. It is important to understand all of the various influences on the degree of ionization if one wants to obtain good chromatographic selectivity and reproducible retention.

In this chapter, we discuss these influences in great detail. In our treatment, we rely on the latest information available; for example, the state-of-the-art knowledge on the influence of an organic solvent on the pH of the mobile phase and the pK_a values of analytes. This information has become available in recent years, largely through the investigations of the Barcelona group of researchers [2]. In addition, we discuss buffers used in classical HPLC as well as buffers useful for LC/MS applications. A few rules on the dependence of retention on the functional groups of the analytes are included as well.

1.3.2
Main Section

1.3.2.1   Ionization and pH
The retention of an ionizable analyte depends on its degree of ionization. For simple substances, one can state that the non-ionic form of the analyte always has a much higher retention than the ionic form. If the analyte has multiple stages of ionization, the form with the higher degree of ionization usually exhibits lower retention. The degree of ionization of the analyte depends on the pH of the solution and on the pK values of the ionization stages of the analyte.

The dependence of the retention on the degree of ionization is shown in Fig. 1 for both an acidic and a basic analyte. For both compounds, the retention changes by more than an order of magnitude. This is typical: for most compounds, the change in retention between the ionized and the non-ionized form is of the order of 10- to 30-fold. The ionized form always has lower retention in RP-chromatography. Therefore, the retention is lowest under acidic conditions for a basic analyte and under basic conditions for an acidic analyte. On the other hand, high retention is observed when the analyte is in its neutral form, i.e. under acidic conditions for acidic analytes and under basic conditions for basic analytes. There is a transition

HPLC Made to Measure: A Practical Handbook for Optimization. Edited by Stavros Kromidas
Copyright © 2006 WILEY-VCH Verlag GmbH & Co. KGaA, Weinheim
ISBN: 3-527-31377-X

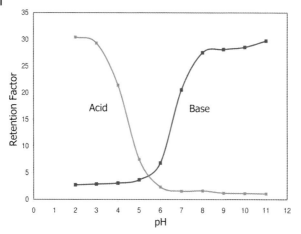

Fig. 1. Dependence of the retention factor of an acidic and a basic analyte on the pH value of the mobile phase. Column: XTerra® RP$_{18}$, 3.9 mm × 20 mm.

range between the two stages, in which the retention depends on the degree of ionization of the analyte. The following equation describes this behavior [1, 3]:

$$k = \frac{k_0 + k_1 \cdot d}{1 + d} \tag{1}$$

Here, k is the retention factor of the analyte, k_0 is the retention factor of the protonated form of the analyte, and k_1 is the retention factor of the deprotonated form; d is the degree of deprotonation and is defined as follows:

$$d = 10^{pH - pK_a} \tag{2}$$

In this equation, pH is the pH value of the solution and pK_a is the pK value of the relevant dissociation step of the compound. For multiply-charged compounds, the equation can be expanded in a simple way. This is shown in the following equation for a doubly-charged compound, such as an acid or base with two dissociation steps or a zwitterion:

$$k = \frac{k_0 + k_1 \cdot d_1 + k_2 \cdot d_1 \cdot d_2}{1 + d_1 + d_1 \cdot d_2} \tag{3}$$

The retention factors usually decrease with the degree of ionization. For example, a zwitterion is less retained than the singly-charged forms of the same analyte. Such a case is illustrated in Fig. 2. The sample is fexofenadine, a compound with a carboxylic acid group and a tertiary amino group. In the strongly acidic pH range, the carboxylic acid is protonated and therefore not charged. The compound therefore bears just one positive charge. In the intermediate pH range, both the carboxylic acid group and the amino group are ionized. Due to the fact that there is now a dual charge on the molecule, the retention is lower. In the alkaline pH range, the amino group is deprotonated, the molecule bears only a single, negative charge, and the retention increases again.

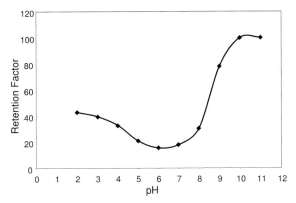

Fig. 2. Dependence of the retention of a zwitterionic sample on the pH of the mobile phase [1]. Column: XTerra® RP$_{18}$, 3.9 mm × 20 mm. Mobile Phase: 20% acetonitrile, 80% 30 mM buffer (from [1] with permission of Elsevier Science B.V.).

The pK values of the analyte determine the pH range in which the charge, and therefore the retention, changes. If the pH value is outside ± 2 pH units around the pK_a of the compound, the analyte is either 99% dissociated or 99% undissociated. This means that outside this pH range the retention does not change with a change in pH. However, within this range, especially within ± 1.5 pH units around the pK_a, the retention changes markedly. For the practice of chromatography, this means that good pH control is absolutely essential in order to achieve reproducible retention times. We will discuss this in more detail a little later in this chapter.

Aliphatic carboxylic acids have pK_a values around 5. For example, ibuprofen has a pK_a of 5.2 [4, 5]. Typical pK_a values for amines are around 9. The pK_a value of amitriptyline, for instance, is 9.4. Phenols, on the other hand, are very weak acids (the pK_a of phenol is 10.0). Similarly, anilines are very weak bases (the pK_a of aniline itself is 4.7). These data are good reference points for an estimation of the pK values of unknown compounds in the absence of measured data. It should be mentioned that the pK values of aromatic compounds need to be estimated by taking into account the conjugation with other groups through the aromatic ring(s).

1.3.2.2 Mobile Phase and pH

In Fig. 1, we have shown that the retention of an analyte may depend strongly on the pH of the mobile phase. In order to maintain reproducible pH values, one needs to use buffers in all pH ranges except at very acidic or strongly alkaline pH values. Buffers are solutions of ionogenic compounds that contain a conjugated pair of a proton donor and a proton acceptor. Thus, they stabilize the pH against the addition of small amounts of acid or base [6]. Let us discuss an acetate buffer as an example. This buffer contains an equimolar amount of acetic acid, the proton donor, and acetate, the proton acceptor. The pH of such a solution in water is 4.75. If one adds small amounts of acid or base (below the total buffer concentra-

Table 1. HPLC buffers.

(a) Buffers based on acids (neutral or anionic proton donors)

Name	pK_a
Trifluoroacetate	0.5
Phosphate I	2.15
Phosphate II	7.20
Phosphate III	12.38
Acetate	4.75
Formate	3.75
Hydrogencarbonate	10.25
Borate I	9.24

(b) Buffers based on bases (cationic proton donors)

Name	pK_a
Ammonium	9.24
Trimethylammonium	9.80
Triethylammonium	10.72
Pyrrolidinium	11.30

tion), the pH changes very little. If one prepares a solution of dihydrogenphosphate in water, one will also obtain a solution with a pH of around 4.5. However, this solution is not a buffer, because it does not stabilize the pH. If one adds a small amount of acid or base to the dihydrogenphosphate solution, the pH will change instantaneously to a more acidic or more neutral value. This does not happen with the acetate buffer. As we can see, a buffer is a solution that stabilizes the pH. If it does not do this, it is not a buffer, merely a salt solution.

The pK_a values of some commonly used HPLC buffers are shown in Table 1. The table is divided into two parts: buffers based on acids, and buffers based on a base. Of course, this list can be expanded at will. The buffers selected here cover the entire pH range of interest to the chromatographer. All organic buffers as well as hydrogencarbonate and ammonium buffers are volatile and are therefore compatible with MS detection, provided that a suitable counterion is used.

1.3.2.2.1 Buffer Capacity

Buffer capacity is a measure of the strength or quality of a buffer. It is defined as the reciprocal value of the slope of the pH curve (or titration curve) of the buffer [6]. It depends on two factors: (1) the concentration of the buffer; (2) the distance between the pH and the pK_a of the buffer. In Fig. 3, we show the buffer capacities of acetate buffers at concentrations of 5 mM, 10 mM, 20 mM, and 40 mM. The maximum of the buffer capacity is always at the pK_a, i.e. 4.75 for the acetate buffer. The maximum buffer capacity increases in direct proportion to the buffer concentration. At the same time, the pH range over which a particular buffer capacity is achieved increases with increasing concentration. For example, a 10 mM acetate buffer has a buffer capacity in excess of 0.005 in the pH range from

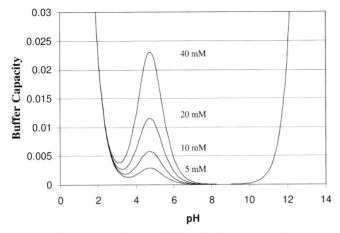

Fig. 3. Buffer capacities of acetate buffers of various concentrations. From top to bottom: 40 mM, 20 mM, 10 mM, 5 mM.

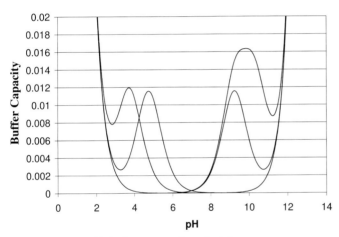

Fig. 4. MS-compatible buffers, 20 mM; from the left: formate, acetate, ammonium, ammonium hydrogencarbonate.

4.5 to 5.0, while a 40 mM buffer has the same buffer capacity between pH 3.5 and pH 6.0. This is important for understanding the quality of a buffer. Figure 3 also shows the buffer capacities of the H^+ and OH^- ions at the extremes of the pH scale.

Figure 4 shows the buffer capacities of a few MS-compatible buffers: formate, acetate, ammonium and ammonium hydrogencarbonate, all at a concentration of 20 mM. Formic acid or ammonium formate are preferentially used at acidic pH. Acetic acid and acetate are useful for a slightly higher pH range, but they are less frequently employed. In the alkaline pH range, one can use ammonia or ammonia together with a volatile ammonium salt such as formate, acetate or hydrogencarbonate. Ammonium hydrogencarbonate is the preferred buffer for

LC/MS applications in the alkaline pH range. The reason for this is the overlapping and additive buffer capacities of the ammonium cation and the hydrogencarbonate anion. Above 60 °C, ammonium hydrogencarbonate decomposes into water, ammonia, and carbon dioxide.

1.3.2.2.2 Changes of pK and pH Value in the Presence of an Organic Solvent

The pH and pK_a values of buffers as well as the pK_a values of analytes found in the literature have commonly been measured using water as the solvent. It is important to understand that both the pH values and the pK values change when an organic solvent is added [2]. Generally, the pK_a values of acids increase, while the pK_a values of bases decrease with the addition of the organic modifier. This phenomenon has been described in detail in the literature [2]. For the practitioner, it is important to understand that the buffer capacity does not change in the presence of the organic solvent. A buffer that is a good buffer in water is also a good buffer in the presence of an organic solvent. Contrary to recommendations in the literature [2], we do *not* recommend the measurement of the pH of the buffer *after* the addition of the organic solvent. The most important value for the practitioner of HPLC is the buffer capacity, and this does not change with the addition of the organic solvent, while the pK_a value of the buffer and the optimal pH do change. The true pH value after the addition of the organic solvent is certainly of interest to the theoretician, but for the practitioner the only relevant value is the buffer capacity, and this value can be estimated without difficulty with reference to the aqueous pK and pH values. The aqueous pK values of buffers can be found in textbooks.

It is nevertheless of value to review the phenomena described in Ref. [6]. If one is interested in an understanding of the changes in retention time occurring at different pH values in the presence of an organic solvent, it is absolutely necessary to measure the pH after the organic solvent has been added. To do this, two pH scales can be used. For the first pH scale, the pH meter is calibrated with standard buffers in the presence of the organic solvent. This pH scale is called the $_s^s$pH scale. The upper index designates the method used for the measurement, and the lower index the method of calibration. If one calibrates the pH meter with standard buffers in water and measures the pH after the addition of the organic solvent, one obtains pH values on the $_w^s$pH scale. The difference between the two scales is the activity coefficient of the hydronium ion γ_H^0:

$$_w^s\mathrm{pH} = {}_s^s\mathrm{pH} - \log\left(_w^s\gamma_H^0\right) \tag{4}$$

The difference between the pH value in water and in a mixture of an organic solvent with water is caused by a change in the autoprotolysis constant of water due to the addition of the organic solvent.

Figure 5 shows the change in the pK values of several buffers for methanol/water mixtures. The data for this figure have been taken from the literature [7, 8]. As mentioned above, one can see that the pK_a values of acids, e.g. formic acid, increase upon the addition of the organic solvent. For buffers generated from bases, the pK values change to more acidic values over a large part of the solvent

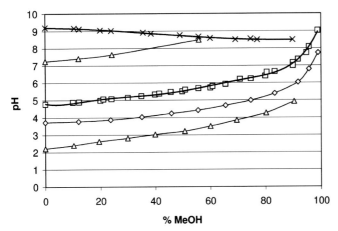

Fig. 5. $_s^s pK_a$ values of several buffers in methanol/water mixtures:
△ phosphate (aqueous pH 2 and 7), ◇ formate, □ acetate, × ammonium.
(Adopted from [8] with permission from the authors).

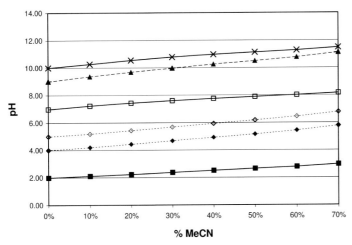

Fig. 6. $_w^s pH$ values of several buffers in acetonitrile/water mixtures:
□■ phosphate (aqueous pH 2 and 7), ◇♦ acetate (aqueous pH 4 and 5),
△▲ borate, × hydrogencarbonate.

composition. This means that the addition of the organic solvent results in a weakening of both acids and bases. Figure 6 shows the changes in the pH of buffered mobile phases based on acetonitrile measured in our laboratory. All the buffers shown in this chart are based on acids. Therefore, the pH values shift towards alkaline pH as the concentration of acetonitrile is increased. An interesting situation is obtained when the pK_a values of the acidic and basic components of a buffer overlap, as is the case for an ammonium hydrogencarbonate buffer (Fig. 7). The $_w^s pH$ value of this buffer does not change much as the concentration of the organic solvent is increased. This is due to the fact that the pK_a value of the

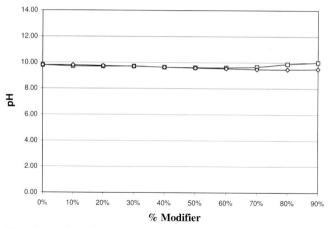

Fig. 7. ^s_wpH values of an ammonium hydrogencarbonate buffer with an aqueous pH of 9.8: ◇ methanol, ☐ acetonitrile.

hydrogencarbonate shifts towards more basic pH, while the pK_a value of the ammonium ion shifts towards more acidic pH.

The pK_a values of analytes shift upon the addition of an organic solvent in exactly the same way as described for the buffers. We will discuss this subject in more detail below in the discussion of the influence of pH on retention.

1.3.2.3 Buffers

1.3.2.3.1 Classical HPLC Buffers

Which buffers are preferred for an application depends primarily on the choice of detector. For this reason, we distinguish between classical HPLC buffers, which are used with UV detection, and MS-compatible buffers, which are highly volatile. We discuss first the classical buffers.

Phosphate buffers are the preferred buffers for UV detection. Their pK_a values are 2.15 and 7.20. They can be used without difficulty at low UV wavelength, for example 210 nm. At weakly alkaline pH, buffers based on the ammonium ion (pK_a = 9.24) with a suitable counterion with low UV absorption can be used. Borate buffers can be used in the same pH range and also have a low UV absorption. In the more alkaline pH range, some simple amines can be used: pyrrolidine has a pK_a of 11.3, and triethylamine a pK_a of 10.7, and both can be used at low UV wavelengths. With all amines, their purity is an important factor for their use at such UV wavelengths.

Other buffers often exhibit a substantial UV absorption below 215 nm. However, for the normal UV detection range around 254 nm a large number of buffers is available. Acetate with its pK_a of 4.75 is probably the most popular buffer, since its pK is exactly intermediate between the first and second dissociation constants of phosphate. Ammonium hydrogencarbonate is a preferred buffer for the pH range around 9 to 10, since the buffering ranges of the ammonium ion (pK_a = 9.24) and the hydrogencarbonate ion (pK_a = 10.25) overlap.

1.3.2.3.2   MS-Compatible pH Control

We have already mentioned some of the commonly used mobile phase additives for MS-compatible pH control. The most important attribute is the volatility of all buffer components. The most frequently used mobile phase additives are formic acid and acetic acid, together with the true buffers formic acid/ammonium formate and acetic acid/ammonium acetate. In the alkaline pH range, the preferred additive is ammonia, and the buffer of choice is ammonium hydrogencarbonate. One can also use the ammonium ion with volatile counterions such as formate or acetate to establish a true buffer at the pK_a of the ammonium ion ($pK_a = 9.24$). One finds sometimes in the literature the reported use of ammonium formate or acetate at neutral pH. It needs to be pointed out that this has little to do with pH control, since these salt solutions have no buffer capacity at pH 7!

Trifluoroacetic acid is used in the very acidic pH range. It should be mentioned that this additive often contributes substantially to ion suppression, and it should therefore only be used when there are no viable alternatives. In our laboratory, we have also used semi-volatile buffers such ammonium phosphate at pH 7. However, a very low concentration should be used, and one should plan for a more frequent clean-up of the MS inlet and source.

1.3.2.4   Influence of the Samples

The retention of an analyte strongly depends on its ionization. As mentioned above, the non-ionized form of the sample has a much higher retention, up to a factor of 30 higher, than the ionized form. However, the difference is not the same for all analytes. For some analytes, the retention changes more, for others less. This is shown in Fig. 8, where we have plotted the retention times under gradient conditions for more than 70 samples under basic *versus* acidic conditions.

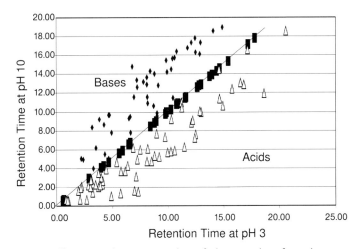

Fig. 8. Differences in the retention values of a large number of samples under gradient conditions as a function of pH.
■ neutrals, ◆ basic compounds, △ acidic compounds.

The x-axis corresponds to the retention times at pH 3, and the y-axis to those at pH 10. In the center of the figure we have drawn a line through the data points for non-ionizable compounds for which retention does not change with pH. Analytes with an acidic functional group are neutral under acidic conditions and thus are retained more. Conversely, basic analytes are neutral at basic pH, and therefore exhibit a higher retention in the alkaline pH range.

1.3.2.4.1 The Sample Type: Acids, Bases, Zwitterions

How much the retention of an analyte changes with pH depends both on the degree of ionization as well as on structural parameters. Sulfonic acids are very acidic. Benzenesulfonic acid has a pK_a of 0.7. The pK values of aliphatic sulfonic acids are higher. Aliphatic carboxylic acids commonly have pK_a values around 4.5 to 5. Aromatic carboxylic acids, on the other hand, are more acidic. Sulfonamides are weakly acidic. The pK_a values of phenols can be found in the mildly alkaline pH range (phenol: pK_a = 9.99) and depend strongly on the substituents on the aromatic ring.

Basic analytes are protonated at acidic pH, and lose their charge under alkaline conditions. Anilines are very weak bases: the pK_a of aniline itself is 4.6. Pyridine too is a very weak base with a pK_a of 5.2. On the other hand, aliphatic amines are rather strong bases: methylamine pK_a = 10.6, dimethylamine pK_a = 10.8, trimethyl-amine pK_a = 9.8. An analysis of the retention time shifts of the strong bases in Fig. 8 has shown that the retention of tertiary amines changes more than the retention of secondary amines, when the data at acidic pH are compared with those at alkaline pH. This may be due to the fact that the charge on a protonated amine creates a strongly bound hydration shell around the amino group, which "hides" the adjacent methyl or methylene groups. On the other hand, the methyl groups do contribute to retention when the amino group is not protonated. Another possibility could be that under the measurement conditions only the tertiary bases were completely deprotonated.

The retention of zwitterions also changes predictably with pH. We have already highlighted the example of fexofenadine earlier in this chapter. The retention of this analyte decreased by a factor of three on going from the acidic pH range to neutral pH, and increased again by a factor of seven on going from neutral pH to alkaline pH. For analytes that contain an aniline and a phenol function, one obtains the opposite result. Under acidic conditions, the sample is positively charged. In the intermediate pH range, the analyte loses its positive charge, the molecule becomes neutral, and the retention increases. In the alkaline pH range, the phenol group becomes negatively charged, and the retention decreases again. As one can see, one can obtain significant changes in the retention and thus in the selectivity of a separation by manipulating the pH of the mobile phase.

1.3.2.4.2 Influence of the Organic Solvent on the Ionization of the Analytes

In Section 1.3.2.2.2, we established that the pH of a buffer changes upon the addition of an organic solvent. So does the pK_a value of an analyte. It is important to keep these things in mind when developing HPLC methods. The combined effect of the pH shift of the buffers and the pK_a shift of the analytes can yield surprising results. For example, let us look at the retention of amitriptyline in a phosphate buffer in 65% methanol [9]. The pH value of the phosphate buffer as measured in water was 7.0. The pK_a value of amitriptyline in water is 9.4. Since there is a difference of 2.4 pH units between the pK_a of amitriptyline and the pH of the mobile phase, one would expect amitriptyline to be completely protonated in this buffer. However, measurements of the retention behavior have shown that under the conditions of the measurement in 65% methanol only one-third of the amitriptyline molecules are ionized. Under these conditions, the second pK_a of phosphoric acid is around 9 [8]. This means that the pK_a of amitriptyline shifted from 9.4 in water to around 8.5 in 65% methanol. The combination of both effects has the result that amitriptyline is only partially protonated under the measurement conditions.

Consequently, it is important to keep these pH and pK shifts in mind when developing a new method. A good control over the degree of ionization of the analytes is an important requisite for method reproducibility and robustness.

1.3.3
Application Example

As we have seen, the pH value plays a major role with regard to the separation of ionizable compounds. This means that the pH is the most important factor in method development, and one should explore its effect at the beginning of the method development effort. We have devised a procedure that does exactly that: it demonstrates relatively early on which combination of pH and mobile phase composition is most promising for a further fine-tuning of the method [10]. The foundations of this principle had already been developed earlier [11].

In this method, a separation is explored using two pH values, two organic modifiers, and several columns with significant selectivity differences. If one uses a pH-stable hybrid packing such as XTerra®, one can combine an alkaline pH (pH 10 with ammonium hydrogencarbonate) with an acidic pH (e.g., pH 3.5 with an ammonium formate buffer). Such a method can be used with MS detection [10]. With classical silica-based columns and a UV detector, it is proposed that phosphate buffers at pH 2.5 and 7.0 are used [11]. In both cases, the large change in pH results in a change in the ionization of the analytes and in a striking change in the selectivity of the separation.

Acetonitrile and methanol are used as the organic solvents. Methanol is a proton donor, and acetonitrile is a proton acceptor. Significant selectivity differences can therefore be achieved with these two solvents. In addition, the solvent selectivity can be varied continuously, and a fine tuning of the selectivity can thus be obtained using mixtures of the two solvents. In classical method development, THF was used as well, but in most cases this is not necessary. Moreover, one can often use a fine-tuning of the pH, i.e. around the pK of the buffer, to improve the separation. However, this is only possible if the pK values of the analytes are not too far away from the pH value.

In addition, this method permits a rapid exploration of the selectivity of the stationary phases. We have generally used three different XTerra® columns: XTerra® MS C$_{18}$, with a trifunctionally bonded C$_{18}$ chain; XTerra® RP$_{18}$, with an incorporated carbamate group; and XTerra® Phenyl, with an aromatic group. This column selection maximizes the selectivity differences within the given family of packings. Alternatively, one can use the silica-based packings Symmetry® C$_{18}$ and SymmetryShield™ RP$_{18}$ with acidic and neutral buffers. The relevant selectivity differences between these two packings are discussed in Chapter 2.1.2 of this book.

In the following, we discuss the separation of nine diuretics as an example of our method development strategy. In order to rapidly evaluate the influence of pH, organic solvent, and column chemistry, the same gradient (from 0 to 80%B in 15 min) was run under acidic (pH 3.64) and basic (pH 9.0) conditions on three short 4.6 mm × 50 mm XTerra® MS C$_{18}$, XTerra® RP$_{18}$, and XTerra® Phenyl columns, using both methanol and acetonitrile as organic modifiers. The separations are shown in Fig. 9. One can clearly see the influence of the pH on the selectivity of the separation. Generally, the separation extends over a larger range of the solvent composition at acidic pH than at alkaline pH. If the final separation can be a gradient separation, then one should select the acidic pH range. Indeed, it was possible to optimize the separation without much additional effort using the XTerra® MS C$_{18}$ column under acidic conditions. If one wishes to achieve an isocratic separation, the alkaline pH range is more promising. However, in this case it is necessary to use solvent selectivity, i.e. mixtures of acetonitrile and methanol, to achieve the final optimized isocratic method.

Such a separation is shown in Fig. 10. We used a mixture of 13% acetonitrile and 4% methanol and an ammonium hydrogencarbonate buffer at pH 9.0. A short 5 cm XTerra® MS C$_{18}$ column was sufficient for this separation. All peaks are cleanly separated. This example demonstrates the value of pH adjustment for the separation of ionogenic compounds. This method development strategy can be employed generically for all separation problems involving ionogenic analytes.

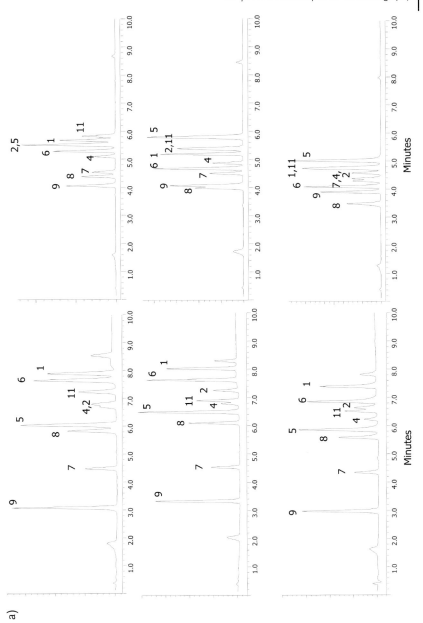

a)

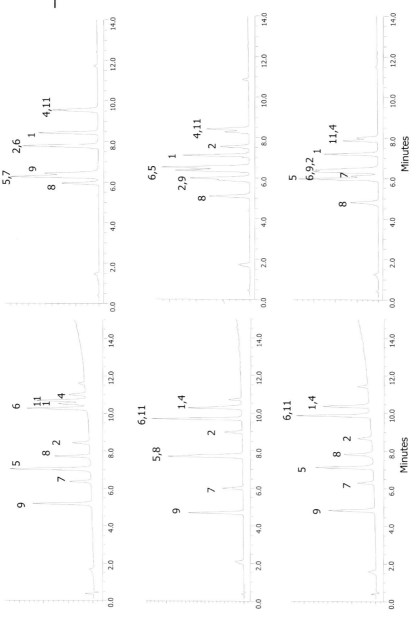

Fig. 9. Gradient separation of nine diuretics with acetonitrile (a) and methanol (b) as organic modifiers.
Left: pH 3.64; right: pH 9.0.
Top: XTerra® MS C₁₈; Middle: XTerra® RP₁₈; Bottom: XTerra® Phenyl.
Samples: bumetanide (1), benzothiazide (2), canrenoic acid (4), althiazide (5), probenecid (6),
chlorothalidone (7), furosemide (8), triamterene (9), and ethacrynic acid (11)
(from ref. [11] with permission from Elsevier Science B.V.).

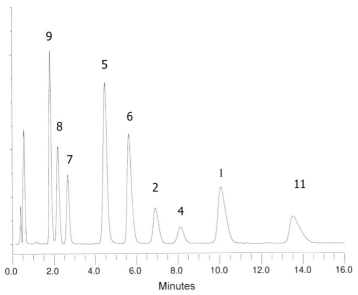

Fig. 10. Isocratic separation of nine diuretics. Column: XTerra® MS C$_{18}$, 4.6 mm × 50 mm, 3.5 µm. Mobile phase: 13% acetonitrile, 4% methanol, 73% water, 10% 100 mM ammonium formate, pH 9.0. Detection: 254 nm. Peak designation as in Fig. 9 (from ref. [11] with permission from Elsevier Science B.V.).

1.3.4
Troubleshooting

1.3.4.1 Reproducibility Problems
When separating ionizable compounds, new problems emerge that are unknown for simple non-ionic molecules. As shown in this chapter, the retention times of analytes in reversed-phase chromatography change drastically with their level of ionization. In order to obtain reproducible results, it is therefore important to maintain a constant level of ionization. The degree of ionization is stable outside of ± 1.5 pH units around the pK_a of the analyte. If the analytes are simply bases or acids, the best results are obtained at either high or low pH values. Virtually all bases are completely ionized in a phosphate buffer at pH 2. Under the same conditions, most, but not all acids are not ionized. On the other hand, all acids are completely charged in an ammonium hydrogencarbonate buffer at pH 10, while most, but not all bases are uncharged. If the charge of all the compounds is well-defined at the pH value under consideration, then it is often not necessary to control the pH with great precision. As an example, the use of trifluoroacetic acid (pK_a = 0.5) is often sufficient to achieve a uniform ionization at acidic pH.

If the analytes contain several ionizable groups, then it is often impossible to find a pH range in which the dissociation does not change upon small changes in the pH value. This is also true for all substances in the intermediate pH range.

Under these circumstances, it is absolutely necessary to have excellent control over the pH value. The consequence of this is that true buffers must be used in order to achieve and control the desired pH value. If this is not done, problems with the reproducibility of retention times are almost inevitable, and often problematic peak-shape phenomena such as fronting or tailing occur as well. In addition to the use of a true buffer, it is also essential to measure and control the pH with high precision to obtain reproducible separation patterns. On the one hand, a change in the pH value is the best tool to manipulate and optimize the separation of ionizable compounds. On the other hand, careful pH control is mandatory.

1.3.4.2 Buffer Strength and Solubility

In order to achieve good reproducibility of a separation, it is important that the buffers used have a sufficient buffer capacity. A good buffer concentration is around 50 mM. However, with a high proportion of organic solvent in the mobile phase, such a concentration may result in precipitation of the buffer. Due to the multitude of buffers and solvent compositions, there are no tables that describe the solubility of the buffers used in reversed-phase chromatography. If one needs to work with a new buffer/mobile phase combination, it may be advisable to quickly test its solubility in the presence of the organic solvent. This can be done without difficulties and very rapidly, and it saves a lot of trouble with precipitated salts in the check valves of the pump.

In LC/MS, the typical buffer concentration used today is around 10 mM. This results in a reduced control over retention compared to classical HPLC. At the same time, the specificity of the mass spectrometer compensates for the sloppiness on the chromatography side.

1.3.4.3 Constant Buffer Concentration

Today, the HPLC instrument is commonly used to mix the aqueous buffer solution with the organic component of the mobile phase. For an isocratic analysis, this is only a convenient option, but for gradient separation there is no way around this. Under normal circumstances, such a procedure implies that the ionic strength of a buffer changes with the concentration of the aqueous component in the mobile phase. However, there is another way to do this, and with this method the concentration of the buffer can be kept constant over the gradient. The HPLC instrument needs three solvent inlets to the gradient mixing chamber to accomplish this. One of the inlets is connected to water, the second to the organic component of the mobile phase, and the third inlet is used for the buffer. The buffer is then added throughout the gradient at a constant rate of 10% of the flow rate. The buffer concentration is ten times higher than the desired final concentration in the mobile phase. With respect to the organic solvent, the gradient is mixed as usual. Commonly, there are four inlets to the gradient mixing chamber. If one is used for the addition of methanol, and another for the addition of acetonitrile, one has already part of a basic set-up for automated method development.

1.3.5
Summary

The majority of the separation problems in reversed-phase chromatography require buffered mobile phases. We have shown in this chapter the possibilities that exist, and what complications are likely to be encountered. While knowledge about the properties of buffers in the presence of organic solvents has increased in the last decade, one can still expect an improvement of this knowledge in the future. We hope to have pointed out the major problems in this chapter.

References

1 U. D. Neue, C. H. Phoebe, K. Tran, Y.-F. Cheng, Z. Lu, "Dependence of Reversed-Phase Retention of Ionizable Analytes on pH, Concentration of Organic Solvent and Silanol Activity", *J. Chromatogr. A* 925 (2001), 49–67.

2 M. Rosés, E. Bosch, *J. Chromatogr. A* 982 (2002), 1–30.

3 Cs. Horváth, W. Melander, I. Molnár, *Anal. Chem.* 49 (1977), 142.

4 W. O. Foye, T. L. Lemke, D. A. Williams, "Principles of Medicinal Chemistry", Lippincott Williams & Wilkins, Philadelphia, 1994.

5 J. N. Delgado, W. A. Remers, "Wilson and Gisvold's Textbook of Organic Medicinal and Pharmaceutical Chemistry", Lippincott-Raven, Philadelphia, 1998.

6 F. Seel, "Grundlagen der analytischen Chemie", Verlag Chemie, 1968.

7 E. Bosch, P. Bou, H. Allemann, M. Rosés, *Anal. Chem.* 68 (1996), 3651–3657.

8 I. Canals, J. A. Portal, E. Bosch, M. Rosés, *Anal. Chem.* 72 (2000), 1802–1809.

9 U. D. Neue, E. Serowik, P. Iraneta, B. A. Alden, T. H. Walter, *J. Chromatogr. A* 849 (1999), 87–100.

10 U. D. Neue, E. S. Grumbach, J. R. Mazzeo, K. Tran, D. M. Wagrowski-Diehl, "Method development in reversed-phase chromatography", Handbook of Analytical Separations Vol. 4 (Ed.: I. D. Wilson), Elsevier Science, Amsterdam, 2003.

11 U. D. Neue, "HPLC Columns – Theory, Technology, and Practice", Wiley-VCH, New York, 1997.

1.4
Selecting the Correct pH Value for HPLC

Michael McBrien

1.4.1
Introduction

Mobile phase pH is one of the most powerful variables for changing selectivity in high-performance liquid chromatography. It is common to observe k changes of an order of magnitude with changes in mobile phase pH. Additionally, pH can have a profound effect on peak broadening and tailing.

The strong influence of pH on chromatographic performance has implications for the robustness of chromatographic methods as well. The sensitivity of retention time to small changes in pH (see the pH region close to the pK_a in Fig. 1) leads to concerns about the reproducibility of separations. Systematic approaches to pH choice are particularly important when stable, reproducible results are sought. Fortunately, the practical range of pH values available to the chromatographer has recently been expanded considerably with the addition of new base-stable stationary phases.

Effective chromatographic pH values can be determined on a purely experimental basis, but this process can be facilitated greatly through the use of any knowledge that the chromatographer has regarding expected structure(s) and/or functional groups of the compounds of interest.

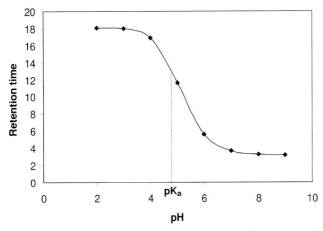

Fig. 1. Retention of an acidic compound as a function of pH in reversed-phase chromatography. At low pH, the neutral, protonated form predominates; the compound has a relatively long retention time. At high pH, the compound is present in ionized form and is weakly retained.

HPLC Made to Measure: A Practical Handbook for Optimization. Edited by Stavros Kromidas
Copyright © 2006 WILEY-VCH Verlag GmbH & Co. KGaA, Weinheim
ISBN: 3-527-31377-X

1.4.2
Typical Approaches to pH Selection

Changes in reversed-phase chromatographic performance as a function of pH can be due to solvation sphere changes, stationary phase modification, or changes of ionization state of the analyte. For the purpose of this chapter, the main focus will be the effects of changes in hydrophobicity due to changes in ionization state. The change in hydrophobicity of a compound as a function of pH can be quite complex depending on the type and number of ionization sites for that compound. For this reason, optimization of pH cannot be treated easily with the same systematic software optimization tools that are commonly applied to optimization of other parameters, such as concentration of organic modifier in the mobile phase and column temperature [1]. Figure 2 shows the change in *k* as a function of pH for norfloxacin, fleroxacin, and ofloxacin. It is clear that a linear (or simple curve-based) relation of pH to retention time will be effective only over very small pH ranges. For this reason, it is wise to consider chemical structures when optimizing separations. These can give a good starting point for method development. If further consideration of pH is desired, the range of optimization can be chosen to be suitably small, enabling accurate experimental optimization.

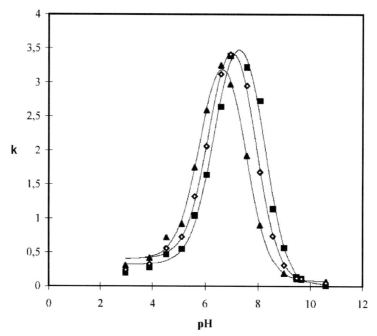

Fig. 2. The change in capacity factor for multiply ionizable species with change in pH: □ norfloxacin, ▲ fleroxacin, ○ ofloxacin [2].

1.4.3
Initial pH Selection

In Chapter 1.3, options for buffer content were discussed. There are a number of ways to utilize any structural information available to assist with selections between available buffers. These can be grouped into two main approaches:

1. Application Databases

The global body of work in chromatography is simply astounding. Recently, there have been a number of efforts to compile at least a subset of this knowledge into a mineable resource [3]. Application databases can offer very fast insights into pH and other method choices, particularly when chemical structures are stored with the applications. Even when only partial structures are known, substructure searches on key (especially ionizable) functional groups give a good estimation of methods that previously worked for similar compounds. A drawback of this approach is that subtleties in the chemical structures can change pK_a values significantly, and invalidate these choices. However, this is a very fast approach to retrieval of reasonable starting points for method development, and this, in conjunction with subsequent optimization over a small pH range, can be effective.

2. pK_a Look-up and Prediction

If we consider Fig. 1 in more detail, it is clear that the largest changes in retention time occur when the pH of the mobile phase is close to the pK_a of the analyte. In addition, the secondary equilibria that result in peak broadening can also be emphasized at this pH. There is a fairly simple solution to this problem: where analyte pK_a values are known, the chromatographer can simply retrieve these values and be sure to work at least 2 pH units above or below the pK_a. Chromatographic mechanisms should be relatively free from secondary equilibria effects in this area; i.e., the analytes will exist in either fully ionized or fully unionized form. This should result in better peak shape and more reproducible chromatography.

In addition to the familiar CRC Handbook [4], there are software-based resources for the retrieval of pK_a values [5]. The advantage of these is that they are typically also substructure-searchable, such that the resource is useful even for novel compounds. Substructure searches for ionizable groups should give a reasonable idea of pK_a values for these groups present in novel species.

Approximate pK_a values for individual functional groups can be obtained from any standard reference system. Chromatographers can, with some confidence, estimate pK_a values for new species based on these compounds. However, errors will arise when considering secondary effects of other functional groups. A more systematic approach is to use prediction software that is now readily available. Predicted pK_a values can be obtained from several resources, including on-line [5], as a standalone package [6], or as part of chromatographic prediction software [7]. All of these resources offer predictions based on an approach similar to the one described below.

1.4.4
Basis of pK$_a$ Prediction

The ACD/pK$_a$ algorithm has not been reported in the literature to date. The general structure of the algorithm is classification followed by a linear free energy (Hammett equation) relationship, using the well-known sigma constants as descriptors of the electron withdrawal and/or donation potency of substituents electronically connected to the ionization center. Several additional complexities need to be considered, however: ionic forms for polyelectrolytes, reference compounds, and transmission effects for subclasses of compounds with distal substituents (see Fig. 3 for an illustration). The classes, subclasses, reference compounds, sigma parameters, and transmission coefficients were derived by study of nearly 16,000 compounds with over 30,000 pK$_a$ values. Advanced topics such as tautomeric equilibria, proton migration, covalent hydration, vinylology, ring-breaking approximations, ring-size correction factors, steric effects, and variable charge effects are also taken into account, but are beyond the scope of this description.

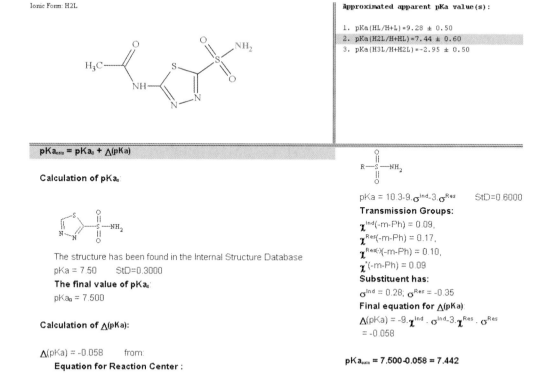

Fig. 3. Illustration of the computation of pK$_a$ for acetazolide; ACD/pK$_a$ DB 8.0.

1.4.5
Correction of pH Based on Organic Content

Organic mobile phase solvents in reversed-phase chromatography influence the selection of pH in two ways. The first is that there are differences in the pK_a values of compounds in the presence of organic solvent as opposed to the pure aqueous phase. The second issue is that the effective pH of the mobile phase changes due to the presence of organic solvent (see also Chapter 1.3). This pH shift is due to a change in the ionization of the buffer, not changes in buffer concentration or buffer capacity. Due to this pH shift, a correction factor should be applied prior to selecting the mobile phase pH based on pK_a values [8, 9]. The correction factor is of the order of 0.2 units per 10% modifier for acetonitrile, and 0.1 units per 10% methanol, depending on the buffer system, and is well-behaved in the range 0–60% in each case.

Correction of pH based on the presence of organic solvent should be done in two simple steps. The first accounts for the effect of the solvent on the buffer. The second accounts for the effect on the ionizable species.

Notice that if the buffer and the analyte are both acid or both base, the terms will cancel out to a large extent. In cases where the analyte is basic and the buffer is acidic, or vice versa, the effect will be very pronounced.

An illustration may be helpful in this case. If we consider an acidic analyte/ acidic buffer, we can, in general, use the aqueous pK_a as an indicator of the pH at which to prepare the aqueous buffer. The changes in pK_a for the buffer and analyte

Table 1. Correction factors for acidic buffer, acetonitrile, and mobile phase pH.

Acetonitrile %	Acid analyte correction factor	Base analyte correction factor	Acidic buffer correction	Total correction factor
10	+0.2	–	−0.2	0
20	+0.4	–	−0.4	0
30	+0.6	–	−0.6	0
40	+0.8	–	−0.8	0
50	+1.0	–	−1.0	0
60	+1.2	–	−1.2	0
10	–	−0.2	−0.2	−0.4
20	–	−0.4	−0.4	−0.8
30	–	−0.6	−0.6	−1.2
40	–	−0.8	−0.8	−1.6
50	–	−1.0	−1.0	−2.0
60	–	−1.2	−1.2	−2.4

will cancel. We need only make sure that we consider the effective pH gradient that we will see. In the case of a basic compound with an acidic buffer, if the pK_a of the species indicates that a pH range of 3 to 5 is unusable and we anticipate working at 50% acetonitrile, we then must consider the effective change. Looking at Table 1, we can see that the correction factor will be about 2 pH units. This means that our effective unusable pH range is actually 1 to 3 instead. Provided that we prepare the mobile phase with a pH greater than 5, we can expect stable retention times and reasonable peak shapes.

Particularly interesting is the mixture of an acid and a base, wherein we must then explicitly model the change in pK_a of the base and mobile phase. Initially, we consider the "aqueous" systems that are shown in Fig. 4a and b. These are the pH/retention factors for an acid and a base with no acetonitrile. Based on this, the pH choices could be either less than 2.5, or more than 7.5. Most chromatographers would elect to choose a pH of 2.5 or so. However, if we are going to use an appreciable concentration of an organic modifier, the picture changes. If we assume that we are going to prepare an acid-based buffer, the elution profile for the acid remains the same; however, for the base, the pH shift will be in opposite directions for buffer and analyte. The elution profile will shift.

Based on this result, a pH of more than 6.5 is acceptable.

To summarize: The presence of organic modifiers affects the ionization of both analytes and buffers. The effects are documented and well-behaved for acetonitrile and methanol. Mobile phase pH correction is particularly important when analyte and buffer are not of the same type; i.e., when bases are studied with an acidic buffered mobile phase. In these cases, the effective change is twofold, with the pK_a of the base shifting down on the pH scale, and the pK_a of the acid shifting up.

1.4.6
Optimization of Mobile Phase pH Without Chemical Structures

Chromatographers will likely be familiar with standard systematic approaches to optimization of well-behaved variables such as concentration of organic modifier in the mobile phase and column temperature in chromatographic method development. This concept is covered in Chapter 1.1. It is possible to approach the problem of pH optimization in a similar manner, provided that it is done with caution. Figure 2 shows the response of a series of ionizable compounds to changes in mobile phase pH. It is clear that the response of these compounds across the entire pH range cannot be modeled based on a few simple experiments. However, due to the tremendous power of pH in changing the retention times of ionizable compounds, optimization of pH can still be helpful in designing a robust separation. Where methods will be applied repeatedly, such as the quality control laboratory, it is wise for the method development chromatographer to anticipate the needs of validation, and to simply perform an optimization of pH over a narrow range (say, three or four experiments over a single pH unit) to examine the effect on the suitability of the separation.

0% MeCN

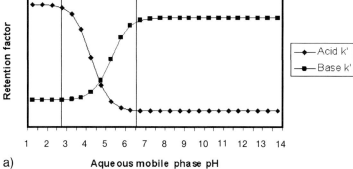

a)

50% MeCN

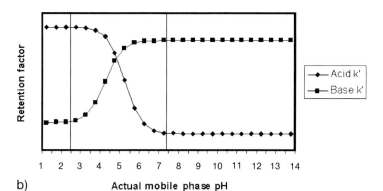

b)

50% MeCN

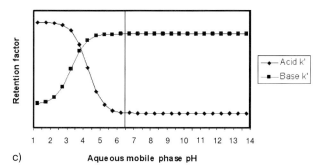

c)

Fig. 4. Retention of acid and base as a function of pH.
(a) pH ranges for robustness and peak shape.
(b) With the addition of acetonitrile as a modifier, the inflection points of the
acid and the base are moved in opposite directions. This assumes that the
user has prepared a buffer that corrects for organic modifier content.
(c) Assuming that buffers are made up in aqueous solution, this is the profile
that would result. The acceptable aqueous range is above pH 6.5.

This non-structural approach provides an effective complement to experimental design based on pK_a values. Solvation spheres and effects of pH on columns will inherently be taken into account, as is the ionization of unknown species. The key to this step is the small pH range; extrapolation over a large range of pH values has little hope of being successful.

1.4.7
A Systematic Approach to pH Selection

It is a relatively simple process to select the pH for analysis of a given sample, even in cases when one or more compounds are not known. In general, it can be expected that compounds for which no functionality is known will be at trace levels, and thus peak shape will be less critical. In addition, unknown compounds will often tend to have similar functionality to other compounds in the sample. For this reason, it is reasonable to use the pK_a values of known compounds as a starting point wherever possible.

An effective approach is as follows:

1. Collect all structural information on the sample in question.
2. Obtain or predict pK_a values for all species. In cases where only partial structures are known, consider ionizable functional groups.
3. Divide the pH scale into areas of "use/don't use" based on keeping each compound in the unionized state. This implies 2 pH units from the pK_a. Where conflicts arise, choose the major components first, and move to areas that push the equilibria for other compounds to be fully ionized where possible. Estimate the solvent strength required and correct pH as detailed in Section 1.4.4. Note that slight adjustments may still need to be performed.
4. Perform a slow gradient experiment at the indicated pH. Estimate the solvent strength at which compounds elute and correct the pH accordingly, if desired.
5. Prepare the new buffer, and re-inject.
6. Where necessary, optimize pH across a small range, and/or perform optimization of other parameters. Note any broad peaks in the separation. If they are part of the critical pair (two closest eluting significant peaks), then this may be a concern. Remember that more than just pH will vary on subsequent separations, and look for a separation that is very stable. If the suitability is sensitive to the pH, consider a slight modification to a more reasonable pH area. If this is not possible, then it may be necessary to control other factors very carefully, including column parameters and temperature.
7. When validating the final method, prepare buffers based on gravimetric preparation rather than adjusting pH using a pH meter and "titration" with acid or base. pH meter maintenance is often neglected, resulting in inconsistencies of measurement.

1.4.8
An Example – Separation of 1,4-Bis[(2-pyridin-2-ylethyl)thio]butane-2,3-diol from its Impurities

An example of this approach to pH selection is shown for an example compound, 1,4-bis[(2-pyridin-2-ylethyl)thio]butane-2,3-diol (see Fig. 5) and three unknown impurities.

The predicted most acidic pK_a of this species is 5.1 (ACD/LC Simulator v8.04, Advanced Chemistry Development, Inc.). It can be concluded that pH ≤ 3.1 should be reasonable for examination of this species. If we knew the structures of impurities we could also predict for these, of course. Our next step is to design a gradient experiment around this pH and inject.

Figure 6 shows the resulting chromatogram. Based on the elution time of the known compound, we could apply pK_a correction factors based on solvent content. For the sake of simplicity, we will not do this here. Note that we have reasonable resolution of the four components, but it is not yet clear if this will be a robust separation or if we can achieve better resolution. Prior to optimization of other parameters, we will study the pH further; we inject at pH 2.5 and 3.8 in order to generate an optimization system. At this stage, we can determine an optimal pH by using experimental optimization software.

Figure 8 shows the resolution map describing the response of the system versus pH. Indeed, despite the fact that we had reasonable resolution of all components in our first experiment (pH 2.9), it is clear that a small change in pH (Fig. 8) would result in co-elution of components 3 and 4. pH 2.6 appears to offer a good solution; we can proceed to further optimization, examining for an isocratic and/ or faster separation from here.

This approach has enabled us to target a pH range that is relatively stable to small changes, but it has also speeded the process of optimization while decreasing the likelihood of validation difficulties. Further optimization of other variables should lead to a high quality method.

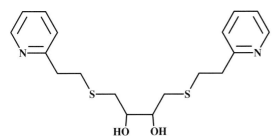

Fig. 5. Chemical structure of 1,4-bis[(2-pyridin-2-ylethyl)thio]butane-2,3-diol.

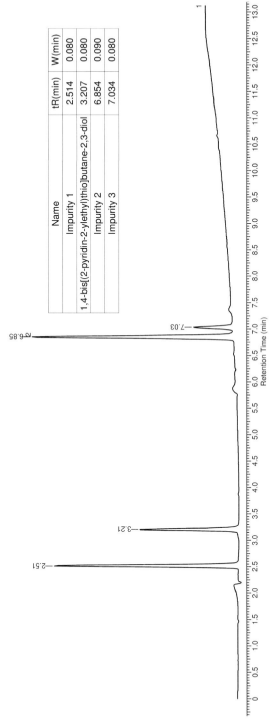

Name	tR(min)	W(min)
Impurity 1	2.514	0.080
1,4-bis[(2-pyridin-2-ylethyl)thio]butane-2,3-diol	3.207	0.080
Impurity 2	6.854	0.090
Impurity 3	7.034	0.080

Fig. 6. Initial "scouting gradient" for 1,4-bis[(2-pyridin-2-ylethyl)thio]butane-2,3-diol and three unknown impurities. Conditions: 11 min gradient from 10 to 80% acetonitrile; buffer was 50 mM HCOOH/HCOONH$_4$.

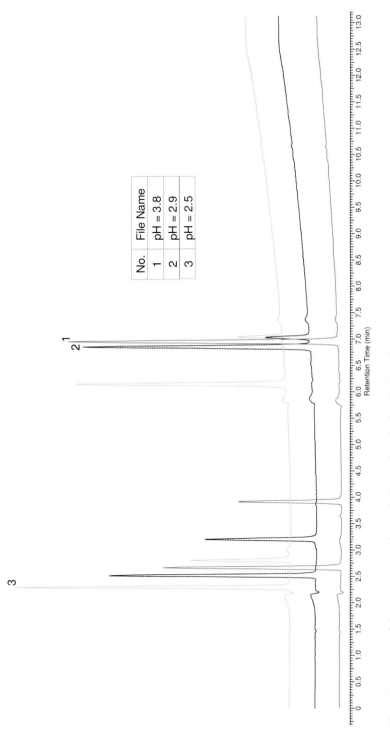

No.	File Name
1	pH = 3.8
2	pH = 2.9
3	pH = 2.5

Fig. 7. Comparison of chromatograms with varied pH; sample and other HPLC conditions as noted in Fig. 6.

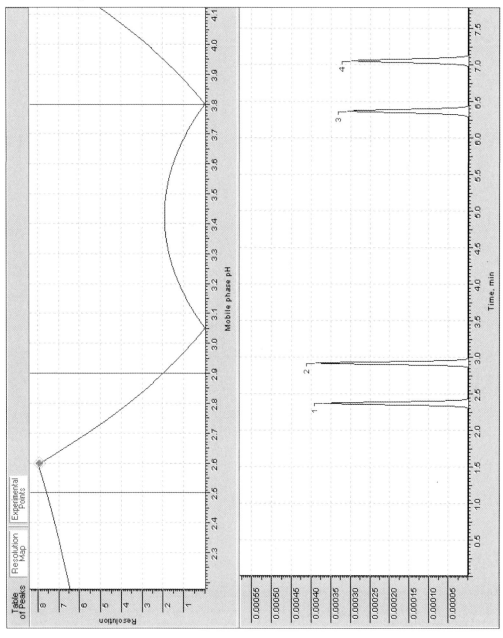

Fig. 8. Optimization of pH for 1,4-bis[(2-pyridin-2-ylethyl)thio]butane-2,3-diol and three unknown impurities; ACD/LC Simulator v8.04, Advanced Chemistry Development, Inc.

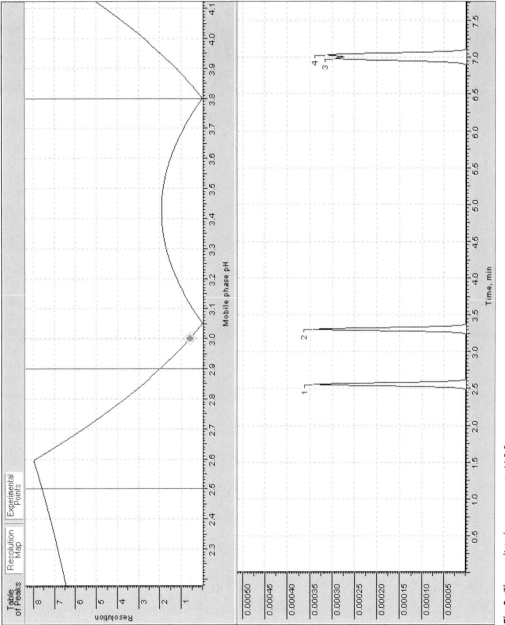

Fig. 9. The predicted response at pH 3.0.
Note that components 3 and 4 can be expected to co-elute.

1.4.9
Troubleshooting Mobile Phase pH

Due to the tremendous sensitivity of selectivity to changes in mobile phase pH, problems with separations are often a result of relatively small changes in effective pH. This may be due to the usual causes, including temperature changes and incorrect buffer preparation, or more subtle effects. The effective design of robust methods should help to alleviate these problems. When the methods break down, it can often be useful to revisit the original development work.

It can be helpful for the troubleshooter to revisit optimizations of pH that were performed during method development. For example, an experimental chromatogram from a failed suitability test can be reconciled with the predicted chromatogram based on, for example, a 0.3 unit difference in pH from the original optimization, leading to the conclusion that mobile phase pH may have been the culprit. Where these optimizations are not available (and where chemical structures *are* available), structure-based systems can be generated quickly, combining the pH/hydrophobicity curves with experimental retention times. The familiar resolution maps contain extrapolated chromatograms that can be quickly examined and reconciled to observed chromatograms. The reader is reminded that small pH changes may alleviate the problem at hand, but this sensitivity should be noted. Again, the addition of column temperature control or gravimetric mobile phase preparation to the method may help to reduce future failures in the method.

1.4.10
The Future

In order to estimate biological activity of given compounds, it is important to consider the ionization state in the medium of interest. Solubility, in particular, is of great interest, and highly dependent on pK_a. For this reason, interest in effective pK_a prediction is increasing in many areas of chemistry. Initiatives are being put in place to measure pK_a values for drug candidates, for example. The pK_a prediction algorithm described in Section 1.4.4 depends in part on the experimental data that forms its basis. The more comprehensive and chemically relevant the measurements are, the better the predictions can be expected to be. It is not unreasonable to expect that these pK_a measurement initiatives will continue. It is possible to make these measurements available to the entire organization, reducing pK_a prediction to "look-ups" in some cases, or in the worst-case scenario, giving prediction algorithms access to experimental values for more relevant compounds, thus increasing accuracy. This accumulation of chemical knowledge both within and outside organizations should be an objective for chemists in the future. In addition, the simple rule-based correction of pH for organic solvent content described in Section 1.4.5 is a problem that should be solvable within pK_a prediction software, particularly software designed for support of chromatographic method development.

1.4.11
Conclusion

Mobile phase pH is one of the most powerful tools for modifying selectivity of compounds in HPLC. A systematic selection process for pH can make the difference between a fast development of a robust method and a long development process that may result in a separation that cannot be reproduced. While it is usually impractical to measure pK_a values of species prior to method development, the combination of literature searches, investigative experiments, and software tools can result in an effective pH choice in a very short time frame. A primary goal in buffer selection should be a successful separation that is stable with respect to small changes in pH. Where this is not possible, developers can explicitly note the sensitivity of the method and create stringent experimental specifications.

References

1 A number of experimental optimization programs are available, including DryLab 2000 (www.lcresources.com), ACD/LC Simulator 8.0 (www.acdlabs.com), and ChromSword (www.chromsword.com).

2 J. Barbosa, B. Rosa, V. Sanz-Nebot, *J. Chromatogr. A* 823 (1998).

3 Examples are: CP ScanView, ChromAccess, and ACD/ChromManager.

4 CRC Handbook of Chemistry and Physics, 84th Edition (Ed.: D. R. Lide), CRC Press, Boca Raton (2003).

5 Chemical Abstracts Service (www.cas.org), the Merck Index Online, MDL® Comprehensive Medicinal Chemistry (www.mdl.com), ACD/ILab (http://www.acdlabs.com/ilab/), and more.

6 ACD/pKa DB 8.0 (www.acdlabs.com), and pKalc 3.1, Compudrug (www.compudrug.com).

7 ACD/LC Simulator 8.0 (www.acdlabs.com)

8 R. LoBrutto, A. Jones, Y. Kazakevich, H. McNair, *J. Chromatogr. A*, 913 (2001) 173–187.

9 S. Espinosa, E. Bosch, M. Roses, *Anal. Chem.*, 74 (2002) 3809–3818.

1.5
Optimization of the Evaluation in Chromatography

Hans-Joachim Kuss

1.5.1
Evaluation of Chromatographic Data – An Introduction

Chromatography is a very important part of instrumental analytics and has some unique features. Optimization means improvement or best utilization of the given conditions. Thus, one of the aims of an analytical chemist is the successful optimization of a chromatographic technique. For the optimal use of chromatography, it is also necessary to optimize the evaluation so as not to lose any information. Integration software provides a quick and flexible means of evaluating data. However, decisions can only be made by the user; only he or she can assess the method.

The most important information is contained in the chromatogram. It would seem obvious that the person responsible for the analysis of results should inspect each chromatogram. This has to be done by a chromatography specialist, not by a computer specialist or statistician.

Clearly, the potential offered by computer techniques should be exploited. The individual values of the internal standard in a serial analysis should always be visualized graphically, so as to recognize any possible trend. An automated calculation should be used to determine the area/height ratio for each peak and to reassess the (graphically) established extreme values in the chromatogram. If possible, two detection signals should be monitored simultaneously, for example, two UV wavelengths or signals from two independent detectors such as UV and fluorimeter, FID and NFID, two mass traces, etc. The two signals should be estimated independently and it should be assessed whether the results are in a predefined window. This twofold analysis can lead to a marked improvement in the quality of results.

1.5.2
Working Range

The measured signals can only be calculated and assessed after prior calibration. First, the working range WRx has to be defined, which is limited by the lower concentration x_L and the upper concentration x_U.

$$WRx = \frac{x_U}{x_L}$$

On the one hand, one wishes to take advantage of the sensitivity of the detector by measuring low concentrations. On the other hand, in a few cases it may be possible that high concentrations are present. A value can then only be quantified

HPLC Made to Measure: A Practical Handbook for Optimization. Edited by Stavros Kromidas
Copyright © 2006 WILEY-VCH Verlag GmbH & Co. KGaA, Weinheim
ISBN: 3-527-31377-X

with much additional and troublesome work because it is above the upper concentration x_U of the working range. If the value for very few samples exceeds x_U, one should consider the possibility of dilution with fresh matrix [1].

Quantification is limited on the lower side by the limit of quantification (LOQ). In the case of very sensitive measurements, one may encounter problems with the LOQ. Therefore, it would seem advantageous to estimate the LOQ early in the development of a method.

$$x_L > LOQ$$

In chromatography, the LOQ can be simply estimated by calculation of the signal-to-noise ratio of a peak with known concentration. The height of the peak H_x corresponds to the signal intensity and will be given directly in μV. One then looks for a region of, for example, one minute in the chromatogram that corresponds to baseline only and measures the noise N in μV as lower apex to upper apex. The LOQ is the concentration with a 10-fold signal-to-noise ratio – in practice, one considers a 20-fold signal-to-noise ratio, in order to compensate for any slight instrumental deteriorations.

$$LOQ = \frac{20 \cdot x \cdot N}{H_X}$$

In principle, any peak with known (preferably low) concentration can be used to estimate the LOQ by calculating the height in relation to the noise. This method should always be used in chromatography, provided that it conforms to existing guidelines.

1.5.3
Internal Standard

In the next step in the preparation of a validation, three (or preferably more) samples with the envisaged lowest concentration x_L after sample preparation considering the estimated recovery rate should be analyzed. If it is at all possible, an internal standard should be added to the samples. A final decision as to whether to use a final method with or without internal standardization can then be taken later. An internal standard is generally the best quality control possibility. Each analysis sample will then be individually controlled with virtually no additional effort.

With three measurements, it is possible to calculate the mean My, the standard deviation SDy, and the coefficient of variation cv(%) using the method of external and internal standards. If linear regression leads to a straight line with an intercept not significantly different from zero – which is normal in chromatography – then the cv(%) also holds for the relative deviation in the concentration axes.

$$\frac{SDy_L}{My_L} = cv(\%) = \frac{SDx_L}{Mx_L} \approx \frac{MU_L}{1.5 \cdot t}$$

Here, MU_L is the measurement uncertainty at x_L and t is Student's t-factor, which depends on the number of calibration points.

Measurement uncertainty MU_L can be roughly estimated from the product of $1.5 \times t \times$ cv(%). If the governing rules allow a maximal MU_L of 25%, 33% or 50%, the highest tolerated cv(%) can be readily obtained through division by $1.5 \times t$.

For the internal standard method, the areas (heights) of the compound peaks divided by the areas (heights) of the related internal standard peaks in the same chromatogram should be used, instead of using the areas (heights) directly. It can immediately be seen if the internal standard decreases the cv(%), which is the purpose of the exercise.

1.5.4
Calibration

After tentative predefinition, the working range must be verified with several calibration concentrations in order to establish a linear relationship between measurand (y) and concentration (x).

$$y = b \cdot x + a$$

The calculation of the intercept a and slope b for the straight line and the back-calculation of the concentration for the analysis samples according to

$$x = \frac{y - a}{b}$$

is an essential feature of each integration program. It may be the case that some chromatographers use weighted regression giving scant attention to its origin. In the usual integration programs it is possible to weight with $1/y$ or $1/y^2$. This is confirmed by a mouse click; thereafter, subsequent workings are hidden and so this can be forgotten. If the values are carried forward into a validation program – which often does not allow for weighting – anomalies will inevitably be programmed. Using high working ranges, weighting often generates an advantage, but this needs to be demonstrated.

For validation purposes, some further parameters have to be calculated, including measurement uncertainty MU from the calibration line. To date, this has usually been included in validation programs, but not in integration programs. Unfortunately, without a basic understanding, it can be hard to see what the meaning of some calculated values really is. The principle train of thought is simple, but often unnecessarily complicated.

1.5.5
Linear Regression

Linear regression is also called the method of least squares. This refers to the squared residues R, which are the perpendicular (y)-distance of the calibration

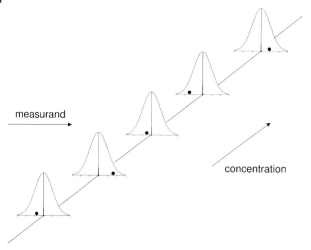

measurand

concentration

Fig. 1. Linear Regression is based on the assumption of the same standard deviation at all calibration points. Then all deviations are considered to be of the same importance – they have the same weight. The variation in the residues is thought to be random.

points y from the calculated points yx on the straight line. It is assumed that the imprecision in the concentration is sufficiently low that it need not be taken into account when calculating the inaccuracy of the measurand.

$$R = y_k - yx_k \qquad\qquad R^2 = \text{min}$$

y_k = measurand at the calibration point k
yx_k = value calculated from the straight line for calibration point k

This calculation relies on the equality of the residues R and assumption of the same standard deviation at different concentrations (Fig. 1).

If this assumption is not fulfilled, the residues with higher standard deviations, due to their higher averaged amount, exert a greater influence on the location of the straight line than do residues at calibration points with lower standard deviations (Fig. 2). In these cases, it is necessary to minimize:

$$\sum \frac{(y_k - yx_k)^2}{SD_k^2} = \sum w(y_k - yx_k)^2$$

This may re-establish the equality of the residues. In the standard form, the residues are individually multiplied by a weighting factor w, which is normally adjusted for $\Sigma w = n$. This can only be done in two steps. First, a temporary weighting g is calculated, which is adjusted using the previously unknown Σg.

$$g = \frac{1}{y^{\text{WE}}} \qquad\qquad w = \frac{g \cdot n}{\sum g}$$

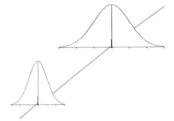

Fig. 2. Knowing from the *f*-test (variance quotient) that the assumption of the same standard deviation does not apply, we have to balance the disequilibrium by a factor – the weighting factor. Otherwise the calculation will be wrong. The 2-fold standard deviation of the upper point averaged induced 2-fold residues. Thus, the straight line will be (2^2) 4-fold more influenced by the location of the upper than by the lower calibration point. Therefore the weighting factor has to weight the variances at the lower point 4-fold.

The calculation method known as linear regression is based on the equality of the standard deviations in the whole working range. In chromatography, we often have the same percentage deviations. In a defined concentration region a method has a precision of a few percent. At lower concentrations, the imprecision will increase to a limit that cannot be tolerated – for instance 33% *MU* – and the LOQ is undercut.

Let us consider a working range of 100, and assume that the cv in the middle of the working range is 1%. According to the (unweighted) linear regression, the residual standard deviation appears as a nearly constant band spanning both sides of the straight regression line. The cv(%) at x_L must be 10% and cv(%) at x_U must be 0.1%. The cv values should be multiplied by 1.5 times Student's *t*-value to give a rough estimate of the confidence interval. Therefore, 10% is higher than allowed for the LOQ. To reach a cv of 0.1% in chromatography is unrealistic. Often it is hard to reach even 1%. The conclusion that can be drawn is that using a working range higher than 10 in chromatography, linear regression without weighting is a model that is most probably inadequate.

Let us assume that in a chromatographic study with multiple measurements of the same sample the cv(%) is constant in the working range from x_L to x_U. Then:

$$\frac{SDx_k}{My_k} = cv(\%) = const.$$

This means that over the whole calibration line there is a constant ratio between the measurand y_K and the related standard deviation SDy_K. It is then possible to replace $w = 1/SD^2$ with $w = 1/y^2$. The standardization on *n* compensates for differences in the values of *y*. The unknown standard deviation is estimated from the known measured values.

A weighting with $1/y^2$ is exactly equivalent to the case when the cv is constant in the working range. Experimentally, this may be found only in approximation. Therefore, it is a model assumption. However, a model is necessary to calculate a weight for the measured values of unknown analysis samples. Measuring the

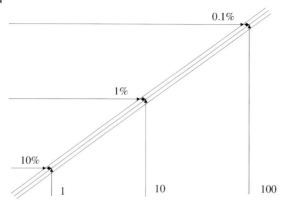

Fig. 3. The same standard deviation at all concentrations (1, 10, 100) leads to great differences in the coefficients of variation (10%, 1%, 0.1%).

standard deviations with great effort at each calibration point only allows interpolation at the measured analysis values with unavoidable inconsistencies.

1.5.6
Weighting Exponent

In $1/y^2$, the weighting exponent WE is 2. With WE = 0, we have the ordinary linear regression because all weighting factors are set to 1 ($y^0 = 1$).

Let us first look at these two extreme cases.

In chromatography, we have on the one hand the "instrumental variance", which is not dependent on the concentration, and which is related only to the noise of the detector or the oscillations of the pump. This variance is the same at high and at low concentrations and is described by WE = 0 ($w = 1$ at all concentrations). On the other hand, we can postulate a "sample preparation variance" with the same percentage deviations at all concentrations (characterized by WE = 2).

The dominating volume effects (the error in volume does not "know" the concentration) of the injection volume variability or of the volume transferring steps in sample preparation have the effect that the standard deviation increases linearly with the concentrations and the measured values (SDy/y = const.). The chromatographic reality can be between the two extremes, depending on the dominant effects. This means that all values between 0 and 2 are possible. Using a very simple sample preparation a WE near 0 can be expected; using multiple volume transfer steps a WE near 2 can be expected.

$$w = \frac{1}{y^{WE}} \qquad WE = 0 \text{ to } WE = 2$$

The broader the working range ABx, the earlier the deviation from ordinary linear regression can be detected. The F-test is used to discriminate between homogeneous and heterogeneous variances.

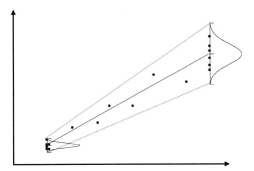

Fig. 4. The weighting factor interpolates the standard deviations between the measured values at the lower and the upper ends of the working range.

$$F = \frac{SD_U^2}{SD_L^2}$$

It is possible to weight variances in such a way that F will be 1. Therefore, WE must be calculated from:

$$WE = \frac{\log F}{\log AB\gamma} = \frac{\log SD_U^2 - \log SD_L^2}{\log \gamma_U - \log \gamma_L}$$

The quotient of the variances is used for this calculation. Therefore, the method might be called weighting by variance ratio.

1.5.7
In Real Practise

In quality control, it is usual to use a working range of 2. Assuming the same cv(%) over this range, the variance at the upper calibration point is $2^2 = 4$ times the variance at the lower calibration point. Measuring these two concentrations six times, no significance is reached at 95% confidence. This means that the differences in such a small working range are too low to be of practical importance.

In a working range of 10 and with the assumption of the same cv(%), F will be calculated as 100, which means heterogeneity of variances. In DIN 32645 [4] it is pointed out that in a working range of 10 homogeneity of variances can be expected. In chromatography, this is not normally the case. As ever, theory must bow to reality. If heterogeneous variances are identified, it is necessary to calculate them by weighting variances or to decrease the working range.

1.5.8
Drug Analysis

When measuring drug concentrations, a working range of 10 is normally too small. Often, a factor of 100 is necessary. Scattering of the calibration points at equal distances is then not possible. It is unclear as to whether in some governing rules a logarithmic arrangement of concentration steps is forbidden, even if this makes common chemical sense. This does not influence the calculated reference values, with the exception of the process coefficient of variation (Mx is lowered; see below).

In reality, the very powerful chromatographic methods can reach a working range of 100 or 1000. The equations of LOQ, PI, and MU include the RSD and, without weighting, give far too high values in the lower concentration range. These complicated equations give the impression of high precision, but they are only valid under defined preconditions.

The FDA guidelines for bioanalytical method validation [5] not only point out a minimal precision of 15%, but also a sensible choice for maximal deviation RE(%) of the back-calculated calibration concentrations of 15%. This level of accuracy is not normally obtainable in chromatography with a working range above 10 without weighting, because the lower calibration points, due to their low weight, often show systematic deviations. The FDA guideline accepts the use of a weighting, assuming that an explanation is given.

$$RE(\%) = \frac{x_{calc} - x_k}{x_k}$$

It has been suggested [6] that one should create a diagram showing the relative percentage deviation RE(%) plotted against the logarithm of the concentration. This prevents the low concentrations from collapsing graphically into one point by taking into account the constancy of the relative and not the absolute deviations typically found in chromatography. In this way, the possible advantage of weighted regression can be illustrated.

1.5.9
Measurement Uncertainty

By measuring usually six equal samples at x_U and x_L, the measurement uncertainty at these concentrations will be known. By interpolation between these two concentrations, a reasonable estimate of the measurement uncertainty is possible. This estimate is certainly better than sticking with a constant RSD and MU over the whole working range, as in the case of ordinary linear regression, when the readings (significant F-test) show otherwise.

MU at x_L can be used to estimate LOQ: For MU below 33%, x_L is above the LOQ. Essentially no more information is necessary. The statistical considerations for LOD, identification limit, and LOQ are partly inconsistent and may create irritation and confusion.

It seems doubtful whether neologisms such as "uncertainty manager" and "budget of measurement uncertainty" will be able to increase quality in daily practice. Analytical jargon contains too much theory and formalism [7].

Using the linear regression equation, it is simple to calculate yx-values at the calibration concentrations. The residues R are the differences of measurand y and calculated yx.

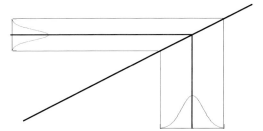

Fig. 5. The measurand of an analysis sample has a 99% prediction interval, which is reflected through the straight line in a 99% measurement uncertainty at the concentration axis.

$$RSD = \sqrt{\frac{\sum R^2}{n-2}}$$

The sum of the squared residues increases with the number n of calibration points. The square root of the sum of the squared residues divided by $n-2$ is called the residual standard deviation RSD, which represents a mean value for the residuals. Using ordinary linear regression, the RSD nestles at both sides against the straight line as a constant band, according to a random scatter of the residues.

More useful than the deviation in measurand direction is the directly resulting deviation of the concentrations. Graphically, this is a view on the same RSD band vertically or horizontally. The standard deviation on the concentration axis is the process standard deviation VSD, which is calculated by division of RSD by the slope b. VSD is the central goodness criterion [3], which can be used to calculate the process coefficient of variation pcv(%).

$$VSD = \frac{RSD}{b} \qquad pcv(\%) = \frac{VSD}{Mx}$$

So far, we have only considered standard deviations, which means mean deviations containing in a normal distribution 66% of the individual values. To get the 95% confidence interval of the y-values (PI) or the x-values (MU), it must be multiplied by Student's t-factor.

Prediction interval: $\qquad \pm PI(95\%) = t(95\%) \cdot RSD \cdot \sqrt{1}$

Measurement uncertainty: $\quad \pm MU(95\%) = t(95\%) \cdot VSD \cdot \sqrt{2}$

The complete (shown in the following inclusive of weighting) equations can be found in Ref. [8] or papers cited therein. Here, the two square roots should be replaced by a factor of 1.2, which leads (without weighting) to the approximation of two constant confidence lines parallel to the straight line. Visualized, these confidence lines would be elastically fixed at the ends and slightly pressed in the middle. This is exactly the effect of the real square-root term, for which using ordinary linear regression w_A (w_K) must be set to 1 and $\Sigma w = n$.

$$\sqrt{}_1 = \sqrt{\frac{1}{w_K} + \frac{1}{\sum w} + \frac{(x_K - Mxw)^2}{\sum w \cdot (x - Mxw)^2}}$$

$$\sqrt{}_2 = \sqrt{\frac{1}{w_A} + \frac{1}{\sum w} + \frac{(y_A - Myw)^2}{b_w^2 \sum (x - Mxw)^2}}$$

Limit of Quantification (LOQ). The LOD according to DIN 32645 can be estimated as $4 \times$ VSD, because Student's t-value for eight degrees of freedom and 1% error probability is 3.35 and the square root can be approximated by 1.2. The LOQ is approximately three times the LOD.

$$LOD = t \cdot VSD \cdot \sqrt{1 + \frac{1}{n} + \frac{Mx^2}{\sum (x - Mx)^2}}$$

$$LOQ \approx 3 \cdot LOD \approx 12 \cdot VSD \qquad \frac{VSD}{LOQ} \approx 8\%$$

Therefore, cv(%) must be below 8%. Let us assume a constant VSD in the working range (homogeneous variances). Then, VSD > SDx_L, because the location of Mx_L is not exactly on the straight line and there is an additional scatter about the straight line. Let us additionally assume that VSD decreases from x_U to x_L (variance heterogeneity). Then VSD must be much greater than SDx_L. Because VSD > SDx_L, it has realistically to be expected that cv(%) values above 5% are due to the LOQ being exceeded. In the case of variance heterogeneity this can occur much earlier (perhaps 1%). This can lead to serious problems in trace analysis.

The obvious solution would be to use SDx_L directly. The best estimate of the standard deviation at x_L is a direct measurement, rather than a complicated calculation. Dividing the maximum tolerated deviation of 33% by Student's t-value of 3.3 gives a maximum tolerated cv(%) of 10%. A method can break down as a result of the difficulty of obtaining a cv(%) of 5% instead of 10%.

To estimate the LOQ according to DIN 32645, a linear regression with ten concentrations in the region of the LOQ needs to be performed; the price is thus a very considerable working effort. In chromatography, the integration systems are unfortunately unable to give area values for peaks at concentrations near the LOQ. Therefore, this method must be skipped. The calibration method should only be used if neither the signal noise nor the blank sample method are possible.

Great calculation expenditure gives an impression of increased security. The problem is not the increased (computer) effort, but the unnecessary intransparency in using complicated equations. Each quality demand is useless when it is suffocated in formalism. It is carried out because rules have to be fulfilled. It is well meant, but intransparent perfectionism can be demoralizing.

1.5.10
Calibration Line Through the Origin

What advantage can be gained in chromatography by using a calibration line with an intercept? In a blank sample, no peak should appear in the chromatogram at the retention time of the substance to be analyzed. If at this stage no baseline is found, the method would seem to be inadequate. In each chromatographic analysis series, some blank samples should be included to establish this. In this way, it is possible to recognize memory effects, but the possibility of interfering underlying peaks at the same retention time in a chromatogram cannot be excluded. Fitting a straight line with an intercept does not prevent this. Only measurements with two different detectors can largely exclude the occurrence of interferences.

For non-chromatographic analyses with a measurable blank value, a straight line with an intercept is unavoidable. In cases where the intercept has a physical importance, results are invalidated by neglecting the intercept.

Conversely, results are also invalidated by considering an (insignificant) intercept in chromatography, even though it has no importance either physically or statistically. The intercept corresponds to a systematic deviation that must be considered (especially important at low concentrations) in the calculation, although by chance it can take positive or negative values. The back calculation of the signal values for the calibration concentrations clearly shows this.

An analytical chemist who is able to optimize his or her sample preparation and HPLC analysis can very well decide whether to use an internal standard, whether to calculate the straight line with or without weighting, and whether or not to consider an intercept. Obviously, it is mandatory that one can give reasons for these decisions.

The rapid advent of LCMS(MS) has changed liquid chromatography. It is regrettable that changes in the governing rules have lagged behind for a decade. Let us hope that future guidelines will take into account the idea of giving the practitioner real help and will allow the necessary flexibility in the different analytical applications.

References

1 V. P. Shah et al., *Pharm. Res.* 17 (2000), 1554.
2 P. Haefelfinger, *J. Chromatogr.* 218 (1981), 73–81.
3 H.-J. Kuss, in *Handbuch Validierung in der Analytik* (Ed.: S. Kromidas), Wiley-VCH, Weinheim, 2000, p. 182.
4 DIN 32645, Beuth Verlag, Berlin, 1994.
5 Guidance for Industry: Bioanalytical Method Validation;

http://www.fda.gov/cder/guidance/Index.htm.
6 E. L. Johnson et al., *J. Chromatogr. Sci.* 26 (1988), 372–379.
7 R. M. Lequin, *Clin. Chem.* 50 (2004), 977–978.
8 H.-J. Kuss, in *More Practical Problem Solving in HPLC* (Ed.: S. Kromidas), Wiley-VCH, Weinheim, 2004, p. 236, ISBN 3-527-31113-0.

1.6

Calibration Characteristics and Uncertainty – Indicating Starting Points to Optimize Methods

Stefan Schömer

1.6.1

Optimizing Calibration – What is the Objective?

The goal of any optimization is the most efficient use of the available resources (time, devices, staff). For calibration, this may mean:

- minimizing efforts in measuring, without compromising in terms of precision or accuracy, or
- improving precision and/or accuracy, without increasing measurement efforts.

In this chapter, six issues are addressed with a view to providing guidelines for the optimization of calibration procedures, as well as of routine analytical methods in general:

1. Does enhanced sensitivity mean improved methods?
2. A constant variation coefficient – is it good, poor or just an inevitable characteristic of method performance?
3. How to prove effects due to matrices – may the recovery function be replaced?
4. Having established matrix effects – does spiking prove necessary in every case?
5. Testing linearity – does a calibration really need to fit a straight line?
6. Enhancing accuracy – obtaining 'robust' calibration functions with weighting.

Considerable potential for optimization will arise just by analyzing routine procedures. The essential objective of any calibration, which is clearly ensuring the accuracy of results, should always be kept in mind. Hence, the expenditure of some effort in calibrating a method is undoubtedly unavoidable.

Appraising trueness and accuracy has to proceed with an evaluation of precision in order to finally ensure and prove the traceability of analytical results to known standards. Thus, in the examples given herein, the focus is on calibration data and characteristics in order to derive an accurate estimate of the precision and trueness of the method.

Why should we optimize the calibration of a method?
What benefit should we expect from optimizing calibration, in view of the relatively small error ranges due to calibration that arise in routine analytical methods? To answer simply, we look for improvements in precision and accuracy to enable efficient use of resources; this holds true not only for calibrating, but also more generally with regard to efficiency for daily routine analysis. Clearly, we have to consider different starting points for optimization and to look for the optimal balance between individual, possibly conflicting, objectives.

HPLC Made to Measure: A Practical Handbook for Optimization. Edited by Stavros Kromidas
Copyright © 2006 WILEY-VCH Verlag GmbH & Co. KGaA, Weinheim
ISBN: 3-527-31377-X

1.6.2
The Essential Performance Characteristic of Calibration

The method's standard deviation (DIN 32645, DIN 38402 T51) represents the key parameter of performance. It is calculated by dividing the residual standard deviation by the sensitivity of the method and it essentially determines the maximum precision for evaluation of analytical results. The standard deviation of a method provides the scale to estimate the accuracy of a method early on the basis of its calibration characteristics.

Due to the software that is widely available for evaluating calibration characteristics, calibration formulas are not explicitly written out here. Instead, we refer to plots of the calibration function and especially of the residuals, which provide a suitable basis for discussion by transparently showing the expected range of possible but still reliable results. More so than the confidence range of the calibration function, the prediction interval enables an evaluation of the precision and trueness. For this reason, we included the prediction limits in the residuals plot in a manner first published in Ref. [1].

1.6.3
Examples

1.6.3.1 **Does Enhanced Sensitivity Improve Methods?**
At first sight, striving for enhanced sensitivity seems to be worthwhile in any case to achieve optimization. A better response, or more exactly an increased change in signals as a result of changes in concentrations, should enable more precise measurements. For example, changing reagents to obtain a more intense color, exchanging detector systems or choosing an alternative detection wavelength are possible courses of action.

We wish to optimize the detecting components of an established HPLC method. Which parameter will provide a reliable characterization of the improvements achieved?

The example will give us the opportunity to introduce reliable characteristic data as well as some criteria for visually evaluating the calibration function and the respective, specially designed residuals plot [1, 2]. We will be able to recognize an efficient improvement at one glimpse.

The measurements of the calibration series to be considered are given in Table 1. The response of method 2 clearly seems to be the better one. The detector used was based on additional electronic amplification of input signals.

Sensitivity is given by the slope of the calibration function applied for the desired operating point. Because we are also going to consider nonlinear calibrations, with their sensitivity being dependent on the operating point, we focus on the middle of the working range. This will enable sensitivity data to be compared numerically and visually for any calibration function that may apply.

The characteristics and especially the ranges of uncertainty that prove approximately equal for the concentration evaluated, are given numerically as well as

Table 1. Measurements provided in two calibration series by methods of different sensitivity.

No.	Values 'x' concentration [mmol]	Method #1 values 'y' signal [arb.]	Method #2 values 'y' signal [arb.]
1	0,15	9,17	22,32
2	0,30	20,75	44,21
3	0,45	31,58	61,43
4	0,60	39,09	85,19
5	0,75	52,50	101,64
6	0,90	59,54	124,98
7	1,05	69,82	142,47
8	1,20	82,11	163,00
9	1,35	88,66	175,61
10	1,50	100,21	198,70

Table 2. Some essential method characteristics of the two calibration series.

	Method 1	Method 2
Sensitivity [arb./mmol]	66.53	129.71
Method standard deviation [mmol]	0.0214	0.01980
Uncertainty at the operating point [mmol]	0.825 ± 0.0753	0.825 ± 0.0700

graphically. Hence, efficiently increased sensitivity does not automatically ensure better analytical results. The residuals plot shows the sensitivity enhanced by a factor of about two. However, this is in fact compensated by an equally increased sensitivity for interference effects. The range of random errors increases by a factor of two on the signal scale as well. The two effects cancel each other out almost completely.

Thus, we should avoid interference effects and any errors additionally arising in the optimization of detector systems. Our example proves amplifying as a stand-out measure to fail the desired improvement of methods.

Conclusion

Increasing response does not automatically lead to better analytical methods. This fact is surely well-known in practice. Nevertheless, most laboratories cannot help but take immense efforts in measuring series of results for repeatability and reproducibility in the belief of verifying actual improvements. Such effort may be confidently left out.

Calibration, which is anyway available and required for both methods, provides objective evidence through the standard deviation of methods, which will indicate the method of significantly lower standard deviation to be the better one. Also, the coefficient of variation of the method (method 1 at 2.59%, method 2 at 2.41%) may help to verify that an improvement has actually been achieved.

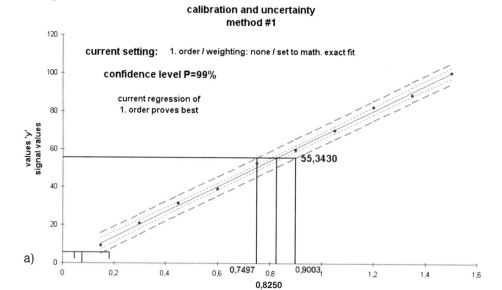

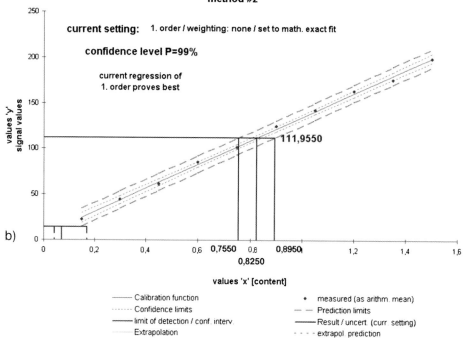

Fig. 1. Comparison of calibration functions and their residuals; uncertainty of measurements as a criterion for success in optimizations.

c)

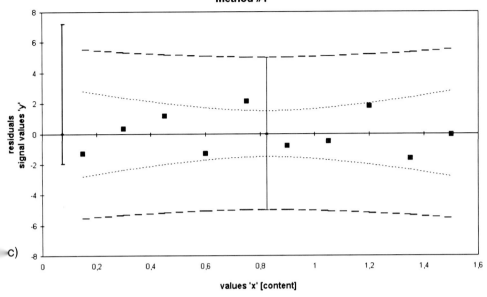

d)

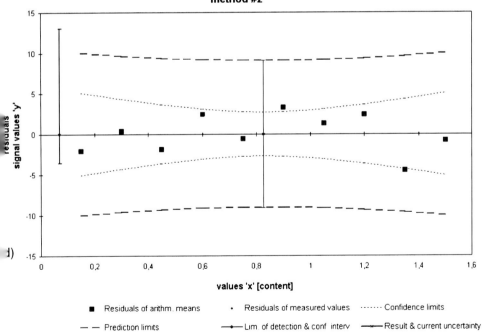

| ■ Residuals of arithm. means | ◆ Residuals of measured values | ⋯⋯ Confidence limits |
| – – Prediction limits | ━◆━ Lim. of detection & conf. interv | ━✖━ Result & current uncertainty |

Fig. 1. (continued)

However, variation coefficients only provide valid results for comparisons of methods that both refer to the same concentration range, thus applying the same middle of the operating range as reference to calculate this parameter.

Alternatively, simply by asking one's supplier, one should obtain sufficient information about sensitivity and signal precision of devices to be applied for the purpose of comparison. Even if the manufacturer does not apply the same analytical method for qualification purposes of devices, the performance characteristic will nevertheless provide criteria suitable for your decision and will help to avoid calibration efforts of your own. Again, one should be careful when relying on information about variation coefficients only, since any reported percentage portion is easily changed to the desired value just by referring to a higher concentration.

Additional extensive test series to complete one's own calibration characteristics will provide no further essential information to assist in correctly choosing the "better" method. This effort is only suitable for validation purposes, in order to provide the most realistic information about the performance of the method of one's choice to be established in a routine application.

1.6.3.2 A Constant Variation Coefficient – Is it Good, Poor or Just an Inevitable Characteristic of Method Performance?

As introduced in the previous example, the standard deviation of the method evaluated on the basis of calibration data already provides sufficient information to describe the expected range of random error. It should apply to the whole operating range considered.

Recording additional measurements of repeatability is absolutely no longer required for this purpose.

In chromatographic practice, we expect variation coefficients (relative standard deviations) to be constant in the operating range of concentration. Thus, the uncertainty of measurements, scaled in absolute concentration units, will increase directly proportionally with increasing concentration of samples. What are the consequences of the variation coefficients being constant, if we want to apply calibration characteristics in order to recognize some starting point for optimization as early as possible? In the following example, the aim of optimization is an expansion of the intended operating range by about one order of magnitude of concentrations.

At first sight, the calibrating function appears suitable for the purpose of quantifying samples in the desired operating range. However, in particular those points labeled by their residual values in the residuals plot give an indication that the standard deviation of the method provided by the calibration data is no longer able to fit the uncertainty that is actually involved. The precision that actually applies proves to increase with concentration. Thus, we are faced with an increasing risk of measurements exceeding prediction limits provided by calibration characteristics. Referring to the so-called conventional true result, the method increasingly encounters systematic errors as we approach the upper operating range. Consequently, quantifying samples will sometimes result in too low or too high a concentration. The risk of encountering systematic errors that prove

calibration and uncertainty

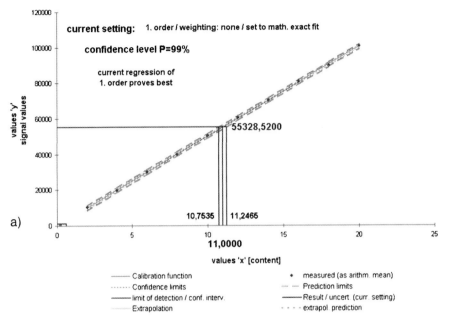

a)

residual chart of calibration function and uncertainty

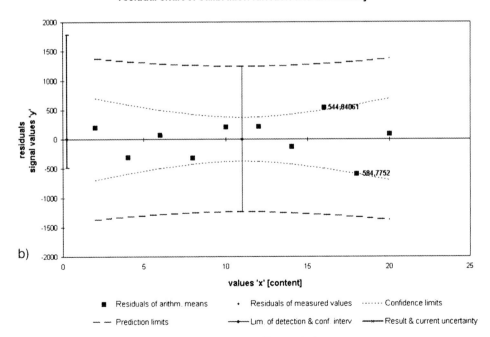

b)

Fig. 2. Ten points of calibration at constant variation coefficient, single measurements at each concentration.

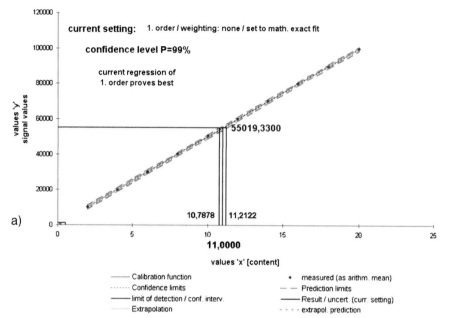

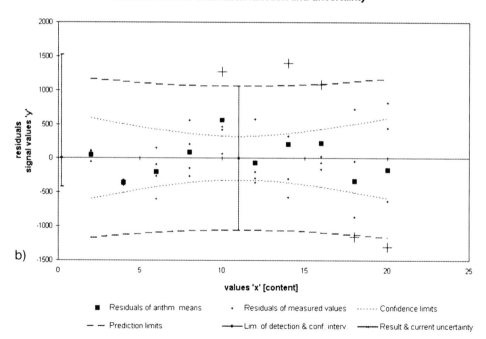

Fig. 3. Ten points of calibration, four repeated measurements at each standard, assuming a constant variation coefficient.

to be caused by random but underestimated effects may easily escape one's attention if only single measurements are applied [3], as is surely commonly the case.

This behavior of a method at constant variation coefficient becomes clearer if we recalibrate the method, this time applying fourfold repeated measurements at each standard concentration. The result is shown in Fig. 3.

The five values marked in the residuals plot evidently reveal the risk of wrong quantification, even for the same calibration standards treated as if they were unknown samples.

Keeping in mind our original purpose of minimizing measuring efforts, the following examples will go on expanding the operating range up to four orders of magnitude of concentrations. Choosing such a widely spanned operating range is quite common practice in testing the purity of products, applying one chromatographic run to simultaneously scan for the target substance (content set to 100%) as well as for impurities (down to 0.01%) [4].

Operating Range 0.01–100% with Homogeneous Variances

Even with the calibration spanning four orders of magnitude of concentration, Fig. 4 evidently shows the calibration data to provide valid performance characteristics. The actual behavior of measurements will be correctly described by the calibration. The obtained standard deviation of the method applies to the whole operating range, as long as we can ensure that the variances are homogeneous. The residuals reveal that just one single result (out of 60 measured) exceeds the estimated limits of prediction (i.e., results just meet the initially set error level and thus prove statistically acceptable).

As can be seen, it is simply impossible to introduce an optimization strategy that should be applicable for the entire desired operating range. We do not have to refer to statistics or calibration characteristics. Looking at Fig. 4, the question obviously arises as to why anyone should calibrate by further diluting standards to below 1% anyway.

One does not need to be an expert to recognize the obvious way of proceeding with the optimization, considering the multitude of highly condensed overlaying calibration points in the range below 1%.

Applying considerably fewer calibration measurements, i.e. one single concentration measured below 1%, we may instantly introduce a 50% reduction in measuring efforts without any loss of quality. With one single calibration standard remaining in this range – let us choose 0.5% for example – we will of course no longer be able to quantify samples at concentrations of less than 1% at the precision required. To reach this conclusion is just a matter of common sense and is not affected by any restricted boundary conditions given by guidelines or regression statistics, as some authors continue to claim. Sometimes even a semi-logarithmic scale to plot calibration data is recommended in order to conceal the facts that we have discussed. However, we feel responsible to serve our reader's interests by clearly rejecting to open some boxes of tricks or to introduce possible conversions. These are nothing more than fudge factors. In the final analysis, moreover, concentrations will not be reported on a logarithmic scale either.

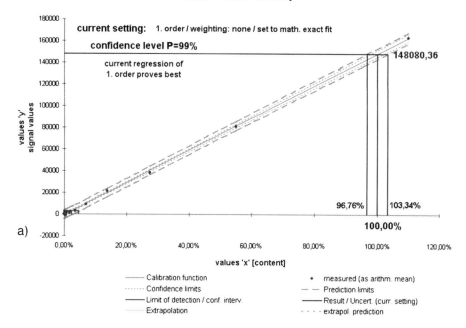

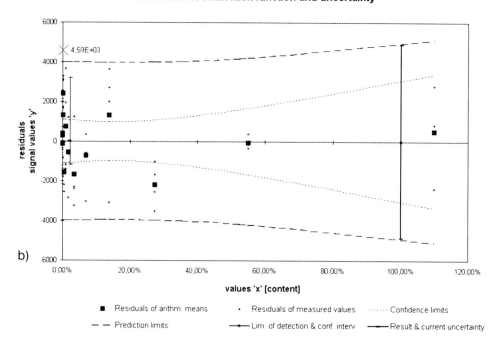

Fig. 4. Homogeneous variances in the operating range 0.01% to 100%.

The uncertainty in the operating point at 100%, as is clear from Fig. 4, is in fact unsymmetrical and correctly indicated [1, 5]. The formulas given in standards and guidelines (i.e., ISO 8466, DIN 38402, DIN 32654) to calculate uncertainties in the concentration scale show deviations hereafter and are no longer strictly valid.

Operating Range 0.01–100% with a Constant Variation Coefficient

The effects and risks involved as previously discussed increase dramatically if we take variation coefficient to be constant in the operating range. Obtaining a valid statement in quantifying samples on the basis of a calibration spanning four orders of magnitude of concentration is associated with a serious and great risk of encountering systematic errors. In principle, it is impossible that an analytical assay, whatever clever and optimized procedure is applied, will enable valid characterization of such wide spanning operating ranges by calibration, in spite of constant variation coefficients occurring at the same time. We also feel responsible to reveal and not to conceal limiting conditions for optimization strategies and requests that sometimes may exceed even the capabilities of modern powerful analytical assays. The result of a calibration applying these boundary conditions is illustrated in Fig. 5.

No regression model in the world will be able to accurately describe the behavior of an analytical assay spanning more than four orders of magnitude, nor will any calculation provide valid and realistic characteristics. The reason is simple and easy to understand. Any evaluation of calibration characteristics is designed to provide one single parameter for precision, i.e. one single residual standard deviation and hence one single standard deviation of the method. In mathematical terms, the specified purpose definitely requires the homogeneity of normal distribution applying to the desired operating range of the calibration as a whole, whereas taking a variation coefficient to be constant means just the opposite.

Trying to ignore these requirements will cause various problems in the practice of future routine analysis. Here, it should be mentioned that we are going to encounter "out of specification" (OOS) results again and again. These are only due to random effects, and only seldom are they ascribed to the real reason of precision being estimated too optimistically.

Fortunately, this fact is of little consequences in practice, because common operating ranges are limited to span only one order of magnitude of concentration. In this case, the precision will be estimated quite correctly, especially if we consider the middle of the operating range.

Evaluating the following example demonstrates that the weighting of the calibration function does not provide a valid solution, but increases the problem. Any basics required for calculating are stressed over their limits.

The correct estimation of performance characteristics, i.e. the standard deviation of the method, will be further distorted by any weighting applied. Consequently, false analytical results are unavoidable, occurring at higher frequency than valid results. More than three-quarters of the operating range show this effect. Focussing on concentrations between 0.01% and 1% in Fig. 6 d), one can see that one runs

calibration and uncertainty

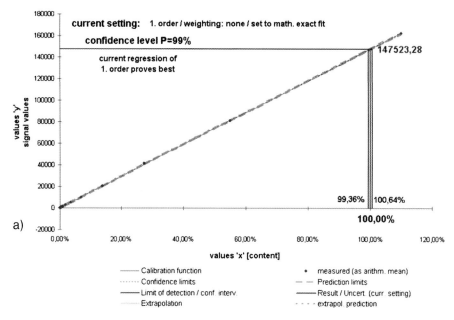

residual chart of calibration function and uncertainty

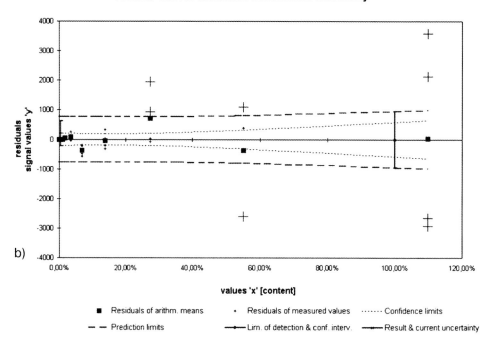

Fig. 5. Constant variation coefficient in the operation range 0.01% to 100%.

the risk of obtaining false analytical results. Even quantifying the calibration standard at concentration 0.27% fails in the case of three out of four measurements.

Solely a purely empirical model, introduced by Kuss [6], applying set points of precision at the boundaries of the operation range, enables adjustment of the calculation and evaluation to the set points of the precision specified but requires the precision at each of the relevant concentrations to be known already. Calibration, which we formerly expected to provide this information, in fact fails. Hence, additional repeated measurements are required, contrary to our original objective, which was to allow less effort in terms of measurements. Unfortunately, the calibration model provides no reliable precision characteristics, but in this case it is no longer required.

Conclusion

Constancy of the variation coefficient for chromatographic assays is equivalent to the standard deviation of the method, depending on the concentration. Consequently, the standard deviation of the method provided by the calibration model does not apply. By introducing this performance parameter as essential for our optimization strategies, we actually prove that optimizing an analytical assay, which appears with constant variation coefficients, will never apply to the operating range as a whole. By the way, this is state-of-the-art and has always been reflected in various guidelines and technical standards.

In any case, one should define exactly the section of one's operating range to be optimized in order to evaluate a standard deviation that will really apply for this section. Only repeated measurements will provide the precision data depending on concentration that will actually apply at each respective operating point. Any further calculation as well as weighting has to refer to these repeatability precision measurements and to take into account the measured precision data as boundary conditions in order to provide reliable and realistic confidence limits of analytical results as a function of concentration. Without any knowledge provided by repeated measurements, evaluation of pure calibration characteristics can easily yield arbitrary and even wrong results. The possibilities for optimizing such methods are considerably restricted, since calibration data alone are unable to provide valid precision characteristics.

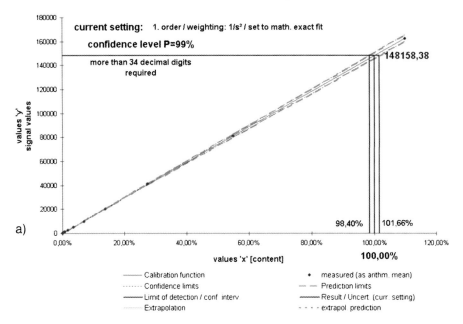

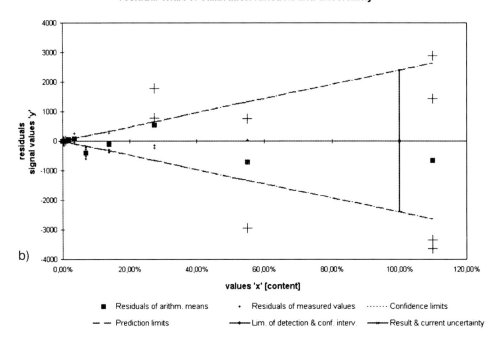

Fig. 6. Weighted calibration; weighting $1/s^2$ applies standard deviations s proportional to concentration; the average variation coefficient of the method is 1.88%.

calibration and uncertainty

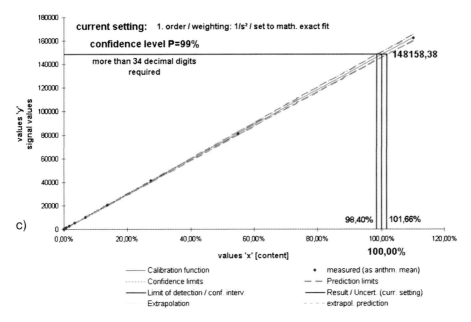

residual chart of calibration function and uncertainty

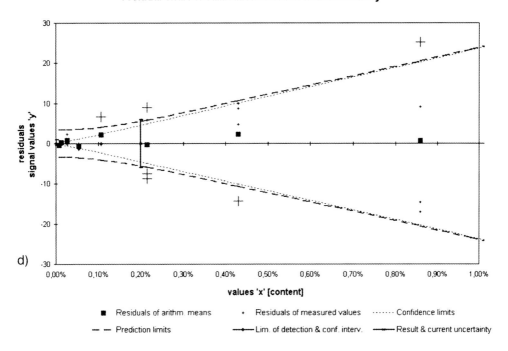

Fig. 6. (continued)

calibration and uncertainty

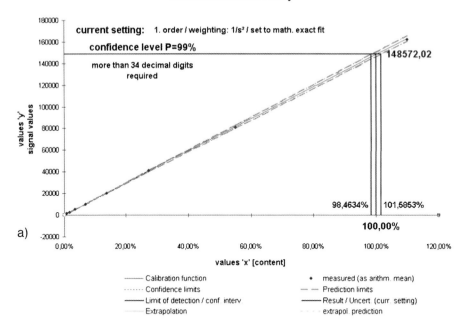

current setting: 1. order / weighting: 1/s² / set to math. exact fit

confidence level P=99%

148572,02

more than 34 decimal digits required

98,4634%   101,5853%

100,00%

a)

values 'x' [content]

———— Calibration function   ♦   measured (as arithm. mean)
·········· Confidence limits   — — Prediction limits
———— Limit of detection / conf. interv. ———— Result / Uncert. (curr. setting)
———— Extrapolation   ······ extrapol. prediction

residual chart of calibration function and uncertainty

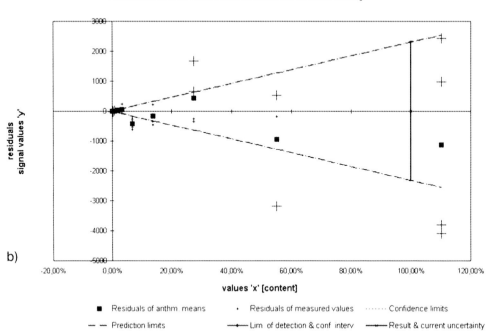

b)

values 'x' [content]

■ Residuals of arithm. means   · Residuals of measured values   ······ Confidence limits
— — Prediction limits   ——◆—— Lim. of detection & conf. interv.   ——■—— Result & current uncertainty

Fig. 7. Weighting with concentrations 'x' applies $1/x^2$ with absolute units of concentration portion 'x'. The result does not really differ from the one in Fig. 6.

1.6.3.3 How to Prove Effects Due to Matrices – May the Recovery Function be Replaced?

Determining the recovery function surely requires some additional measurement efforts. Table 3 sets out some possible ways of proceeding following procedure A or procedure B.

Procedure B already reduces measurement efforts down to two steps. To achieve this 33,3% reduction, the recovery function, i.e. plotting the concentration of individually prepared standards measured in the matrix against the known concentration, is required to reveal no significant deviations from expected slope '1' and intercept '0'. If significant differences are detected, even procedure B leads to a third step. It is then necessary to prepare recalibration standards in the matrix, since the standard sample applied for calibration should be prepared individually and separately. The existing control samples should not be used for this purpose.

Possibilities for Optimization with a Specified Degree of Reduction of Efforts

In our example, measurement efforts are minimized to a maximum of two calibration series. If the recovery function is not explicitly required by guidelines that apply to one's practice, one may skip the preparation and independent measurement of series of control samples. An instant comparison of calibrations in the matrix and in pure solvent will suffice. The objective evidence of this statistically sound comparison proves absolutely equivalent to evaluating the recovery function.

Result of the Graphical Comparison

If the calibration function of the pure solvent without matrix (lower residual standard deviation expected) does not intersect the confidence limits of the calibration function applied in blank matrix spiked with standards of known concentration (higher residual standard deviation expected), or completely exceeds

Table 3. Efforts to evaluate the recovery function applying procedure A or procedure B.

Procedure A (admittedly, a somewhat circumstantial way!)	*Procedure B* (established, quick, and at first sight efficient)
1. Calibrate in pure solvent	1. Calibrate in pure solvent
2. Calibrate including matrix	2. Control samples, prepared independently from calibration samples, evaluated based on the current calibration function
3. Control samples: samples in matrix, prepared expressly independently from calibration standards, are evaluated applying the respective calibration function in each case	3. Calibrate including matrix (only required if the recovery does not meet the slope '1' and intercept '0')

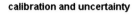

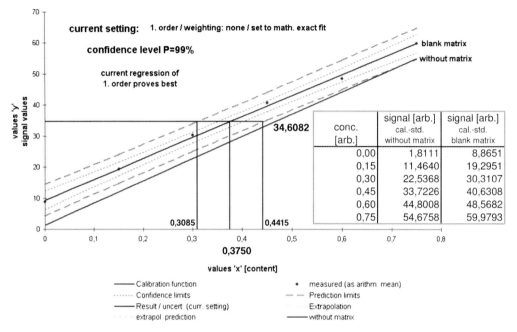

Fig. 8. Exemplary data and graphic of a calibration – a constant matrix effect is evident and we include a blank sample for clarity.

the confidence ranges, an effect caused by the matrix is significant. Figure 8 illustrates this result graphically. If no blank matrix is available, one may also apply the same principle to a reference matrix (sample matrix), as long as the effect due to the pure blank matrix is known. It should be explicitly pointed out that including a blank sample in a calibration series does not comply with common guidelines relating to calibration, but we apply a blank here to emphasize transparency of results.

The sum of squared differences of each standard to the middle of the operating range (Qxx) and the squared standards (Sxx), both given on the concentration scale, are the only parameters that were additionally required to proceed with comparing linear calibration functions directly [5]. All other input refers to common calibration characteristics. On the important left-hand-side of the spreadsheet, one can see the calculation proceeding step-by-step in the way which is commonly known in chemical laboratories for validation purposes [7], i.e. explicitly comparing precision data and mean values applied to two measured series under repeatable conditions (F-test, t-test).

Some software, such as validation or chromatographic software, enables t-testing of the intercept or the slope of the recovery function compared to theoretically expected set points '0' or '1'. However, this strategy for a statistical test procedure does not apply for our purposes and results may be unreliable.

Table 4. Results – statistically sound comparison of linear calibration functions [7].

Wshstve3.xls [Schreibgeschützt]

Comparing two Calibrations (applies to straight lines only, regression of 1st. order)

calibration series #

values given in units [x] characteristics	1st. calibration		2nd. calibration	
	slope	intercept	slope	intercept
number of cal.-points:	6		6	
coefficient:	71.52757 [y/x]	1.34568 [y]	67.37339 [y/x]	9.34317 [y]
standard deviation:	0.83061 [y/x]	0.37722 [y]	1.42809 [y/x]	0.64856 [y]
residual std.-deviation:	0.5212 [y]		0.89612 [y]	
sum of squares Q_x:	0.39375 [x]²		0.39375 [x]²	
sum of squares S_{xx}:	1.2375 [x]²		1.2375 [x]²	
df of calibration:	4		4	

comparing precisions / testing for homogeneous variances

	residual variance	precision of the slope	precision of the intercept
conf. level P=99.90% F-test:			
s^2_{max}:	0.80303 [y]²	2.03944 [y·x]²	0.42063 [y]²
s^2_{min}:	0.27165 [y]²	0.14229 [y·x]²	0.14229 [y]²
df (s_{max}):	4	4	4
df (s_{min}):	4	4	4
F_{crit}:	15.97709343	15.97709343	15.97709343
F_{test}:	2.956081711	14.33251739	2.956081711
result:	variances prove homogeneous (ok)	proved homogeneous (ok)	proved homogeneous (ok)
approval:	Res. std.-dev. homogeneous ●Yes ○No	s² homogeneous (slopes) ●Yes ○No	s² homogeneous (intercept) ●Yes ○No
	F-test passed	F-test passed	F-test passed

Comparing coefficients requires homogeneous residual variances!

comparing coefficients / testing homogeneity

only applies for homogeneous variances

	slope	intercept
conf. level P=99.90% t-test:		
df_{test}:	8	
$s^2_d = s^2_{test}$:	2.72935 [y/x]²	0.56293 [y]²
t_{crit}:	3.355380613	3.355380613
t_{test}:	2.514524209	10.65926659
result:	slopes prove the same	intercepts differ
approval:	homogeneous slopes ●Yes ○No	homogeneous intercepts ○Yes ●No
	t-test passed	t-test: homogeneity rejected

Conclusion

Recording a maximum of two calibration series enables proof of a matrix effect being significant and at the same time provides a calibration series in the matrix should this be required. This is further discussed and verified in the following example.

Measuring and evaluating a recovery function does not provide any further information. Measuring efforts to determine the recovery function may be cut in this step of the procedure without replacement.

Further proof of the previous statements will be obtained by submitting control samples (low, medium, high concentrations) to medium- and long-term validations or by evaluating long-term observations supported by control charts.

In summary, our aim is to apply to the full the existing elements that quality management provides anyway and to combine the information so as to succeed in optimizing, i.e. reducing efforts in measurements to the minimum required to obtain objective evidence.

1.6.3.4 Having Established Matrix Effects – Does Spiking Prove Necessary in Every Case?

In the previous example, we discussed a deviation of the calibration function, which was proved to be constant and due to an interfering matrix effect. Of course, proportional deviations may occur, either alone or in addition to existing deviations. The procedure introduced also applies in all these cases. The actual uncertainty proves relevant in reporting analytical results and again provides a valid criterion for decision. Applied to some boundary case, one will achieve the goal of cutting the immense efforts of spiking and even gain improved precision and accuracy of the method.

Figure 9 illustrates the calibration function of the spiking experiment shifted to a higher signal level and involving an enhanced uncertainty compared to the calibration applying a blank matrix shown in Fig. 8. This result is expected due to the finite but unknown concentration of the sample. We note an additional uncertainty in Fig. 9, which is solely due to the evaluation of the spiking series. Evaluating needs extrapolation exceeding the actual operating range of concentrations of the spiked standards.

Extrapolation is allowed here by referring to the exact regression formulas [1, 5]. Applying exact calculation formulas ensures compliance to existing standards and is required by extrapolating, whereas evaluations exceeding the operating range have to be rejected if we refer to calibration formulas given by standards. This is because these formulas represent approximations that are restricted to the actual operating range only and will no longer be valid as we exceed the operating range. From the residuals plot of Fig. 9, one obtains the overall combined uncertainty in units of measured signals at a variation coefficient of 7.73%.

In spite of the asymmetric uncertainty on the concentration scale, the uncertainty in signal units is sure to still provide symmetrical confidence limits.

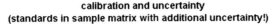

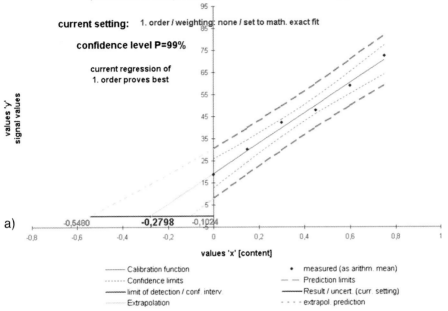

a)

Calibration function	• measured (as arithm. mean)
Confidence limits	— — Prediction limits
limit of detection / conf. interv.	Result / uncert. (curr. setting)
Extrapolation	· · · · extrapol. prediction

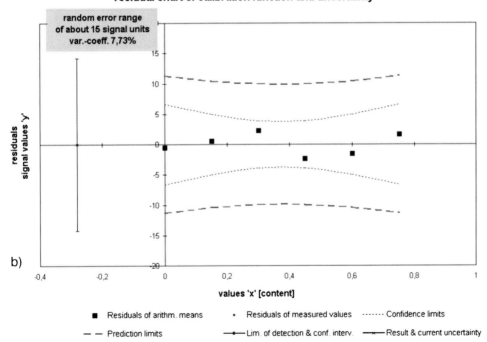

b)

| ■ Residuals of arithm. means | ✦ Residuals of measured values | ·····Confidence limits |
| — — Prediction limits | Lim. of detection & conf. interv. | Result & current uncertainty |

Fig. 9. Calibration by spiking procedure (standard addition method).

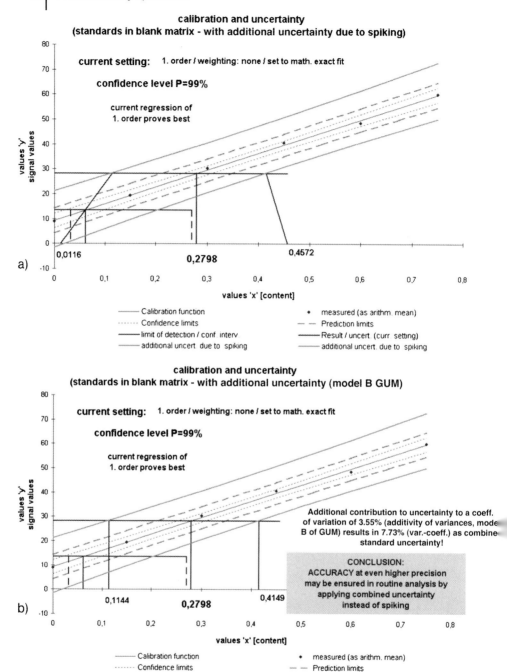

Fig. 10. Calibrations in a blank matrix including enhanced uncertainty given by one single and initial spiking experiment.

We just need to proceed with one single spiking series to get information about the actual uncertainty involved in order to apply this enhanced uncertainty level to any further calibration. Hence, routine calibration requires only a blank matrix solvent and calibration will be possible without spiking.

The enhanced uncertainty due to extrapolation is illustrated in the left plot of Fig. 10 for comparison purposes. In fact, we do not need to consider this additional component of uncertainty (trapezium). We thus get results of higher precision because future evaluations do no longer need extrapolations according to DIN 32633 to calculate the analytical result.

Conclusion

In our example, spiking is no longer required in spite of matrix effects being evident. The additional effort would be nothing but a waste of resources. By comparing calibrations and their respective uncertainty shown in the two plots of Fig. 10, we even improve the quality of our analytical result in spite of reduced measurement efforts. The trueness of the result remains unchanged, but we are able to report an improved precision.

The described strategy provides an even better method at essentially reduced measurement effort. In this way, we even succeed in the simultaneous realization of two competitive and conflicting objectives in optimization (precision versus number of measurements).

1.6.3.5 Testing Linearity – Does a Calibration Really Need to Fit a Straight Line?

Testing the linearity of a measured calibration is an established part of any calibrating. Software provided by modern instrumental analytical methods, such as HPLC, of course supports evaluation of nonlinear calibration. Unfortunately, only a few standards and guidelines deal with basic requirements to evaluate such calibration functions. In principle, there is no reason to deny nonlinear functions. Even the limits of detection and of quantification may be provided by nonlinear calibration and will comply with the sense of existing standards. As long as the analytical assay is not applied to quantify concentrations approaching the limit of detection, the limits of detection and of quantification that result on the basis of nonlinear calibration functions will surely provide sufficient evidence to prove fitness for use, i.e. to meet requirements in validation. Although the focus of standards has hitherto been in terms of straight-line calibration, we refer, for example, to ICH guidelines, which explicitly allow the application of alternative and further calculations, as long as they prove statistically sound.

In our example, we will take this proposal for granted. Again, we focus attention on the expected uncertainty of analytical results to reach our objective in optimizing methods.

We suppose that the method is going to be used for purposes of monitoring and approval at given limits of specifications. The task in optimizing is to reach the best precision available. Thus, when specified limits are approached, we will provide enhanced margins for decision thresholds at an unchanged significance level. The calibration data are presented in Table 5.

Table 5. Calibration measurements provided to test linearity.

Values 'x' (initially set)	Measurand values 'y' arith. means
0,15	0,0448
0,3	0,0886
0,45	0,1317
0,6	0,1716
0,75	0,2179
0,9	0,2556
1,05	0,2939
1,2	0,3388
1,35	0,3764
1,5	0,4124

Tests for linearity are designed to find a calibration function that fits best, but at the same time it should be as simple as possible. Although commonly applied, neither correlation coefficients nor residual standard deviations provide valid criteria to choose the optimum order of a calibration function. We will apply comparison of F-statistics of both regressions involved to provide results that exactly meet the sometimes better known Mandel test. This test supports practical help for decision-making. The result for the data considered suggests the application of a linear regression of first order. However, merely by visually checking the residuals it can be quickly realized that we may just as well choose a calibration function of the second order and ignore the results of numerical linearity testing.

To the trained eye, the clearly bent course of the residuals is sufficient reason to doubt the results of linearity testing, and to proceed to look for an essentially enhanced precision of future analytical results. As Fig. 12 illustrates, we reach a considerably improved precision simply by applying a squared function for calibration.

Conclusion

Simply by applying a calibration function of second order, we enable optimization of the method. The improvement of precision of the analytical assay reaches considerable 35.2% and provides enlarged margins of uncertainties to be tolerated in routine production in our example. In boundary cases, only the precision improved in the way that we have described will enable validation and prove the method's fitness for use, whereas controlling of given specification limits with the poorer precision associated with a first-order calibration will fail. In fact, we enable validation of the method: Even with respect to the demanding requirements of high precision, some extended margins as action limits at specified tolerances are opened up.

calibration and uncertainty

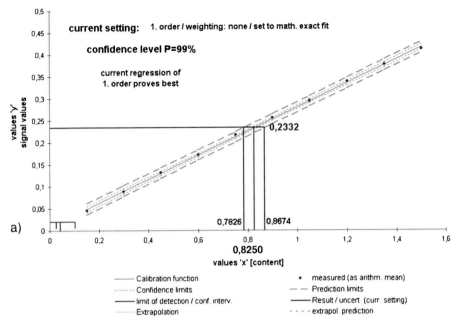

residual chart of calibration function and uncertainty

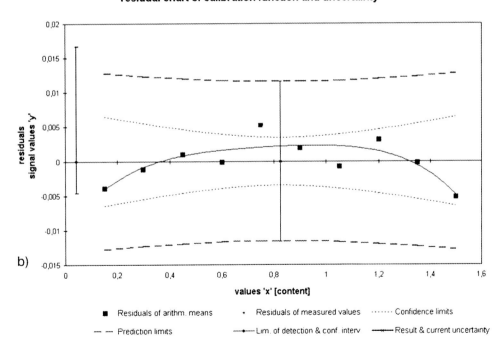

Fig. 11. Calibration of first order, including residuals.

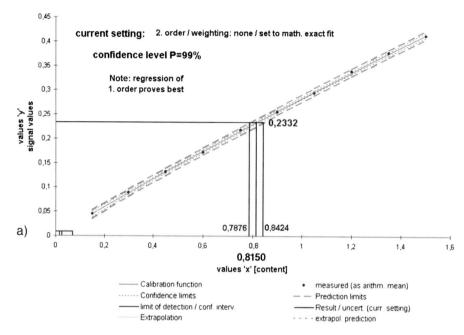

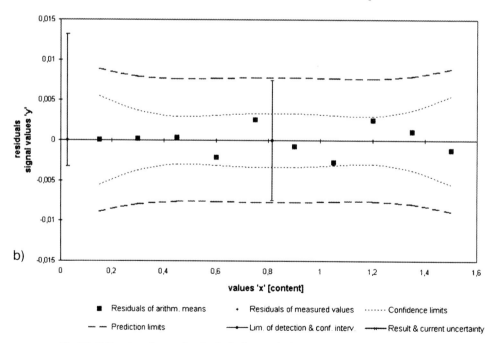

Fig. 12. Calibration of second order, including residuals.

1.6.3.6 Enhancing Accuracy – Obtaining 'Robust' Calibration Functions with Weighting

Calibration guarantees the trueness and ultimately the traceability of analytical results. Errors encountered in calibration give rise to a risk of affecting all future results through a systematic propagation of errors (biases). For this reason, the German DEV[1]-39, 1997 "Strategien für die Wasseranalytik" recommends replacement of the total calibration if outliers of regression are evidently encountered.

Following this recommendation would necessitate a considerably enhanced effort in calibrating. It is surely preferable to check for reasons beforehand and to avoid them in future routine chemical analysis. However, we have to be aware that high sensitivity of methods often implies low robustness at the same time and such methods are not immune from being sensitively affected by interference effects. Whether interference effects are due to the sample matrix itself or to a varying handling of analytical procedures is immaterial.

Especially if we consider the development of methods, it is of no use to track down suspicious results already in the first development cycles. The same applies to wasting time recalibrating again and again until no further suspicious calibration point occurs. After all, it is an essential task to identify possible failure modes and effects as development progresses so that significant errors can be avoided in routine analysis later on. Nevertheless, we should in any case avoid the effect of suspicious calibration points that may result in wrong calibration functions and hence in biasing of results.

Even the question as to whether calibration points outside of the expected range are best eliminated or left unchanged will impede any progress in development the same way as recalibrating does for regression outliers.

Table 6. Calibration data, residuals, and weighting factors (method development).

Values 'x' (initially set)	Measurand values 'y' arith. means	Weighting factor $1/s^2$	Residuals unweighted
0,15	10,92	1	2,21
0,3	18,81	1	−0,76
0,45	31,67	1	1,25
0,6	40,38	1	−0,89
0,75	51,6	1	−0,52
0,9	60,43	1	−2,54
1,05	77,58	3,76	3,76
1,2	79,36	1	−5,31
1,35	90,41	1	−5,12
1,5	114,3	7,92	7,92

---

[1] DEV denotes "Deutsche Einheitsverfahren", i.e. "German Standard Procedures", dedicated to the chemical analysis of water, waste water, and sludge; the relevant paper is entitled "Strategies for chem. analysis of water".

calibration and uncertainty

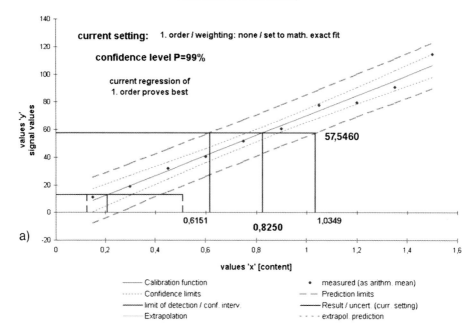

a)

—— Calibration function   •   measured (as arithm. mean)
······· Confidence limits   — — Prediction limits
—— limit of detection / conf. interv.   ——Result / uncert. (curr. setting)
—— Extrapolation   - - - extrapol. prediction

residual chart of calibration function and uncertainty

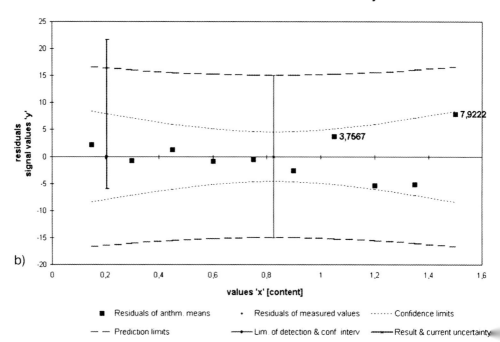

b)

■ Residuals of arithm. means   •   Residuals of measured values   ······· Confidence limits
— — Prediction limits   —◆—Lim. of detection & conf. interv   —✕—Result & current uncertainty

Fig. 13. Unweighted calibration including suspicious regression outliers.

calibration and uncertainty

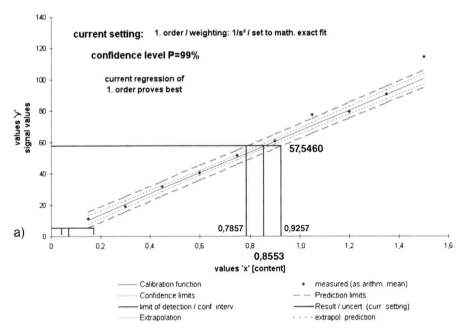

a)

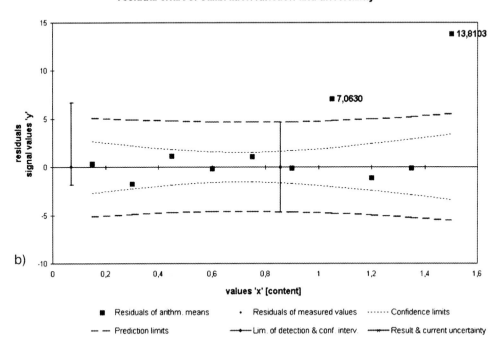

b)

Fig. 14. Weighted calibration (weighting factors are 1/(squared distance)).

Therefore, a weighting of regression is applied in our final example to ensure accuracy of the calibration. Weighting will ensure that calibration is no longer affected by outliers.

We quickly realize that the uncertainty of the calibration is quite high and unsatisfactory, but typical in such cases. The residuals of Fig. 13 clearly reveal the calibration points #7 and #10 to be suspicious stragglers. Both prove to be outliers by testing for regression outliers according to DEV. The residuals of the eight remaining calibration points are supportive assuming the slope of the "true" calibration to be essentially lower.

Instead of eliminating the straggling values, we proceed by applying "weighted regression" [5, 8]. With the current calibration data that lack any individual precision information, only the residuals provide a suitable measure of distances, which we apply as reciprocal squares for weighting. The calibration points that remain unsuspicious are each assigned a distance of "1" for weighting, which corresponds to no weighting and results in normal unweighted regression for these values. Any weighting other than the reciprocal squares of distances should be rejected in principle, since the units of calculation only remain consistent in this way. This helpful principle is worth remembering for any scientific calculation. Besides the sound theoretical background, this principle has considerable practical consequences. If we should ignore it, we easily run the risk of encountering some arbitrary results in formally applied calculations.

Conclusion

Weighted regression proves to be robust and is not affected by suspicious calibration outliers. Weighting ensures valid evaluation of the "true" calibration function, also providing the "true" precision as well. "Weighting out" suspicious calibration points leads to a reduction in uncertainty by a factor of three, which represents a 300% improvement in precision. The measurement of repeatability standard deviations will confirm the better precision, once the method is validated and established for routine analysis. Again, we emphasize the requirement cited in the primary literature [5], to apply squared distances for any weighting. As already mentioned, all other weighting factors run the risk of providing nothing more than arbitrary results. In particular, weighting that refers to concentrations may be ascribed to misunderstanding, since the abbreviation assigned to the weighting factors in primary literature is "c", which in no case means concentration, but is well defined and applied as $1/c = 1/s^2$, with s assigned to standard deviations. Hence, any estimates of s by other measures of distance will be adequate.

Once development and validation of the method has been completed, most effects that triggered regression outliers will have been eliminated anyway. Hence, the profit we obtain by weighted regression is a reliable and true operating calibration provided early in the first steps of development. We avoid posing the question of elimination of calibration points too early as we do not need to recalibrate too often. We thereby proceed to develop the method without any disturbance. The strategy will apply systematically when searching for and recognizing undesired influences that still affect the method. Finally, in the development of methods one should not waste time in eliminating individual measurements but efficiently achieve the goal of establishing a valid method that is fit for use by efficiently eliminating interference effects.

References

1 St. Schömer, ProControl® 4.0 manual – Applied Statistics for Laboratories and Industry, 2004.

2 St. Schömer, *Pharma International* 2 (2001), 55.

3 St. Schömer, *GIT Fachz. Lab.* 9 (1996), 904.

4 M. M. Kiser, J. W. Dolan, *LCGC Europe* 17(3) (2004), 138–143.

5 U. Graf, H.-J. Henning, K. Stangl, P.-T. Wilrich, Formeln und Tabellen der angewandten mathematischen Statistik, 3. Auflage, Springer-Verlag, Berlin, 1987.

6 H.-J. Kuss, *LCGC Europe* 16(12) (2003), 819–823.

7 St. Schömer, Kalibrierung chemisch-analytischer Prüfverfahren, training course, 1996–2004.

8 St. Schömer, *GIT Fachz. Lab.* 9 (2000), 1043.

2
Characteristics of Optimization in Individual HPLC Modes

HPLC Made to Measure: A Practical Handbook for Optimization. Edited by Stavros Kromidas
Copyright © 2006 WILEY-VCH Verlag GmbH & Co. KGaA, Weinheim
ISBN: 3-527-31377-X

2.1
RP-HPLC

2.1.1
Comparison and Selection of Commercial RP-Columns

Stavros Kromidas

2.1.1.1 Introduction

The diversity of RP-phases is a well-known fact and the topic of many publications. In this chapter, results of our own investigations concerning the comparison of commercially available columns and their properties are reported in condensed form. The focus of the explanations is on experimental results and on conclusions from the point of view of the user. Three topics are addressed, namely:

1. Are there general rules allowing informed pre-selection of columns depending on the type of compounds to be separated?
2. Which commercially available columns have similar properties and how can this be tested?
3. What should be in my personal column tool kit ("column portfolio")?

2.1.1.2 Reasons for the Diversity of Commercially Available RP-Columns – First Consequences

Usually, an RP-phase is made by chemical modification of the surface of a suitable matrix, which in most cases is silica gel. The degree of conversion is 50–70%, except in specific cases such as polymerized phases.

It is known that the physico-chemical properties of the silica gels used, which have a chromatographic relevance, can be very different; see Table 1, which lists the physico-chemical properties of some commercially available columns. Values have been obtained partly from the author's own measurements and partly from the literature.

Because the surface of the silica gel plays an active role in RP-HPLC since 30–50% of it is available to the analyte, the type of silica gel (or other matrix) used is an important variable in the chromatographic process. In addition, one can see that for a "simple" RP-phase such as the C_{18}-phase, the modification is by no means constant. Each manufacturer has essentially its own special "recipe", e.g., reaction of silica gel with silanes with or without catalyst, traces of water in the toluene-based reaction medium, purity of the silanes used, reaction temperature, mono-, di- or trifunctional bonding with the alkyl chain, etc. Considering the large number of possible combinations of different silica gels and different types of chemistry of the manufacturers, the large number and the diversity of commercially available RP C_{18} columns and even RP columns can readily be appreciated.

HPLC Made to Measure: A Practical Handbook for Optimization. Edited by Stavros Kromidas
Copyright © 2006 WILEY-VCH Verlag GmbH & Co. KGaA, Weinheim
ISBN: 3-527-31377-X

Table 1. Physico-chemical properties of some commercial RP-phases.

Number	Name	Manufacturer/ supplier	Particle form	Particle diameter [μm]	Packing density [g/ml]	Endcapped?	Pore diameter [Å]	Specific surface [m²/g]	Carbon content [%C]	Pore volume [ml/g]	pH-value of the silica matrix	Coating degree [μmoles/m²]	Coating density [μmol/g]
60	Aqua	Phenomenex	spherical	5		yes (hydrophilic)	125	320	15	1,05	4,8–5,1	2,42	774
1	Bondapak	Waters	irregular	10	0,40	yes	125	330	10	1	Porasil 5,4	1,42	469
61	Chromolith Performance	Merck	silica road			yes	130	300	18	1		3,43	1029
2	Discovery Amid C$_{16}$	Supelco	spherical	4		yes	180	200	11,34	0,94		2,55	510
3	Discovery C$_{18}$	Supelco	spherical	4		yes	180	200	12,31	0,93		2,92	584
4	Fluofix IEW	Neos/Dr. Maisch	spherical	5		yes	120	300		1			
5	Fluofix INW	Neos/Dr. Maisch	spherical	5		no	120	300		1			
6	Gromsil AB	Grom	spherical	5	0,60	yes	100	200	11	0,5	8,7	2,40	480
7	Gromsil CP	Grom	spherical	5	0,55	polymer coating	120	320	15	0,8	4,4	2,80	896
8	Hypercarb	Thermo Electron		5			250	120		0,75			
9	Hypersil BDS	Thermo Electron	spherical	4,62		yes	136	169	11,1	0,65		3,20	541
11	Hypersil ODS	Thermo Electron	spherical	5		no	120	170	10	0,7		2,86	486
62	HyPURITY Advance	Thermo Electron	spherical	5			180	190		1	7		
10	HyPURITY C$_{18}$	Thermo Electron	spherical	5		yes	183	193	12,4	1		3,19	616
12	Inertsil ODS 2	GL Sciences/MZ	spherical	5	0,58	yes	150	320	18,5	1,2	6,5	2,13	682

Table 1. (continued)

Number	Name	Manufacturer/ supplier	Particle form	Particle diameter [μm]	Packing density [g/ml]	Endcapped?	Pore diameter [Å]	Specific surface [m²/g]	Carbon content [%C]	Pore volume [ml/g]	pH-value of the silica matrix	Coating degree [μmoles/m²]	Coating density [μmol/g]
13	Inertsil ODS 3	GL Sciences/MZ	spherical	5	0,58	yes	100	450	15	1,1		1,30	585
14	Jupiter	Phenomenex	spherical	5	0,46	yes	300	157	13,65	1,5	5,7	4,40	691
15	Kromasil	Akzo Nobel/MZ	spherical	5		yes	100	340	19	0,9	5,6 Si	3,09	1051
16	LiChrosorb	Merck	irregular	5	0,50	no	100	300	16,2	1,1	6,5	2,70	810
17	LiChrospher	Merck	spherical	5	0,50	no	100	350	21	1,25	3,5	3,62	1267
18	LiChrospher Select B	Merck	spherical	5	0,60	no	60	360	11,5	0,9	3,5	3,55	1278
19	Luna	Phenomenex	spherical	5	0,56	yes	108	440	18	1	5,7	3,00	1320
20	MP-Gel	Omnichrom	spherical	5	0,64	yes	120	340	13,3	1,05	5,5±0,5	2,92	993
21	Nova-Pak	Waters	spherical	4,6	0,80	yes	60	120	7	0,3	5,1	2,67	320
22	Nucleosil 100	Macherey-Nagel	spherical	5		yes	100	350	14	1,1	5,9	2,50	875
23	Nucleosil 50	Macherey-Nagel	spherical	5		yes	50	400	14,5	0,8	6,3	2,30	920
24	Nucleosil AB	Macherey-Nagel	spherical	5		yes	100	350	24	1,1	5,5	4,80	1680
25	Nucleosil HD	Macherey-Nagel	spherical	5		yes	100	350	21	1,1	5,6	3,80	1330
63	Nucleosil Nautilus	Macherey-Nagel	spherical	5		yes	120	350	16	0,5		3,67	1285
26	Nucleosil Protect 1	Macherey-Nagel	spherical	5		yes	100	350	11	1	5,8	2,27	795
27	Platinum C$_{18}$	Alltech	spherical	5	0,66	yes	100	207	6,2	0,516	6,2	1,33	275
28	Platinum EPS	Alltech	spherical	5	0,66	no	100	207	4,8	0,516	6,2	1,03	213
29	Prodigy	Phenomenex	spherical	5	0,56	yes	100	450	15,5	1,2	5,7	1,80	810

Table 1. (continued)

Number	Name	Manufacturer/supplier	Particle form	Particle diameter [µm]	Packing density [g/ml]	Endcapped?	Pore diameter [Å]	Specific surface [m²/g]	Carbon content [%C]	Pore volume [ml/g]	pH-value of the silica matrix	Coating degree [µmoles/m²]	Coating density [µmol/g]
64	ProntoSil ACE	Bischoff	spherical	5	0,59	yes	120	300	18,5	1	5,9	2,87	861
30	ProntoSil AQ	Bischoff	spherical	5	0,59	yes (hydrophilic)	120	300	13	1	5,3	2,17	651
31	ProntoSil C$_{18}$	Bischoff	spherical	5	0,59	yes	120	300	17	1	5,3	3,03	909
32	Purospher	Merck	spherical	5	0,50	yes	120	340	18,5	1,1	4,0	3,14	1068
65	Purospher Star	Merck	spherical	5		yes	125	330	17	1,05		2,90	957
33	Repro-Sil AQ	Maisch	spherical	5		yes (hydrophilic)	120	300	15	1	4	2,01	603
34	Repro-Sil ODS 3	Maisch	spherical	5		yes	120	300	17	1	4	3,20	960
35	Resolve	Waters	spherical	5	0,60	no	90	200	10	0,5		2,39	478
36	SMT OD C$_{18}$	ICT	spherical	5		no (polymer coating)	100	340	24	0,9	6,0	7,40	2516
37	Spherisorb ODS 1	Waters	spherical	5	0,60	no	80	220	6,2	0,5	6,1	1,56	343
38	Spherisorb ODS 2	Waters	spherical	5	0,60	yes	80	220	11,5	0,5	6,1	2,84	625
39	Supelcogel TRP	Supelco		5			100						
40	Supelcosil ABZ plus	Supelco	spherical	5		yes	120	170	12	0,6		3,21	546
41	Superspher	Merck	spherical	4	0,40	no	100	350	21	1,25	3,5	3,62	1267
42	Superspher Select B	Merck	spherical	4	0,50	no	60	360	11,5	0,9	3,5	3,55	1278
43	Symmetry C$_{18}$	Waters	spherical	5	0,45	yes	100	335	19	0,9		3,14	1052

Table 1. (continued)

Number	Name	Manufacturer/ supplier	Particle form	Particle diameter [μm]	Packing density [g/ml]	Endcapped?	Pore diameter [Å]	Specific surface [m²/g]	Carbon content [%C]	Pore volume [ml/g]	pH-value of the silica matrix	Coating degree [μmoles/m²]	Coating density [μmol/g]
66	Symmetry Shield C$_{18}$	Waters	spherical	5		yes	100	335	17,5	0,9		3,24	1085
80	Synergi MAX RP	Phenomenex	spherical	4		yes	80	475	15	1,05		2,36	1121
81	Synergi POLAR RP	Phenomenex	spherical	4		yes (hydrophilic)	80	475	11	1,15		2,09	993
44	TSK	TosoHaas	spherical	5	0,40	yes	80	198	15	0,6	3,5–4,0	1,50	297
45	Ultrasep ES	Sepserv	spherical	5		no	120	280		0,8	6,2	3,20	896
46	VYDAC	Vydac	spherical	10		no	90	275	12	0,65		2,15	591
67	XTerra	Waters	spherical	5		yes	125	175	14,95	0,68		2,32	406
68	XTerra MS	Waters	spherical	5		yes	125	175	15,45	0,68		2,35	411
47	YMC ODS AQ	YMC	spherical	5		yes (hydrophilic)	120	300	14,6	1	5,5±0,5	2,50	750
48	YMC Pro C$_{18}$	YMC	spherical	5			120	320	16,8	1,06	5,5±0,5	2,59	829
69	Zorbax Bonus RP	Agilent Technologies	spherical	5		yes	80	180		0,5		3,00	540
70	Zorbax Extend	Agilent Technologies	spherical	5		yes	80	180		0,5		3,80	684
49	Zorbax ODS	Agilent Technologies	spherical	5	1,00	yes	70	300	17	0,48		2,80	840
50	Zorbax SB C$_{18}$	Agilent Technologies	spherical	5	1,00	no	80	180	10	0,45		2,08	374
51	Zorbax SB C$_8$	Agilent Technologies	spherical	5	1,00	no	80	180	5,5	0,45		2,00	360

2.1.1.2.1 On Polar Interactions

The authors mentioned in Refs. [1–5], as well as other research groups, have shown, through a lot of measurements under different experimental conditions, that steric aspects and, above all, polar interactions are the most important factors for selectivity in RP-HPLC, besides the generally assumed and expected hydrophobic interactions. Polar interactions in particular play by far the most important role. The reason is the diversity of possible polar groups on the surface of a typical C_{18} phase. One might speak about a "polar heterogeneity", which should not be underestimated: absolute concentration of silanol groups in dependence on the degree of coverage of the phase surface, type of remaining silanol groups (free, vicinal, geminal), concentration and type of metal ions in the silica gel framework or at the silica gel surface (alkali, alkaline earth, heavy metal ions, and Al, which in type A silica can have a strong influence as an acidic center [6]), siloxane bonding, residual amino and other polar groups, etc. Therefore, the differences between various C_{18}/C_8 phases can be huge. The diversity of polar groups inevitably causes a corresponding diversity of possible polar interactions: ion-exchange, hydrogen-bonding, dipole-dipole, and ion-dipole interactions, London forces, π-π interactions or complexation. Such polar interactions between analytes and a stationary RP-phase are the norm since there are relatively few molecules that do not contain any oxygen, nitrogen or a hydroxy group, disregarding substance classes such as alkanes, which play a minor role in HPLC.

In the case of more modern phases, there are also some other polar functionalities: additional polar groups to enhance the polar selectivity (e.g., Platinum EPS, **E**nhanced **P**olar **S**electivity, SynergiFUSION RP, Nucleodur Sphinx), polar and even ionic groups in the alkyl chain (EPG, **E**mbedded **P**olar **G**roup, e.g., Symmetry Shield, Primesep A), polar terminal groups on the alkyl chains (e.g., SynergiPOLAR RP), hydrophilic endcapping (e.g., ProntoSIL AQ, YMC AQ), short fluorinated chains (e.g., Fluofix), and many more. On the other hand, hydrophobic/lipophilic interactions are more or less uniform. Therefore, only a minor differentiation between analytes is possible, if these dominate during the separation process.

2.1.1.2.2 First Consequences

These facts lead to the following consequences:

1. If hydrophobic interactions dominate a separation, only small differences in selectivity are observed between the phases.
2. The more polar the phase is, the larger is the range of selectivity (α values) in the case of the separation of polar substances.

These statements are illustrated by two examples:

a) In Fig. 1a, the factors controlling the separation of ethylbenzene/toluene (methylene group selectivity, "pure" RP mechanism) on ca. 50 commercial stationary phases are shown. With the exception of very polar phases, quite different phases can also have similar selectivities. The α values are between

1.35 and 1.45. This experiment was repeated using fairly new, varied stationary phases as well as some phases from earlier experiments which provided a kind of internal standard. The observation mentioned above has been confirmed (see Fig. 1b).

b) In Fig. 2a, the factors controlling the separation of ethylbenzene/fluorenone are shown. In this case, polar interactions are possible because of the additional C=O group of fluorenone; the second difference to the above example is the higher molecular weight of fluorenone in comparison to that of toluene. In this case, the phases can "show" their individual character and the result is that the separation factors are between 1.00 and 1.65. In the case of polar phases Spherisorb ODS1, Fluofix, and SynergiPOLAR RP, even an inversion of the elution order is observed. Measurements with the new stationary phases had the same results (see Fig. 2b). The selectivity bandwidth lies between 1 and 1.75. In polar phases Acclaim PA C_{16}, XBridge Shield and Primesep C 100 reversed elution has been observed.

Stationary phases in the left part of Figs. 2a and 2b have good hydrophobic selectivity. Such phases can separate analytes with very little difference in their hydrophobic characteristics, e.g. differences in alkyl chain length ("methylene group selectivity"), homologous series, aldehydes, ketones. Furthermore, they are also selective for analytes that differ in their polar characteristics. This could be a difference such as the presence/absence of a polar group (such as keto- or hydroxy groups), or it could be isomerism. However, where polar/ionic interaction is dominant or separation is only possible using the difference in the ionic characteristics of the analytes, stationary phases from the right section of Figs. 2a and 2b are a better choice. They feature good polar/hydrophilic and aromatic selectivity, i.e. they are suitable for the separation of molecules such as phenol and caffeine, aromatic compounds, small bases in ionic form such as amines, polar degradation products or metabolites.

As explained above, one should bear in mind that even in unbuffered systems exclusive hydrophobic interactions – i.e., an "ideal" RP mechanism – rarely dominate. Even large, hydrophobic, unsubstituted aromatic substances are capable of polar interactions through induced dipoles. In Fig. 3a and b, the retention and separation factors (bar and line chart) for the separation of chrysene/perylene are depicted.

In addition to phases with a polymer layer on the phase surface, polar RP-phases in particular show very good aromatic selectivity. Modern hydrophobic phases, on the other hand, which are hardly capable of polar interactions, show only moderate selectivity.

As a final example of the separation of aromatic compounds, the separation of planar/nonplanar molecules based on steric selectivity is described. In the 1990s, a lot of attention was paid to steric selectivity ("shape selectivity") (Wise, Sanders, Tanaka). This concerns the capability of a phase to differentiate between planar and nonplanar molecules and is known as "the molar recognition mechanism".

Fig. 1a. Separation factors of ethylbenzene/toluene on about 50 customary commercial RP-phases in 80/20 (v/v) methanol/water.

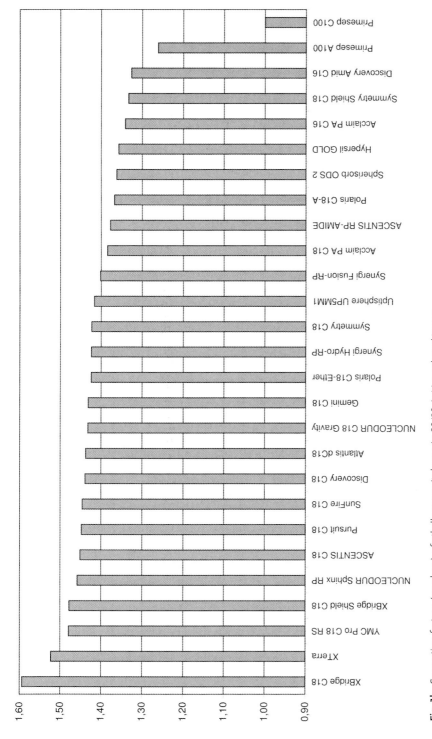

Fig. 1b. Separation factors (α-values) of ethylbenzene/toluene in 80/20 (v/v) methanol/water on 27 new commercial RP-phases.

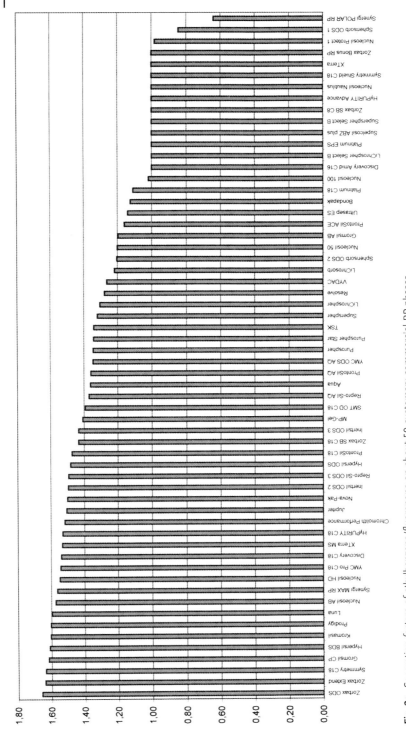

Fig. 2a. Separation factors of ethylbenzene/fluorenone on about 50 customary commercial RP-phases in 80/20 (*v/v*) methanol/water.

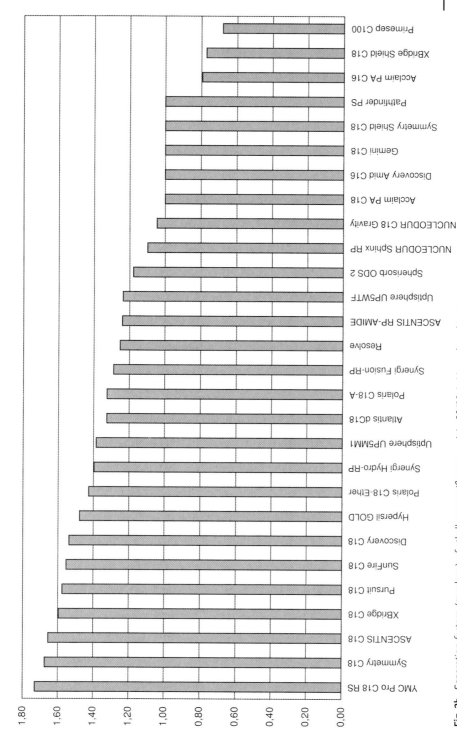

Fig. 2b. Separation factors (α-values) of ethylbenzene/fluorenone) in 80/20 (v/v) methanol/water on 27 new commercial RP-phases.

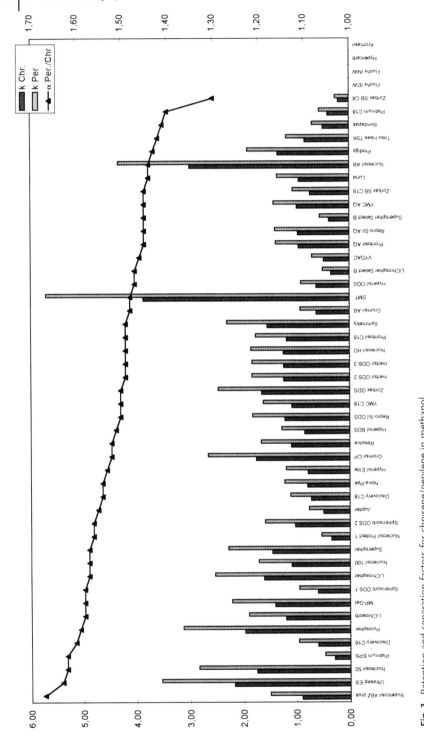

Fig. 3. Retention and separation factors for chrysene/perylene in methanol, (a) on about 50 customary commercial phases.

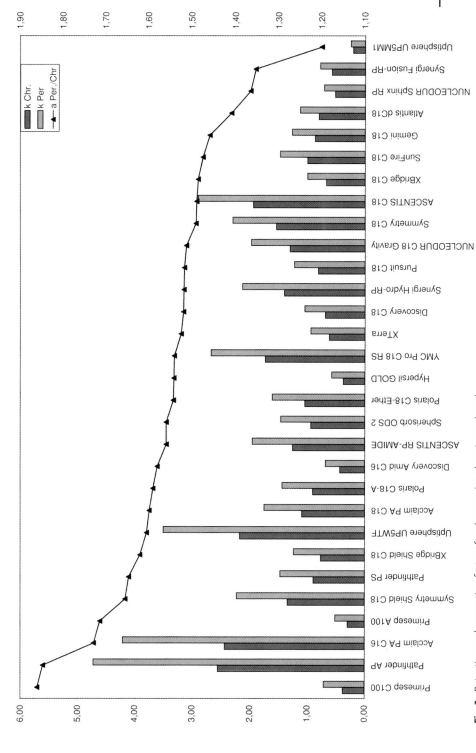

Fig. 3. Retention and separation factors for chrysene/perylene in methanol, (b) on 27 new commercial RP-phases. For details see text.

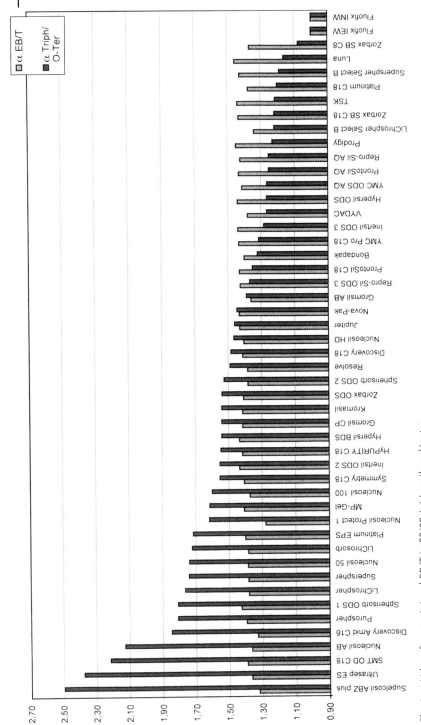

Fig. 4. α-Values for o-ter/tri and EB/T in 80/20 (v/v) methanol/water, (a) on about 50 customary commercial phases.

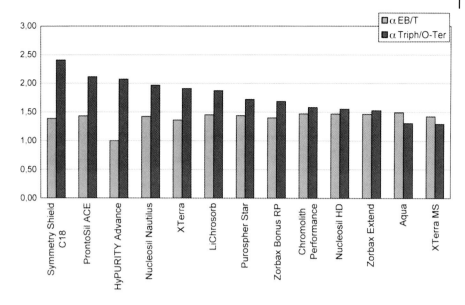

Fig. 4a. (continued)

According the authors named above, good steric selectivity should correspond to the polymeric character of the phase. There should be a correlation between the steric selectivity and the polymeric character of the phase or the degree of coverage.

As model substances, polycondensed aromatic substances and, above all, triphenylene/o-terphenyl, were frequently used. The last named analytes have the advantage in this context of having the same molecular weight and being nearly the same size.

Whereas triphenylene has a planar structure, o-terphenyl has a distorted structure. Figure 4 shows the α values for o-ter/tri and for comparison also the α values of EB/T for a couple of RP columns.

As a matter of fact, two phases with a polymer layer (Nucleosil AB, SMT) show excellent selectivity for o-ter/tri. Phases with polar properties show particularly good selectivity as well (left part of Fig. 4a and b). This finding is at variance with the conclusions of the authors named above.

A possible explanation stems from the fact that o-ter/tri are aromatic substances. As in the case of chrysene/perylene, polar interactions of the π-electron systems are possible here as well. (A chemometric analysis carried out with nine molecular properties for a lot of very different substances showed the strong similarity of these four aromatics [7]). The difference in conformation results in a difference in the aromaticity of the two analytes and as a consequence in the polar character as well, and this ultimately results in a different affinity for the polar groups on the stationary phase.

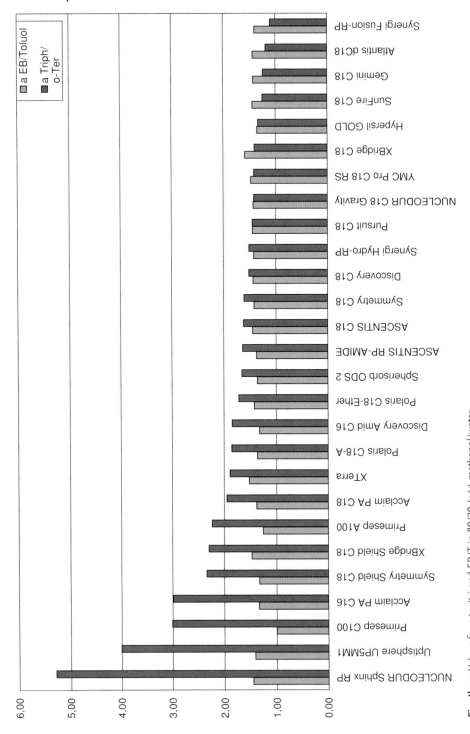

Fig. 4b. α-Values for *o*-ter/tri and EB/T in 80/20 (*v/v*) methanol/water, (b) on 27 new commercial RP-phases. For details see text.

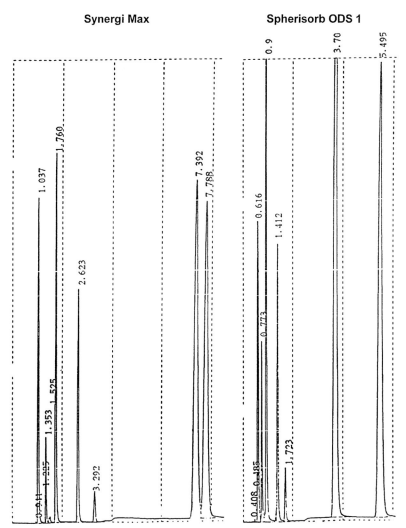

Fig. 5. Separation of uracil, methyl and ethyl benzoates, toluene, ethylbenzene, triphenylene, and *o*-terphenyl in 80/20 (*v/v*) methanol/water using two C$_{18}$ columns; for comments, see text.

Fig. 5 shows the separation of a mixture of uracil, an unknown contaminant, methyl and ethyl benzoates, toluene, ethylbenzene, triphenylene, and *o*-terphenyl using firstly a modern very well covered hydrophobic phase (SynergiMAX RP) and secondly a non-endcapped, less covered, "old" phase with acidic silanol groups (Spherisorb ODS1). On SynergiMAX RP, the small mononuclear aromatic substances are separated very well, but the selectivity of the *o*-ter/tri separation is only just good enough (last two peaks). In the case of the strongly silanophilic, less covered, and non-endcapped Spherisorb ODS1, the observed selectivity is exactly the opposite.

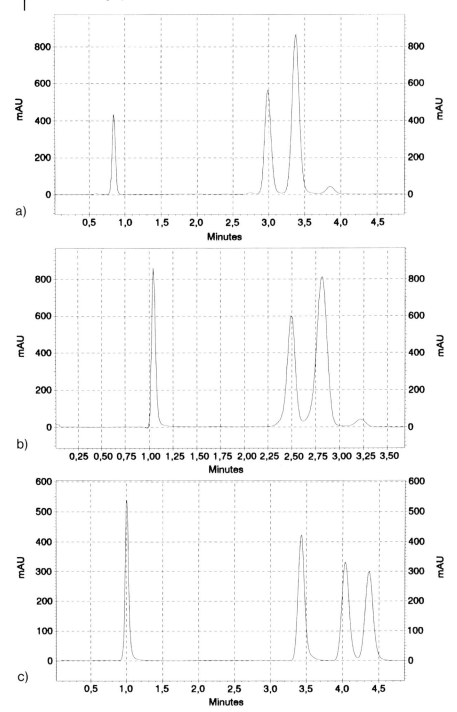

Fig. 6. Separation of steroids on several RP materials; for comments, see text.

This finding has been verified by several experiments: phases with polar properties show a good steric selectivity (see Fig. 6). Thus, isomeric steroids (double-bond isomerism) cannot be separated selectively using hydrophobic phases (co-elution of peaks 2 and 3 in the upper and middle chromatograms), but they can if polar groups are present on the surface of the stationary phase, as in the lower chromatogram in Fig. 6. The stationary phase is Resolve, a non-endcapped material.

An RP mechanism can be enforced by the pH value. If a uniform state of ionization of the analytes that are to be separated is generated by a given pH value (here, suppression of ionization), then their individual polar character recedes into the background, and the separation runs purely according to their different organic character. As mentioned above, apolar interactions result in a narrow range of selectivities, and the α values are small and similar. Figure 7 shows the retention and separation factors for the separation of tricyclic anti-depressants in an acidic phosphate buffer on several different phases.

Despite quite different retention factors (bars), which depend on the hydrophobic character of the phases, one observes very similar separation factors (line). Under such chromatographic conditions, the stationary phase decreases to a secondary factor in the optimization process (see Fig. 8); there is a large range of retention factors (retention time), and a very small range of separation factors (selectivities).

Figure 9 serves to demonstrate this "equalizing" of the stationary phases in the presence of buffers even for non-ionic analytes. In Fig. 9a, the separation of the isomers of nitroaniline on four rather different stationary phases with the help of an alkaline acetonitrile buffer is shown. Apart from small differences in the retention time, the separation of the three peaks looks rather similar on each of the four columns. Fig. 9b shows the separation of the nitroanilines on Symmetry Shield and on Zorbax Bonus in a methanol/water mixture. The chromatograms look absolutely different; even an inversion of the elution order is observed. This means that to exploit the individual properties of the stationary phases in the realm of "ultimate" selectivity, one should dispense with buffers, which is not easy to realize in routine work, where reproducible retention times are required. Nevertheless, one should remember this in the case of orthogonal tests; see below. These phenomena are observed even with simple, polar, non-ionizable analytes such as ketones (see Fig. 10).

On the other hand, if the analytes or a part of them are in ionic form by virtue of the pH of the eluent, then the differentiation is easier. If different interactions are possible, then there is a better chance for discrimination of (similar) analytes, i.e., better selectivity.

The fact that hydrophobic well-covered phases do not facilitate polar interactions, which is often the case in the context of the neutralization of ionizable analytes by the pH value, can result in a good peak symmetry, but frequently also in inferior selectivity (see Fig. 19 in Chapter 1.1). The following risk thus arises: in the case of well-covered, hydrophobic phases, a good peak symmetry frequently gives a false impression of similarly good selectivity.

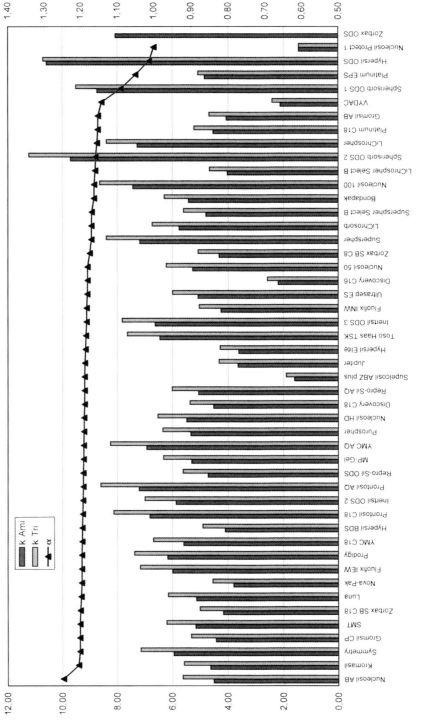

Fig. 7. Retention and separation factors for tricyclic anti-depressants in acetonitrile/acidic phosphate buffer.

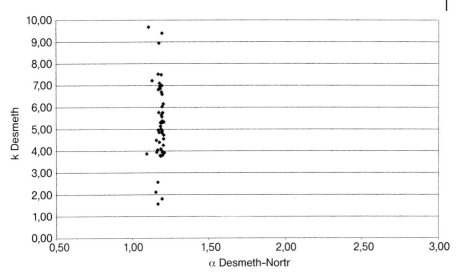

Fig. 8. On the bandwith of retention and separation factors by the separation of metabolites of tricyclic anti-depressants in acetonitrile/acidic phosphate buffer.

Summing up, we may conclude that:

- If polar interactions are possible, or if such interactions can be made possible by adjusting the pH, for example, one invariably finds a better selectivity.
- A uniform mechanism results in fast kinetics and inferior differentiation; this means good peak symmetry and frequently poor selectivity.
- A dual mechanism frequently results in slow kinetics and therefore in poor peak symmetry, but in most cases in good selectivity.

Therefore, for the separation of very polar analytes (metabolites, degradation products) for example, polar interactions are acceptable, but should not correspond with slow kinetics. The latter situation arises when undissociated silanol groups (free and so acidic) and/or other polar groups are bound directly on the surface. The problem with the free, "aggressive" silanol groups in the case of a lot of older materials (type A silica) is that they are "hidden" under the C_{18} groups and are therefore poorly accessible. After interaction with these groups, subsequent desorption of a polar/ionic analyte is a kinetically slow process (ion-exchange mechanism). Silanol groups, on the other hand, as well as other polar groups, which are freely accessible, do not generate tailing (e.g., HILIC on silica gel, as in Atlantis HILIC or SynergiFUSION RP). Possible solutions could then be:

Polar groups in the alkyl chain, which result in fast kinetics (EPG phases), steric protection, and either "no" free silanol groups on the surface of the stationary phase or a large number of sterically unhindered silanol groups/other polar groups, which are not overloaded in the case of interaction with the analytes, e.g., Zorbax SB C_8, Fluofix, SynergiPOLAR RP.

a) **ACN/phosphate buffer (32/68), pH = 7.6**

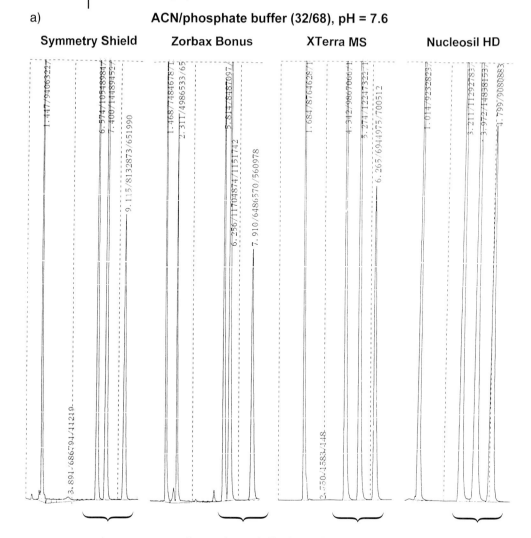

Fig. 9. Separations of nitroanilines in buffered (a) and unbuffered eluent. For details, see text.

b)

MeOH/H₂O (40/60)

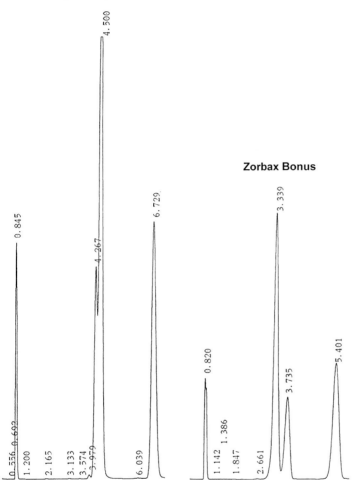

Fig. 9. (continued)

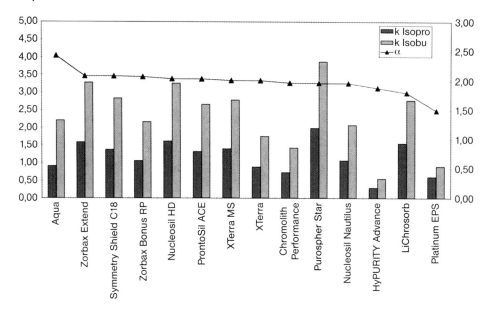

Fig. 10. Retention and separation factors of isobutyl/isopropyl ketones; for comments, see text.

Therefore, one can conclude that it is not the ionic interactions per se that are responsible for the chemical tailing. However, if as a consequence of some mechanisms – in this case ionic interactions – the kinetics of the desorption of the analytes from the surface of the stationary phase is slowed down, then tailing will result.

2.1.1.3 Criteria for Comparing RP-Phases

2.1.1.3.1 Similarity According to Physico-chemical Properties

Physico-chemical properties and chromatographic behavior are suitable criteria for stating the similarity of phases. The latter is the more expressive, of course.

The most important physico-chemical data are the specific surface area, the pore diameter, and the carbon content (more precisely, the degree of coverage); see Table 1. In the light of much chemometric analysis carried out by the author's group [7], it has become apparent that these three properties have the largest influence in terms of retention and separation factors. Although the packing density and the pore volume are essential for a differentiated comparison, because the absolute amount of stationary phase in a column can be calculated from these, on which, in turn, the retention time depends, the following can be stated. If two phases have comparable values of specific surface area, pore diameter, and degree of coverage, and if both also belong to the same type (e.g., "hydrophobic"/ "silanophilic"), one can expect quite similar properties.

2.1.1.3.2 **Similarity Based on Chromatographic Behavior;**
          Expressiveness of Retention and Selectivity Factors

The characteristic chromatographic parameters that describe the chromatographic behavior of stationary phases and that therefore are suitable criteria for comparison purposes, are the plate number (measure of the bond broadening), the retention factor (measure of the strength of interactions of an analyte under defined conditions), and the separation factor (measure of the capability of the chromatographic system to separate two analytes under defined conditions). In addition, the asymmetry factor is frequently used.

Which of these four factors is the "best", if these parameters are considered separately?

In a lot of tests in the literature for the characterization of RP-phases, retention factors of often "ideal" analytes are used to rank the phases according their hydrophobic character. Irrespective of the analytes used, one should consider the principle question as to whether retention factors are a suitable criterion for comparison at all. The retention behavior (i.e., a measure of the hydrophobicity) is, from the practical point of view, only of secondary interest. In the case of comparison tests, the aim is to identify columns that separate any compounds similarly well, and a measure of the separability is the selectivity. Besides the retention factor, also the third parameter, the plate number, seems to be less suitable. It depends on so many factors that one needs a very large-scale standardization procedure to relate detected differences in plate numbers exclusively to differences in the quality of the packing procedure of several columns. One should only consider the simple case that the next column is better/worse packed. It is quite labor-intensive to obtain statistically relevant results in this context. Also in the case of the declaration of asymmetry factors, a correlation is generally assumed between asymmetry and silanophilicity. This undoubtedly exists, but not exclusively. Also, if a comparison of columns is made correctly, of what use is the information that two columns are similarly well-packed and have similar silanophilic character in relation to two standard bases as model compounds?

We may thus conclude that if single chromatographic data are used as a criterion for the comparison of columns, the most expressive from the user's point of view may be the separation factor.

Refined Comparison Criteria

Remark: The following section (up to Section 2.1.1.4) goes into much detail; this text is intended for readers who are especially interested in column tests and their evidence. Readers who are rather interested in the key points should direct their attention in the next pages only to the sections headed "conclusion", "expressiveness of the presented tests" and "simple tests for the characterization of RP phases".

Recently, the author's group has been concerned with the question as to whether combined data concerning chromatographic parameters perhaps result in more differentiated statements. In the following, some results are given in condensed form for the first time:

1. Plate number and retention factor

The plate number depends, among other things, on the retention time of the analytes used or the retention factor. Under defined experimental conditions and if the retention factors are by chance similar, the resulting plate numbers of different columns can be compared directly. If the retention factors are different (perhaps as a result of different properties of the columns under consideration), a direct comparison would seem problematical. As an alternative, the following standardization can be applied: plate number divided by the k value or plate number divided by the square root of the k value. Without consideration of the exact correlation between k and N, one could state that the application of a relation of these two values results in a more substantiated comparison of the columns compared to when plate numbers alone are considered (see Figs. 11 and 12).

A column is more efficient, the lower the k value required to reach a defined plate number.

Example: The plate number for perylene using column X is 10,000 at a k value of 1.35, while the plate number using column Y is 10,000 at a k value of 2.21. Column X is hence the more efficient one. The quotient $N_{analyte}/k_{analyte}$ is ostensibly a measure of the efficiency of *this* analyte at *this* retention factor on *this* column.

Let us consider column A and column B in another example.

With the same particle size of 10 μm and a similar plate number for uracil of about 3000 for both columns, only the quotient $N_{perylene}/k_{perylene}$ shows that A, with a value of 4689, is more efficient (this column is better packed) than B with a value of 3983.

It should be pointed out that "better" efficiency means better in the actual case only. Thus, here, for the analyte "perylene" a better efficiency using A compared to B is found. For another analyte, e.g. ethylbenzene, B shows a better efficiency. These examples show how important the specific interactions between an analyte and a stationary phase are when considering the resulting plate number. This means that the mechanism and consequently the actual kinetics has a strong influence on the plate number, as expected. From a practical point of view, or in terms of "marketing optimized" strategies, this means that judicious choice of the analyte and the other experimental conditions can render a column of average quality highly efficient.

High values of the quotient N/k at the same k value are suggestive of small particle sizes or well-packed columns. Remarkably small quotients indicate a retarded mass transfer (C term in the van Deemter equation).

Finally, a third numerical example:

Let us consider the parameters for ethylbenzene and *o*-terphenyl on two columns A and B. Both show comparable k values for ethylbenzene (0.99 and 0.92) and for *o*-terphenyl (2.68 and 2.64). In the case of ethylbenzene, the quotient N/k for B (6699) is larger than that for A (4954). In the case of *o*-terphenyl, on the other hand, the quotients are similar, just a little larger for A (3696 compared to 3516).

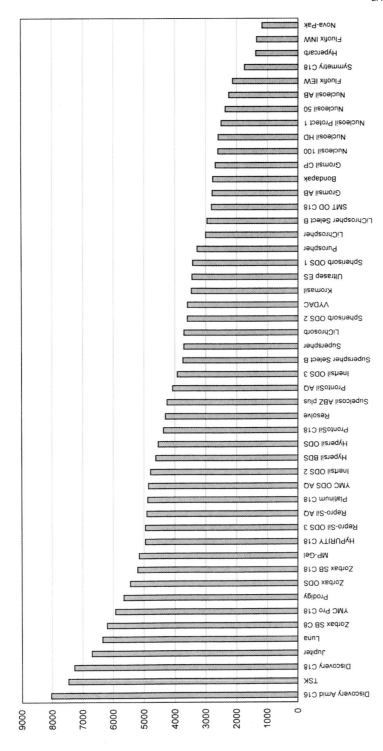

Fig. 11. "Normed" plate number of ethylbenzene.

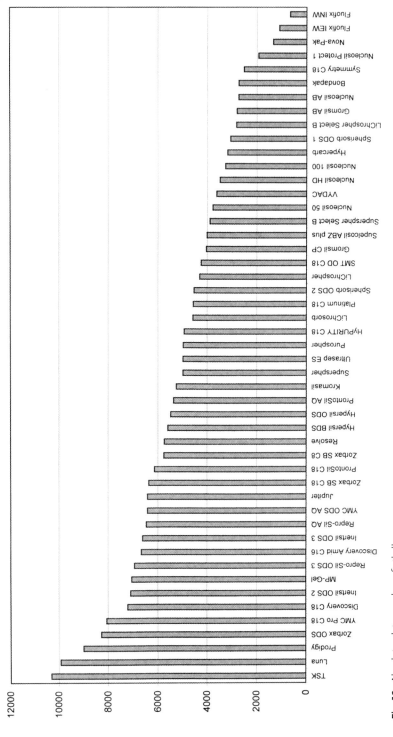

Fig. 12. Absolute plate number of ethylbenzene.

Obviously, the efficiency of column B is better, which can be established using "unproblematic" analytes such as ethylbenzene. However, if a specific effect results in slow kinetics, as in this case because of the spatially quite demanding structure of o-terphenyl, the corresponding quotient, in this case $N_{o\text{-terphenyl}}/k_{o\text{-terphenyl}}$, decreases strongly.

Conclusion: The direct comparison of plate numbers of several columns, in spite of different retention factors, is general practice, but seems problematical because of the aforementioned reasons, because wrong conclusions on the quality of packing can be made. A standardization under consideration of retention factors (N/k) could be an acceptable compromise. The comparison of N with N/k or of N_1/k_1 with N_2/k_2 yields interesting indications of a special affinity of an analyte to a given surface (C term in the van Deemter equation). Thus, a column might yield an excellent plate number for particular types of analytes, but for other analytes it might be very poor because of ionic interactions or steric effects. Other columns with a similar plate number invariably show a fast mass transfer for different analytes and therefore consistently good efficiency. Such a column is a better choice – based on comparable selectivity – than a competitor product.

2. Plate number and pressure

In practice, the following aspect is also important: one prefers to work with columns that have a high plate number but that also generate a moderate pressure. In Fig. 13, the standardized pressure and the standardized plate numbers of some columns are shown. Some columns generate quite a high pressure at a moderate plate number.

3. Retention and separation factors

As is well known, a good selectivity is not the same as a "good" separation in all cases. In addition to a low efficiency (band broadening, tailing), which can result in an insufficient resolution, a large retention factor (run time) also proves a disadvantage. As a criterion for the "effectivity" of a separation, the quotient of retention and separation factor could be used; see the example in Fig. 14.

Separations are optimal when run within a robust range with the best possible selectivity. This is the case here with Hypersil BDS and Nucleosil AB. In a robust k value range of ca. 2–4, both columns show α values larger than 3. In the case of isocratic separations and aiming for as short as possible run times, these quotients should be in a range between ca. 1.3 and 2.5. For difficult separations and/or a large number of peaks, this value can be up to 3–4. Other interesting diagrams, which are not presented here, show resolution versus retention time, or versus retention factor, respectively.

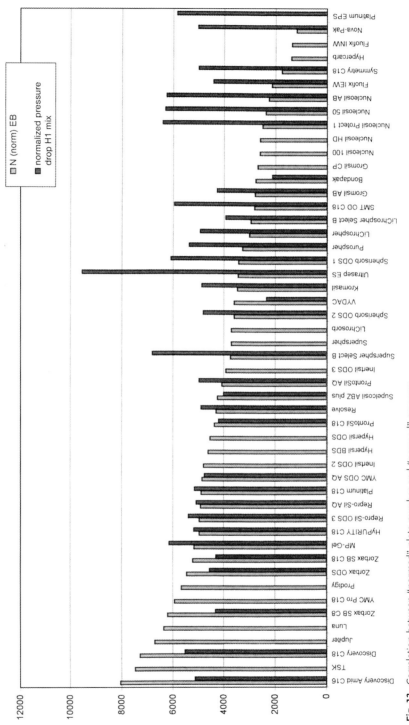

Fig. 13. Correlation between "normed" plate number and "normed" pressure.

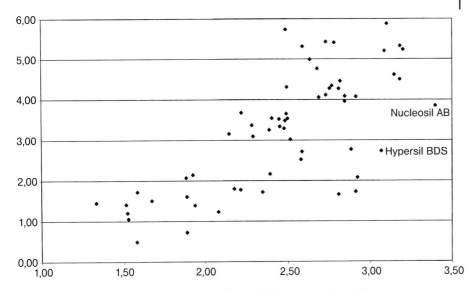

Fig. 14. Correlation between the retention factors for fluorene (*y*-axis) and the separation factors for fluorene/fluorenone (*x*-axis); for details, see text.

2.1.1.3.3 Tests for the Comparison of Columns and Their Expressiveness

The chemical character of a phase, i.e., if it appears more polar or more apolar, is regarded as the most important criterion for the comparison of RP columns. Therefore, the aim of more or less all tests reported in the literature is to check the silanophilicity or the hydrophobicity. In the 1980s and early 1990s, Walters, Sander, and Tanaka also took steric aspects into consideration. The efficiency was thereby further checked. In the following, the expressiveness of such tests of "silanophilicity"/"hydrophobicity" is discussed in short form and some results are presented. Finally, of some recently developed tests [5], one is described that yields information about the homogeneity of the surface.

Silanophilicity

In recent years, several tests have been developed to characterize the silanophilic properties of RP phases. These tests are usually based on the behavior of the phases towards basic substances, as, for example, in the case of the Engelhardt test, whereby acidic, neutral, and basic compounds are used. As a measure of the silanophilicity, one uses either the elution of a base compared to phenol (an earlier elution than phenol indicates a low silanophilicity of the phase) or the asymmetry factor of *p*-ethylaniline in a methanol/water eluent. Recent results show that a classification by these criteria can only be quite rough: "silanophilic"/"hydrophobic".

The lack of differentiation between acidic and "mild" silanol groups, the neglecting of (almost invariably found) mixed mechanisms, and the use of quite weak bases results in a doubtful similarity of the phases. Some phases, denoted as silanophilic according to the less rigorous criteria of the Engelhardt or similar tests, are very similar to hydrophobic phases. Therefore, it is not permissible to speak about *the* silanophilicity without a more differentiated consideration. An example illustrates this: Zorbax ODS has very acidic silanol groups. In the presence of metal ions on silica gel, such silanol groups are dissociated even in acidic eluents. The result is a strong tailing of the peaks of bases. On the other hand, Zorbax ODS has a quite pronounced hydrophobic character, because of the relatively high degree of coverage. This has been verified by several tests. Therefore, it seems meaningful to make at least the following distinction, if the point of interest is the relation of phases according to the criterion "silanophilicity" (see also Table 1 in Sectoin 2.1.3.1):

1. Concentration of acidic silanol groups. These are groups which are dissociated even in an acidic environment and which are able to interact with protonated bases (ion exchange).
2. Total concentration of available silanol groups; i.e. "good" shielding of the surface, activity of total silanol groups (hydrogen bonding, dipole-dipole interactions, etc.).

In Ref. [5], newer tests are described, which as well as making such a distinction possible also show the differences between different polar groups, but these tests will not be considered here.

Therefore, one should decide consciously in the laboratory if one really wishes to classify according to the criteria "hydrophobicity/silanophilicity" as some column producers do for understandable reasons. One may construct x/y diagrams, in which the k value of a neutral component as a measure of the hydrophobicity is plotted on the y-axis and the asymmetry factor of a base is plotted on the x-axis. In such cases, the phases are compared using values (the asymmetry factor and the retention factor) that characterize different properties of a phase without differentiation of enthalpic and entropic contributions, and this for different analytes. This could result in wrong conclusions. Even with a "correctly" performed test, one obtains only the following type of information: "the retention factor of the hydrophobic compound A on column C is x times larger/smaller than on column C'" and "the asymmetry factor of a standard base is x times larger/smaller", and this for the given conditions. There is seldom any information about the applicability of the column for a separation problem.

Hydrophobicity

The term "hydrophobicity" is not used consistently. We shall use the following definition:

- *Hydrophobicity:* The strength of the interaction between an apolar component and the stationary phase. As a measure, the retention factor of toluene can be used, for example.
- *Hydrophobic selectivity:* The ability of a phase to separate two apolar components.

Undoubtedly, apolar is an "flexible" term, especially if two analytes are involved, which is true in the case of separation factors. We use in our work a total of six quite different pairs of analytes for the investigation of the hydrophobic selectivity of phases. Without going into too much detail, after a lot of measurements and chemometric analysis, it was established that the representation in Figs. 15, 2b and 16 represents the selectivity of phases quite well. Here, some commercially available phases are listed according to decreasing hydrophobic selectivity, or polar selectivity, respectively (Fig. 16).

Considering the retention factors of several hydrophobic analytes, the hydrophobic selectivity, and the degree of coverage, commercially available columns can be divided into "more hydrophobic" and "more polar" (see Table 2).

Table 2. Examples of hydrophobic and polar RP-phases.

Hydrophobic RP-phases	Polar RP-phases
Gromsil CP	Synergi POLAR RP
SMT OD C18	HyPURITY Advance
Nucleosil AB	Platinum EPS
Nucleosil HD	Fluofix IEW
Luna	Fluofix INW
Prodigy	Nucleosil Protect 1
Zorbax Extend	Zorbax SB C8
Zorbax ODS	LiChrospher Select B
Synergi MAX RP	Superspher Select B
Symmetry C18	Supelcosil ABZ plus
Kromasil	Spherisorb ODS 1
Hypersil BDS	Zorbax Bonus RP
HyPURITY C18	LiChrosorb
YMC Pro C18	Discovery Amid C16
Discovery C18	Symmetry Shield C18
Jupiter	Nucleosil Nautilus
Repro-Sil ODS 3	XTerra
ProntoSil C18	Spherisorb ODS 2
Chromolith Performance	ProntoSil ACE
XTerra MS	Resolve
YMC Pro C18 RS	Acclaim PA C16
Ascentis C18	Nucleodur Sphinx RP
Pursuit C18	Primesep C100
SunFire C18	XBridge Shield
Hypersil GOLD	Uptisphere UP5MM

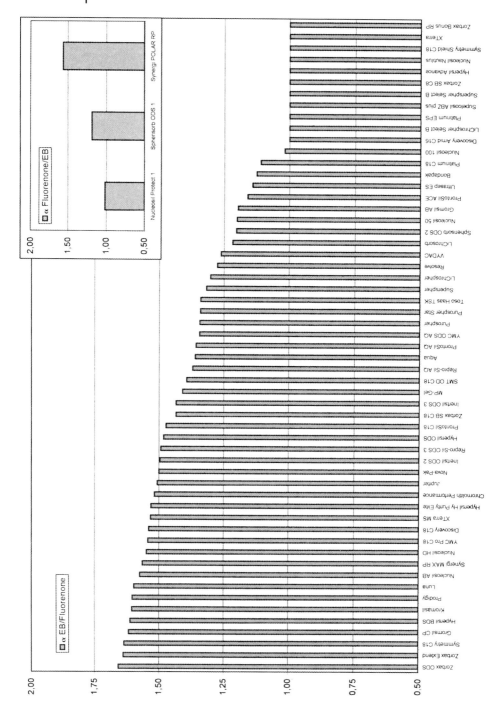

Fig. 15. Ranking of commercial RP-phases according decreasing selectivity

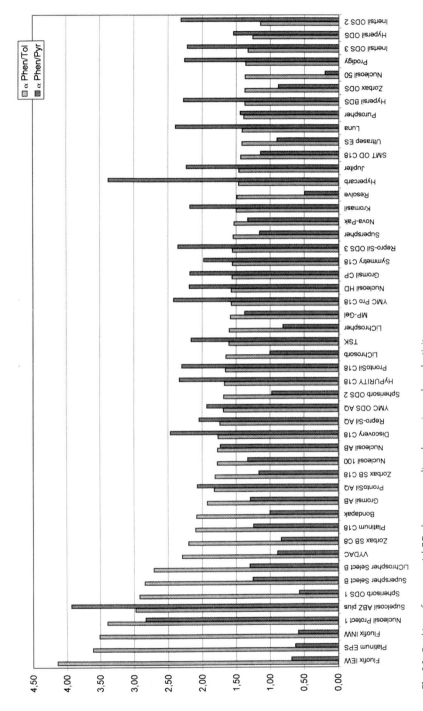

Fig. 16. Ranking of commercial RP-phases according decreasing polar selectivity

Table 3. Chromatographic data from different separations on columns
on basis of the same silica matrix; comments, see text.

Name	k toluene	α ethylbenzene/toluene	k Prop.benzene	α Prop.benzene/Eth.benzene	k tri	α tri/o-ter	k steroid C	α steroid B/C	k steroid D	α steroid D/E
Zorbax SB C 18	1,04	1,43	0,78	1,44	5,29	1,22	3,36	1,08	6,27	1,19
Zorbax SB C 8	0,50	1,44	0,50	1,32	1,70	1,08	3,35	1,09	4,91	1,08
Discovery C 18	0,70	1,41	0,46	1,44	4,03	1,48	1,55	1,06	3,44	1,25
Discovery Amid C 16	0,52	1,33	0,38	1,34	3,20	1,85	1,45	1,00	5,83	1,00
Spherisorb ODS 2	1,15	1,38	0,88	1,38	6,74	1,52	3,17	1,08	4,72	1,09
Spherisorb ODS 1	0,56	1,43	0,56	1,29	3,44	1,81	3,01	1,00	4,94	1,04
Second measurement										
Zorbax Bonus	0,89	1,40	0,73	1,97	3,33	1,69	3,17	1,00	12,58	1,00
Zorbax SB C 18	1,23	1,46	0,53	2,41	6,56	1,20				
Zorbax SB C 8	0,63	1,37	0,38	2,00	2,18	1,09				
Zorbax Extend	1,61	1,46	0,57	2,35	7,43	1,53	3,26	1,07	7,27	1,27
Spherisorb ODS 2	1,27	1,39	0,53	2,21	7,87	1,58	2,94	1,08	6,59	1,06
Spherisorb ODS 1	0,63	1,33	0,31	2,38	3,94	1,97	2,38	1,00	5,06	1,06
XTerra	0,89	1,36	0,60	1,81	3,15	1,91	2,53	1,00	7,85	1,05
XTerra MS	1,23	1,42	0,35	1,65	4,94	1,29	2,76	1,00	6,21	1,23
Reprosil AQ	1,22	1,42	0,91	1,41	6,47	1,42	3,81	1,05	7,81	1,18
Reprosil Pur	1,38	1,43	0,94	1,41	7,83	1,43	3,41	1,06	7,41	1,21

Table 4. Polar and hydrophobic versions of RP-Phase based on the same silica gel.

Long/short alkyl chain	Zorbax SB C_{18}/Zorbax SB C_8
Endcapped/non-endcapped	Spherisorb ODS 2/Spherisorb ODS 1
Classical endcapping/hydrophilic endcapping	Reprosil Pur/Reprosil AQ
Classical coverage/embedded phases (Zorbax Bonus: plus steric protection)	XTerra MS/XTerra, Zorbax Extend/Zorbax Bonus
Classical coverage/"embedded phase" plus shorter alkyl chain Discovery C_{18}/Discovery Amide C_{16}	Discovery C_{18}/Discovery Amide C_{16}

The hydrophobicity/hydrophobic selectivity also needs to be considered from another point of view. By means of the values in Table 3, the influence of the coverage on the separation behavior of different phases is shown, which are prepared using the same silica gel matrix. In Table 4, the retention and separation factors of several analytes are listed. In all cases, hydrophobic and polar versions of the phases are used.

Results

- Retention factor
  In the case of a "pure" RP separation, such as ethylbenzene/toluene, the larger retention factor is invariably obtained using hydrophobic phases. For Zorbax SB C_{18}/Zorbax SB C_8 the factor is ca. 2; in the case of Reprosil Pur/Reprosil AQ it is only negligibly larger. The larger the quotient of the retention factors "apolar phase" to "polar phase", the more one can speak about "pure" RP interactions. A small quotient indicates similar hydrophobic properties of the two stationary phases.

- Separation factor
  In the case of "pure" RP interactions (e.g., a difference of only an additional CH_2 group, "methylene group selectivity"), the stationary phase is largely unimportant. The choice of the column is of secondary importance. Therefore, one obtains only a small difference in the separation factors between Zorbax Bonus and Zorbax Extend, which are two completely different phases. If the analyte has polar groups, as in the case of hydroxybenzoates, a differentiation of columns is more easily possible and the separation factors cover a wider range. In the case of triphenylene/o-terphenyl (additional steric effect), the range is even wider. If the polar phase can display its capacity for polar interactions, a better selectivity is observed in all cases. If the difference is only a shorter alkyl chain, as in the case of Zorbax SB C_8, then no distinct positive influences are observed.

"Steric selectivity" ("shape selectivity")

Using the analytes triphenylene and o-terphenyl, two molecules with the same molecular weight and nearly the same size but different spatial arrangements (planar/helically deformed), several columns were investigated for their ability to separate steric complex molecules. It was found that besides phases with a "polymer layer" (e.g., SMT OD, Nucleosil AB/HD, Gromsil CP), polar phases show a good "steric" selectivity.

Homogeneity of the surface – indication of different active groups

As is known, there exists a correlation between log P and the k value. The better this correlation (large correlation factor) the more uniform is the mechanism. Therefore, a good correlation equates to a uniform mechanism, for which *one* functional group is essentially responsible. If one deliberately uses a mixture of analytes that differ in chemical nature, then the retention should be controlled by several functional groups of the in this case heterogeneous surface. In other words, the different functional groups show a discriminatory selectivity for different types of analytes. It should be borne in mind that silanophilic *and* hydrophobic properties of a phase do not have to be in contradiction. One should consider only the well covered and metal ion containing silanophilic Zorbax ODS and LiChrospher, the very hydrophobic C_{30} alkyl chain and simultaneous hydrophilic endcapping of Develosil, or the Acclaim PA C16 with an embedded group and a 16–18% carbon load.

The correlation factor can be calculated for a linear and for a quadratic regression.

The larger the difference between the two values, the stronger the indication of different functional groups on the surface, which are of course relevant for the types of analytes used.

This was checked for several samples on several columns under different conditions. In Fig. 17, the column-specific differences for the separation from the experiment "hydrophobicity" are denoted. The subject is the separation of neutral components (ethylbenzene and toluene) and aromatic planar/nonplanar molecules (triphenylene and *o*-terphenyl).

At the top end of Fig. 17a and b there are phases in which polar interactions are dominant. Polar groups responsible for this ability are either intentionally introduced (embedded phases, Platinum EPS, Ascentis RP-Amid, Uptisphere UP5MM) or result from the preparation process (Supelcosil ABZ PLUS). Alternatively, they may be simply the residual silanol groups of not or not completely endcapped phases. As a matter of fact, such phases show very good "shape selectivity". At the lower end of Fig. 17, phases are denoted which do not offer any specific interactions to this mixture. These are phases with a rather homogeneous, well-covered surface which therefore shows only a restricted "shape selectivity". The same applies to Fig. 17b: In this graph, there is not a single phase – e.g. between Spherisorb ODS 2 and Synergi Fusion RP – that features a distinctly polar group. Such phases are hardly able to show ionic/$\pi...\pi$ interactions, which could be an advantage in some separations (see below).

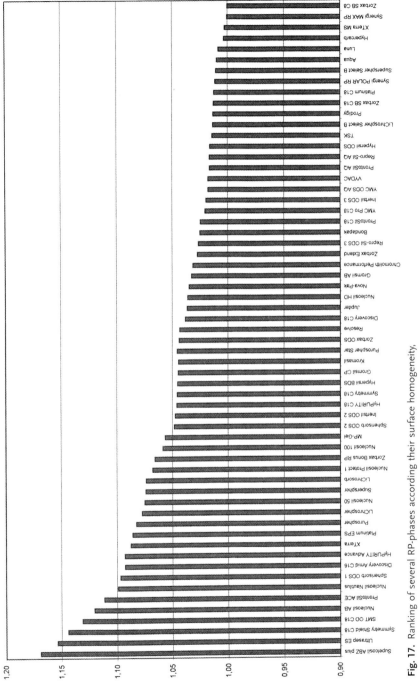

Fig. 17. Ranking of several RP-phases according their surface homogeneity, (a) on about 50 customary commercial phases. For comments, see text.

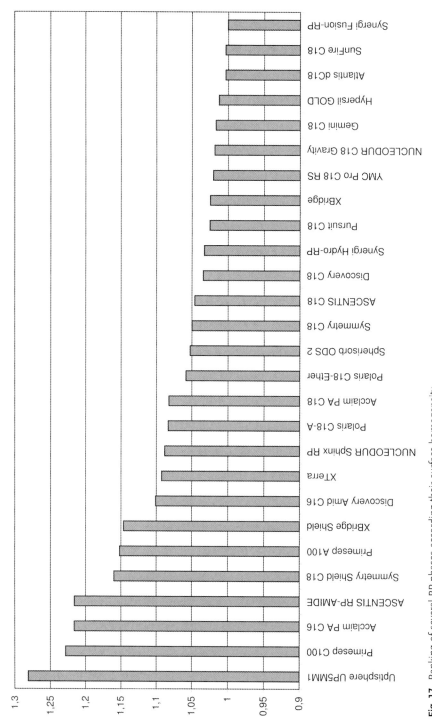

Fig. 17. Ranking of several RP-phases according their surface homogeneity, (b) on 27 new commercial phases. For comments, see text.

Expressiveness of the Presented Tests

The relevance of the results of these tests for one's own work depends on the actual aim under consideration. The following example is illustrative:

- The injection of stronger bases (e.g., benzylamine) in an unbuffered methanol/water eluent is a very stringent test: only by using very well covered phases can a stable retention time as well as a symmetric peak be obtained.
- On the other hand, there is hardly a "bad" column on the market that does not give wonderful peaks with toluene, ethylbenzene, and anthracene.
- A further test could be injection in an acidic buffer, in which only strongly acidic silanol groups show an effect. The silanol groups of, for example, Zorbax ODS, Resolve, Hypersil ODS or Spherisorb ODS1 are so "aggressive" (acidic) that they are also dissociated in the acidic range. This also has the following consequence. If a phase that perhaps performs poorly in these tests is used in an acetonitrile/phosphate buffer with sufficient buffer capacity and with the "right" pH value, then it can yield very good results for weaker organic bases, i.e. very good selectivity at sufficient peak symmetry.

Consider the following (simplified):
Using a silanophilic phase, large retention and separation factors are obtained with low peak symmetry in the case of the separation of protonated bases. Using hydrophobic phases, low retention and separation factors are obtained with good peak symmetry unless because of the pH value there exist almost pure hydrophobic interactions. The compromise is in all cases like traversing peaks and troughs. Considering the choice of the column, the result of this compromise can be influenced very strongly by the pH value. The "optimal" test does not exist; only more (e.g., primary/secondary amines and unbuffered eluents) or less (e.g., ternary amines and buffered eluents) stringent tests are available. An important factor with regard to the choice of the test is the purpose.

Whether such tests are sufficient for one's requirement or whether it is better to investigate the columns for suitability for the analytical problem at hand using the actual sample, can only be decided in each specific case. From a practical point of view, the latter approach is probably better.

Simple Tests for the Characterization of RP Phases

Although it is possible to carry out laborious tests, as described above, and obtain good RP phase characterization, this is far too extravagant for most laboratories. It was therefore a matter of finding out if it was possible to carry out two quick and simple tests (isocratic runs, no buffers, retention time under approx. 10 min), to obtain a maximum of information.

Data from previous measurements were reexamined with a view on the requirements stated above, and it seems that only two simple tests are needed to obtain information ranging from column characterization (hydrophobic/polar character?), establishing groups of columns with similar properties to their suitability for the separation of specific analyte groups. In order to verify this statement, 25 fairly new columns with diverse chemical surface properties were tested alongside with previously examined columns which served as a sort of internal standard. The two tests require two runs and two eluents:

Test 1:

Eluent: 80/20 methanol/water
Flow: 1 mL/min
Temperature: 35 °C
Detection: 254 nm
Injection of ethylbenzene und fluorenone

Information from Test 1: The graphical presentation of the ethylbenzene/ fluorenone separation factors shown in Figs. 2a, 2b and 15 highlights the hydrophobic characteristics of the phases. The phases in the left half of each figure are hydrophobic in character (more precisely: they have good hydrophobic selectivity), whereas the phases in the right half could be described as polar.

Test 2:

Eluent: 40/60 methanol/water
Flow: 1 mL/min
Temperature: 35 °C
Detection: 254 nm
Injection of phenol, pyridine, benzylamine

Information from Test 2:

1. The longer benzylamine takes to elute in comparison to phenol, the stronger the polar character of the phase (more precisely: the capacity of strong polar/ ionic interaction). Co-elution with phenol indicates a hydrophobic surface, and elution even before phenol indicates a very hydrophobic RP surface with good coverage (see Fig. 17d).

2. The peak shape of benzylamine may provide a second criterion, as long as the solution injected has been freshly prepared. The injection volume should not exceed 10 µL. This is a rigorous test, as only very few columns will achieve good peak symmetry for benzylamine (see Figs. 17c and 17d). Fig. 17c shows the injection of uracil, polar contamination, pyridine, phenol, benzylamine, and an unknown contamination on YMC Pro C_{18}. Fig. 17d shows the injection of benzylamine on XBridge C_{18}, with good peak symmetry, in spite of the fact that the column has been deliberately overloaded.

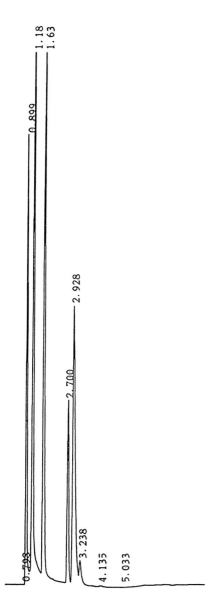

Fig. 17c. Injection of polar compounds in methanol/water on YPC-Pro C18; for comments, see text.

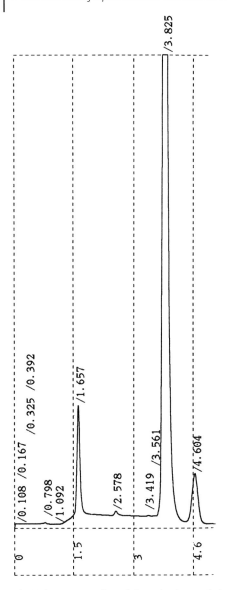

Fig. 17d. Injection of uracil, benzylamine, and phenol in methanol/water on XBridge; for comments, see text.

Information derived from the data of both tests: Plotting the ethylbenzene/ fluorenone separation factor (Y-axis) against the ethylbenzene/phenol selection factor results in a selectivity map or chart: The columns can be subdivided into groups with similar characteristics, see relevant explanations for Figs. 19 and 20.

Note: In Figs. 19a and 19b as well as in Fig. 20, the abscissa shows the phenol/ toluene separation factor, but using the ethylbenzene/phenol separation factor will yield identical results.

Thus, two simple runs can characterize RP-columns as follows:
• hydrophobic characteristics,
• expected selectivity regarding specific compound categories (see above and section 2.1.1.5),
• similarities with other RP columns (column comparison).

If time is at a premium, a selectivity chart need not be established – just injecting ethylbenzene/fluorenone at 80/20 methanol/water and phenol/benzylamine at 40/60 methanol/water may sometimes yield quite enough information. Figures 2 and 15 show the hydrophobic characteristics and indirectly the suitability of the phases for certain compound categories, while the peak shape of benzylamine in connection with its elution compared to phenol reveals whether the column in question carries strongly polar/ionic groups on its surface or not.

For Test 1 you must be prepared to test several columns, as hydrophobic characteristics do not exist as an absolute, but are defined by comparison. In a new column, it is also possible to compare the ethylbenzene/fluorenone separation factor obtained with the figures given in Figs. 2a and 2b. A rough comparison with approx. 90 commercial columns regarding hydrophobic characteristics is thus possible. By contrast, Test 2 can also be carried out for single columns. Elution of benzylamine after phenol – probably as a broad, tailing peak – points to the polar characteristic of the phase. Co-elution or even elution before phenol, however, is seen in hydrophobic phases with good coverage.

2.1.1.4 Similarity of RP-Phases

As explained above, in the author's opinion, for a separation technique such as HPLC selectivity is the most important and most expressive criterion for the comparison of columns or for establishing the similarity of RP-phases. Retention factors should only be considered as a secondary issue. To verify this, a chemo-metric analysis (see below) was performed with a large number of data (very different separation conditions and different analytes); see Fig. 18. The further the variables (here, k and α values) are from the origin on the y-axis, the stronger is their influence on the differentiation of objects, which here means columns. As can easily be seen, the α values are indeed suitable for this. If, nevertheless, one wants to use k values, then "problematical" compounds are best suited for this, e.g., polar analytes in unbuffered eluents (e.g., k Cos, caffeine) or strong acids in buffered eluents (k Phs, phthalic acid and k Tes, terephthalic acid).

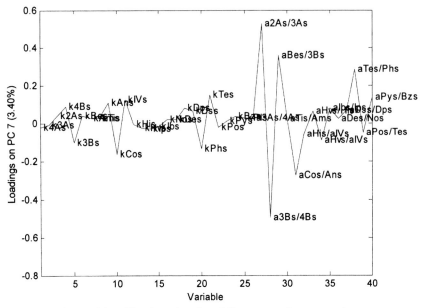

Fig. 18. On the suitability of k and α values for a differentiation of stationary phases; for comments, see text.

The data obtained from such investigations can be depicted in various ways. Good visualization tools are selectivity maps, selectivity plots, selectivity hexagons, and dendrograms. In the following, the similarity of some commercially available phases will be discussed with the aid of some examples.

2.1.1.4.1 Selectivity Maps

Selectivity maps primarily serve to find similar phases by means of the selectivity of these phases for given analytes under defined conditions. For this, the separation factors for the separation of different analytes are plotted against each other. Phases which are close together in these diagrams are similar with respect to the selectivity of the types of analytes under consideration. In the following, this tool will be presented by means of an example. The more different the pairs of analytes used are, the more accurately the similarity should be found. See Chapter 2.1.6 for an alternative tool suitable for establishing column similarities.

Example: Neutral Molecules, Unbuffered Eluents

Consider Figs. 19 and 20. The y-axis shows the selectivity of columns for the separation of hydrophobic/polar analytes, i.e., the selectivity for analytes that have a difference in their size and/or show a difference in their polar character, e.g., by a C=O group (ethylbenzene/fluorenone). The x-axis shows selectivity for polar/hydrophobic analytes (phenol/toluene). Large α values on the x-axis indicate polar phases; large values on the y-axis indicate hydrophobic phases.

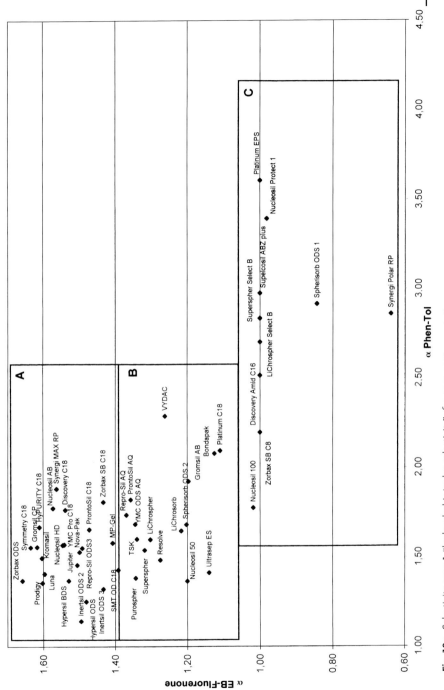

Fig. 19a. Selectivity map 1 "hydrophobic/polar selectivity"; for comments, see text.

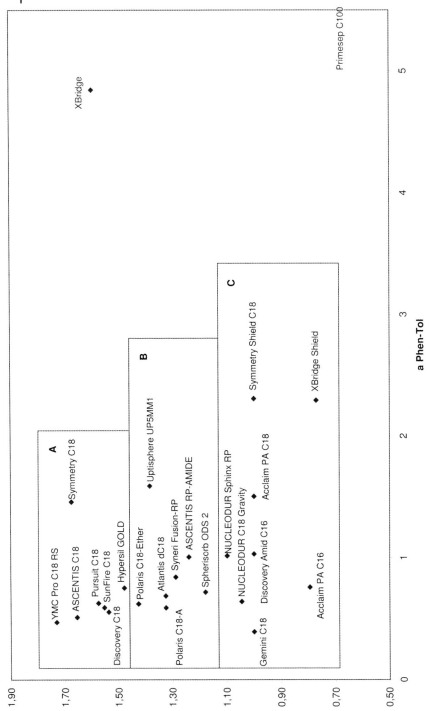

Fig. 19b. Selectivity chart 2 ("hydrophobic selectivity"); for comments, see text.

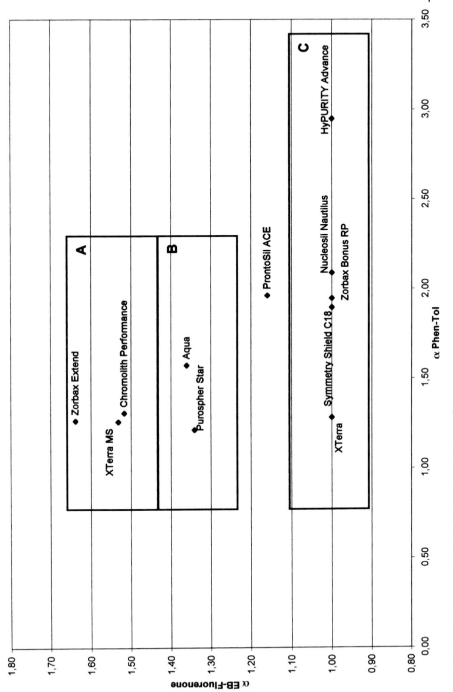

Fig. 20. Selectivity map 2 "hydrophobic/polar selectivity"; for comments, see text.

The columns can be loosely classified in three clusters:

A) Hydrophobic phases

 On these phases, the hydrophobic character dominates. This results, for example, from a strong coverage, a metal ion-free silica, or from a polymer layer.

B) Hydrophobic phases with distinct polar character (middle polar phases)

 The observed polar character of these phases results from low coverage, free silanol groups, or additional polar groups on the surface.

C) Polar phases

 The strong polar character, which is characteristic of these phases, results from short alkyl chains, polar groups on the surface, or embedded polar groups in the alkyl chains.

Application of a Selectivity Map

If, for example, in the case of a column of group A an unsatisfactory selectivity is obtained, for the next trial one should not use a column from the same part of the selectivity map as this column. It can be expected to have similar properties and thus be equally unsuitable for the test substances. A better choice would probably be a column of group B or C with completely different properties. Various cross experiments have shown that selectivity charts such as Figs. 19 and 20 very well represent the similarity of stationary phases.

2.1.1.4.2 Selectivity Plots

Interesting results are also obtained by plotting separation factors (α values), obtained from supposedly similar phases, against each other.

 The more similar the phases, the nearer the different values should be to the bisector of the angle (diagonal). In a series of experiments, 68 different pairs of analytes were separated under deliberately different chromatographic conditions (nine different eluents) and the obtained α values were plotted against each other. Because of the large number of values and the different experimental conditions one obtains rather reliable conclusions about the similarity. An example of such a plot is depicted in Figs. 21 and 22. In this case, Symmetry C_{18} and Purospher e are compared.

Some Comments

From the small diagram in which all values are contained, it can be assumed that both columns are in fact similar. This was to be expected, because both are hydrophobic C_{18} phases with classical high coverage on a base of very pure silica gel. To obtain a more differentiated picture, another type of graduation was used for the plot of the 68 values (large diagram).

 The seemingly anomalous 21 is remarkable. The α value corresponds to the separation of 2,2'-/4,4'-bipyridyl after 4 weeks of using the column. A large α

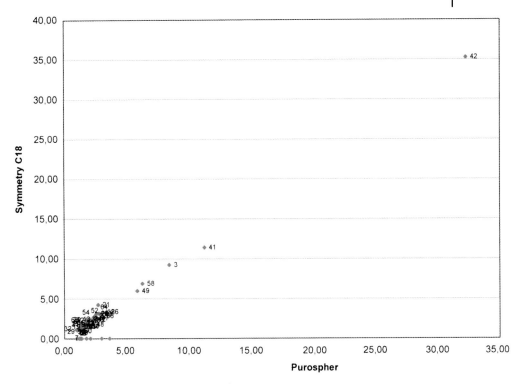

Fig. 21. Selectivity plot Symmetry C$_{18}$/Purospher e.

value can be taken as an indication of the presence of heavy metal ions, because 2,2'-bipyridyl is a complexing agent, but 4,4'-, in contrast, is not. In the course of the use of the column, metal ions clearly accumulate on the silica gel surface of Symmetry and the tendency for complexation increases. In addition, the somewhat greater hydrophobic character of Symmetry is remarkable; see separations "31" (phenol/pyridine) and "64" (fluorene/fluorenone). Purospher, on the other hand, shows a rather better selectivity for the separation of organic acids, in this case HVS/HIES (hydroxyvanillic acid and hydroxyindoleacetic acid) ("48"); its additional polar character is remarkable.

Such plots are a suitable tool for quite accurately classifying the similarity of columns in terms of their separation behavior. Selectivity plots give a reliable visualization of for which substances two columns show similar selectivity and for which they do not. The construction of several selectivity plots, of which just a few examples are depicted here, yielded the following results:

In general, one can make the following observation: differences between phases are most obvious if ionic interactions are possible between the analytes under consideration and the stationary phase. This was clearly recognizable in all of our selectivity plots. The consequence of this is the following: for a reliable investigation

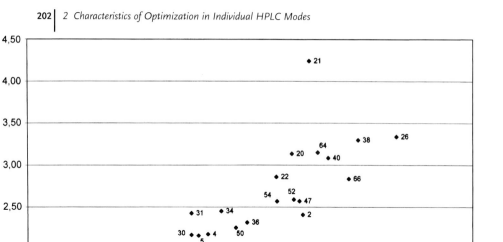

Fig. 22. Selectivity plot Symmetry C$_{18}$/Purospher e, different scale to that in Fig. 21.

of the "charge to charge" reproducibility of a phase or for the investigation of the stability of a column in routine use or for the comparison of two supposedly similar columns, polar, or better ionic, analytes should be used. The stringency of the tests and consequently the relevance of the result increases if unbuffered methanol/water eluents are also used.

- Prontosil and Reprosil are very similar phases (see Fig. 23).

- Superspher and LiChrospher are nearly identical (see Fig. 24).

- Hypersil ODS and Hypersil BDS mostly behave similarly. The deactivation of the phase surface in the case of Hypersil BDS ("weak" silanol groups) shows an effect with regard to the peak form only in relation to the separation of basic substances.

- The highly covered phases Zorbax ODS and Hypersil BDS show similar selectivity behavior for a lot of separations. The consequence of a deactivation of the phase surface (reduction of the silanol group activity) in the case of Hypersil BDS is clearly evident for the separation of basic compounds (see Fig. 25).

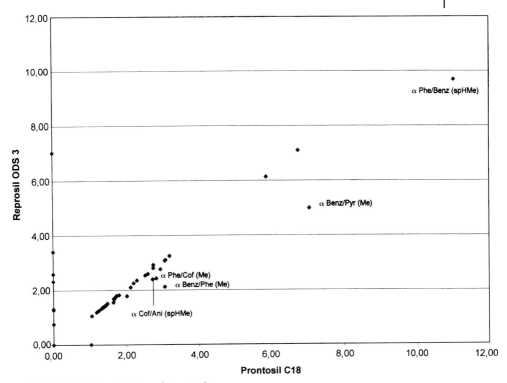

Fig. 23. Selectivity plot Reprosil/Prontosil.

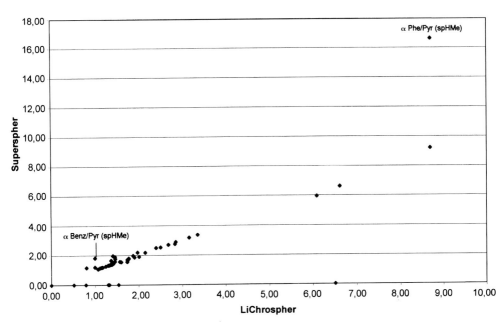

Fig. 24. Selectivity plot LiChrospher/Superspher.

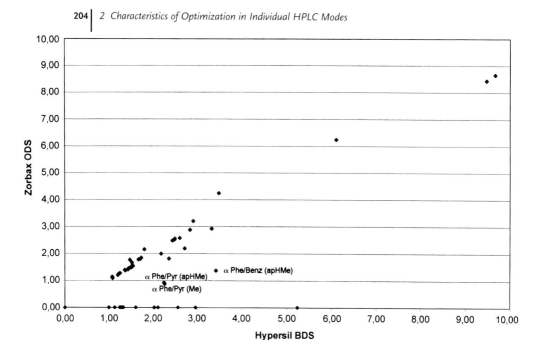

Fig. 25. Selectivity plot Zorbax ODS/Hypersil BDS.

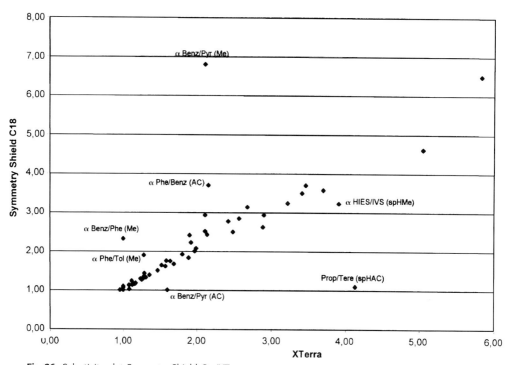

Fig. 26. Selectivity plot Symmetry Shield C_{18}/XTerra.

- Hypersil ODS obviously has more "acidic", "aggressive" silanol groups than the (highly covered) Zorbax ODS material. The interactions with basic compounds are stronger.

- Zorbax Bonus is a more "hydrophobic" embedded phase than HyPURITY ADVANCE.

- Symmetry Shield and XTerra show, as expected, comparable selectivity behavior because of the shielding effect of the carbamate group. The difference in the matrix, synthetic silica gel in the case of Symmetry as opposed to hybrid material in the case of XTerra, shows an effect for polar pairs of analytes, as expected. XTerra is indeed a low covered material, but the activity of this matrix towards bases is low (see Fig. 26).

2.1.1.4.3 Selectivity Hexagons

Selectivity hexagons are an additional visualization tool, which nicely show the similarity of different phases. Several such hexagons have been prepared for a quick decision about the applicability of a phase for the separation of a given class of substances in a given eluent (methanol/acetonitrile) or at a given pH value (acidic/basic phosphate buffer). In addition, this type of diagram shows which column has similar or completely different properties compared to other columns for a specific application under given conditions. In these diagrams, normalized α values measured for given pairs of analytes are denoted at the corners of hexagons.

Consider, for example, six different pairs of analytes, all containing an aromatic ring (bases, acids, isomers, etc.) which were separated under different conditions (methanol, methanol/water, methanol/buffer). The more symmetrical the hexagon, the more applicable is the corresponding stationary phase for the separation of aromatic substances in general. In addition, it is easy to decide which column is suitable, e.g., for the separation of quite strong aromatic acids (phthalic acid/terephthalic acid, "Tere/Phthal"), which is best for weaker aromatic acids (3-hydroxy-/4 hydroxybenzoic acid, "3/4-OH"), and which is best for planar/nonplanar aromatic substances (triphenylene/o-terphenyl, "Triph/o-Ter").

Finally, the similarity of phases can be recognized at first glance because of the condensed information; as with pictograms, e.g., for traffic signs, characteristic pictures are produced; see below.

In the following, three figures containing a lot of representative hexagons (columns) for different pairs of analytes and eluents are shown and brief comments are made with regard to the similarity of the columns.

Note: In some hexagons, certain α-values are missing. The reasons for this are various problems during the measurements (i.a. air bubbles, plugged injection needle). In view of the aim to compare the similarity of the columns under really identical conditions, repeat measurements were dispensed with.

Example 1: Aromatic Compounds in Methanol/Water

Consider Fig. 27. As mentioned above, the most important distinctive feature of stationary RP-phases is their capacity for polar/ionic interactions. If this ability is missing or not needed, as is frequently the case when separating neutral analytes in unbuffered eluents, then a lot of stationary phases show a similar behavior. Other properties then come to the fore, such as the degree of coverage.

Results

- Inertsil ODS 2 is more suitable for polar interactions than Inertsil ODS 3; viz. the better selectivity for the planar/nonplanar pair triphenylene/*o*-terphenyl.
- LiChrospher, Superspher, and Purospher e show similar selectivity patterns because of the similar high degree of coverage and the presence of polar groups on the surface. The doubtless existent metal ions and free silanol groups on the surface of LiChrospher and Superspher have no effect in this case. Also similar are Symmetry and Zorbax ODS, two phases that are markedly different in the case of the separation of ionic analytes.
- In the case of the following stationary phases, a fine differentiation is possible. In the following sequence, the apolar character increases: HyPURITY C_{18}, Hypersil BDS ≈ Inertsil ODS 3, Luna 2.
- From numerous experiments, it is known that Hypersil ODS and Zorbax ODS have very acidic silanol groups. If no basic analytes are to be separated, then the more hydrophobic character of Zorbax ODS compared to Hypersil ODS comes to the fore. In addition, it can be seen that while both phases bear acidic silanol groups, the total concentration of silanol groups is low. Other stationary phases have a considerably higher total concentration of polar groups, e.g. LiChrospher, Superspher, Discovery Amide, Spherisorb ODS 1. This becomes noticeable through their very good selectivity in the separation triphenylene/*o*-terphenyl.
- The similarity of the "C_8 phases" LiChrospher Select B, Superspher Select B, and Zorbax SB C_8 is clearly evident.
- Phases with a polymer layer on the surface show the best selectivity for aromatic compounds per se (SMT OD, Nucleosil AB), and secondly for this comes Ultrasep ES.
- The phases Bondapak, Gromsil AB, and Platinum C_{18} show strong similarity.
- XTerra MS resembles Purospher Star, except in the polar groups of the latter. It is these that give rise to the good selectivity of Purospher in separating triphenylene/*o*-terphenyl.
- The higher hydrophobicity of XTerra MS compared to XTerra is clearly evident.
- Nucleosil Protect 1, Spherisorb ODS 1, and Platinum EPS are the most polar phases here (and in other cases too).

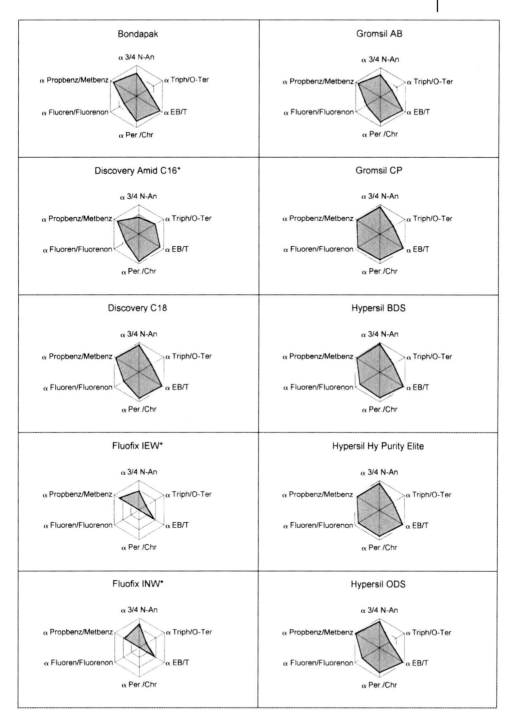

Fig. 27. Selectivity hexagons: "different aromatic compounds in methanol/water".

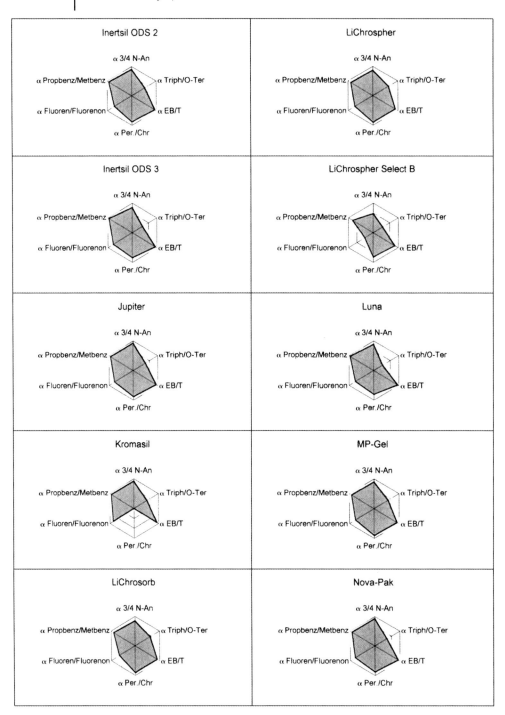

Fig. 27. (continued)

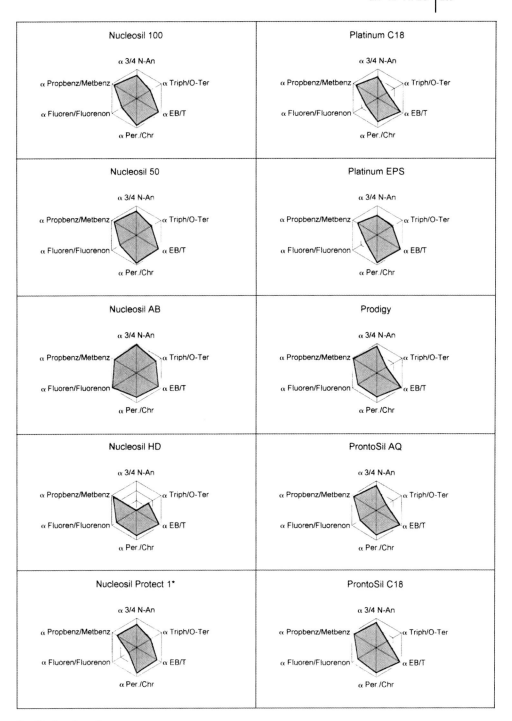

Fig. 27. (continued)

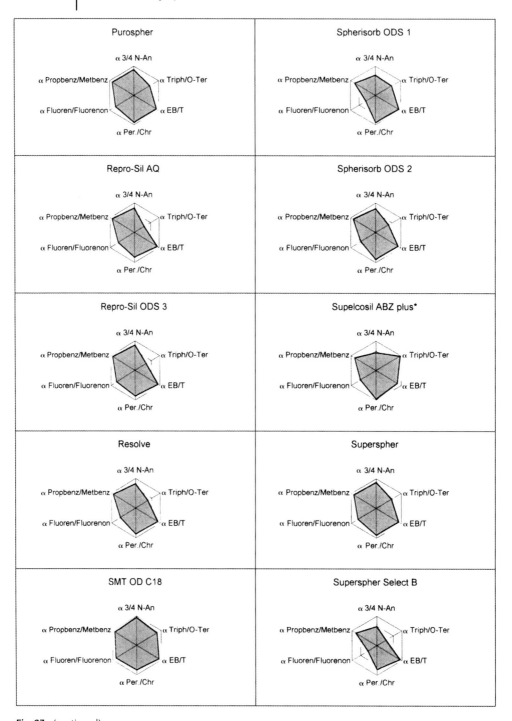

Fig. 27. (continued)

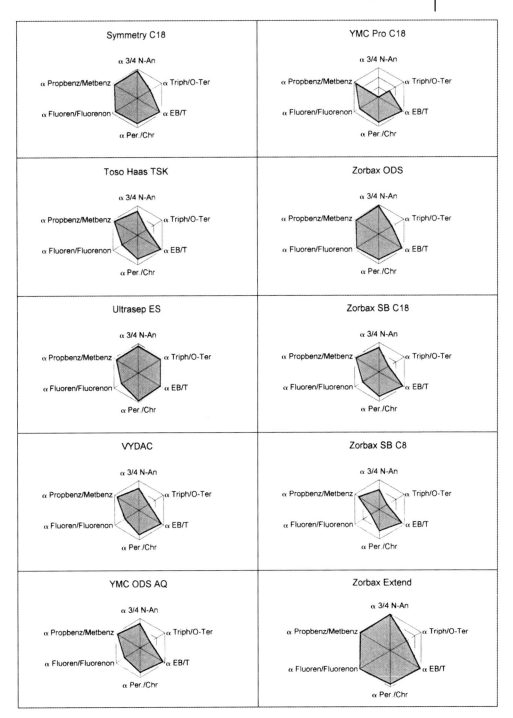

Fig. 27. (continued)

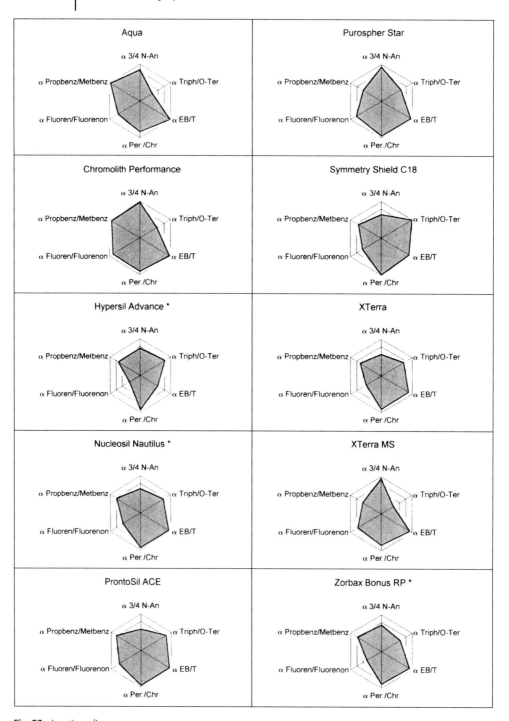

Fig. 27. (continued)

**Example 2: Neutral and Acidic Aromatic Compounds in Unbuffered Eluent/
in Acidic Phosphate Buffer**

Consider the hexagons in Fig. 28.

Results

- Prontosil AQ, which is more hydrophilic than Prontosil C_{18}, is better at separating the polar pair phthalic acid/terephthalic acid. Otherwise, these two phases are quite similar.
- Reprosil AQ and Prontosil AQ are nearly identical phases.
- The polar XTerra material (embedded phase) efficiently separates triphenylene/ o-terphenyl and phthalic acid/terephthalic acid, while the apolar XTerra MS (classical coverage) is more selective for 3-/4-nitroaniline and 3-hydroxy-/ 4-hydroxybenzoic acid.
- As is known, LiChrospher and Superspher are essentially identical, the only difference being in their particle size (5 µm compared to 4 µm). The hexagons are nearly identical. In the case of LiChrospher Select B and Superspher Select B, they are again nearly the same.
- The surface of the Zorbax SB material is sterically well protected because of the bulky diisopropyl groups. Therefore, it shows a lower selectivity in the case of the separation of triphenylene/o-terphenyl.

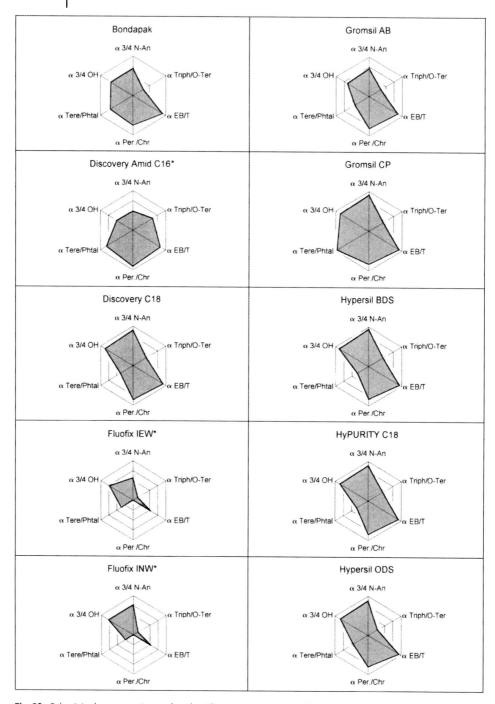

Fig. 28. Selectivity hexagons: "neutral and acidic aromatic compounds in unbuffered eluent/acidic phosphate buffer".

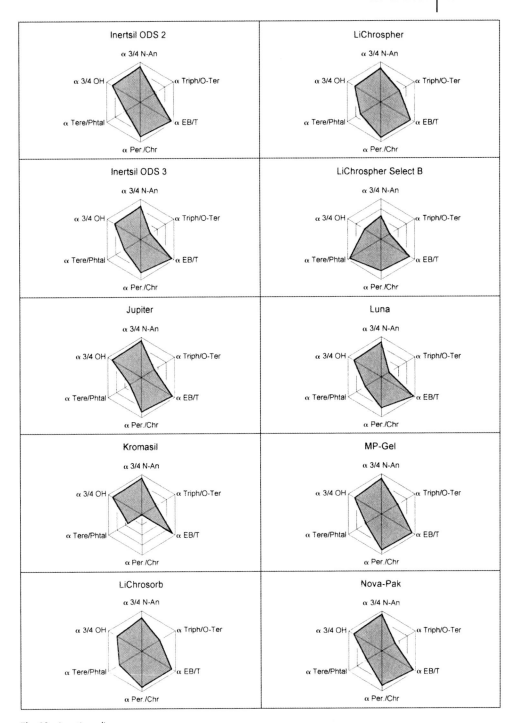

Fig. 28. (continued)

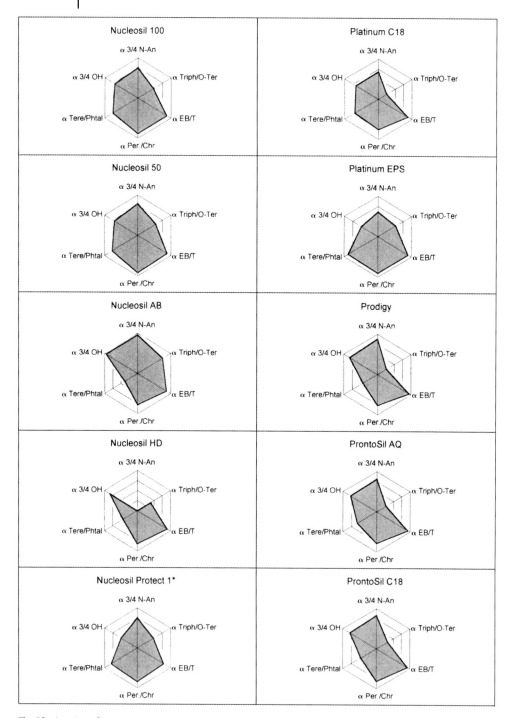

Fig. 28. (continued)

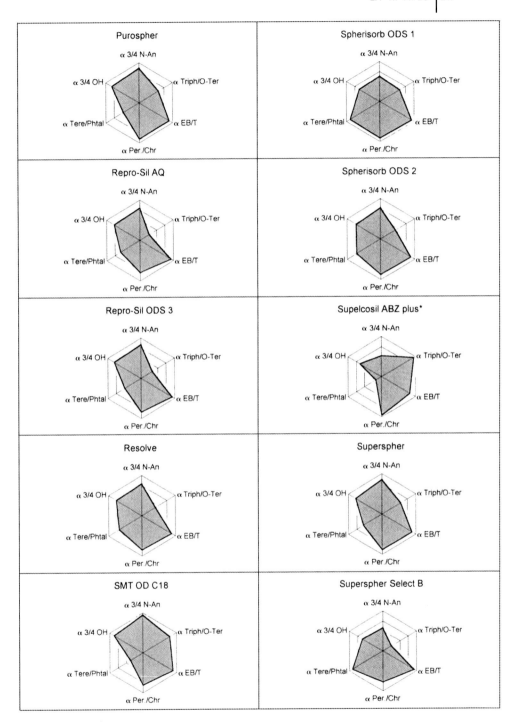

Fig. 28. (continued)

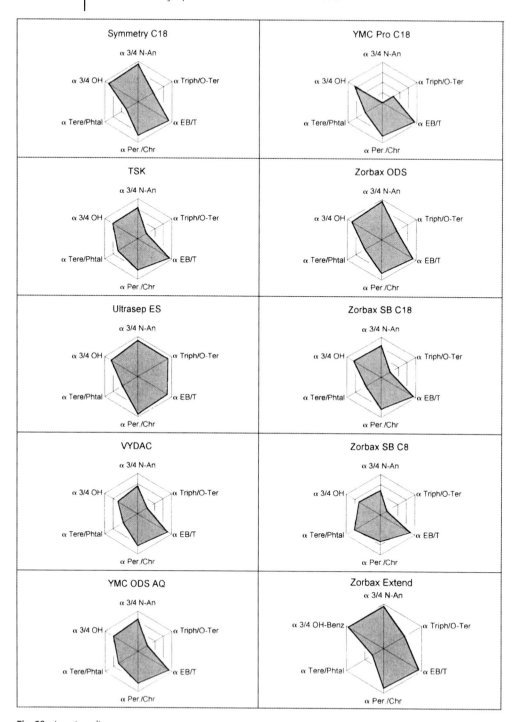

Fig. 28. (continued)

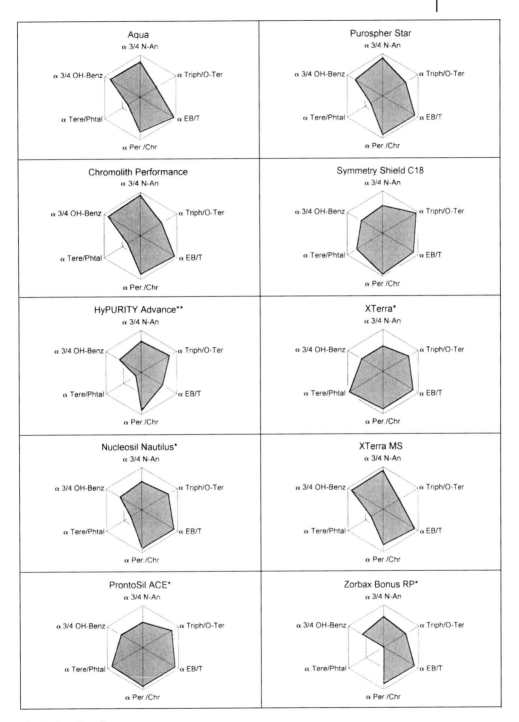

Fig. 28. (continued)

Example 3: Aromatic Acids in Unbuffered Eluent / in Acidic Phosphate Buffer

Consider the hexagons in Fig. 29. For the construction of these hexagons, very different pairs of analytes (e.g., hydrophobic/polar) and very different eluents (methanol/water, acidic buffer) were deliberately used. In this way, the selectivity of columns for different analytical problems may be demonstrated, i.e., the application of a column as a "universal" column for acidic aromatic compounds.

Results

- For good selectivity in separating planar/nonplanar molecules, polar groups are needed on the surface. However, merely the polar character of the stationary phase, as in the case of short C_8 alkyl chains, is not enough.
- Inertsil ODS 2 has a higher degree of coverage, but shows more polar character (polar groups/"impurities" on the surface?) than Inertsil ODS 3, which has a lower but "better" degree of coverage.
- Purospher e seems to be the "most polar" of common modern hydrophobic phases. It is marginally more apolar (in the absence of bases) than LiChrospher/Superspher.
- If the eluent used contains no more than ca. 90% water/buffer, Prontosil AQ is barely more polar than Prontosil, and Reprosil AQ barely more polar than Reprosil. As also shown by other experiments, these materials are very similar. Prontosil ACE is by far the most polar of the Prontosil family. YMC AQ, on the other hand, has different properties to YMC Pro C_{18}. This is indeed understandable, because the silica matrix is different.
- SynergiMAX RP and Luna 2 are both strongly hydrophobic RP materials.
- Symmetry, a classic material based on silica gel, is more selective in the separation of triphenylene/o-terphenyl than the hybrid material XTerra MS. The density of the silanol groups on the surface of the latter is obviously lower.
- Symmetry Shield is more hydrophobic than the low covered XTerra (see the values for isobutyl/isopropyl methyl ketone and 3-/4-nitroaniline), but, like Symmetry, has more surface silanol groups than XTerra.
- Zorbax Bonus is more hydrophobic than Nucleosil Nautilus, and this in turn is more hydrophobic than Prontosil ACE. The most polar embedded phase with a long alkyl chain is Symmetry Shield, with carbamate as the embedded polar group.
- The most polar embedded phases of all are such phases with even shorter alkyl chains, e.g. HyPURITY ADVANCE and SynergiPOLAR RP.

Some hexagons from Fig. 29 are depicted directly side-by-side in Fig. 30. In the left-hand column, the enhanced hydrophobicity of Discovery C_{18} compared to Discovery Amide C_{16} is clearly distinguishable: good selectivity in the separation of weaker acids (α-3,4-OH-benzoic acid). Discovery Amid C16 verhält sich genau umgekehrt. But a poor selectivity for more dissociated acids (α-HIES/VS). Symmetry Shield, which is a silica gel-based stationary phase, has a higher affinity for phenol than the less covered hybrid material XTerra (middle column). In the right-hand column, the transition from hydrophobic Inertsil ODS 2 (note the similarity with Discovery C_{18}) through Superspher to the quite polar Spherisorb ODS phases can clearly be recognized.

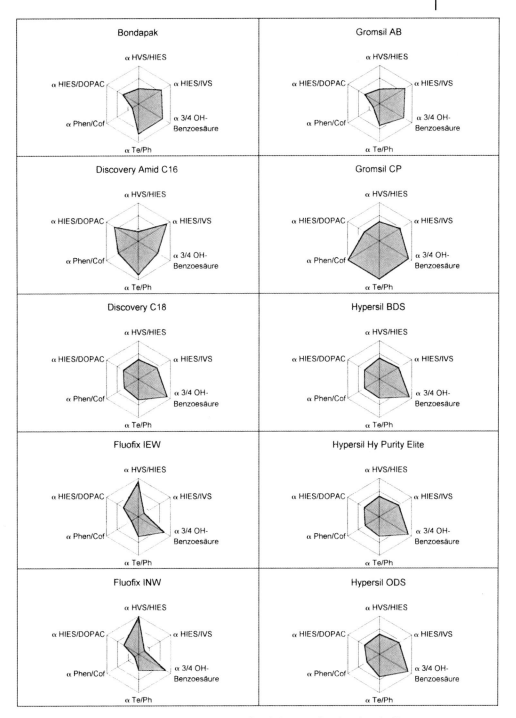

Fig. 29. Selectivity hexagons: "aromatic acids in unbuffered eluent/acidic phosphate buffer".

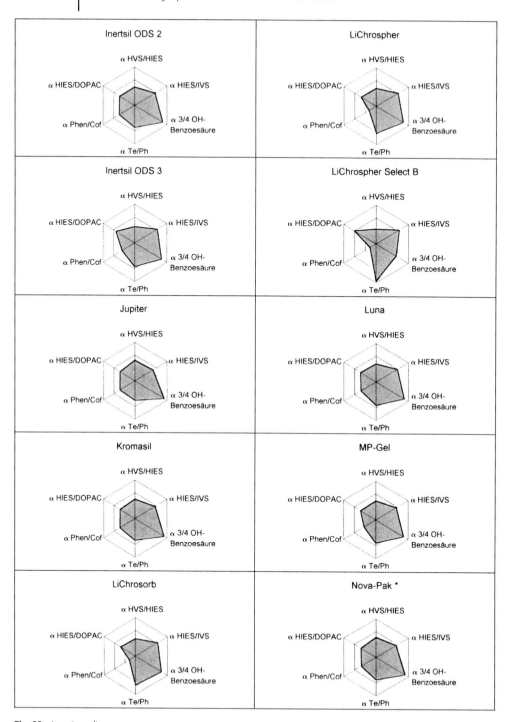

Fig. 29. (continued)

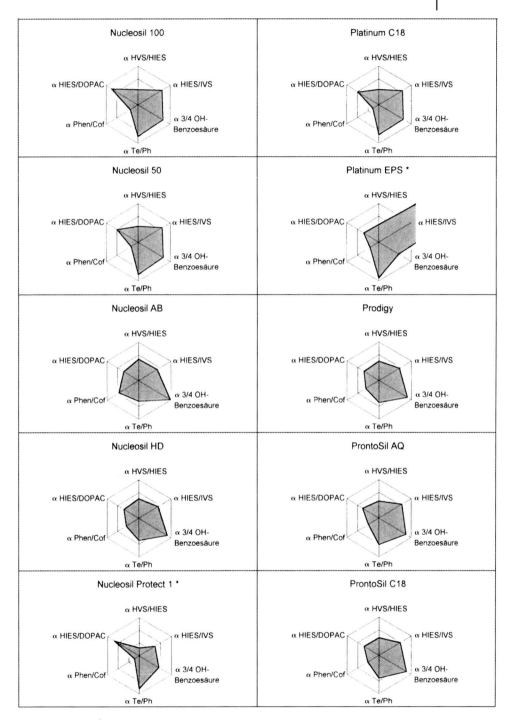

Fig. 29. (continued)

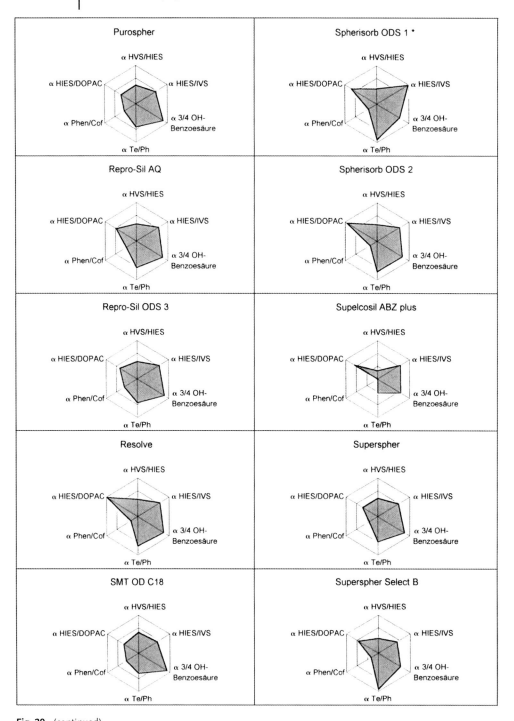

Fig. 29. (continued)

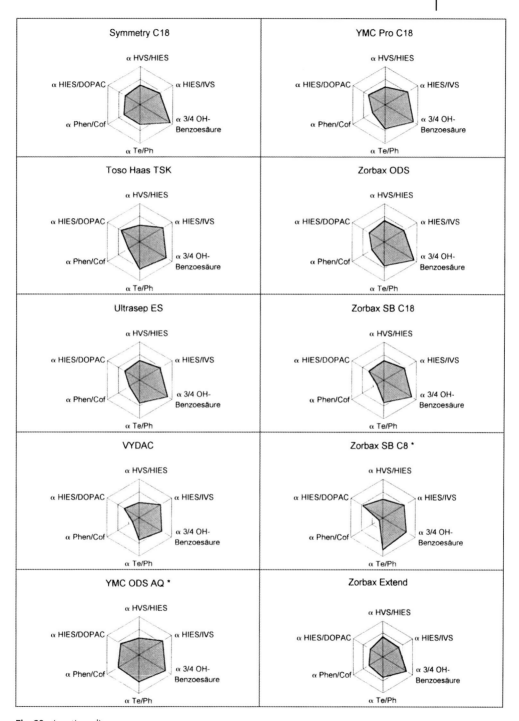

Fig. 29. (continued)

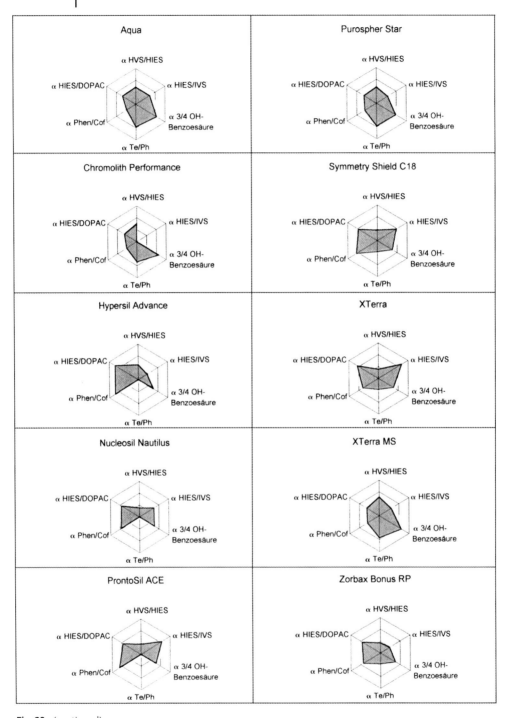

Fig. 29. (continued)

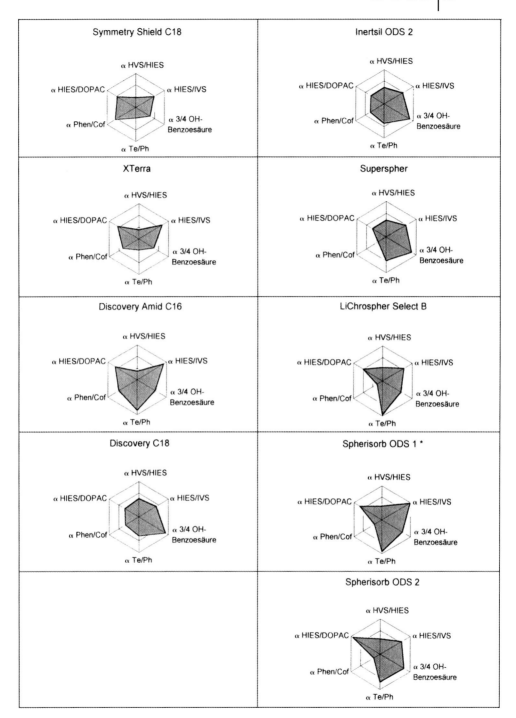

Fig. 30. Some selected hexagons from Fig. 29; for comments, see text.

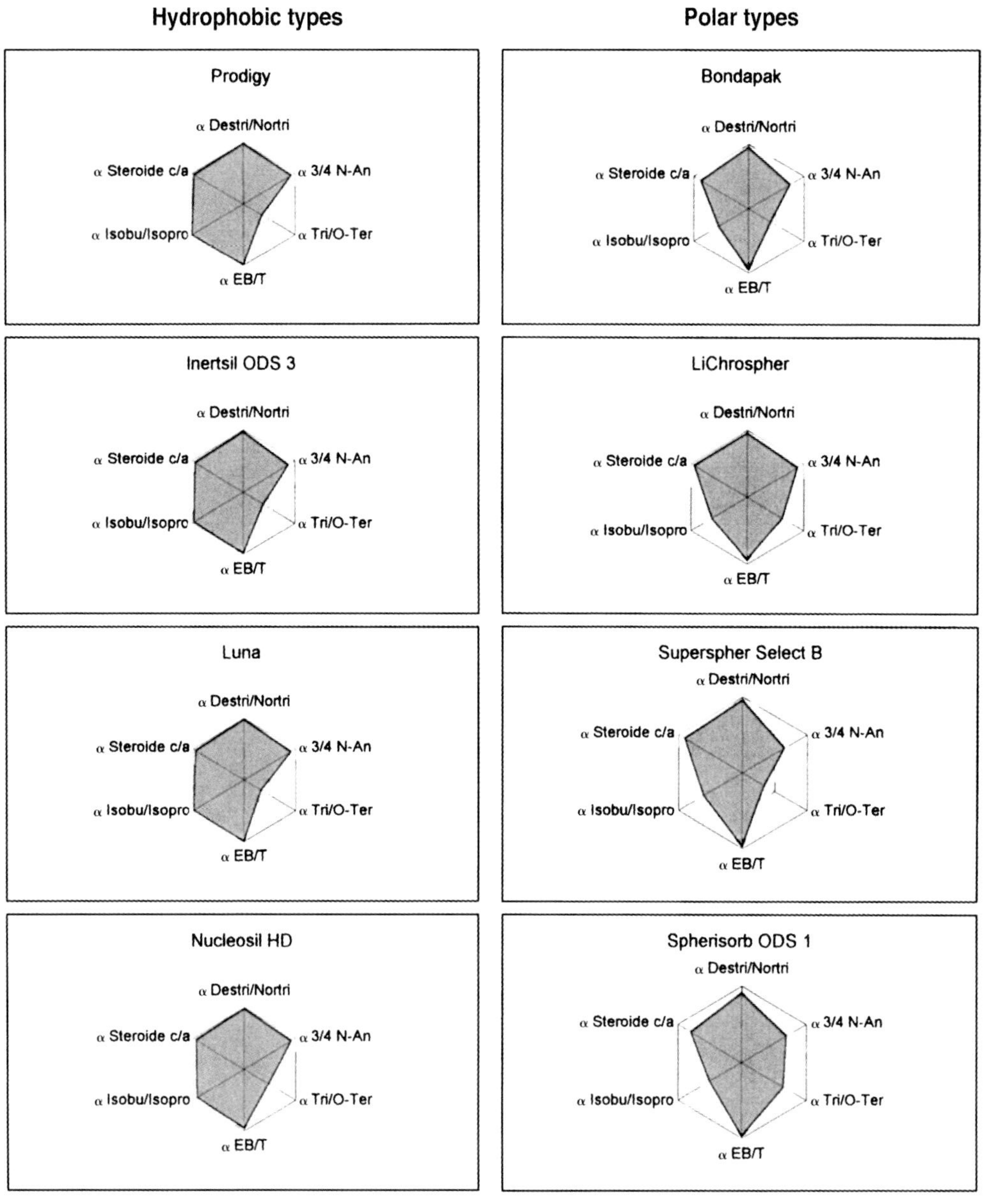

Hydrophobic types

Prodigy

Inertsil ODS 3

Luna

Nucleosil HD

Polar types

Bondapak

LiChrospher

Superspher Select B

Spherisorb ODS 1

Fig. 31. Hexagons referring to "hydrophobic" and "polar" RP-phase types.

Finally, in Fig. 31 one can see characteristic hexagons of the "hydrophobic" and "polar" types. Such pictograms can serve for a quick optical classification due to the selectivity behavior of the columns for given compounds under given separation conditions.

2.1.1.4.4 Chemometric Analysis of Chromatographic Data

Chemometric analysis is a tool that reveals the similarity of objects – in this case columns – by means of characteristic data (properties). Its principal advantage is the handling of a large number of data in a multidimensional space. In addition, it makes possible a weighting of the criteria (factors) that are used to test for similarity. The application of these tools in HPLC is described in detail in Chapters 2.1.3 and 2.1.4. In the following, some results of the author's own investigations are presented. Variables characterizing the different chromatographic properties of the phases were used.

A short description of the tools used can be found in the Appendix. For more details on the background of chemometrics/multivariate analysis, one should refer to the additional literature. First of all, a data analysis is briefly considered, which was kindly performed by Prof. Dr. Heinz Zwanziger, Merseburg, Germany.

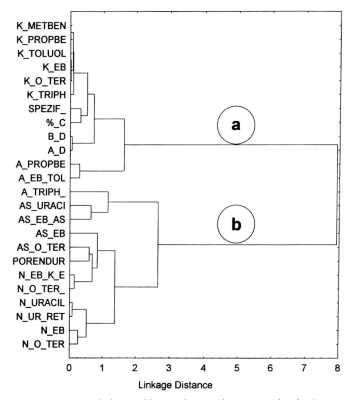

Fig. 32. Clusters with the variables "analytes", "chromatographic data" (i.e., retention factor) and "physico-chemical properties of the stationary phases"; for comments, see text.

Even if this analysis only takes the data of one experiment into account ("hydro-phobicity"), the results obtained show the expressiveness of chemometric tools.

The clustering of 24 variables of 30 columns is depicted in Fig. 32. Two clusters can be discerned. The first cluster ("a") includes k values, values for the specific surface area, and the carbon content. The other variables, e.g. the asymmetry factor (AS), the plate number (N), and the diameter of the pores are in the second cluster ("b").

Findings/Remarks

- The k value depends primarily on the specific surface area and on the carbon content.
- While the specific surface area and the carbon content are responsible for the retention behavior, these properties of the phases are not significant with respect to the selectivity, even in the case of "typical" RP analytes such as ethylbenzene/toluene, and for triphenylene/*o*-terphenyl ("shape selectivity") they are not relevant at all.
- The diameter of the pores is responsible for the symmetry of helically deformed structures such as *o*-terphenyl.
- There is not necessarily a correlation between plate number and asymmetry factor.

Cluster analysis of 30 columns further led to the following results:

- Hypersil BDS and Hypersil ODS react similarly, if no ionic interactions play a role.
- In addition, the similarity of Prontosil and Reprosil, as well as of Luna and Prodigy, is obvious.

In total, several thousand values were used for the chemometric analysis. Some revealing diagrams, which will be referred to below, are depicted in the Appendix).

Short Remarks

- It has become apparent that the similarity of columns depends on the basic parameter (parameter under consideration). For example, comparing the columns according to the criterion "similarity of retention properties" (see Appendix Fig. Chem. 1, retention factors), a different pattern is obtained compared to that obtained by comparing them according to the criterion "similarity of selectivity properties" (see Chem. 2, separation factors). If "only" the selectivity is in the foreground, dendrograms like Chem. 2 should be considered. If retention properties (run time) are an additional criterion, dendrograms like Chem. 3 should be used for comparison.

- In an alkaline environment, the phases are "more similar" with respect to selectivity. Although small differences in separation factors are obtained, the ramifications in the dendrograms start quite early. On the other hand, the differences are enhanced in a neutral or acidic eluent. The sequence is $\Delta \alpha_{acidic}$ $> \Delta \alpha_{neutral} > \Delta \alpha_{alkaline}$. Therefore, in an alkaline buffer the material of the column plays a less important role, and the choice of the column is of secondary importance. This applies for Chem. 5 and 7, Chem. 15 and 19, Chem. 23 and 22, as well as Chem. 26 and 25. Consider, for example, Platinum EPS and HyPURITY ADVANCE. Both are very polar phases, which is verified in a lot of experiments. Their exceptional position is clearly apparent in Chem. 4. There, the similarity of the columns under consideration of all variables is demonstrated: retention and separation factors in acidic/alkaline, methanol/acetonitrile-phosphate buffer. Working in neutral or alkaline as opposed to acidic media, the phases cannot show their "individuality", i.e. the phases under consideration all behave similarly and are members of a "family"; see Chem. 6. Nevertheless, a differentiation between HyPURITY ADVANCE and Platinum EPS is possible in neutral media; see Chem. 11. Platinum EPS is the only one of the phases denoted in Chem. 11 that bears a large number of polar groups on the surface. Its role of an outsider is obvious. Chem. 8 (similarity in acidic and neutral media) strongly resembles Chem. 4. This is an additional proof that the alkaline range does not show any variability; in other words: separation tests using buffers in the neutral/weak alkaline range are not suitable for making statements about the similarity of stationary phases.

- As a rule, the bandwidth of the selectivity in methanol is larger than that in acetonitrile; cf. Chem. 21 and 24. This means the following: to obtain the maximal spectrum of selectivity of a stationary phase, one should work in methanol/water or in a methanol/acidic buffer.

- It should be emphasized once more that different similarities are obtained in methanol vs. acetonitrile and in acidic vs. alkaline media. The following example illustrates this: Nucleosil HD and XTerra MS show similar behavior in acidic media; see Chem. 7. In neutral media, on the other hand, Nucleosil HD shows a strong similarity to Zorbax Extend (Chem. 9). This is apparent from the following: considering only separation factors as the criterion, a similarity is found (Chem. 9), and considering both separation and retention factors together, a similarity is also obtained (Chem. 10). This means that if two columns show similarity in a neutral eluent, then these columns do not necessarily have to show similarity in an acidic buffer as well. In practice, this could have the following consequence: let us assume that column A tested using eluent A is not suitable, and the same is true for column B. Therefore, a "negative" similarity exists. If one changes the eluent, e.g. eluent B (neutral or acidic), then it is worth testing both columns using the new eluent. If a similarity is also obtained using the second eluent (the criterion is the chromatogram), the two columns are probably very similar. This finding results in the following effective procedure for optimization. In the first step, one performs a separation with the combination column A and eluent A (e.g., a hydrophobic phase and a methanol/water eluent). In a second run, one uses a completely different combination, e.g. column B and eluent B (e.g., a polar phase and an acetonitrile/water eluent). If the number of peaks remains the same, as verification one could replace 5% acetonitrile by 5% tetrahydrofuran, or in a buffered eluent shift the pH by ± 0.5 pH units. If the number of peaks still remains constant, it can be safely assumed that there were no hidden peaks that went unnoticed. (For explanations of the "orthogonal" principle, see also Chapters 1.1 and 2.1.6).

- Biplots are very useful to delineate the weighting of different variables in the case of similarity considerations of objects. In our case, the corresponding question is: "which pair of analytes in particular is suitable to show the differences of phases as exactly as possible?". For this, see Chem. 13.

Results

1. Hypercarb and the two Fluofix columns show obvious differences to the other columns, which is not surprising because these types of phases are not "classical" RP-phases based on silica.
2. Expressive pairs of analytes are benzylamine/pyridine, 4,4'-/2,2'-bipyridyl, and *o*-terphenyl/toluene.

Consequently, for an effective comparison of columns, analytes should be used for which the separation/retention strongly depends on the properties of the phases that are particularly responsible for the differences between the phases. Such analytes are, for example, bases (benzylamine/pyridine), complexing agents (2,2'-bipyridyl) or even pairs of analytes with interactions that are based on different mechanisms, e.g. spatial size/van der Waals interactions (*o*-terphenyl/ethylbenzene).

In the following, some phases are presented that have proved to be quite similar under different conditions.

However, a comment on the expected clustering of two different types of phases is first necessary. In Chem. 24, two clusters are obvious. The upper cluster, Symmetry Shield to Nucleosil Nautilus, contains embedded phases. The lower cluster, XTerra MS to Chromolith Performance, contains phases with a classical coverage.

The results presented in this section are of course primarily valid for the investigated pairs of analytes and the experimental conditions. A transfer of the results to other analytes and eluents introduces uncertainty.

Similar phases:

- Nucleosil AB and Gromsil CP – two phases with a "polymer layer" (Chem. 12 and 17).
- Symmetry and Purospher e (Chem. 20) – Purospher also shows similarity to MP-Gel for specific separations. The latter two phases are highly covered phases, which also show polar character (Chem. 18).
- Prodigy and Inertsil ODS 2 show similarities (and to some extent resemble Purospher and MS-Gel), and have a certain polar character.
- SMT and Ultrasep ES – two phases of "polymer type".

- Prontosil and Reprosil.
- Prontosil AQ and Reprosil AQ, and secondary similarity to YMC AQ.
- Zorbax ODS and LiChrospher, and secondary similarity to Ultrasep ES (Chem. 14 and 20). These are highly covered phases with additional free silanol groups. Hypersil ODS, which also has acidic silanol groups, is not a member of this group because of its low coverage.
- Gromsil AB, Bondapak, Plantinum C_{18}, and Vydac (Chem. 14 and 16).
- HyPURITY C_{18}, Discovery C_{18}, and Jupiter (Chem. 16). The similarity is caused by the hydrophobic character and the quite large pore diameter. Working in neutral environments, the importance of an "extreme" pore diameter as a special property of the selectivity clearly increases, and therefore Novapak (60 Å) joins the three aforementioned columns in this case (Chem. 14).
- Zorbax Extend, Nucleosil HD, XTerra MS, and Purospher Star (Chem. 24).
- Hypercarb (Graphite), Nucleosil 50 (high coverage together with a small pore diameter), and Fluofix INW and IEW (short alkyl chains with fluorine atoms) are "oddities".

2.1.1.5   Suitability of RP-Phases for Special Types of Analytes and Proposals for the Choice of Columns

2.1.1.5.1   Polar and Hydrophobic RP-Phases

At the beginning of this section, it was explained that different interactions may dominate because of different chemistry of the phase surfaces and different analyte structures. Of course, these can be readily influenced by the composition or pH of the eluent. The resulting mechanism determines the selectivity of the actual chromatographic system and the kinetics, i.e., the peak form as well. It is clearly interesting which physico-chemical properties of the substances to be separated have the greatest chromatographic relevance. To test this, we performed a chemometric analysis of the following properties of 42 quite different substances: molar mass (MM), heat of formation (BW), dipole moment (DM), molecule area (FA), molecule volume (VO), heat of hydration (HE), log P value (LP), polarization (PO), and molar refraction (MR). It was then investigated which of these properties are the most important for chromatographic discrimination. It was found that the chromatographically most relevant properties are the dipole moment, the heat of formation, the hydration energy, and the log P value (see Figs. 33 and 34).

What do these results mean? The greater the difference in the corresponding numbers of the compounds, the more easily these compounds can be separated. For example, two substances with similar properties but different dipole moments and different log P values are probably easy to separate on polar phases, whereas a greater difference between the heats of formation of two compounds could indicate a good selectivity for their separation on hydrophobic phases.

In the following, "polar" and "apolar" RP-phases are presented in brief and thereafter their suitability for the separation of specific classes of substances is discussed. Finally, some proposals with regard to column choice are given.

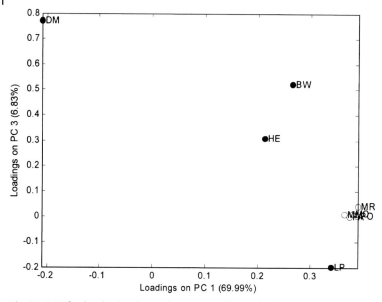

Fig. 33. PCA for the physico-chemical properties of several substances; for comments, see text.

Polar RP-phases

For various reasons, the following types of phases show a more or less polar character:

- non-endcapped, with metal ion contaminated phases
- polar (residual) groups on the matrix surface
- low coverage and therefore high total silanol group concentration
- embedded polar/ionic groups, zwitterions
- hydrophilic endcapped phases
- fluorinated phases
- short alkyl chains
- combinations, e.g. short alkyl chain + embedded polar groups + polar end groups + hydrophilic endcapping
- sterically protected phases
- classical polar phases, e.g. phenyl, nitrile, diol

Considering several criteria and a weighting, which need not be explained here in detail, the following phases can be denoted as some of the most polar types. These phases also show a low affinity for hydrophobic analytes and an enhanced selectivity for strong ionic pairs of analytes in buffered eluents.

Fluofix INW	Spherisorb ODS1	Acclaim PA C16
Fluofix IEW	Nucleosil Protect1	XBridge Shield
Platinum EPS	Supelcosil ABZ PLUS	Primesep C100
Hypersil ADVANCE	Zorbax SB C$_8$	
SynergiPOLAR RP	Zorbax Bonus	

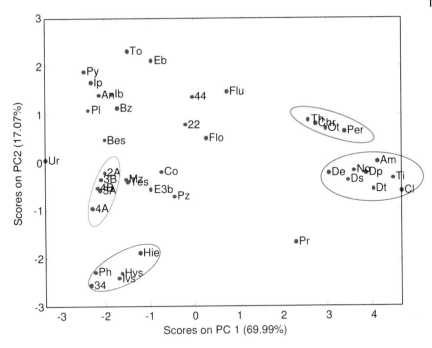

Name	Abb.
2,2'-Bipyridyl	22
2-Nitroaniline	2A
3,4-Dihydroxyphenylacetic acid	34
3-Hydroxybenzoic acid	3B
3-Nitroaniline	3A
4,4'-Bipyridyl	44
4-Hydroxybenzoic acid	4B
4-Nitroaniline	4A
5-Hydroxyindole-3-acetic acid	Hie
Amitriptyline	Am
Aniline	An
Benzoic acid	Bes
Benzylamine	Bz
Chrysene	Chr
Clomipramine	Cl
Caffeine	Co
Desipramine	Ds
Desmethyldoxepine	De
Desmethylmaprotyline	Dp
Desmethyltrimipramine	Dt
Ethyl 3-hydroxybenzoate	E3b

Name	Abb.
Ethylbenzene	Eb
Fluorene	Flu
Fluorenone	Flo
Homovanillic acid	Hvs
Isobutyl methyl ketone	Ib
Isohomovanillic acid	Ivs
Isopropyl methyl ketone	Ip
Methyl 3-hydroxybenzoate	Mz
Nortriptyline	No
o-Terphenyl	Ot
Perylene	Per
Phenol	Pl
Phthalic acid	Ph
Propranolol	Pr
Propyl 3-hydroxybenzoate	Pz
Pyridine	Py
Terephthalic acid	Tes
Toluene	To
Trimipramine	Ti
Triphenylene	Th
Uracil	Ur

Fig. 34. Score plot with the "variable" physico-chemical properties of several substances; for comments, see text.

Of course, RP-phases with a polar group such as CN or diol also show a strong polar character.

Comments

1. Hydrophilic endcapping ("AQ", "AQUA") imparts a phase with only slight polar character compared to the analogous phase without hydrophilic endcapping. Such phases represent the most polar types of hydrophobic phases.

2. Polar groups directly on the surface, as in the case of Spherisorb ODS1, Platinum EPS, and Supelcosil ABZ PLUS, and perhaps in combination with short alkyl chains as in the case of SynergiPOLAR RP, impart the strongest polar character, which is not possible by the introduction of a polar group into the alkyl chain (embedded phases).

3. The most important property for the differentiation of C_{18} phases is their capacity for polar/ionic interactions with polar analytes. If this property of differentiation is lost, then the phases behave ever more similarily. Therefore, Hypersil BDS and Hypersil ODS on the one hand, and Superspher, LiChrospher, and Purospher on the other hand, as well as Spherisorb ODS1 and Spherisorb ODS2, show quite similar separation properties in unbuffered systems and in the absence of basic analytes.

4. Low coverage together with "good" endcapping (e.g., Atlantis d C_{18}) or the presence of additional short chains with polar groups (e.g., Synergi FUSION RP) does not result in very polar character. Such phases occupy an interesting intermediate position.

Hydrophobic Phases

Hydrophobic character results largely from the following three factors:

1. High coverage. A metal-free matrix enhances the hydrophobic character; compare, for example, the highly covered but metal ion containing Zorbax ODS and LiChrospher with SunFire and SynergiMax RP on the one hand, and Ascentis C_{18} on the other.
2. Polymer layer on the surface, e.g. SMT OD, Nucleosil HD/AB.
3. Hydrophobic matrix, e.g. graphite or polymer.

Through numerous experiments, we could conclude that the following RP-phases show significant hydrophobic character:

Luna 2	YMC Pro C_{18} RS	Hypersil GOLD
SynergiMAX RP	Ascentis C_{18}	SMT OD
Hypersil BDS	Pursuit C_{18}	Gromsil CP
Symmetry C_{18}	SunFire C_{18}	Zorbax Extend
Nucleosil HD	XBridge C_{18}	

2.1.1.5.2 Suitability of RP-Phases for Different Classes of Substances

A series of experiments under different experimental conditions yielded the following results:

1. Organic, neutral analytes, "typical" RP separations

(e.g., small, mononuclear aromatic compounds, "methylene group selectivity", aldehydes, ketones, esters)

For these, apolar phases are suitable, although as a rule the differences in selectivity are rather small (see Fig. 1).

2. Large, hydrophobic, unsubstituted aromatic substances

(e.g., chrysene, perylene)

A good selectivity for large, hydrophobic, unsubstituted aromatic compounds is found primarily using phases with a polymer layer on the surface and polar phases; more precisely, with phases containing polar groups. Polar groups on the surface or on the alkyl chains (embedded phases) can interact with the π-electron system of the aromatic substances. A decrease of the hydrophobicity, e.g., by shortening of the alkyl chain (e.g., Zorbax SB C_8, LiChrospher Select B, Superspher Select B) or by a reduced carbon content (Bondapak, Platinum C_{18}) does not result in this case in selective interactions. Well-covered, hydrophobic phases occupy an intermediate position. Phases with specific steric properties, e.g. small or large pore diameters, show equally good aromatic selectivity.

3. Planar/nonplanar molecules, steroids

("steric selectivity", "shape selectivity")

Good selectivity is achieved using phases containing a polymer layer or phases containing polar groups; see Fig. 6.

4. Organic bases

Example: Small molecules, e.g. mononuclear, primary – tertiary amines

- The best selectivity is found in unbuffered methanol/water eluents, and the best peak symmetry in acetonitrile/acidic buffer eluents.
- An excellent peak form of benzylamine and propranolol is also found in unbuffered eluents when using modern phases ("pure" silica, good coverage of the surface).
- If one decides to opt for a buffer because of the robustness and the better peak symmetry and for the alkaline range because of the better selectivity, then acetonitrile seems more suitable than methanol. Here, non-endcapped phases show a better peak form compared to an unbuffered eluent, although not a satisfying one.
- A good peak symmetry is observed using hydrophobic, well-covered phases and embedded phases. Indeed, the chromatograms of different phases of this type ("uniform" surface) are quite similar.

- The more distinct the polar/ionic character of the analytes to be separated, the more selective are the phases that are suitable for such interactions. Therefore, phases with strongly acidic silanol groups have the best selectivity for bases. Spherisorb ODS1 and Resolve, for example, are able to discriminate strong bases even in an acidic buffer with methanol as organic solvent. In contrast, the bases elute inert or are even excluded in the case of acetonitrile. However, for good selectivity one often has to accept a stronger asymmetry.

In practice, the following procedure could be useful. One separates basic components as usual with acetonitrile/phosphate buffer and/or ion pair reagents. In the case of important (critical) peaks, the peak homogeneity can subsequently be checked using methanol/water on a polar phase. For this purpose, the (likely worse) peak form would be of secondary importance, but the reliability of the result would increase; see Fig. 16 in Chapter 1.1.

Example: "Large" organic bases, e.g. tricyclic antidepressants

If the selectivity is sufficient, acidic buffers and modern, well-covered phases are the first choice for the separation of organic bases. In the alkaline range, a better selectivity is frequently obtained at longer retention times.

Example: More polar bases, e.g. metabolites of tricyclic antidepressants

Hydrophobic phases are selective for the separation of less polar metabolites. Polar phases are selective for the separation of polar metabolites. This shall be demonstrated with three examples.

Using the modern, well-covered Inertsil ODS 3 (Fig. 35, top), polar metabolites of tricyclic antidepressants cannot be separated optimally, but using the older Inertsil ODS 2 (Fig. 35, bottom), which is probably "contaminated" with polar groups, they can (see also Fig. 36). Using the polar Discovery Amide C_{16} (top), a better selectivity is obtained than with the hydrophobic Discovery C_{18} (bottom). In the front of a chromatogram from a polar CN phase (see Fig. 37) there is enough space for polar components, but weakly hydrophobic analytes are only partially separated. Simplistically, one can state the following: one should use polar phases for polar analytes (early peaks) and apolar, hydrophobic phases for apolar analytes (later peaks).

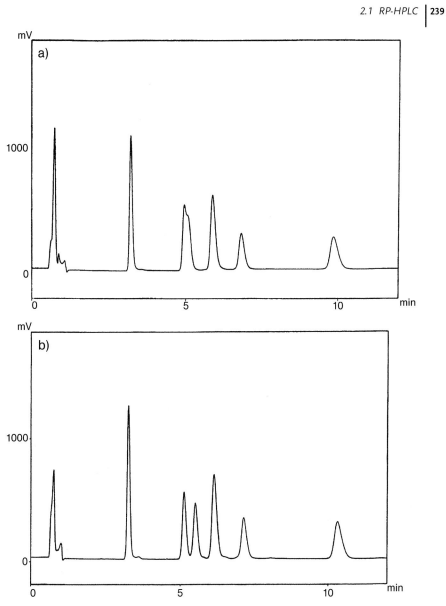

Fig. 35. On the selectivity of polar and hydrophobic RP-phases for the separation of polar and apolar compounds; for comments, see text.

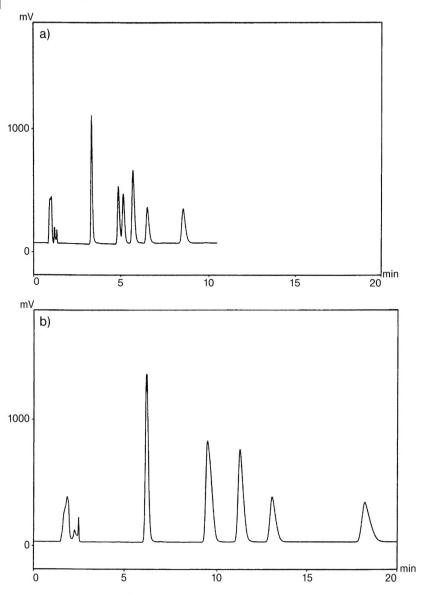

Fig. 36. On the selectivity of polar and hydrophobic RP-phases for the separation of polar and apolar compounds; for comments, see text.

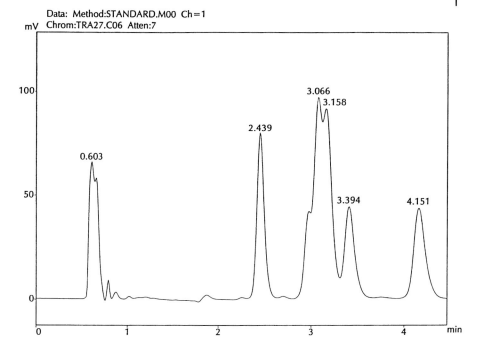

Fig. 37. On the selectivity of polar and hydrophobic RP-phases for the separation of polar and apolar compounds; for comments, see text.

5. Organic acids

In principle, the separation of ionizable components strongly depends on their dissociation state. The dissociation, in turn, depends on the pK_a/pK_b value of the acid/base and on the actual pH of the eluent (see Chapters 1.3 and 1.4).

In addition, the "polarity" of the substance plays a role. The log P value gives an indication of this. In this context, it should be noted that on the one hand nitrogen-containing components show remarkably low log P values compared to similar structures without nitrogen. On the other hand, the polarity of a substance depends on the degree of solvation, i.e., ultimately on the eluent.

Thus, the relationships are quite complex and we are only just beginning to understand them.

According to one proposal [5], the degree of ionization of an acid/base at a defined pH is a better measure of the "strength"/"weakness" of an acid/base than the pK_a or the log P value.

The degree of ionization can be calculated by the following equation:

$$\text{Degree of ionization (\%)} = \frac{100}{(1 + 10 \, e^{pK-pH})}$$

While the pK_a values of the acid compounds listed below differ only marginally, there are remarkable differences in the degrees of ionization:

Benzoic acid	3.1%
4-Hydroxybenzoic acid	1.3%
3-Hydroxybenzoic acid	4%
Homovanillic acid	2%
Isohomovanillic acid	2.1%
Terephthalic acid	14%
Phthalic acid	36%

Despite the use of an acidic buffer, the dissociation of "weaker" acids remains below 10%.

The first (simple) question is therefore: "is the dissociation of the acids in my sample higher or lower than 10%?". With values lower than 10%, the acids behave more like neutral organic compounds. Classical RP conditions exist in this case: hydrophobic, classical RP phases show the best selectivity (see Fig. 38) because in the acidic buffers typically used for such separations the polar differences of the phases recede into the background (see above) and their hydrophobicities come to the fore. This is true, for example, for the separation of 4-hydroxy-/3-hydroxy-benzoic acids. Both remain largely undissociated at pH 2.7.

Conversely, this means that polar/embedded phases are not very selective in this case. It should be briefly mentioned once more that methanol shows a better selectivity at a longer retention time and frequently a lower efficiency than acetonitrile. On the other hand, if strongly acidic components are dissociated, i.e. ionic, then phases with additional polar properties show a better selectivity.

Three examples of this are the following:

Consider Fig. 39: separating 4-hydroxy-/3-hydroxybenzoic acids, an insufficient selectivity is observed using Discovery C$_{16}$ Amide (upper chromatogram, peak 2). Using Discovery C$_{18}$, on the other hand, the selectivity is excellent (lower chromatogram).

In the case of the separation of phthalic/terephthalic acids, Discovery C$_{16}$ Amide shows a good selectivity (upper chromatogram in Fig. 40), but Discovery C$_{18}$ shows just sufficient selectivity (lower chromatogram in Fig. 40). Strong acids (eluted early in Fig. 41) are better separated using the polar phase Symmetry Shield C$_{18}$. Weaker acids, which react like neutral molecules and are eluted later, are better separated using the hydrophobic phase SynergiMAX RP. Strong acids are separated with difficulty on modern, hydrophobic RP-phases (five peaks on Symmetry Shield vs. four on SynergiMAX RP).

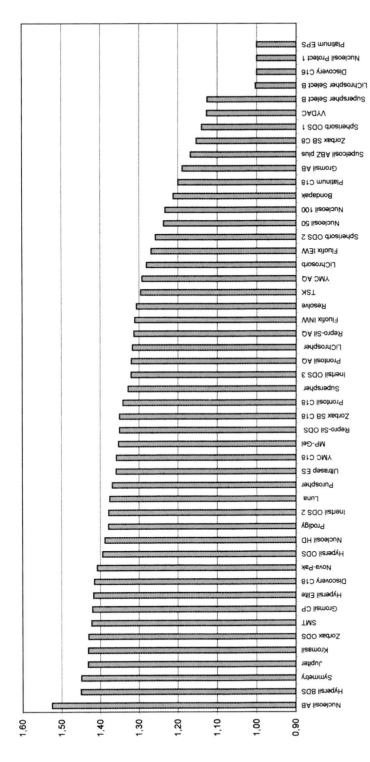

Fig. 38. Separation factors of 4-hydroxy-/3-hydroxybenzoic acid in a 40/60 (v/v) methanol/20 mmol phosphate buffer, pH 2.7.

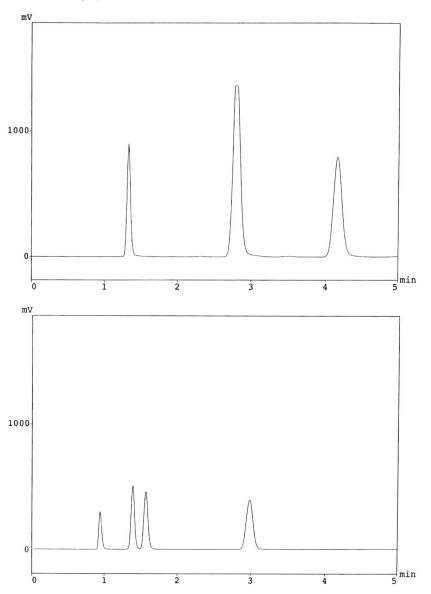

Fig. 39. Separation of 4-hydroxy-/3-hydroxybenzoic acid on a polar (upper chromatogram) and on a hydrophobic RP-phase (lower chromatogram); for comments, see text.

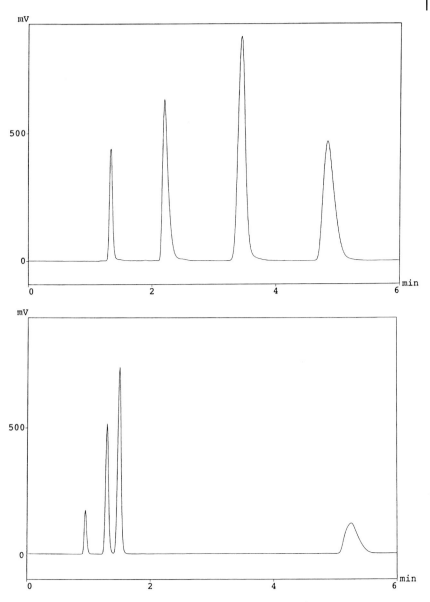

Fig. 40. Separation of phthalic/terephthalic acids on a polar (upper chromatogram) and on a hydrophobic RP-phase (lower chromatogram); for comments, see text.

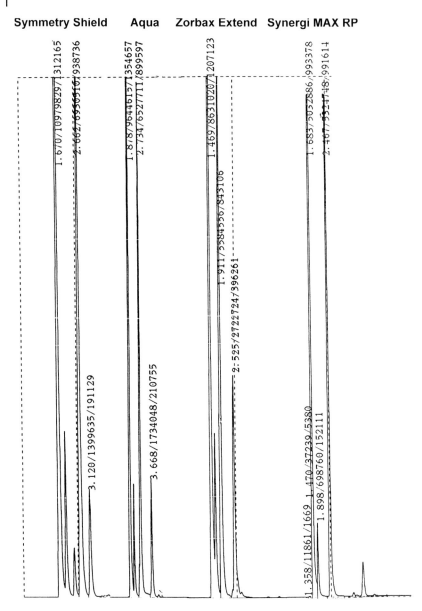

Fig. 41. Separation of weak and strong aromatic acids on various RP-phases; for comments, see text.

Conclusion

The following rules have been established from these investigations:

Suitability of polar phases

1. Unbuffered eluents
   - hydrophobic, unsubstituted "large" aromatic substances
   - planar/nonplanar aromatic substances
   - isomers (positional, double-bond isomerism)

2. Buffered eluents
   - basic substances; the poor peak form often overshadows the good selectivity
   - moderately strong (dissociated) acids
   - strongly polar metabolites, decomposition products, polar contaminants

Suitability of hydrophobic phases

1. Unbuffered eluents
   - polar to apolar, small, neutral organic molecules (aldehydes, hydroxy-benzoates, mononuclear aromatic substances)
   - compounds of different polarity; the difference of the polar character can also be caused by a group ($C=O$, CH_2, etc.) or by isomerism

2. Buffered eluents
   - weak acids (undissociated)
   - organic, weak bases

In summary, we can conclude that:

- Apolar aromatic substances can be selectively separated using polar phases. As the difference in the polar character of the analytes increases, apolar phases are needed.
- Polar substances can be efficiently separated in unbuffered eluents using apolar phases. This is also true for compounds that show a difference in their polar character because of isomerism or a substituent, e.g., OH or H vs. CH_3 or CH_2 vs. $C=O$.
- In buffered eluents, polar substances can be efficiently separated using polar phases and apolar compounds using apolar phases.
- In water-rich eluents, hydrophobic interactions influence the selectivity greatly; in organic solvent-rich eluents the importance of polar interactions (i.e., hydrogen bonding) increases.

In the case of an unbuffered eluent, a contrast between the character of the analyte and the stationary phase is necessary, while in the case of a buffered eluent a similarity is needed to obtain good selectivity.

Remark: "Suitability" in this context relates only to the selectivity and not to the peak symmetry. Of course, when separating neutral or, through the pH, neutralized molecules on a routine basis, an RP-phase with a hydrophobic, well-covered, homogeneous surface is advantageous, if it offers selectivity. In the following cases, the suitability should always be critically assessed and the selectivity should be checked by "cross-experiments" using polar phases (orthogonal tests):

- dissociated acids and bases
- strongly polar metabolites and degradation products
- "difficult" isomers
- polynuclear hydrophobic aromatic substances (limited)
- planar/non-planar molecules

2.1.1.5.3 Procedure for the Choice of an RP-Column

With regard to the choice of a column, several factors play a role, of course, such as price, stability in routine work, "charge-to-charge" reproducibility, availability in the event of transfer of the method, possible different column sizes, if necessary unproblematic scale-up, reliability of suppliers, etc. Here, we should consider first of all the principal question of selectivity.

"Universal Column?"

Of course, a "universal column" does not exist, and will probably never be realized. However, one may speculate about which properties that a fictitious, ideal RP-column with "universal" selectivity would need to have.

Very simplistically, one would need any type of modern LiChrospher or Zorbax ODS based on an alkali-resistant silica gel in two versions, 50 and 300 Å.

Necessary features would be:

- An average degree of coverage of ca. 2.5 mol m^{-2} to obtain sufficient retention.
- Polar groups, to facilitate selective interactions with polar compounds.
- A smaller (e.g., 50–80 Å) or larger (e.g., 180–300 Å) pore diameter to facilitate a possibly necessary steric effect.

Explanations

As noted above, the following three factors are relevant for selectivity: hydrophobicity, silanophilicity, and steric aspects. A phase possessing these three properties would have the necessary pre-conditions for a separation of nearly all types of analytes in RP-HPLC.

- "modern": synthetic silica gel with a low metal ion content; good "charge-to-charge" reproducibility possible;
- "alkali-resistant": if necessary, separation in alkaline eluent is possible;
- 50, 300 Å: provision for any necessary steric aspects (steric selectivity);
- column such as LiChrospher, Zorbax ODS:
  - highly covered phases: hydrophobic interactions possible
  - non-endcapped materials: ion-exchange interactions possible
  - high total silanol group concentration: polar interactions such as hydrogen-bonding and dipole-dipole interactions possible.

The interactions would be individually adjusted for each class of compound by the eluent, e.g. +5% tetrahydrofuran, n-butanol, isopropanol, etc., for neutral components, +/−0.5 pH units for ionic components. The final step of optimization under special consideration of the run time would involve a change of the gradient (starting and end percentages of B, steepness, gradient volume) and the temperature.

Because no single column is likely to possess all of the relevant properties for "universal" selectivity, one should consider the following three main types of RP-phases for the assembly of an individual column tool kit for use with unknown samples:

- "polar" phases – these are phases with quite a low degree of coverage and/or additional polar functionalities, e.g., Spherisorb ODS1, Atlantis d C_{18}, Platinum EPS, SynergiPOLAR RP, etc.
- "moderately polar" phases with sufficient hydrophobic as well as polar character, e.g., LiChrospher, Nucleosil 100, SynergiFUSION RP, Acclaim PA, etc.
- "hydrophobic" phases – these are phases with decidedly hydrophobic character, imparted by a high degree of coverage, and are mostly free from polar groups, e.g., Discovery C_{18}, Symmetry C_{18}, SynergiMAX RP, Hypersil GOLD, YMC Pro C18 RS, etc.

It would be optimal to have in addition one phase with a small pore diameter (e.g., NovaPak, Nucleosil 50) and one with a large pore diameter (e.g., Zorbax SB 300, Jupiter). If one anticipates a need for separations in alkaline media, one should consider XBridge/XTerra, Gemini, Zorbax Extend, Pathfinder, or perhaps also a Hypercarb, a zirconium dioxide or a polymer phase.

One may wish to remain with one column supplier and to select a personal tool kit of five columns. With these five columns, a test run should be performed overnight with the unknown sample to be separated with the help of a column selecting valve. For this, one may use overall gradients at several pH values, different gradient volumes, and, if necessary, different modifiers. Indeed, most of the major column suppliers offer a wide range of stationary phases to tailor an individual column kit. Alternatively, one can rely on fully complete column kits (method development kit). If a special type of stationary phase is lacking in the offer of one's regular supplier, one can/should of course use a column of another supplier. In the following, such column combinations are listed as a possible representative kit for some suppliers. It should be emphasized that this is by no means a recommendation, but merely an example of columns with different properties. Alternative columns and any specific columns are given in brackets.

- **Phenomenex**
  SynergiMAX RP/SynergiHydro, Jupiter, SynergiFUSION RP, SynergiPOLAR RP, Luna Phenyl Hexyl (LiChrospher, in alkaline pH: Gemini)

- **Waters**
  SunFire, Symmetry C_{18}, NovaPak, Symmetry Shield, Atlantis d (Resolve, in alkaline pH: XTerra/XTerra MS/XBridge)

- **Macherey-Nagel**
  Nucleosil AB/HD, Nucleodur C18 Gravity, Nucleosil 100, Nucleosil C_{18} Pyramid, Nucleodur Sphinx (Nucleosil Protect 1, where necessary: Nucleosil 50)

- **Agilent**
  Zorbax SB C_{18}/300, Zorbax Eclipse XDB, Zorbax ODS, Zorbax SB C_8, Zorbax Bonus (Zorbax SB AQ, in alkaline pH: Zorbax Extend)

- **Supelco**
  Discovery C_{18}, Discovery Amide C_{16}, Ascentis C_{18}, Ascentis RP-Amide, Supelcosil ABZ PLUS/Discovery HS F5

- **Varian**
  Pursuit C_{18}, Polaris C_{18} A, Polaris C_8-Ether, Polaris C_{18} Amide, Polaris NH_2 (MetaSil Basic, MetaSil AQ)

- **YMC**
  YMC-Pro C_{18} RS, YMC-Pro C_{18}, YMC AQ, Hydrosphere C_{18}, YMCbasic

Almost all of the above columns are alkyl phases, i.e. C_{18}/C_8. For very polar compounds, one should also consider CN, phenyl or diol phases or even silica. In combination with RP eluents, the latter often yields very interesting selectivities. At last, stationary phases with zwitterion functionality such as ZIC-HILIC or, in the alkaline region, ZIC-pHILIC, are well suited for separating strongly polar compounds.

In the following two tables, principle recommendations for the choice of columns are given, in case one would like to use columns of different suppliers.

As explained in Chapter 1.1, a column selecting valve, perhaps together with a computer program for optimization, is a very efficient tool for the development of methods or the optimization of poor separations. A six-way valve seems to be the most popular version. To get an idea of which six columns should be preferred for the first experiments, see Tables 5 and 6. In Table 5, phases are listed that represent different retention mechanisms in RP-HPLC. Table 6 shows a broader spectrum of hydrophobic to polar RP-phases, with some typical examples for each case.

The listed items are not recommendations for the specific quoted columns, but examples of the corresponding types of phases.

Remarks About Judicious Column Choice

The assembly of a column tool kit depends on the actual analytical problem, of course. Therefore, the number of polar phases in Table 5, for example, can be reduced for the benefit of additional hydrophobic phases.

If it is known, for example, that if the sample consists of small organic molecules, polar phases can easily be waived. On the other hand, if it is known that the sample contains components of similar polarity but different structure, classical, hydrophobic RP materials will be less useful. In this case, it is better to test a non-endcapped or any "polar" RP-C_{18}-phase, considering several types of "polar" phases, e.g. Develosil (C_{30} chain *plus* hydrophilic endcapping). If the sample contains

Table 5. Different types of RP-phases I.

Selectivity feature	Examples of columns	Column characteristics
Steric aspect	Jupiter Nucleosil 50	Large pore diameter Small pore diameter
Hydrophobic surface	YMC-Pro C18 RS Gromsil CP	"Classical" coverage Polysiloxane layer
Free silanol groups	Spherisorb ODS 1 Zorbax ODS	Acidic silanol groups and low coverage Acidic silanol groups and high coverage
Polar groups in the alkyl chain (embedded phases)	HyPURITY ADVANCE Prontosil ACE	"Polar embedded phase" "Hydrophobic embedded phase"
Strong polar surface	SynergiPOLAR RP Platinum EPS	Short alkyl chain, embedded polar group, polar terminal phenyl group, hydrophilic endcapping Polar groups on the surface
Free selection as needed	e.g. Hypercarb Fluofix IEW/INW Gemini/XBridge ZirChrom-MS Primesep A Acclaim PA ZIC-HILIC Pathfinder	 Very hydrophobic Very polar fluorine atoms Alkaline-stable Other matrix chemistry Ionic embedded group Hydrophobic and polar character Alkyl chain with zwitterions Polymer layer, temperature-stable

Table 6. Different types of RP-phases II.

Hydrophobic phases (including steric aspects in some cases)	"Polar" phases from the hydrophobe group	"Hydrophobic phases" from the polar group	Polar phases
Luna 2 (Inertsil ODS 3, Kromasil)	Purospher	Spherisorb ODS 2	HyPURITY Advance
SMT (Gromsil CP, Nucleosil HD)	Synergi Fusion RP	Platinum C_{18}	SynergiPOLAR RP
SynergiMAX RP (Zorbax Extend, XTerra MS)	Acclaim PA C18	Nucleosil Nautilus	XBridge Shield
Nucleosil 50 (Novapak)	Zorbax ODS	Symmetry Shield	Acclaim PA C16
Zorbax SB 300 (Nucleodur C18 Gravity)	Reprosil AQ	Zorbax Bonus	Fluofix INW
Discovery C_{18} (YMC Pro, Hypersil BDS)	LiChrospher	Polaris C_{18}-Ether	Platinum EPS

alkaline compounds, one could take directly for the first experiment three hydrophobic, two embedded phases, and one polar phase as "watcher" (see Fig. 19 in Chapter 1.1).

The more is known about the sample, the better directed the selection of columns will be, of course. If no information is available, which should rarely be the case, it would be informative to test one column of each of the different types.

A 12-way valve could, for example, be loaded in the following way:

- endcapped/non-endcapped
- Si 60 Å/Si 300 Å
- sterically or chemically protected/hydrophilic endcapped ("SB"/"embedded", "AQ")
- ca. 8% C/ca. 20% C
- diol/phenyl
- amine/nitrile

Finally, some inherent rules should be briefly summarized:

- Hydrophobic, modern, well-covered phases are suitable for neutral molecules, for molecules that can be neutralized by adjusting the pH, and for molecules that show a distinct polar character because of different substitution or isomerism.
  Remember: Hydrophobic phases are unproblematic and should always be the first choice, especially for a routine method, after the selectivity has been checked by a "cross"-experiment.

- Hydrophobic phases are also useful for the separation of weakly basic components, as well as for separating strong bases that have been neutralized/"masked" by ion-pairing reagents. For these, one can always expect good peak symmetry. Frequently, sufficient selectivity for basic components with distinct organic character is obtained even in acidic media. Polar phases are more selective but yield lower peak symmetry if basic components are present in the eluent in protonated form, and if these components are small molecules such as mononuclear amines.
  Remember: Good selectivity *and* fast kinetics are frequently mutually unobtainable if different mechanisms are operative.

- Phases possessing polar/ionic/ionizable groups are selective if the analytes that have to be separated are in an ionic state. Therefore, rather polar phases are of interest for acidic components with a high degree of ionization.
  Remember: For "difficult" separations, make polar interactions available.

- It is advisable to always consider steric aspects. For this, one should test, e.g., a hydrophobic phase with a large pore diameter (Jupiter, Zorbax SB 300, Discovery C_{18}, Hypersil BDS) as well as a phase with hydrophobic and/or polar properties and a small pore diameter, e.g., Nucleosil 50, Novapak, Superspher Select B.
  Remember: Even in the case of "small" molecules, the selectivity can be influenced by hindered/unhindered diffusion in different sized pores.

- If polar *and* apolar interactions are necessary for separation (e.g., components with amino and acidic groups), phases may be encountered that show sufficient hydrophobicity but which are also capable of polar interactions because of additional polar groups on the surface, e.g., Spherisorb ODS 2, LiChrospher, Polaris C18-A, Ascentis RP-Amide, Purosphere, Uptisphere UP5MM, Nucleodur Sphinx RP, Atlantis d C_{18}, SynergiFUSION RP, and Acclaim PA. The "fine adjusting" for a better selectivity can be most effectively achieved by means of the pH of the eluent *and* by variation of the acid/base used [8]. A possible lack of robustness of the method has to be accepted in such systems.

  Remember: When separating polar components, good selectivity is frequently associated with poor robustness.

References

1   J. J. Gilroy, J. W. Dolan, L. R. Snyder,
     J. Chromatogr. A 1026 (2004), 77.
2   M. R. Euerby, P. Petersson,
     J. Chromatogr. A 994 (2003), 13.
3   U. D. Neue, "Column Technology",
     in Encyclopedia of Analytical Science,
     P. Worsfold, A. Townshend, C. Poole
     (Eds.), Vol. 5, pp. 118–125, Academic
     Press, Amsterdam, 2005.

4   C. Stella, lecture presented at
     "HPLC 2003", Nice.
5   S. Kromidas, "Eigenschaften von
     kommerziellen C_{18} Säulen im Ver-
     gleich", Pirrot Verlag, Saarbrücken, 2002.
6   H. Zobel, GIT Spezial Separation, 2000.
7   S. Kromidas, U. Panne, unpubl. results.
8   S. Kromidas, "More Practical Problem
     Solving in HPLC", Wiley-VCH,
     Weinheim, 2005.

2.1.2
Column Selectivity in RP-Chromatography

Uwe D. Neue, Bonnie A. Alden, and Pamela C. Iraneta

In this section, we explain the principles that influence the selectivity of a reversed-phase column. The first parameter is the hydrophobicity of the stationary phase, which can be measured with purely hydrophobic probes. The second value describes the silanol activity, which is of special importance for basic analytes. Naturally, the silanol activity is best measured using a basic compound, using a correction for the hydrophobic contribution of the structure of the analyte to its retention. The third value is the polar selectivity, which measures the formation of hydrogen bridges between analytes and the stationary phase. With these values, one can create selectivity charts that can help in the selection of the best stationary phase for a particular separation problem. These charts help in method development, whether one would like to find a stationary phase that is drastically different or one that is rather similar to one that is available or in use. At the end of the chapter, we briefly touch on the subject of stationary phase reproducibility.

2.1.2.1 Introduction
When one develops new reversed-phase (RP)-HPLC methods, one usually uses the selectivity of the mobile phase as the primary method development tool. The chromatographic separation can be influenced by the choice of the organic solvent (mainly methanol and acetonitrile), or by variation of pH or buffer type. Schemes for method development using these parameters have been described in the literature [1, 2]. Most important are the selectivity changes caused by pH changes, which are well-understood and easily predictable [3]. It is well known that the stationary phase influences the selectivity as well, but this effect is often not very well understood. The primary reason for this is the fact that reliable methods for the description of the stationary phase selectivity have only become available fairly recently. In the last few years, several papers have been published that deal with the subject of selectivity in a fundamental way [4–9] or represent a data collection based on older methods [10–15]. In this chapter, we describe in detail the method used in our laboratory. We then look at our selectivity charts and discuss our results. It needs to be pointed out in advance that selectivity charts only accurately represent the properties of a stationary phase under the conditions of the measurement. If we depart from the mobile phase composition of the test, the relationships between different columns will change, since selectivity arises from a combined effect of the mobile phase and the stationary phase.

2.1.2.2 **Main Section**

2.1.2.2.1 **Hydrophobicity and Silanol Activity (Ion Exchange)**

Reversed-phase packings have several properties that determine their interactions with sample molecules. The first property is the hydrophobicity of the stationary phase. This property is best determined by using analytes without any polar groups, i.e. purely hydrophobic compounds. This means that useful candidates are hydrocarbons, preferentially benzene derivatives with aliphatic side chains, or simple aromatic compounds. We have chosen naphthalene and acenaphthene as hydrophobic reference compounds in our method. We have shown that the logarithms of the retention factors of both compounds correlate linearly, if one compares a large number of stationary phases [13]. This means that the hydrophobicity of a phase can be determined simply by using the retention factor of acenaphthene. In the literature, the methylene group selectivity is often used as a measure of the hydrophobicity of a phase. However, this value specifically excludes the contribution to the retention of the specific surface area of a packing. Therefore, we believe that the retention factor is a better criterion for hydrophobicity, and we use the retention factor as our measure of hydrophobicity.

Most stationary phases used in RP-chromatography are based on silica or closely related hybrid packings. Such stationary phases contain silanol groups, which can influence the retention of basic analytes. The strength of this effect depends on the mobile phase, as well as on the stationary phase and the analyte itself. The pH of the mobile phase has the largest influence. Under acidic conditions, silanols are protonated, and do not interact at all or interact only weakly with basic analytes. Under conditions of neutral pH, the silanols are negatively charged and interact with positively charged basic analytes by way of an ion-exchange mechanism. This means that the pK_a of the base plays an important role, since uncharged analytes are not subject to this interaction. At neutral pH, more than half of the silanol groups on a packing are ionized. Complete ionization is only possible at alkaline pH, but these conditions are unsuitable for a column test since silica-based bonded phases are not stable at alkaline pH values. For these reasons, we selected pH 7.0 as our reference value for the measurement of the silanol activity of a packing. We chose potassium phosphate at a concentration of 30 mM as our buffer [11]. The tailing of basic analytes depends on the nature of the probe. The differences are most likely caused by the steric accessibility of the basic function. For this reason, we selected specific compounds that are known to exhibit significant tailing at pH 7, which indicates a strong interaction of the analyte with the surface silanol groups. We selected tricyclic antidepressants, such as amitriptyline, and other strong bases such as propranolol. Today, one finds that tricyclic antidepressants are used in many column tests. The silanol activity is the desired measure, and we determine this from the retention factor of the base amitriptyline after correction of this value for the hydrophobic contribution to retention [13]:

$$S = \ln(k_{\text{amitriptyline}}) - 0.7124 \times \ln(k_{\text{acenaphthene}}) + 1.9748 \qquad (1)$$

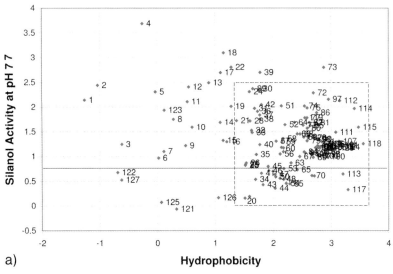

a)

Hydrophobicity

Fig. 1. Plot of silanol activity versus hydrophobicity of the stationary phase. *Designations:* (1) Nova-Pak CN HP, (2) Waters Spherisorb CN RP, (3) Hypersil CPS CN, (4) Waters Spherisorb Phenyl, (5) Keystone Fluofix 120N, (6) YMC-Pack CN, (7) Ultra PFP, (8) Zorbax SB-CN, (9) Hypersil BDS Phenyl, (10) Inertsil 3 CN, (11) Fluophase RP, (12) Hypersil Phenyl, (13) Zorbax SB-Aq, (14) YMC-Pack Ph, (15) YMC Basic, (16) Ultra Phenyl, (17) Inertsil Ph3, (18) Platinum EPS C_{18}, (19) Synergi Polar-RP, (20) XTerra RP_8, (21) Nova-Pak Phenyl, (22) Zorbax SB-Phenyl, (24) Zorbax Rx C_8, (25) XTerra MS C_8, (26) Prodigy C_8, (27) Zorbax Eclipse XBD Phenyl, (29) Zorbax SB C_8, (30) µBondapak C_{18}, (31) YMC J'Sphere L80, (32) Supelcosil LC DB-C_8, (33) ZirChrom PBD, (34) Discovery RP Amide C_{16}, (35) Hypersil BDS C_8, (36) HydroBond AQ, (37) Lichrospher Select B, (38) Allure Ultra IBD, (39) Platinum C_{18}, (40) Nova-Pak C_8, (41) Capcell Pak C_{18}, (42) Alltima C_8, (43) Discovery RP AmideC_{16}, (44) XTerra RP_{18}, (45) Symmetry300 C_{18}, (46) Spectrum, (47) Zorbax Bonus RP, (48) Supelcosil LC-ABZ Plus, (50) SymmetryShield RP_8, (51) Lichrosorb Select B, (52) PolyEncap A, (53) Prism, (54) Supelcosil LC-ABZ+, (55) Supelcosil LC-ABZ, (56) Luna C_8(2), (57) Inertsil C_8, (58) Kromasil C_8, (59) Zorbax Eclipse XDB C_8, (60) Symmetry C_8, (63) Hypersil HyPurity Elite C_{18}, (64) Hypersil ODS, (65) Polaris C_{18}-A, (66) Luna Phenyl-Hexyl, (67) Hypersil BDS C_{18}, (68) Supelcosil LC DB-C_{18}, (69) Aqua C_{18}, (70) SymmetryShield RP_{18}, (72) Nucleosil C_{18}, (73) Waters Spherisorb ODS-2, (74) Waters Spherisorb ODSB, (75) YMC J'Sphere M80, (77) Zorbax SB-C_{18}, (78) Synergi Max RP, (79) YMC Hydrosphere C_{18}, (80) Nova-Pak C_{18}, (81) PolyEncap C_{18}, (82) TSK-Gel 80Ts, (83) Ace C_{18}, (84) XTerra MS C_{18}, (85) Fluophase PFP, (86) Purospher RP_{18}, (87) Develosil C30 UG 5, (88) Develosil ODS UG 5, (89) Hypersil Elite C_{18}, (90) Zorbax Rx C_{18}, (91) Zorbax Eclipse XDB C_{18}, (92) L-Column ODS, (93) YMC ODS AQ, (94) Prodigy C_{18}, (95) Luna C_{18}(2), (96) Kromasil C_{18}, (97) Allure PFP Propyl, (98) Discovery HS C_{18}, (99) Inertsil ODS-2, (100) Symmetry C_{18}, (101) L-column ODS, (102) Puresil C_{18}, (103) Cadenza CD-C_{18}, (105) Luna C_{18}, (106) Zorbax Extend C_{18}, (107) Inertsil ODS-3, (108) Zorbax Eclipse XDB C_{18}, (110) YMC Pack Pro C_{18}, (111) Purospher RP_{18}e, (112) Alltima C_{18}, (113) ODPerfect, (114) YMC J'Sphere H80, (115) Develosil ODS SR 5, (116) Nucleodur Gravity C_{18}, (117) Inertsil ODS-EP, (118) YMC-Pack Pro C_{18} RS, (119) Discovery HS F5, (120) Atlantis dC_{18}, (121) Discovery HS PEG, (122) Discovery Cyano, (123) Luna CN, (124) Cadenza CD-C_{18}, (125) experimental carbamate-CN packing, (126) experimental carbamate-phenyl packing, (127) Imtakt Presto FT C_{18}, (128) Aquasil C_{18}.

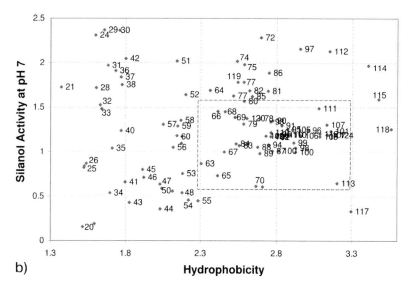

b)

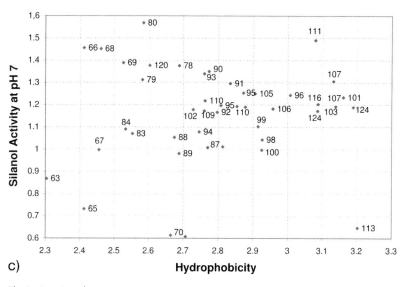

c)

Fig. 1. (continued)

We now have two values that describe the two most important properties of RP-packings: the natural logarithm of the retention factor of acenaphthene as a measure of the hydrophobicity of the packing and the value specified in Eq. (1) for the silanol activity. With these two values, we can now create the first selectivity chart (Fig. 1). Both axes of this figure are logarithmic axes. In Fig. 1a, the *x*-axis covers a 400-fold range of the retention factor of acenaphthene, and the *y*-axis an 80-fold range of the silanol activity; Fig. 1b and c are enlargements of the crowded region of the previous figures. The figures show 117 different commercial RP-packings from manufacturers throughout the world and two experimental phases from our own R&D department.

In Fig. 1a, stationary phases with a low hydrophobicity are found on the left side of the chart. Such phases are the cyano phases, followed by the phenyl packings. An example of a rather polar stationary phase is Nova-Pak CN HP, #1. On the right-hand-side of the graph one finds the C_{18} phases with the strongest hydrophobicity: Develosil ODS SR5, #115, YMC J'Sphere H80, #114, and YMC-Pack Pro C18 RS, #118. Packings with a high silanol activity are found in the upper region of the chart: Waters Spherisorb Phenyl, #4, and Waters Spherisorb ODS-2, #73. Such phases are typically based on older types of silica. The matrices of these older silicas contain non-negligible amounts of metals such as aluminum or iron. These metal ions increase the acidity of the surface silanols and enhance the interaction of the packing with basic compounds. In the lower part of the chart, we find stationary phases with a very low silanol activity. These are for the most part phases with an embedded polar group, such as the experimental carbamate phases #125 and #126, the commercially available XTerra RP_8, #20, which also has an embedded carbamate group, or Discovery HS PEG, #121, with polyethylene glycol as the ligand. The mechanism responsible for this reduced silanol activity is most likely a water layer that is strongly bound by the polar functional groups. Such a water layer would prevent interaction with the surface silanols and with the part of the ligand below the polar functional group. Other mechanisms are conceivable as well.

We have drawn a horizontal line in Fig. 1 to differentiate between classical bonded phases, which can all be found above the line, and phases with embedded polar groups, which lie below the line. Other phases with very low silanol activity can also be found below the line. An example is ODPerfect, #113, which is not based on silica. Beyond the limits of the chart shown here, one finds packings with exceptionally high silanol activities, such as the classical Zorbax C_{18} or Resolve C_{18}, both of which lack end-capping and are based on older silicas.

These comments demonstrate that the classification of the stationary phases is in line with expectation and is consistent with practical experience. The reproducibility of the values surely depends on the phase and the manufacturer. Therefore, one should expect some variation of the values given here. For the retention factor of a hydrophobic probe, one should expect today a relative standard deviation of around 5%. Kele [16] has demonstrated that the relative standard deviation of the batch-to-batch reproducibility achieved with Symmetry C_{18} is as low as 1.3%. The silanol activity can have a standard deviation as high as 15% in the worst case. On the other hand, for a highly reproducible packing such as Symmetry C_{18}, a relative standard deviation of only 2.2% has been measured [17]. One can see that the relative positions of different phases might change, especially for phases with a lower level of reproducibility. Therefore, one should be careful in interpreting the relative positions of the different phases on this chart.

In addition, it is important to understand that the values shown have only an absolute meaning in the context of the mobile phase composition of the test. In other mobile phases with different percentages of organic solvent or even different solvents, the positions of the different phases relative to each other will change. The fundamental reason for this phenomenon is the fact that the selectivity of a separation arises from a combination of the influence of the stationary phase and the influence of the mobile phase. If the composition of the mobile phase is drastically different from the test conditions, one can expect a different position of the different columns relative to each other. It remains correct that Symmetry C_{18} has a lower silanol activity than Spherisorb ODS-2, but whether a Luna $C_{18}(2)$ or a YMC-Pack Pro C_{18} has a higher hydrophobicity or a lower silanol activity surely depends on the details of the measurement conditions.

Let us now examine the first selectivity chart more closely. The position of 119 columns in the chart can be localized without difficulty using the enlargements of the crowded section of Fig. 1a in Fig. 1b and 1c. The packings in the center of Fig. 1c are all RP-packings generated from high-purity silicas. The more hydrophobic phases are found on the right. Examples of columns in the center are the L-Column ODS, #92, Luna $C_{18}(2)$, #95, YMC-Pack Pro C_{18}, #110, Discovery HS C_{18}, #98, Inertsil ODS-2, #99, Symmetry C_{18}, #100, and Luna C_{18}, #105. As one can see, Luna C_{18} and Luna $C_{18}(2)$ are two clearly different materials. On the other hand, the difference between these two packings is smaller than the difference between Symmetry C_{18}, #100, and SymmetryShield RP$_{18}$, #70. Symmetry C_{18} is a monofunctionally bonded phase on a high-purity silica, while SymmetryShield RP$_{18}$ is a phase with embedded polar groups created on the same parent silica. As one can see, the nature of the ligand has a significant influence on the properties of the stationary phase. We see that not only are the hydrophobicities of the two phases different, but also – and more importantly – the silanol activity of Symmetry C_{18} is very much different from that of SymmetryShield RP$_{18}$. These differences can be used in method development, together with the selectivity of the mobile phase, to manipulate and improve a separation.

2.1.2.2.2 Polar Interactions (Hydrogen Bonding)

Other selectivity characteristics of a stationary phase can be measured using the same technique [11]. Some of the phases with an incorporated polar group form hydrogen bonds with either the mobile phase or with analytes. These properties can be measured with suitable probes. For this purpose, we have selected the compound pair dipropylphthalate and butylparaben. Dipropylphthalate represents the polar reference compound. Butylparaben interacts via its phenol function with the polar group of the stationary phase. We have termed this interaction "polar selectivity" or "phenolic activity". Snyder [9] has determined the same or at least a very similar property of packings and called it the hydrogen bond basicity. Eq. (2) describes our method for measuring this parameter:

$$P = \ln(k_{butylparaben}) - 0.8962 \times \ln(k_{dipropylphthalate}) \tag{2}$$

A few other classes of compounds are subject to similar interactions with the stationary phase as phenols. Examples are sulfonamides, and carboxylic acids in their uncharged form, i.e. at acidic pH, with acetonitrile as the mobile phase. This special interaction increases the retention of the compounds affected.

This interaction is shown in Fig. 2. "Polar selectivity" is plotted versus the hydrophobicities of the columns. The designation of the stationary phases is identical to that used in Fig. 1. A horizontal line is shown in this figure. This line separates classical stationary phases (below the line) from packings with an embedded polar group (above the line) and other special phases. The box at the lower right-hand side of the graph includes a group of classical C_{18} phases. Among the special phases above the line, one finds several different types with embedded polar groups: the amide phases Discovery RP Amide C_{16}, #34, and Supelcosil LC-ABZ Plus, #48, the urea phases Spectrum, #46, and Prism, #53, and the carbamate phases SymmetryShield RP_8 and RP_{18}, #50 and #70. Inertsil ODS EP, #117, is based on a diol phase. Slightly above the horizontal line, one finds other special phases such as PolyEncap A, #52, PolyEncap C_{18}, #81, and the version of

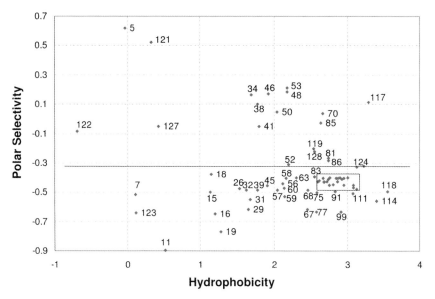

Fig. 2. Plot of polar selectivity versus hydrophobicity of the stationary phase. Column designations as in Fig. 1.
The box encloses the following stationary phases:
78, 82, 88, 89, 94, 96, 97, 100, 102, 103, 105, 107, and 110.

Purospher RP$_{18}$ that is endcapped with an aminosilane. In the left upper part of the chart, one finds Fluofix 120N, #5, and at the lower fringe of the chart there is Fluophase RP, #11. Both are fluorinated phases, but other details of the structures of these stationary phases are not known to us. Another phase found in the upper left-hand corner is Discovery HS PEG, #121, which is formed using polyethylene glycol. Packing #122 is the Discovery Cyano packing.

One can see that the use of the "polar selectivity" creates a grouping of the columns that is distinctly different from the classification by silanol activity. This demonstrates that column selectivity is truly multidimensional. Here, we have discussed three dimensions, while Snyder [4–9] deals with five dimensions. In our treatment, other dimensions are possible as well, for example the "extended polar selectivity" discussed in Ref. [13]. However, we believe that the three dimensions discussed here represent the strongest influences on column selectivity.

How well these selectivity groupings work, and how these similarities translate into the everyday separations of the chromatographer's work is left to the judgement of the reader. There are many separations in many laboratories that can be carried out on more than one stationary phase. The column selection used for the creation of the charts described here is rather large, and the reader will surely find at least some of the columns that he or she is commonly using. With the information provided here, the interested user can then test the described classification scheme and see how well it applies to his or her separation problem. In general, one can be sure that the hydrophobicity of the stationary phase will play the major role, no matter what the analyte is. If the analyte bears basic functional groups, then the silanol activity will surely affect the retention. However, if the analyte does not contain basic functions, then the silanol activity is not very important; see also Chapter 2.1.1. The same is true for the polar selectivity. The similarities and differences of stationary phases depend upon the nature of the analytes. For example, the stationary phases Hypersil Elite C$_{18}$, #89, and Synergi MAX RP, #78, are practically identical with respect to hydrophobicity and polar selectivity, but the silanol activity at neutral pH is much lower for the Hypersil Elite C$_{18}$ column. Therefore, we would expect very similar separations on both columns if the silanol activity does not play a significant role, but more different separations if the opposite is true. On the other hand, if one is working on a new separation and would like to efficiently exploit the selectivity differences of different columns in an automated method development scheme, then the selectivity charts presented here will certainly be useful in the selection of columns of different selectivity properties.

2.1.2.2.3 Reproducibility of the Selectivity

We have pointed out above that the reproducibility of a stationary phase has some limitations. How wide the range is depends on the type of stationary phase, the synthesis conditions, and the manufacturer. We do not want to end this chapter without saying a few words about the reproducibility of methods and stationary phases.

The reproducibility of a stationary phase depends on the experience of the manufacturer and the quality of the test method for the stationary phase. By the latter, we mean the quality of a test of the selectivity of the stationary phase, not a column test method, which is commonly used for measuring the column plate count. The method that we have described here for the characterization of the stationary phases was based on a very sensitive test method for the reproducibility of a packing material. High-quality manufacturers of modern stationary phases now use such highly sensitive methods for the measurement of the batch-to-batch reproducibility of such packings. At the same time, one has learned that batch-to-batch differences are measurable and can be quantified.

If one observes difficulties with the reproducibility of a method, one should not, however, immediately blame the column manufacturer. If the method in question is a new method, i.e., a method that has not yet been used on many different columns, it is not impossible that the column-to-column differences observed are simply related to the age and use of these columns. Often, a brand new column will give different results compared to an old, much used column. In order to test this, it is worthwhile to check whether the two columns in question have been prepared from the same batch of packing material. This information may be contained in the Certificate of Analysis that came with the column, or it can be obtained from the manufacturer. If the batch of packing material in both columns is identical, the selectivity of the separation must be identical. If the selectivity is different, one can be sure that the differences are due to the age and use of the older column. This difficult situation is not the fault of the column manufacturer, and cannot be resolved by the column supplier. There is no other choice but to redevelop the separation from scratch. Unfortunately, many reproducibility problems are of this type. If, on the other hand, the new column is from a different batch of packing material than the older column, then one should inquire whether the manufacturer can supply a new column from the old batch of packing. This is usually possible, unless the "old" column really is several years old.

If this comparison of the selectivity of different columns does indeed demonstrate that the observed difference correlates with a difference in the batch of packing material, then there are two possible solutions to the problem. One possibility is to reserve a quantity of bulk packing or packed columns specifically for the customer or even at the customer's site. This is the best solution if the separation is needed only for a limited amount of time. If, on the other hand, the separation will be needed for many years to come, as is often the case with quality control methods, it is better to redevelop and harden the method. Often, only a small change in the mobile phase conditions is necessary to solve the problem. Now that everybody is aware of the potential differences in the method on different batches of packing, one should check the reproducibility on several new batches of packing. Several column manufacturers offer reproducibility test kits that contain several different batches of packing and enable such studies.

The most difficult situation for everybody involved occurs if a separation that used to run smoothly for several years on many different batches of the packing suddenly does not work anymore. This problem occurs very rarely with modern, highly reproducible packings, but even with the best batch-to-batch reproducibility it is not impossible that the remaining small differences will shift the selectivity of an assay to an unacceptable level. One then faces two questions: how to solve the problem and how to avoid the problem in the future. Because this is a difficult situation, one should first of all make absolutely sure that the interpretation of the results is flawless. An open cooperation between the column user and the column manufacturer is essential. Sometimes, sufficient amounts of older batches are still available. If the batch test procedure is of a sufficiently high quality, it is potentially possible to select suitable older (and subsequently also new) batches based on the batch test results. It is also possible that the generic test procedure does not show the subtle differences that influence this particular assay. Under these circumstances, the manufacturer can provide the user with a test column. However, this is a gamble, and generally a better solution is advisable. Therefore, the less risky, but more involved solution is a modification or even a redevelopment of the assay.

References

1 D. Neue, E. S. Grumbach, J. R. Mazzeo, K. Tran, D. M. Wagrowski-Diehl, "Method development in reversed-phase chromatography", Handbook of Analytical Separations Vol. 4 (Ed.: I. D. Wilson), Elsevier Science, 2003, pp. 185–214.

2 M. Z. El Fallah, in "HPLC Columns – Theory, Technology, and Practice" (Ed.: U. D. Neue), Wiley-VCH, New York, Weinheim, 1997.

3 U. D. Neue, Selectivity Changes with pH, Chapter 1.3 of this book.

4 N. S. Wilson, M. D. Nelson, J. W. Dolan, L. R. Snyder, R. G. Wolcott, P. W. Carr, *J. Chromatogr. A* 961 (2002), 171–193.

5 N. S. Wilson, M. D. Nelson, J. W. Dolan, L. R. Snyder, P. W. Carr, *J. Chromatogr. A* 961 (2002), 195–215.

6 N. S. Wilson, M. D. Nelson, J. W. Dolan, L. R. Snyder, P. W. Carr, L. C. Sander, *J. Chromatogr. A* 961 (2002), 217–236.

7 J. J. Gilroy, J. W. Dolan, L. R. Snyder, *J. Chromatogr. A* 1000 (2003), 757–778.

8 J. J. Gilroy, J. W. Dolan, P. W. Carr, L. R. Snyder, *J. Chromatogr. A*, 1026 (2004), 77–89.

9 N. S. Wilson, J. J. Gilroy, J. W. Dolan, L. R. Snyder, *J. Chromatogr. A*, 1026 (2004), 91–100.

10 M. R. Euerby, P. Petersson, *J. Chromatogr. A* 994 (2003), 13–36.

11 U. D. Neue, B. A. Alden, T. H. Walter, *J. Chromatogr. A* 849 (1999), 101–116.

12 U. D. Neue, "HPLC Columns – Theory, Technology, and Practice", Wiley-VCH, New York, Weinheim, 1997.

13 U. D. Neue, K. Tran, P. C. Iraneta, B. A. Alden, *J. Sep. Sci.* 26 (2003), 174–186.

14 U. D. Neue, "Column Technology", in Encyclopedia of Analytical Science (Eds. P. Worsfold, A. Towneshend, C. Poole), Academic Press, Elsevier, Amsterdam, 2005, pp. 77–122.

15 U. D. Neue, B. A. Alden, P. C. Iraneta, A. Méndez, E. S. Grumbach, K. Tran, D. M. Diehl, "HPLC Columns for Pharmaceutical Analysis", in Handbook of HPLC in Pharmaceutical Analysis (Eds.: M. Dong, S. Ahuja), Elsevier, Amsterdam, 2005.

16 M. Kele, G. Guiochon, *J. Chromatogr. A* 830 (1999), 55–79.

17 U. D. Neue, E. Serowik, P. Iraneta, B. A. Alden, T. H. Walter, "A Universal Procedure for the Assessment of the Reproducibility and the Classification of Silica-Based Reversed-Phase Packings; 1. Assessment of the Reproducibility of Reversed-Phase Packings", *J. Chromatogr. A* 849 (1999), 87–100.

2.1.3

The Use of Principal Component Analysis for the Characterization of Reversed-Phase Liquid Chromatographic Stationary Phases

Melvin R. Euerby and Patrik Petersson

2.1.3.1 Introduction

The most ubiquitous type of stationary phase used within liquid chromatography is of a reversed-phase (RP) nature, with almost 700 differing types being commercially available worldwide. Hence, experienced and novice chromatographers alike are regularly faced with a bewildering choice of stationary phases for any particular application. The situation is further complicated by the many claims and counter claims made by manufacturers. To date, very little work has been done to compare all these stationary phases and to standardize the approach to stationary phase choice. Unfortunately, stationary phase selection appears to be the weak link in the method development cycle for many chromatographers.

Table 1. Description of the chromatographic parameters used in the column characterization studies; for further details, please refer to [2].

Chromatographic parameter	Abbreviation	Description
Retention factor for pentylbenzene	k_{PB}	Reflects the surface area and surface coverage (ligand density).
Hydrophobicity or hydrophobic selectivity	α_{CH2}	Retention factor between pentylbenzene and butylbenzene, $\alpha_{CH2} = k_{PB}/k_{BB}$. This is a measure of the surface coverage of the phase as the selectivity between alkylbenzenes differentiated by one methylene group is dependent on the ligand density.
Shape selectivity	$\alpha_{T/O}$	Retention factor ratio between triphenylene and o-terphenyl, $\alpha_{T/O} = k_T/k_O$. This descriptor is a measure of the shape selectivity, which is influenced by the spacing of the ligands and the functionality of the silylating reagent.
Hydrogen-bonding capacity	$\alpha_{C/P}$	Retention factor ratio between caffeine and phenol, $\alpha_{C/P} = k_C/k_P$. This descriptor is a measure of the number of available silanol groups and the degree of endcapping.
Total ion-exchange capacity	$\alpha_{B/P}$ at pH 7.6	Retention factor ratio between benzylamine and phenol at pH 7.6, $\alpha_{B/P}$ at pH 7.6 $= k_B/k_P$. This is an estimate of the total silanol activity.
Acidic ion-exchange capacity	$\alpha_{B/P}$ at pH 2.7	Retention factor ratio between benzylamine and phenol at pH 2.7, $\alpha_{B/P}$ at pH 2.7 $= k_B/k_P$. This is a measure of the acidic silanol activity of the silanol groups.

In order to assess the chromatographic properties of various stationary phases, a number of parameters must be measured. In our favored approach, which uses a modified Tanaka methodology [1], six chromatographic parameters are evaluated (see Table 1). In our latest publication, in which we report the characterization of 135 different RP materials, this equates to 810 values, which have to be compared [2]. In order to simplify the evaluation of this, and other, large chromatographic databases, the chemometrical tool of Principal Component Analysis (PCA) has been successfully employed [3–10].

2.1.3.2 Theory of Principal Component Analysis

PCA is a chemometrical tool, which allows a simple graphical interpretation of large data tables. A brief description of PCA is outlined below. A more formal algebraic account can be found in Refs. [11] and [12].

A table consist of rows, one row for each object (stationary phase), and columns, one column for each variable (stationary phase characterization parameter). A table with three variables can also be described as a three-dimensional plot, as illustrated in Fig. 1, where each object constitutes a point in a three-dimensional variable space.

Prior to PCA, the data is usually pre-treated, or "autoscaled", in order to give all variables the same importance and thus the same possibility to influence the PCA. For each variable, this is achieved by the division of its standard deviation and subtraction of its average value. Thereby, all variables are given unit variance and an average value of zero.

	Variable 1	Variable 2	Variable 3
Object 1	$x_{1,1}$	$x_{1,2}$	$x_{1,3}$
Object 2	$x_{2,1}$	$x_{2,2}$	$x_{2,3}$
Object 3	$x_{3,1}$	$x_{3,2}$	$x_{3,3}$
Object 4	$x_{4,1}$	$x_{4,2}$	$x_{4,3}$
Object 5	$x_{5,1}$	$x_{5,2}$	$x_{5,3}$
Object 6	$x_{6,1}$	$x_{6,2}$	$x_{6,3}$
Object 7	$x_{7,1}$	$x_{7,2}$	$x_{7,3}$
Object 8	$x_{8,1}$	$x_{8,2}$	$x_{8,3}$
Object 9	$x_{9,1}$	$x_{9,2}$	$x_{9,3}$
Object 10	$x_{10,1}$	$x_{10,2}$	$x_{10,3}$
Object 11	$x_{11,1}$	$x_{11,2}$	$x_{11,3}$
Object 12	$x_{12,1}$	$x_{12,2}$	$x_{12,3}$
Object 13	$x_{13,1}$	$x_{13,2}$	$x_{13,3}$
Object 14	$x_{14,1}$	$x_{14,2}$	$x_{14,3}$

a)

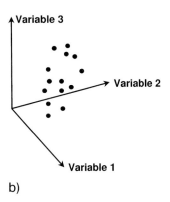

b)

Fig. 1. Different ways of describing the same data set.
(a) A table with 14 rows and 3 columns.
(b) 14 points in a 3-dimensional variable space.

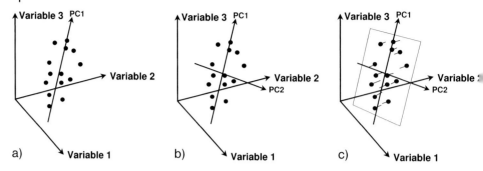

Fig. 2. In PCA the number of variables is reduced by a projection of the objects onto a smaller number of new variables termed principal components (PC).
(a) The PCs are orientated so that the first PC describes as much as possible of the original variation between the objects.
(b) The second PC is orientated in an orthogonal manner to the first PC and is directed to describe as much as possible of the remaining variation. Both PCs go through the average object.
(c) The projection of objects onto a PC describes coordinates called scores. By plotting the scores for two PCs a score plot is obtained. Such plots allow a simple graphical comparison of similarities and differences between objects.

In PCA, the number of variables is reduced by a projection of the objects onto a smaller number of new variables termed principal components (PC). The PCs are orientated so that the first PC goes through the average object and describes as much as possible of the variation between the objects (Fig. 2a). The second PC is orientated in an orthogonal manner to the first PC and is directed to describe as much as possible of the remaining variation between the objects (Fig. 2b). The second PC also goes through the average object.

The projection of objects onto a PC describes coordinates, which are called scores. By plotting the scores for two PCs, a score plot is obtained (Fig. 2c). Using such score plots, it is possible to graphically find similarities and differences between objects (stationary phases). The distance between objects in a score plot shows if they are similar or different. Objects located close to each other are similar and objects located far from each other are different.

An estimate of the contribution of each of the original variables that contributes to a PC is described by so-called loadings, one for each variable and PC. By plotting the loadings for two PCs, it is possible to see which of the original variables are most important (longest distance from the origin) and if any variables are correlated (the same or opposite directions on a straight line through the origin). By comparing the score and the loading plots, it is also possible to explain why an object is located where it is in the score plot (Fig. 3).

In the literature, several examples can be found where PCA has been applied for the characterization of chromatographic columns [2–10, 13]. It should be stressed, however, that PCA is a very powerful tool for analyzing any type of database. The stationary phase datasets described in the following sections have been evaluated by PCA; the examples aim to highlight the ability of PCA to identify

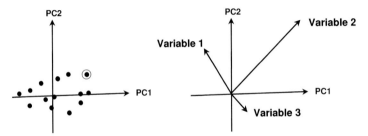

Fig. 3. The contribution of each variable to a PC is described by the term *loadings*. By plotting the loadings for two PCs, it is possible to see which of the variables are most important (longest distance from the origin – in this example variable 2) and if any variables are correlated (the same or opposite directions on a straight line through the origin – in this example variables 3 and 1 are negatively correlated). It is also possible to draw conclusions about why an object is located where it is in the score plot – in this example the encircled object in the score plot is characterized by a high value of variable 2.

phases of orthogonal selectivity or equivalent phases, and to enhance understanding of the retention mechanism and phase bonding technology.

2.1.3.3 PCA of the Database of RP Silica Materials

The diversity of commercially available RP materials can readily be observed by subjecting such large databases of column chromatographic parameters to PCA. In our recent paper, 135 columns equating to 810 data points were successfully evaluated using PCA [2].

The PC1-PC2 score plot of Fig. 4 highlights that there appears to be a large number of similar phases centered near the origin; however, the plot clearly shows that there are many phases which possess fundamentally different chromatographic properties. The score and loading plot for phases in group A indicate that they possess high silanol activity ($\alpha_{C/P}$, $\alpha_{B/P}$ at pH 2.7 and 7.6) and low retentivity (k_{PB} and α_{CH2}) (see Table 1 for an explanation of the chromatographic terms); this group consists mostly of non-C$_{18}$ phases and traditional C$_{18}$ phases based on acidic silica, such as Resolve C$_{18}$ (column no. 87).

In contrast, the group B phases are newer generation C$_{18}$ materials, which exhibit low silanol activity and high retentivity (e.g., Ultracarb ODS 30, column no. 112). Group C phases contain the polar embedded materials (e.g., Suplex pkb 100, column no. 100) and these are differentiated from the other phases due to their high shape-selectivity parameter ($\alpha_{T/O}$).

Since the PC1-PC2 analysis only explained 61% of the variation of the database, the PC1-PC3 plots were evaluated (see Fig. 5). PC3 contributed an extra 15% of the variability of the data and a plot of PC1-PC3 further differentiated between the phases. The acidic phases such as Resolve C$_{18}$, Spherisorb ODS1, and Platinum EPS C$_{18}$ (column nos. 87, 91, and 75) could be grouped together (group E) as these possessed high acidity and total silanol activity ($\alpha_{C/P}$, $\alpha_{B/P}$ at pH 2.7 and 7.6). Within the non-C$_{18}$ phases (group D), the majority of cyano phases could be located as a subgroup H. The PC1-PC3 plot also differentiated the perfluorophenyl

(PFP) phases (group F) on the basis of their shape-selectivity parameter ($\alpha_{T/O}$). The highly retentive phases such as Ultracarb ODS, BetaMax Neutral C_{18}, Gromsil ODS pH7, J'sphere ODS JH, and the Omnisphere C_{18} phases that possess high carbon loads (i.e. > 22%) are grouped together (group G) as they are characterized by high retention factors for pentylbenzene (k_{PB}) and high hydrophobic selectivity terms (α_{CH2}).

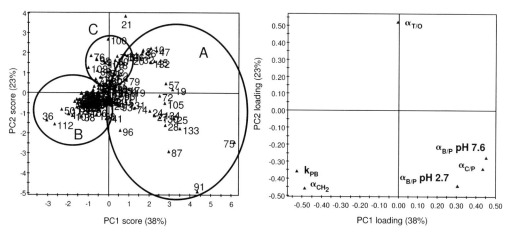

Fig. 4. PC1–PC2 score and loading plots for columns in Ref. [2].
A = mostly non-C18 and traditional acidic (type A) C18 silica phases.
B = mostly non-acidic (type B) C18 silica phases.
C = polar embedded phases.

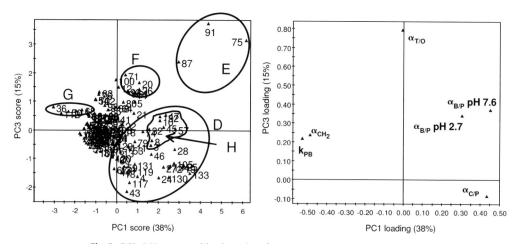

Fig. 5. PC1–PC3 score and loading plots for columns in Ref. [2].
D = non-C18 phases
E = acidic phases
F = perfluorophenyl phases
G = highly hydrophobic phases
H = cyano phases

2.1.3.3.1 PCA of Polar Embedded, Enhanced Polar Selectivity, and AQ/Aqua Phases

Recently, there has been an increase in the number of commercially available phases which the manufacturers claim to be suitable for chromatography using purely aqueous mobile phases. These have been further classified, by the manufacturers, as polar embedded, enhanced polar selectivity, and Aqua phases. Some of the phases in the latter two classifications have been further described by the manufacturers as polar/hydrophilic endcapped phases, depending on the bonding technology employed. Unfortunately, there is a dearth of information regarding the bonding chemistry employed [2].

The PC1-PC2 plot of 31 phases classified by the stationary phase manufacturers as suitable for use under purely aqueous conditions clearly highlights three distinct groups of phases possessing differing chromatographic properties (see Fig. 6 score and loading plots).

- Those possessing enhanced shape-selectivity ($\alpha_{T/O}$) and reduced retention (k_{PB}), as typified by polar embedded phases, e.g., Suplex pk_b 100, HyPURITY ADVANCE, Polaris Amide C_{18}, BetaMax Acidic (column nos. 100, 46, 76, and 9, respectively) – **group C.**
- Those possessing enhanced polar selectivity, as shown by a high hydrogen-bonding capacity (i.e., $\alpha_{C/P}$ values > 1), e.g., Platinum C_{18}18 EPS, Synergi Polar RP, Zorbax SB AQ, and Aquasil phases (column nos. 75, 105, 134, and 5, respectively) – **group A.**
- Those which retain the test analytes on the basis of their lipophilicity, i.e., phases possessing high retention (k_{PB}) and low shape-selectivity ($\alpha_{T/O}$), e.g., Genesis AQ, Hichrom RPB, and Phenonemex AQUA (column nos. 29, 37, and 73, respectively) – **group B.**

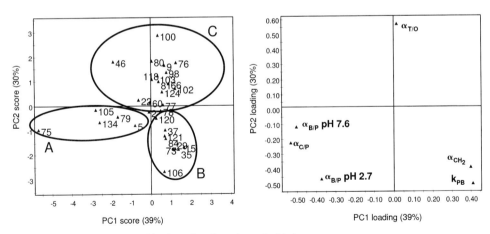

Fig. 6. PC1–PC2 score and loading plots for polar embedded, enhanced polar selectivity and "Aqua" phases
A = enhanced polar selectivity
B = high lipophilic retention
C = polar embedded phases

The enhanced polar selectivity of the group A phases can be attributed to either the presence of polar endcapping functionality (e.g., Aquasil, Synergi Polar RP, column nos. 5 and 105), low surface coverage and non-endcapped phases (e.g., Platinum C_{18} EPS, column no. 75) or the type of silylating reagent that was employed in the bonding (e.g., Zorbax SB AQ, column no. 134).

A number of the phases in group B possess a mixed alkyl phase (e.g., Genesis AQ, Hichrom RPB, column nos. 29 and 37, respectively); this type of mixed bonding is responsible for preventing "phase collapse" (i.e., de-wetting of the pores) in highly aqueous mobile phases. In contrast, phases such as YMC ODS-AQ, Prontosil C_{18}-AQ, Phenonemex AQUA, and Synergi-Hydro-RP (column nos. 121, 84, 73, and 106, respectively) are claimed by their manufacturers to possess a C_{18} ligand plus a hydrophilic/polar endcapping functionality. The PCA suggests that these phases are similar to the mixed alkyl phases and certainly do not possess any additional hydrogen-bonding capacity. The precise nature of these polar endcapping moieties is at present unclear.

2.1.3.3.2 PCA of Perfluorinated Phases

Recently, there has been much interest in the use of perfluorinated phases as they can offer an orthogonal separation mechanism compared to standard alkyl RP materials. As a consequence of this, the number of commercially available phases has increased dramatically. We have recently characterized and performed PCA on ten alkyl and phenyl perfluorinated silica-based stationary phases using the modified Tanaka approach and compared their retention behavior with that of conventional phenyl and alkyl phases [13].

The PC1-PC2 model for the database accounted for 84% of the chromatographic variability within these phases. The score plot in Fig. 7 illustrates that the perfluorinated phases can be categorized into three subgroups.

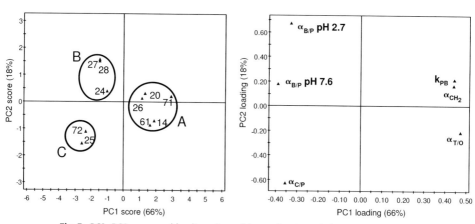

Fig. 7. PC1–PC2 score and loading plots of the perfluorinated phases [13]
A = perfluorophenyl phases
B = perfluoro C6 or C8 phases
C = perfluoro C6 non endcapped and C3 perfluoro phases

Group A containing the perfluorophenyl phases (e.g., Discovery F5 HS and Fluophase PFP, column nos. 20 and 26), Group B the perfluorohexyl branched alkyl endcapped (e.g., Fluofix, column no. 24) and perfluoroctyl phases (e.g., Fluophase RP and FluoroSep RP Octyl, column nos. 27 and 28), and Group C the perfluorohexyl non-endcapped (e.g., Fluofix, column no. 25) and perfluoropropyl (e.g., Perfluorpropyl ESI, column no. 72) phases.

The loading plots indicate that the subgroups possess the following dominant chromatographic properties:

A) High surface area/hydrophobicity, high shape selectivity (k_{PB}, α_{CH2}, and $\alpha_{T/O}$), and low ion-exchange and hydrogen-bonding capacity ($\alpha_{C/P}$, $\alpha_{B/P}$ at pH 2.7 and 7.6).

B) Low surface coverage/hydrophobicity, low shape selectivity (k_{PB}, α_{CH2}, and $\alpha_{T/O}$), and high ion-exchange capacity ($\alpha_{B/P}$ at pH 2.7 and 7.6).

C) Low surface coverage/hydrophobicity, low shape selectivity, low ion-exchange capacity (k_{PB}, α_{CH2}, $\alpha_{T/O}$, and $\alpha_{B/P}$ at pH 2.7 and 7.6), and high hydrogen-bonding capacity ($\alpha_{C/P}$).

2.1.3.4 Use of PCA in the Identification of Column/Phase Equivalency

The issue of column equivalency is of major importance to regulatory bodies such as the Food & Drug Administration (FDA), the United States Pharmacopoeia (USP), and the European Pharmacopoeia (EP) commissions.

The USP general chapter on chromatography (621) lists 52 separate types of HPLC packings (L1–L52), and it has recently been proposed that another seven L-classes should be added to this list [14]. Some of the L-classes are extremely specific, e.g., USP L57 column designation is described as *"Spherical, porous silica gel, 3–5 µm diameter, the surface of which has been covalently modified with palmit-amidopropyl groups and endcapped with acetamidopropyl groups to a ligand density of about 6 µmol per m²"*; to date, only the Discovery Amide C$_{16}$ phase is capable of satisfying these criteria.

In comparison, there are over 200 different varieties of commercial columns worldwide which meet the criteria of the USP L1 column designation *(a packing of octadecylsilane chemistry, chemically bonded to porous silica or ceramic microparticles, 3–10 µm in diameter)*.

This problem has been recognized by the USP in that it was concluded at their recent conference that *"Currently USP classifies all C18 columns as type L1, and USP monographs do not differentiate the myriad differences among columns within this classification"* [15].

PCA of 42 columns (see Table 2) that the manufacturers claim to meet the USP L1 criteria clearly shows that these columns are not equivalent [16] (see Fig. 8a and b). Approximately 82% of the variability of the L1 phases could be described by the first three principal components. It is of interest that the diversity of the columns is more apparent when the PC1–PC3 plot is examined. The difference between the L1 phases is exemplified in their chromatographic performance with a range of hydrophilic bases (see Fig. 9a and b).

Table 2. Table of the USP L1 designated phases and their chromatographic properties; for a detailed explanation of the testing (see [2]).

Column no.	Description	Reference	k_{PB}	α_{CH2}	$\alpha_{T/O}$	$\alpha_{C/P}$	$\alpha_{B/P}$ pH 7.6	$\alpha_{B/P}$ pH 2.7
1	Ace 5C18	HICHROM	4.58	1.46	1.52	0.40	0.47	0.13
5	Aquasil C18	Hypersil	4.14	1.41	1.84	1.18	2.29	0.16
7	Betabasic C18	Hypersil	4.49	1.47	1.56	0.39	0.80	0.12
11	BetaMax Neutral C18	Hypersil	10.62	1.49	1.50	0.40	1.00	0.10
16	Discovery C18	Supelco	3.32	1.48	1.51	0.39	0.28	0.10
17	Discovery C18 HS	Supelco	6.68	1.49	1.55	0.40	0.38	0.10
39	Hypersil C18 BDS	Agilent, Hypersil	4.50	1.47	1.49	0.39	0.19	0.17
40	Hypersil Elite C18	Hypersil	4.76	1.49	1.52	0.37	0.30	0.14
41	Hypersil ODS	Agilent, Hypersil	4.44	1.45	1.28	0.38	1.04	0.64
42	HyPURITY C18	Hypersil	3.20	1.47	1.60	0.37	0.29	0.10
48	Inertsil ODS3	Ansys	7.74	1.45	1.29	0.48	0.29	0.01
50	J'Sphere ODS	YMC	10.60	1.51	1.59	0.39	0.43	0.06
51	Jupiter C18 300A	Phenomenex	2.26	1.48	1.65	0.37	0.47	0.27
52	Kromasil C18	Hypersil	7.01	1.48	1.53	0.40	0.31	0.11
54	Lichrosphere RP18	Agilent, Merck	7.92	1.48	1.73	0.54	1.39	0.19
55	Luna C18	Phenomenex	5.97	1.47	1.17	0.40	0.24	0.08
56	Luna C18(2)	PHenomenex	6.34	1.47	1.23	0.41	0.26	0.06
62	Novapak C18	Waters	4.49	1.49	1.44	0.48	0.27	0.14
64	Nucleosil C18	Agilent, Hypersil	4.80	1.44	1.68	0.70	2.18	0.13
65	Nucleosil C18 HD	Hypersil	6.04	1.48	1.54	0.40	0.47	0.10
74	Platinum C18	Alltech	2.12	1.39	1.23	0.81	2.82	0.21

Table 2. (continued)

Column no.	Description	Reference	k_{PB}	α_{CH2}	$\alpha_{T/O}$	$\alpha_{C/P}$	$\alpha_{B/P}$ pH 7.6	$\alpha_{B/P}$ pH 2.7
75	Platinum C18 EPS	Alltech	0.97	1.31	1.98	2.62	10.11	0.26
82	Prodigy ODS2	Phenomenex	4.94	1.49	1.43	0.37	0.50	0.01
83	Prodigy ODS3	Phenomenex	7.27	1.49	1.26	0.42	0.27	0.09
86	Purospher RP18e	Agilent	6.51	1.48	1.75	0.46	0.34	0.08
87	Resolve C18	Waters	2.40	1.46	1.59	1.29	4.06	1.23
91	Spherisorb ODS1	Hypersil, Waters	1.78	1.47	1.64	1.57	2.84	2.55
92	Spherisorb ODS2	Hypersil, Waters	3.00	1.51	1.56	0.59	0.76	0.23
99	Superspher RP18e	Agilent	5.47	1.47	1.64	0.44	0.42	0.11
101	Symmetry C18	Waters	6.51	1.46	1.49	0.41	0.68	0.01
102	Symmetry Shield RP18	Waters	4.66	1.41	2.22	0.27	0.20	0.04
106	Synergi Hydro-RP	Phenomenex	7.63	1.47	1.47	0.58	0.83	0.25
111	uBondpak	Waters	1.97	1.39	1.28	0.78	1.12	0.15
112	Ultracarb ODS(30)	Phenomenex	13.27	1.52	1.39	0.48	0.73	0.06
116	XTerra MS C18	Waters	3.52	1.42	1.26	0.42	0.35	0.10
118	XTerra RP18	Waters	2.38	1.29	1.83	0.33	0.20	0.07
121	YMC ODS-AQ	YMC	4.44	1.46	1.25	0.57	0.41	0.11
122	YMC ProC18	YMC	7.42	1.53	1.29	0.46	0.26	0.08
125	Zorbax Eclipse XDB-C18	Agilent	5.79	1.50	1.30	0.47	0.35	0.09
126	Zorbax Extend C18	Agilent	6.66	1.50	1.49	0.38	0.20	0.08
127	Zorbax Rx C18	Agilent	5.68	1.57	1.61	0.54	0.55	0.11
128	Zorbax SB-C18	Agilent	6.00	1.49	1.20	0.65	1.46	0.13

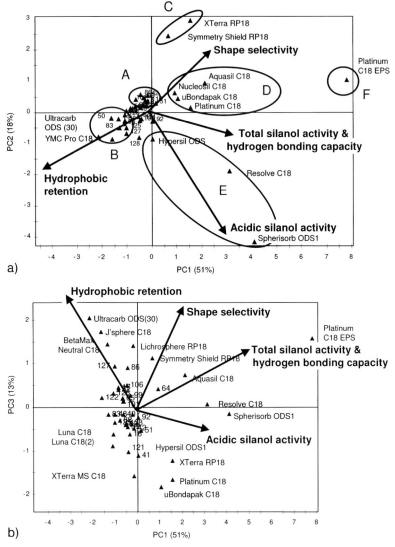

Fig. 8. Combined PC1–PC2 (a) and PC1–PC3 (b) score and loading plots of the USP designated L1 phases.

The loading plots indicate that the subgroups possess the following dominant chromatographic properties:

- **Group A:** Moderate surface area/hydrophobicity and shape selectivity, and low ion-exchange and hydrogen-bonding capacity, as exemplified by the new generation C_{18} silica phases, e.g., BetaBasic C_{18}, ACE C_{18}, Symmetry C_{18}, XTerra MS C_{18}, Purospher RP18e, Superspher RP18e, HyPURITY C_{18}, Discovery C_{18} (column nos. 7, 1, 101, 116, 86, 99, 42, and 16, respectively).

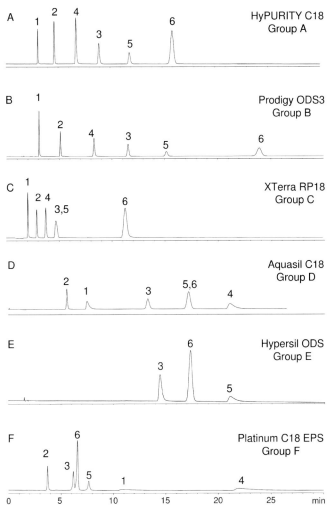

Fig. 9. Comparison of L1 designated phases for the analysis of a range of hydrophilic bases (1 = nicotine, 2 = benzylamine, 3 = terbutaline, 4 = procainamide, 5 = salbutamol, 6 = phenol [neutral marker]). 20 mM KH_2PO_4, pH 2.7 in 3.3 : 96.7 v/v MeOH : H_2O, 1 mL min^{-1} (assumes 150 × 4.6 mm column format), 60 °C, detection 210 nm. For experimental conditions see Ref. [2].

- **Group B:** High surface coverage/hydrophobicity, low shape selectivity, ion-exchange and hydrogen-bonding capacity, as exemplified by high carbon load phases, e.g., Ultracarb ODS 30, YMC Pro C_{18}, J'Sphere ODS, Prodigy ODS3, BetaMax Neutral C_{18}, Luna C_{18}(2), Zorbax Rx C_{18}, Zorbax XDB Eclipse C_{18}, Zorbax Extend C_{18} (column nos. 112, 122, 50, 83, 11, 56, 127, 125, and 126, respectively).

- **Group C:** Low surface coverage/hydrophobicity, high shape selectivity, low ion-exchange and high hydrogen-bonding capacity, as exemplified by the polar embedded phase of Symmetry and XTerra materials, which contains a carbamate as the integral polar moiety [17]; e.g., XTerra RP18 and Symmetry Shield RP18 (column nos. 118 and 102, respectively).

- **Group D:** Moderate surface coverage/hydrophobicity, shape selectivity, and hydrogen-bonding, and low ion-exchange capacity, as exemplified by phases of moderate retentivity and silanol activity, e.g., Nucleosil C_{18}, Aquasil C_{18}, Platinum C_{18}, and μBondapak C_{18} (column nos. 64, 5, 74, and 111, respectively).

- **Group E:** High ion-exchange capacity at pH 2.7, as exemplified by phases possessing a high proportion of acidic silanol groups, e.g., Hypersil ODS1, Resolve C_{18}, and Spherisorb ODS1 (column nos. 41, 87, and 91, respectively).

- **Group F:** High hydrogen-bonding capacity and low surface coverage/hydrophobicity, as exemplified by phases of high polar selectivity, e.g., Platinum C_{18} EPS (column no. 75).

The question of classifying columns is a difficult one as there is often a great diversity and overlap between apparently similar USP designated phases as shown by PCA [16]. The identification of "back-up" columns is required when a chromatographic methodology is registered with a regulatory body. Score plots can be conveniently used to identify equivalent phases; for example, Fig. 10a suggests that Selectosil C_{18} (Phenomenex) and Nucleosil C_{18} (Macherey-Nagel) (column nos. 88 and 64, respectively) are similar in their chromatographic properties [2]. This was confirmed chromatographically when both phases generated similar results when a hydrophilic base mix was chromatographed (see Fig. 10b).

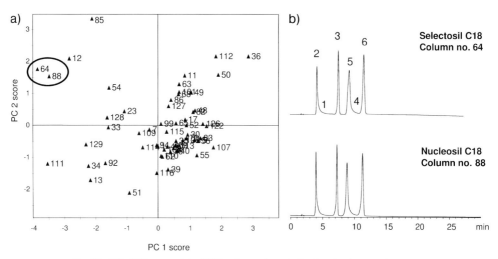

Fig. 10. PC1–PC2 score plot of C18 column characterization database and the analysis of the hydrophilic bases on two similar columns as identified by PCA. For peak assignments and chromatographic conditions see legend of Fig. 9.

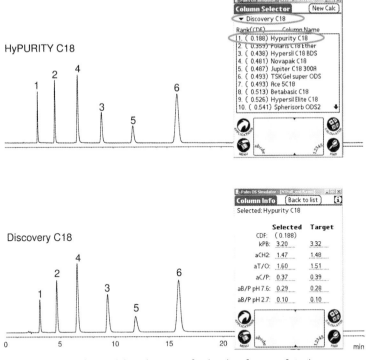

Fig. 11. Example of the web-based program for the identification of similar columns based on the database of Ref. [2] and the analysis of the hydrophilic bases on two similar columns as identified by the downloadable freeware [18]. For peak assignments and chromatographic conditions see legend of Fig. 9.

An alternative approach to the identification of similar columns has recently been described [2]; the authors calculated the distance in the six-dimensional variable space between the column of interest and other columns in the database. This approach and the relevant database has been placed on the worldwide web [18] for open access to all chromatographers in order to allow the identification of similar columns for column equivalence testing and the identification of columns possessing orthogonal retention mechanisms for method development purposes. Figure 11 highlights the usefulness of the approach in identifying an equivalent column to the Discovery C_{18} phase.

2.1.3.5 Use of PCA in the Rational Selection of Stationary Phases for Method Development

PCA of column chromatographic properties is an ideal tool to identify phases with differing selectivities that can be exploited in method development. Figure 12 highlights the stationary phase selectivity differences that can be achieved by selecting phases from the PCA score and loading plots. The selected phases exhibited differing selectivities for the separation of the forced degradation products of the non-steroidal anti-inflammatory drug Ketoprofen.

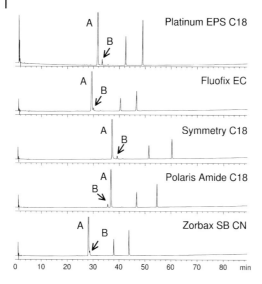

Fig. 12. Differences in stationary phase selectivity on the analysis of degraded Ketoprofen using identical chromatographic conditions. Note the change in elution order for peaks marked (A) and (B) on the polar embedded phase – Polaris Amide C18.

2.1.3.5.1 Proposed Solvent/Stationary Phase Optimization Strategy

1. Use of PCA and column characterization data as an aid to select appropriate stationary phase chemistries and those that are orthogonal in separation mechanism. Phases based on acidic silica would not be suitable for the analysis of basic compounds. In addition, polar embedded phases with residual amino functionalities would not be suitable for analyzing ionized acids.

2. Automated screening of a number of stationary phases of differing selectivity using short columns, applying rapid gradients using a number of organic modifiers at a buffered pH, as suggested from the physico-chemical properties of the analyte, and at low and elevated temperatures.

3. Peak tracking is performed by diode-array detection, determination of peak areas and retention times, and most importantly by mass spectrometry.

4. Optimal isocratic and/or gradient conditions are predicted by computer optimization software by retention time and peak width modeling, e.g., with DryLab (LC Resources, BASi Northwest Laboratory Services, Walnut Creek, CA, USA), LC Simulator (Advanced Chemistry Development, Toronto, Canada), or ChromSword (VWR International, Darmstadt, Germany).

This and similar approaches are being developed into fully automated method development platforms by a variety of instrument manufacturers, e.g., AMDS (Waters Milford, MA), AutoChrom (Advanced Chemistry Development, Toronto,

Canada), ChromSmart MS (Intelligent Laboratory Solutions Inc., Naperville, IL, USA), and ChromSword Auto (VWR International, Darmstadt, Germany). It is envisaged that there will be an increase in the number and flexibility of these systems in the coming years. Inexperienced analysts may then operate the systems in a totally automated fashion starting from the input of the chemical structure of the analyte of interest. For more demanding samples, the more experienced analyst will have access to a fully flexible system, which would allow multiple points of entry and exit from the method development platform.

References

1 K. Kimata, K. Iwaguchi, S. Onishi, K. Jinno, K. Eksteen, K. Hosoya, M. Arki, N. Tanaka, *J. Chromatogr. Sci.* 27 (1989), 721.

2 M. R. Euerby, P. Petersson, *J. Chromatogr. A* 994 (2003), 13.

3 B. A. Olsen, G. R. Sullivan, *J. Chromatogr. A* 692 (1995), 147.

4 M. R. Euerby, P. Petersson, *LC·GC* 13 (2000), 665.

5 D. V. McCalley, R. G. Brereton, *J. Chromatogr. A* 828 (1998), 407.

6 R. J. M. Vervoort, M. W. J. Derksen, A. J. J. Debets, *J. Chromatogr. A* 765 (1997), 157.

7 A. Sandi, A. Bede, L. Szepesy, G. Rippel, *Chromatographia* 45 (1997), 206.

8 R. G. Brereton, D. V. McCalley, *Analyst* 123 (1998), 1175.

9 E. Cruz, M. R. Euerby, C. M. Johnson, C. A. Hackett, *Chromatographia* 44 (1997), 151.

10 T. Ivanyi, Y. Vander Heyden, D. Visky, P. Baten, J. De Beer, I. Lazar, D. L. Massart, E. Roets, J. Hoogmartens, *J. Chromatogr. A* 954 (2002), 99.

11 L. Eriksson, E. Johansson, N. Kettaneh-Wold, S. Wold, Multi- and Megavariate Data Analysis: Principles and Applications, Umetrics AB, Umeå, Sweden, 2001.

12 K. Esbensen, S. Schönkopf, T. Midtgaard, Multivariate Analysis in Practice, Camo AS, Trondheim, Norway, 1994.

13 M. R. Euerby, A. P. McKeown, P. Petersson, *J. Sep. Sci.* 26 (2003), 295.

14 Pharmacopeial Forum, 29 (2003), 170.

15 Output from a discussion group at US Pharmacopoeia Conference, June 2003, Philadelphia, USA.

16 M. R. Euerby, P. Petersson, Is the USP classification satisfactory? – A comparison of manufacturer's USP recommended RP materials with that of a simple RP characterization procedure using principal component analysis. Presented at HPLC 2003, June 2003, Nice, France.

17 J. E. O'Gara, D. P. Walsh, B. A. Alden, P. Casellini, T. H. Walter, *Anal. Chem.* 71 (1999), 2992.

18 www.acdlabs.com/columnselector

2.1.4
Chemometrics – A Powerful Tool for Handling a Large Number of Data

Cinzia Stella and Jean-Luc Veuthey

2.1.4.1 **Introduction**
Reversed-phase liquid chromatography (RPLC) is the method of choice for the analysis of basic compounds. However, because of the problems encountered with these compounds in terms of peak asymmetry and poor reproducibility with conventional supports, "base-deactivated" stationary phases have been developed. The chromatographic behavior of basic substances is closely related to their structure and is easily influenced by many factors, such as pH, molarity, and composition of the mobile phase. Thus, a valuable column evaluation can be obtained by using a set of basic test compounds covering a wide range of physicochemical properties. Principal Component Analysis (PCA) is a helpful chemometric tool in the selection of the set of test compounds possessing different physicochemical and structural properties, such as lipophilicity, pK_a value, accessibility of basic groups, etc. Once the appropriate compounds have been selected, they can be used for the evaluation of different base-deactivated supports according to a particular methodology. Due to the large amount of data generated during a chromatographic test, Principal Component Analysis is a helpful tool in describing differences and similarities between columns. In addition, such a chemometric tool can also be useful in reducing the number of the test compounds that need to be used and in the optimization of the methodology of the chromatographic test.

2.1.4.2 **Chromatographic Tests and Their Importance in Column Selection**
The choice of an appropriate stationary phase is one of the most challenging tasks for a chromatographer. For this reason, many procedures have been developed to classify columns according to their separation mechanisms. Generally, to differentiate stationary phases with regard to their selectivity and peak asymmetry, a chromatographic test is preferable to a physicochemical characterization in which parameters such as carbon loading and specific surface area are determined. The chromatographic approach seeks to examine specific, discrete, chemical interactions with the stationary phase. Therefore, numerous tests have been published in the last 20 years to elucidate the chromatographic behavior of packing materials and thus provide a basis for column selection. For this purpose, a set of molecules, also called test solutes, can be used and, in the case of base-deactivated columns, particular care should be taken in the selection of the appropriate set. The chromatographic behavior of basic compounds is closely related to their physicochemical properties and is easily influenced by many factors, such as pH and mobile phase composition [1, 2].

   After having selected the set of compounds, columns are evaluated with at least two mobile phases at two different pH values (e.g., pH 7.0 and 3.0). For each compound and for each set of chromatographic conditions, chromatographic parameters such as retention factor (k), asymmetry (As), and efficiency (N) are measured.[1]

In order to treat the large amount of data generated and to obtain a clear overview of differences and similarities between chromatographic supports as well as an evaluation of their chromatographic performance, a chemometric tool, such as Principal Component Analysis, is necessary.

2.1.4.3 Use of Principal Component Analysis (PCA) in the Evaluation and Selection of Test Compounds

2.1.4.3.1 Physicochemical Properties of Test Compounds

Selecting the most appropriate set of test compounds for use in the evaluation of stationary phases is a key step of a chromatographic test. Only test compounds possessing quite different physicochemical and structural properties will induce different interactions with the stationary phase, thereby allowing a good evaluation of its chromatographic performance. For this reason, it is highly recommended that a chemometric tool, such as Principal Component Analysis, is used for the selection of the most appropriate compounds.

As already discussed by Euerby and Petersson in Chapter 2.1.3 (theory of PCA), PCA is applied to two-dimensional tables crossing "individuals" (test compounds) and "quantitative variables" (physicochemical properties of test compounds). The main objective of PCA is to evaluate the resemblance between individuals. Hence, two individuals are considered as being similar if their values are closer in comparison to the totality of the variables. In order to better illustrate this concept, an example will be given concerning the selection of the most suitable set of test compounds.

In order to ensure that the selected test compounds (Table 1) are sufficiently different to evaluate chromatographic supports, PCA of some of their physicochemical properties was carried out:

V molecular volume
S_{pol} polar surface area
$\log D^{7.0}$ distribution coefficient at pH 7.0
$\Sigma \alpha$ hydrogen-bond donor capacity
$\Sigma \beta$ hydrogen-bond acceptor capacity

Inspection of the obtained *loading plot* (Fig. 1) and of the contribution of variables (Table 2) to Principal Components (PC) is necessary to evaluate differences and similarities between basic compounds (*scores* in Fig. 2). All variables are well represented in the selected two-dimensional space, as their length is close to 1.0 (see Chapter 2.1.3). By considering the direction of the S_{pol} and $\log D$ vectors and their contribution to PC1, compounds can be ranked along PC1 in relation to

---

[1] When columns of different lengths and particle size are compared, h should be used instead of N. $h = H/d_p = L/(N \cdot d_p)$, where d_p is the particle size, L the column length, and N the efficiency.

Table 1. Physicochemical properties.

Molecule	Abbreviation	V	S_{pol}	log D	Σα	Σβ
Benzylamine	BZ	115.3	23.8	−1.21	0.62	0.16
Chloroprocaine	CL	252.5	22.78	0.37	1.49	0.26
Codeine	CO	277.9	23.24	−0.06	1.99	0.30
D-Amphetamine	DX	149.9	14.75	−1.15	0.62	0.16
Diphenhydramine	DP	265.9	6.03	1.28	1.18	0.00
Fentanyl	FN	345.5	5.57	2.44	1.09	0.00
Mepivacaine	MP	255.2	9.11	1.14	1.4	0.50
Methadone	MT	329.7	6.06	1.63	1.24	0.00
Morphine	MO	261.1	28.43	−0.3	2.02	0.60
Nicotine	NI	168.3	20.31	0.08	1.25	0.00
Nortriptyline	NO	274.9	8.83	1.31	0.7	0.08
Procainamide	PR	243.5	24.87	−1.46	1.87	0.61
Pyridine	PY	82.1	26.17	0.64	0.52	0.00
Quinine	QN	315.1	22.36	1.95	2.21	0.33

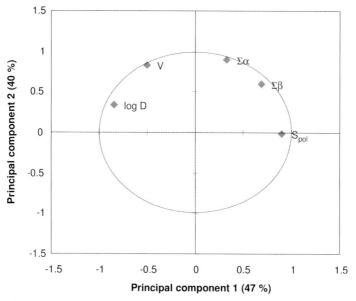

Fig. 1. Loading plot obtained on some of the physical properties of test compounds.

their polarity. In the same way, the direction of the vector V (Fig. 1) and contribution to PC2 (Table 2) indicates that the test compounds can be ranked depending on their size.

In conclusion, the basic test compounds can be clustered in the following groups:

- large polar molecules (morphine, quinine, procainamide, chloroprocaine, codeine, mepivacaine)
- large apolar molecules (methadone, fentanyl, diphenhydramine, nortriptyline)
- small polar molecules (nicotine, benzylamine, amphetamine, pyridine)

and, due to the wide range of physicochemical properties that they exhibit, they represent a valuable set of test compounds.

Table 2. Contribution of variables (%) to Principal Components (PC).

Variable	PC1	PC2	PC3	PC4	PC5
V	10.423	35.247	1.908	12.026	40.396
S_{pol}	34.215	0.014	43.413	0.137	22.221
$\log D$	30.355	5.666	24.643	39.050	0.285
$\Sigma\alpha$	4.677	41.145	7.306	10.245	36.627
$\Sigma\beta$	20.330	17.928	22.729	38.543	0.471

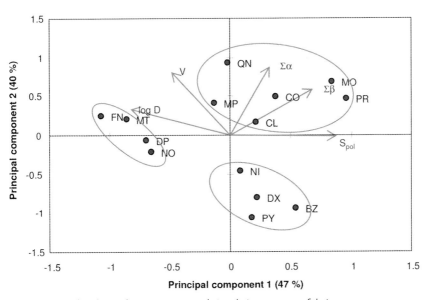

Fig. 2. Scores plot obtained on test compounds in relation to some of their physicochemical properties.

2.1.4.3.2 Chromatographic Properties of Test Compounds

Test compounds can now be used for the evaluation of base-deactivated supports. With this aim, they are injected, in different mobile phases, onto the chromatographic supports to be evaluated. Then, chromatographic parameters such as *retention factor* and *asymmetry value* are measured and used for the evaluation of the performance of the chromatographic support in the analysis of basic compounds.

In the present example, five different chromatographic supports were tested at two pH values (pH 3.0 and 7.0) with three isocratic mobile phases [3, 4]:

- mobile phase 1: acetonitrile – pH 7.0, 0.0375 M phosphate buffer (40 : 60, *v/v*)
- mobile phase 2: methanol – pH 7.0, 0.0643 M phosphate buffer (65 : 35, *v/v*)
- mobile phase 3: acetonitrile – pH 3.0, 0.0265 M phosphate buffer (15 : 85, *v/v*)

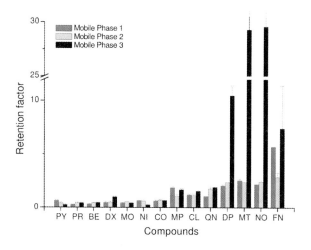

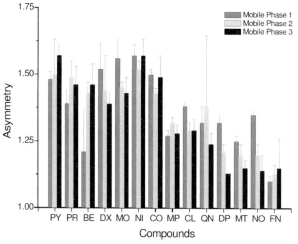

Fig. 3. Distribution of chromatographic parameters.

The 14 pre-selected test compounds were injected individually and their chromatographic parameters measured under each set of chromatographic conditions. For a better understanding of the importance of the selection of the test compounds, let us consider the measured chromatographic parameters.

As shown in Fig. 3, the retention factors and asymmetry values measured on one of the five supports covered a wide range of values, due to the different physicochemical properties of the selected test compounds. For example, some of the test compounds (diphenhydramine, methadone, nortriptyline, fentanyl) were strongly retained, especially in mobile phase 3. This behavior is strictly correlated with their physicochemical properties, since these analytes are large and lipophilic. On the other hand, small and polar molecules were only slightly retained under all of the chromatographic conditions.

Only in this way can a comprehensive evaluation of the chromatographic performances of stationary phases be achieved.

2.1.4.4 Use of PCA for the Evaluation of Chromatographic Supports

Let us now consider how PCA can be used to describe chromatographic similarities or differences between stationary phases. In our example, five deactivated stationary phases (Table 3) were evaluated. According to the chromatographic test methodology described above, selected stationary phases were firstly tested in three mobile phases at two different pH values (pH 3.0 and pH 7.0). It is of interest to know whether these stationary phases exhibit similar (or different) chromatographic behavior in relation to their structures (electrostatically shielded, bidentate, C_{18} carbon chain, etc.) rather than mobile phase composition. This information helps in ascertaining whether it is more important to select the most suitable stationary phase for a particular separation or to work under the optimal chromatographic conditions. Principal Component Analysis can be used to derive this information; in this case, chromatographic columns combined with mobile phases are the *scores* (e.g., ABZ-mobile phase 1).

Table 3. Tested chromatographic supports.

Column	Abbreviation	Manufacturer	Characteristics	Dimensions (mm)	Particle size (µm)
Supelcosil ABZ Plus	ABZ	Supelco	Electrostatically shielded	150 × 4.6 I.D.	5
Discovery C_{16} RP Amide	DIS	Supelco	Electrostatically shielded	150 × 4.6 I.D.	5
Nautilus C_{18}	NAU	Macherey Nagel	Electrostatically shielded	125 × 4.6 I.D.	5
Nucleosil AB	AB	Macherey Nagel	C_{18} carbon chain	125 × 4.6 I.D.	5
Zorbax Extend C_{18}	EXT	Agilent	Bidentate bonding (C_{18})	150 × 4.6 I.D.	5

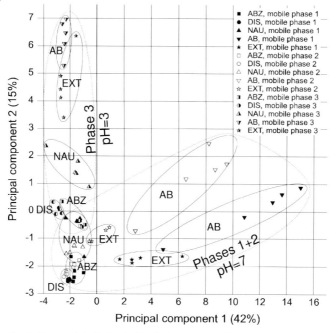

Fig. 4. Scores plot obtained with all chromatographic parameters.

In the obtained *score plot* (Fig. 4), each axis seemingly corresponds to a mobile phase pH: the first axis to pH 7.0 (stationary phases in mobile phases 1 and 2), the second axis to pH 3.0 (stationary phases in mobile phase 3). In addition, cluster positions suggest that the chromatographic behavior of embedded polar group supports (DIS, NAU, ABZ) is less pH-dependent in comparison to the two other supports (AB and EXT). In fact, the pH 7.0 and pH 3.0 clusters of the embedded polar group supports are situated close together. As asymmetry and reduced height are especially involved in the construction of the first axis, score positions along PC1 are correlated to their silanol activity. On the other hand, the second axis ranks the columns especially with regard to their hydrophobicity, as retention is mostly involved in PC2.

Another important point that can be easily assessed in this kind of study is the *batch-to-batch* and *column-to-column* reproducibility. In fact, for each of the studied supports, five columns (three of the same batch and two of different batches) were evaluated and results are reported in the *score plot*. It can clearly be discerned how the batch and column variability can depend on chromatographic conditions.

2.1.4.4.1 Evaluation of Chromatographic Supports in Mobile Phases Composed of pH 7.0 Phosphate Buffer

In view of the previous observations (Fig. 4), chromatographic data should be separately treated in each mobile phase to obtain a fine distinction of stationary phases in relation to their silanol activity (pH 7.0) or hydrophobicity (pH 3.0).

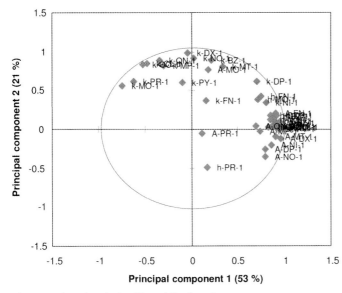

Fig. 5. Loading plot obtained in mobile phase 1.

The determination of silanol activity is an essential step in selecting a suitable stationary phase for the analysis of basic compounds. Silanol activity comprises a number of stationary phase-solute interactions, such as ion-ion (ion exchange), ion-dipole, dipole-dipole (e.g., hydrogen bonding), dipole-induced dipole, and induced dipole-induced dipole interactions (London forces), of which ion-ion and hydrogen bonding are probably the most important [2]. For this reason, chromatographic parameters measured in a mobile phase composed of pH 7.0 buffer (especially asymmetry value) are a good indication of the interactions between charged silanol groups and charged (depending on their pK_a value) basic compounds.

As an example, we will consider only the column evaluation obtained in mobile phase 1, as that obtained in mobile phase 2 is very similar (data not shown).

By inspection of the loading plot obtained (Fig. 5) and the contribution of variables to the selected PC axis (Table 4), we deduce that PC1 mainly consists of asymmetry and reduced height, and PC2 of retention. Thus, differences between NAU, ABZ, and DIS are, for example, mainly due to differences in retention. Nevertheless, most of the variance is explained by PC1, meaning that column positions along this axis can be used as ranking criteria of stationary phases in relation to their silanol masking capacity.

In addition, on the same score plot (Fig. 6), we can observe how some of the tested supports (ABZ, DIS, NAU, ABZ) showed better batch or column reproducibility in comparison to others (AB). These results clearly demonstrate that new generation stationary phases (embedded polar group and bidentate bonding supports) are well adapted to the analysis of basic compounds, even under these "critical" chromatographic conditions.

Table 4. Contributions of variables to PC axes.

Variable	F1	F2	Variable	F1	F2
A-BZ-1	3.972	0.068	h-MP-1	3.516	0.038
A-CL-1	4.171	0.004	h-MT-1	3.634	0.121
A-CO-1	3.920	0.011	h-NI-1	4.085	0.089
A-DP-1	2.864	0.698	h-NO-1	3.286	0.172
A-DX-1	4.198	0.151	h-PR-1	0.129	2.604
A-FN-1	3.746	0.438	h-PY-1	4.082	0.000
A-MO-1	0.154	6.237	h-QN-1	3.142	0.011
A-MP-1	2.474	0.005	k-BZ-1	0.209	8.442
A-MT-1	3.674	0.094	k-CL-1	1.009	7.599
A-NI-1	3.322	0.421	k-CO-1	1.227	7.434
A-NO-1	2.877	1.290	k-DP-1	2.289	4.002
A-PR-1	0.066	0.027	k-DX-1	0.008	10.138
A-PY-1	4.304	0.005	k-FN-1	0.110	1.456
A-QN-1	2.175	0.013	k-MO-1	2.451	3.364
h-BZ-1	3.574	0.262	k-MP-1	0.390	7.227
h-CL-1	3.800	0.001	k-MT-1	0.535	6.922
h-CO-1	3.475	0.021	k-NI-1	2.952	1.250
h-DP-1	3.767	0.345	k-NO-1	0.004	8.893
h-DX-1	3.288	0.331	k-PR-1	1.708	4.036
h-FN-1	2.515	1.848	k-PY-1	0.044	3.902
h-MO-1	2.348	1.541	k-QN-1	0.507	8.485

Table 5. Contribution of variables to PC axes.

Variable	F1	F2	Variable	F1	F2
A-BZ-3	4.805	0.401	h-MP-3	1.500	5.841
A-CL-3	0.076	0.615	h-MT-3	0.002	4.492
A-CO-3	0.036	1.221	h-NI-3	1.733	2.738
A-DP-3	4.872	0.379	h-NO-3	1.930	2.222
A-DX-3	0.296	5.666	h-PR-3	2.971	2.680
A-FN-3	3.926	0.602	h-PY-3	3.064	2.899
A-MO-3	0.996	0.021	h-QN-3	0.000	6.795
A-MP-3	0.105	2.859	k-BZ-3	0.285	6.492
A-MT-3	4.029	0.014	k-CL-3	1.933	1.816
A-NI-3	0.694	0.077	k-CO-3	0.417	4.332
A-NO-3	3.929	0.001	**k-DP-3**	**6.519**	0.011
A-PR-3	3.215	0.974	k-DX-3	3.603	2.168
A-PY-3	0.588	0.243	**k-FN-3**	**6.671**	0.026
A-QN-3	5.191	0.063	k-MO-3	0.466	6.648
h-BZ-3	2.407	2.739	k-MP-3	5.137	0.886
h-CL-3	0.958	6.895	**k-MT-3**	**6.698**	0.008
h-CO-3	0.817	2.380	k-NI-3	3.457	2.978
h-DP-3	0.345	4.806	**k-NO-3**	**6.589**	0.008
h-DX-3	0.018	1.483	k-PR-3	0.261	6.954
h-FN-3	0.000	2.922	k-PY-3	3.837	2.392
h-MO-3	2.644	1.948	k-QN-3	2.979	1.305

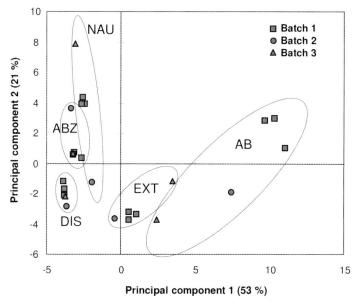

Fig. 6. Score plot obtained in mobile phase 1.

2.1.4.4.2 Evaluation of Chromatographic Supports in Mobile Phases Composed of pH 3.0 Phosphate Buffer

In a mobile phase composed of a pH 3.0 phosphate buffer, silanol groups are mostly unionized, limiting ion-exchange interactions responsible for the problematic peak tailing, and columns will thus be ranked according to their hydrophobic properties.

Contrary to the previously discussed loading plot (Fig. 5), vectors are more spread out in this case (Fig. 7) and some of them are not very well represented on the selected PC axes (length < 1.0). In such cases, as we know that in this mobile phase silanols are mostly unionized and columns are evaluated in relation to their hydrophobicity, we can refer to the direction of the retention vectors of large and apolar compounds. These latter contribute strongly to PC1 and are also highly mutually correlated (Table 5). Thus, chromatographic supports in this direction (AB and EXT in Fig. 8) are certainly more hydrophobic than those in the opposite direction. These findings are also supported by the high retention factors obtained for apolar compounds under these chromatographic conditions. Basic compounds, even if positively charged in a pH 3.0 buffered mobile phase, show strong hydrophobic interactions with the chromatographic support, due to their apolar structures.

As regards the batch-to-batch variability, it is interesting to note that in this mobile phase similar results are obtained both for "new generation" and conventional chromatographic supports. These findings confirm how batch and column reproducibility are strongly influenced not only by the bonding type (more or less adapted for the analysis of basic compounds), but also by the chromatographic conditions.

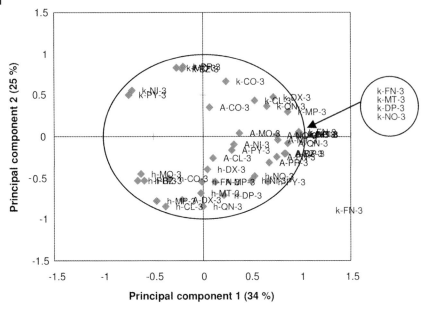

Fig. 7. Loading plot obtained in mobile phase 3.

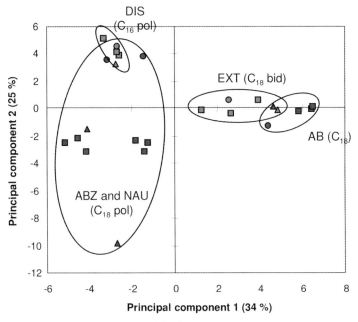

Fig. 8. Scores plot obtained in mobile phase 3.

2.1.4.5 How a Chromatographic Test can be Optimized by Chemometrics

Having demonstrated that the chromatographic test allows a suitable characterization of base-deactivated stationary phases, another important step is the improvement and optimization of its methodology. An easily applied and simple methodology is, in fact, a critical aspect to consider for the quality of a chromatographic test. The maximum amount of information should always be achieved with a minimal number of injections.

2.1.4.5.1 Test Compounds

First of all, the number of test compounds to be employed should be as low as possible. In our ongoing example, we show how a set of 14 test compounds can be successfully reduced to seven, while still maintaining the same quality of column evaluation. The 14 previously selected test compounds possessed different physicochemical properties and could be classified on the basis of size and polarity. If we keep two molecules per physicochemical group (the ones more easily commercially available), the physicochemical domain is still well covered and consequently a valuable column evaluation should be maintained.

A possible way to confirm this hypothesis is to perform column evaluations with both the complete and reduced sets of test compounds and then compare the plots. As an example, we consider the results obtained in mobile phase 1 (Figs. 5 and 9). In both *score plots*, the amount of variance (%) attributable to selected PC axes is similar: 73% (52% for PC1 and 21% for PC2) with 14 test compounds and 71% (53% for PC1 and 18% for PC2) with seven test compounds. Thus, the column evaluation obtained with the reduced set of test compounds is almost identical to that obtained with the whole set.

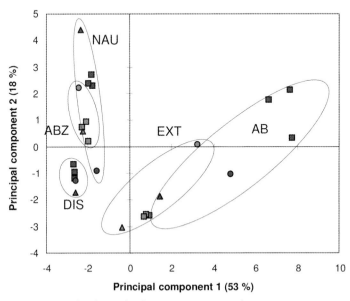

Fig. 9. Scores plot obtained with seven test compounds.

2.1.4.5.2 **Mobile Phases**

Another important issue in the optimization of a chromatographic test is the selection of the most appropriate chromatographic conditions to analyze the test compounds. More specifically, depending on the mobile phase composition, different kinds of information on the chromatographic performance of stationary phases can be gleaned. In a first attempt, columns were tested with two mobile phases composed of pH 7.0 phosphate buffer (one with acetonitrile and the other with methanol) and one composed of a pH 3.0 phosphate buffer.

Column evaluations obtained in mobile phases 1 and 2 were very similar at first sight and both seem appropriate for the evaluation of stationary phases in relation to their silanol activity. Results obtained with the two mobile phases indicated that a slightly better batch variability and asymmetry were achieved when using methanol as organic modifier (data not shown).

This observation indicates that methanol should be preferred to acetonitrile as organic solvent when looking for the optimal conditions (especially in terms of asymmetry factor) in the analysis of basic compounds. Methanol, being a protic solvent, plays an active role in the chromatographic separation, interacting both with free silanol groups and basic compounds. On the other hand, acetonitrile, being an aprotic solvent, does not interact with free silanol groups or basic compounds. For this reason, acetonitrile is the solvent to use in column evaluation, as it allows a better discrimination of stationary phases in relation to their silanol activities (due to secondary interactions between silanol groups and basic compounds).

The third mobile phase, composed of a pH 3.0 phosphate buffer and acetonitrile, allows a fine distinction of stationary phases according to their hydrophobic properties. This is also helpful for selecting the most suitable stationary phase and for this reason it should be retained. In fact, depending on the compound to be analyzed, a more retentive support can, for example, be preferred to a slightly hydrophobic one.

2.1.4.5.3 **Chromatographic Parameters and Batch (Column) Reproducibility**

In a chromatographic test, stationary phases are evaluated on the basis of the measured chromatographic parameters. For this reason, retention, asymmetry, and reduced height should be studied in each mobile phase. Let us first consider the loading plots (Figs. 6 and 10) obtained with mobile phase 1. All asymmetry and reduced height vectors are in the same direction and are well represented (as their length is close to 1.0). This means that, in this particular mobile phase, these two chromatographic parameters are correlated, and so the same information should be maintained by keeping only one of them.

In mobile phase 3, as supports are evaluated according to their hydrophobicity and bonding type, the most important chromatographic parameter is the retention factor. In this acidic mobile phase, silanol groups are mostly uncharged (depending on their pK_a) and secondary interactions with charged basic compounds are unlikely. Nevertheless, it is important to measure asymmetry factors of basic compounds under these chromatographic conditions as well, as they

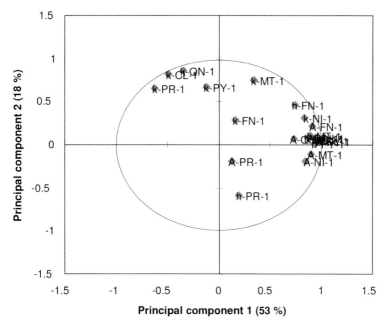

Fig. 10. Loadings plot obtained in mobile phase 1 with seven test compounds.

represent a good indication of stationary phases possessing silanol activity even at low pH.

Column evaluations obtained in mobile phases 1 and 3 (see the loading and score plots obtained with mobile phase 1 in Figs. 11 and 12, respectively) with only retention and asymmetry of test compounds were identical to that previously obtained with the whole set of data.

Concerning the batch-to-batch and column-to-column reproducibility, the results (*score plots* obtained under each set of chromatographic conditions) showed that new generation supports ensure better batch reproducibility compared to conventional ones. This suggests that only inter-batch reproducibility should be assessed, especially when testing new generation base-deactivated supports.

This optimization entails a large reduction in the number of measured chromatographic parameters per stationary phase: 252 (7 compounds × 3 injections × 3 columns × 2 mobile phases × 2 chromatographic parameters) instead of 1890 (14 compounds × 3 injections × 5 columns × 3 mobile phases × 3 chromatographic parameters). Furthermore, in the analysis of basic compounds, only the mobile phase at pH 7.0 is effectively necessary to achieve a good evaluation of silanol masking capacity of chromatographic supports. A second mobile phase (at pH 3.0) may be required only when an assessment of the hydrophobic properties of stationary phases is needed.

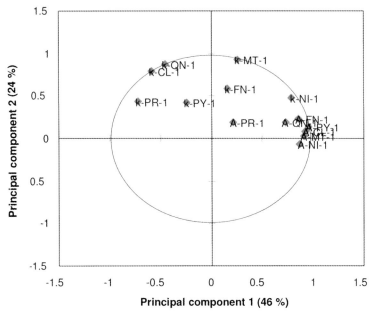

Fig. 11. Loadings plot obtained in mobile phase 1 with only *k* and As of seven test compounds.

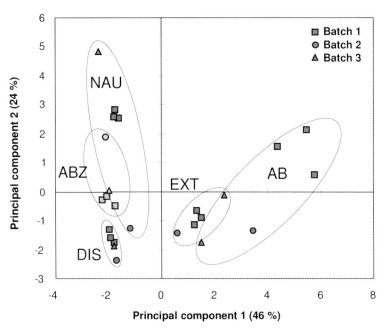

Fig. 12. Score plot obtained in mobile phase 1 with only *k* and As of seven test compounds.

2.1.4.6 Conclusion and Perspectives

Thanks to PCA in this example, and chemometric tools in general, a graphical representation of differences and similarities between individuals (test compounds or chromatographic supports) can be rapidly achieved. We have demonstrated the importance of PCA in selecting the most suitable test compounds, in improving the chromatographic test methodology, as well as in choosing the best column for a given application (e.g., the analysis of basic compounds). This is an important point, especially for the analyst, who rarely has the opportunity to test many columns before finding the "good one". For this reason, the use of computational packages that allow optimization of a particular separation or the selection of the most suitable column is becoming increasingly popular [5].

In the same way, column evaluation obtained with such a chemometric tool certainly influences, and will continue to influence, manufacturer's thoughts about developing new stationary phases adapted to a particular class of compounds (e.g., basic compounds). Databases containing as many chromatographic supports as possible are thus assembled, with a view to improving the development of new stationary phases.

References

1   D. V. McCalley, R. G. Brereton,
       J. Chromatogr. A 828 (1998), 407.
2   J. Nawrocki, *J. Chromatogr. A* 779 (1997),
       29.
3   C. Stella, P. Seuret, S. Rudaz,
       P.-A. Carrupt, J. Y. Gauvrit, P. Lantéri,

J.-L. Veuthey, *J. Sep. Sci.* 25 (2002),
1351–1363.
4   D. V. McCalley, *J. Chromatogr. A* 738
       (1996), 169–179.
5   M. R. Euerby, P. Petersson,
       J. Chromatogr. A 994 (2003), 13.

2.1.5

Linear Free Energy Relationships (LFER) – Tools for Column Characterization and Method Optimization in HPLC?

Frank Steiner

2.1.5.1 Characterization and Selection of Stationary Phases for HPLC

In liquid chromatography, the complexity of parameters contributing to retention is high. This is due to numerous different interactions between the stationary phase and the analyte on the one hand and the mobile phase and the analyte on the other. Moreover, the mobile phase interacts with the stationary phase, thus modulating its properties. Within this multifaceted entity, the stationary phase plays the central role. A schematic illustration of a C_{18} reversed-phase system in a water/acetonitrile eluent is depicted in Fig. 1.

Since a vast number of reversed phases is available on the market, the successful selection of an appropriate column for a given analytical problem is a true challenge. Ignoring simple trial-and-error strategies, it is commonly based on a systematic characterization of column properties. The term column will be used synonymously with stationary phase. Chromatographic column tests are most important, as they directly yield relevant data for later applications. They are mainly

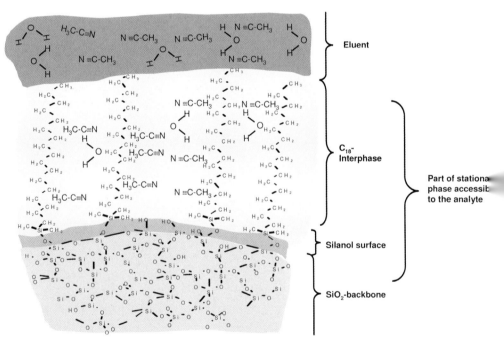

Fig. 1. Scheme of a silica-based C18 reversed phase in contact with a water/acetonitrile eluent according to the concept of an alkyl chain interphase.

based on measuring retention factors of characteristic test substances or, in some cases, peak asymmetries. These primary data need to be plotted in an intelligent way, or even better treated with statistical/chemometric methods such as principal component analysis (PCA) in order to extract further information and to enable an objective classification of columns. Such methods provide the potential to monitor similarities, diametrical differences or orthogonal behavior of systems in a manner far beyond a simple discussion of primary chromatographic results. The chemometric factorial methods, however, do not facilitate the correlation of chromatographic behavior and structural properties. They are hence not convenient for identifying and discussing retention mechanisms.

A different approach to the systematic characterization of stationary phases is the correlation of analyte structure and retention in a given chromatographic system with the help of quantitative structure retention relationships (QSRR), a distinct discipline of linear free energy relationships (LFER). In QSRR, the total retention of an analyte is separated into individual contributions such as dipole-dipole, π-π, acid-base, and hydrophobic interactions. This approach enables strict interrelations on the basis of fundamental mechanistic aspects in order to improve the physico-chemical understanding of chromatographic retention.

2.1.5.2 What are LFERs and Why can they be Profitable in HPLC?

The retention factor k, a common measure of chromatographic retention, depends on the equilibrium constant K describing the partition of the analyte between the mobile and stationary phases and the so-called phase ratio Φ, which is the ratio of the capacities of the two phases to accommodate an analyte (see Eq. 1).

$$k = K \cdot \Phi \tag{1}$$

The retention factor thus follows the rules of equilibrium thermodynamics, whereby the degree of retention is controlled by the change in the Gibbs free energy of the analyte molecule on going from the mobile into the stationary phase. This is in accordance with the general description of chemical equilibria in terms of standard free energies ΔG^0 (Eq. 2) with universal gas constant R and absolute temperature T).

$$\Delta G^0 = -R \cdot T \cdot \ln K \tag{2}$$

The Hammett equation is regarded as the origin of linear free energy relationships (LFER). It correlates the rates of hydrolysis of differently substituted benzoic acid esters with the acidities of the equivalently substituted benzoic acids. A linear relationship between the free energy for the acid dissociation and the free activation energy for the ester hydrolysis is assumed. Both energies are sensitive to the inductive (electron-shifting) effects of a particular substituent, which are embodied in the substituent constant σ, an example of a molecular descriptor. The fundamentals of the Hammett theory are summarized in Fig. 2.

Analogously to the Hammett relationship, and the more complex relationships between structure and reactivity derived from it, chromatographers equally strive for appropriate descriptions of retention with the help of linear free energy

Relationship between the rate of hydrolysis of a substituted ester k_x and that of its non-substituted form k_H and correlation with the corresponding acidity constants K_x and K_H	$\log \dfrac{k_x}{k_H} = \rho \cdot \log \dfrac{K_x}{K_H}$
Definition of the substitutent constant σ_x:	$\sigma_x = \log \dfrac{K_x}{K_H}$
Hammett equation with reaction constant ρ:	$\log \dfrac{k_x}{k_H} = \rho \cdot \sigma_x$

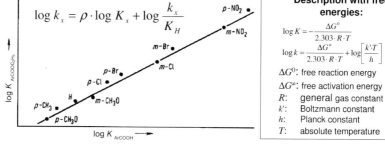

$$\log k_x = \rho \cdot \log K_x + \log \frac{k_x}{K_H}$$

Description with free energies:

$$\log K = -\frac{\Delta G^0}{2.303 \cdot R \cdot T}$$

$$\log k = \frac{\Delta G^\neq}{2.303 \cdot R \cdot T} + \log\left[\frac{k' \cdot T}{h}\right]$$

ΔG^0: free reaction energy
$\Delta G^\neq$: free activation energy
R: general gas constant
k': Boltzmann constant
h: Planck constant
T: absolute temperature

Fig. 2. Derivation of the Hammett equation for the linear correlation of logarithm of rate constant for ester hydrolysis and logarithm of acidity constant of benzoic acids with equivalent substitution. This equation is thus a linear correlation of the free activation energy for hydrolysis with the free energy of acid dissociation.

relationships. A general approach for describing retention with an LFER is given in Eq. (3):

$$\log k = \log k_0 + m(\delta_s^2 - \delta_m^2)V_2 + s(\pi_s^* - \pi_m^*)\pi_2^* + a(\beta_s - \beta_m)\alpha_2 + b(\alpha_s - \alpha_m)\beta_2 \quad (3)$$

with the descriptors for physico-chemical properties of the stationary phase (index $_s$), mobile phase (index $_m$), and the analyte (index $_2$) as explained in the following:

V intrinsic molar volume of analyte
δ Hildebrand solubility parameter for mobile and stationary phases
π^* Kamlet–Taft polarity parameter
α Kamlet–Taft hydrogen-bond acceptor (basicity) parameter
β Kamlet–Taft hydrogen-bond donor (acidity) parameter

$\log k_0$ is the intercept of the multivariate (here four-dimensional) linear regression. It contains the phase ratio of the system, as well as other properties of the chromatographic system that do not depend on the analyte descriptors. The coefficients m, s, a, and b represent the individual contributions of each type of interaction (retention mechanism). It follows that the difference in the stationary phase parameter (e.g., δ_s^2) and the mobile phase parameter (e.g., δ_m^2) multiplied by the analyte descriptor (e.g., V_2) is proportional to the partial change in free energy related to that distinct interaction mechanism. All partial free energy differences are summed to give the total difference in Gibbs free energy, which is proportional to the logarithm of the retention factor k according to the rules of thermodynamics (see Eq. 2). The individual contributions are hydrophobic retention, interactions between electrical dipoles, and the complementary interplay of hydrogen-bonding properties, where the stationary phase can function as an acceptor or as a donor. In reversed-phase chromatography, hydrophobic retention prevails. The related parameters of the chromatographic system are the squares of the Hildebrand solubility parameters, also named cohesion energy densities. The cohesion energy density δ^2 can be calculated from the internal heat of evaporation ΔE over the molar volume V_m of the substance that forms the phase. The molar volume of the solute is considered as an analyte descriptor for hydrophobic retention. It can be calculated theoretically after McGowan from its molecular formula [1].

The rationale for hydrophobic retention on a molecular basis is the transition of the solute from a so-called cavity in the mobile phase to a new cavity that needs to be generated inside the stationary phase, as shown schematically in Fig. 3. It is obvious that the energy needed to create a cavity in a particular phase will increase with the cohesion energy density of the phase and the molecular volume of the solute. In the case of silica-based reversed phases, the stationary phase cavity must be localized within an interphase formed by the solvated alkyl chains bonded to the silica surface (see Fig. 3). When the solute leaves the mobile phase, the former cavity can close-up and energy is liberated. The generation of the new cavity in the stationary interphase requires a smaller amount of energy, since the cohesion energy of the weakly interacting alkyl chains is significantly lower.

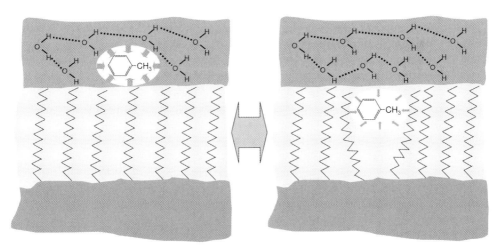

Fig. 3. Retention in reversed-phase chromatography according to the model of interphase and solvophobic effects. Due to the elevated cohesion energy density within the partly aqueous mobile phase, the energy liberated upon closing a cavity therein exceeds the energy required to create a new cavity within the less cohesive alkyl chain interphase.

This is the so-called solvophobic retention model. The analyte is forced to leave a strongly cohesive mobile phase, where it interrupts the internal molecular (mainly hydrogen-bonding) interaction due to its less polar structure, in order to be absorbed into a new cavity in the less cohesive alkyl interphase, which is closer to its own polarity.

2.1.5.3 How to Obtain Analyte Descriptors for the Multivariate Regression

The parameters for a given chromatographic system are determined through multivariate regression from a multitude of measured retention factors of selected solutes with their individual descriptors for relevant chemical properties. Besides the difficulty in defining these relevant properties, the main problem is how to determine the parameters of the chromatographic system, in particular those of the stationary phase.

The strategy of incorporating these parameters into the regression coefficients, as in Eq. (4), is a way of overcoming this problem. One then obtains an expression of the form:

$$\log k = \log k_0 + m' \, V_2 + s' \, \pi_2^* + a' \, \alpha_2 + b' \, \beta_2 \tag{4}$$

where the symbols are as given for Eq. (3). It should be emphasized, however, that the parameters m', s', a', and b' are not identical with the coefficients m, s, a, and b of Eq. (3). The new regression coefficients describe the chromatographic system completely. They can be determined from the retention factors of a set of solutes for a defined chromatographic system, provided that molecular descriptors of these solutes are available.

The solute descriptors such as π^*, α, and β can be extracted from the literature [2–5] and are in most cases empirically determined by spectroscopic measurements. Their calculation is based on the shift of absorption bands of the given solutes due to varying solvatochromic effects when the dipolar or hydrogen-bond donor/acceptor properties of the solvent mixture used for spectroscopy are altered. Hence, they are not true thermodynamic data. Abraham and co-workers also determined descriptors from GC retention data and octanol/water partition coefficients [6–9]. With the help of such descriptors, the so-called solvation equation (Eq. 5) can be set out [10, 11]:

$$\log k = c + r \cdot R_2 + s \cdot \pi_2^* + a \cdot \sum \alpha_2^H + b \cdot \sum \beta_2^H + \upsilon \cdot V_x \tag{5}$$

with descriptors comprising solute excess molar refraction (R_2), solute dipolarity or polarizability (π_2^*), solute overall H-bond acidity ($\sum \alpha_2^H$), solute overall H-bond basicity ($\sum \beta_2^H$), and the McGowan characteristic volume (V_x). The multivariate regression yields a parameter describing the propensity of the chromatographic phases to interact with solute n- or π-electrons (r), one for the dipolarity/polarizability of the phases (s), a measure of the H-bond basicity of the phases (a), one for the H-bond acidity of the phases (b), and one for the hydrophobic properties to describe differences in the energy for cavity formation (υ), besides the intercept of the regression (c).

A concise introduction to LFER strategies in HPLC can be found in the following chapter by Snyder and Dolan.

2.1.5.4 LFER Procedure Using the Solvation Equation

2.1.5.4.1 Comparing Stationary Phases on the Basis of LFER Data

The use of solute descriptors from the solvation equation (Eq. 5) for application in an LFER regression to characterize liquid chromatographic systems in terms of regression parameters is documented in numerous examples in the literature. To illustrate the procedure for determining these parameters and to demonstrate their convenience in phase characterization, an example from the literature, presented in a series of three publications [12–14], is discussed here.

Multivariate linear regression can be carried out with common statistics software and was performed in the given example with Statistica 5.0 (StatSoft). Retention factors were measured for a set of selected test solutes using a set of different columns under one fixed set of eluting conditions (acetonitrile/water, 3 : 7, *v/v*, without addition of a buffer). A buffer was avoided, since Engelhardt and co-workers reported a marked modulation of column properties by salt or buffer additives; this adulterates phase characterization and makes columns look less silanophilic than they actually are [15].

The column hold-up time t_M of all columns was measured with 0.05 mM nitrate solution, which can lead to biased values as anion exclusion effects at dissociated silanol groups are possible. Test solutes (readily available substances) were selected so as to cover a wide range of polarities and diverse molecular structures. Phenolic and weakly basic compounds were included. A list of all test substances, together with the related solvatochromic descriptors, which were extracted from publications of Abraham [6–9], is given in Table 1. Stronger acids or bases were excluded, and anyway these would not elute with regular peaks from a silica-based reversed-phase column without any buffer or salt addition. The multivariate regression yielded six parameters for each chromatographic system (column and eluent) according to Eq. (5). Therefore, a set of 34 linear equations relating to the retention data of the 34 test solutes was used for the regression statistics. Some 15 different stationary phases were selected, covering a reasonable variety of pore widths, specific surface areas, and bonding characteristics. Table 2 lists all of the stationary phases with their characteristic parameters. The set also included phases with very similar properties, e.g., end-capped and non-endcapped versions of the same material. Besides 14 typical reversed phases, partially incorporating polar groups in the alkyl chain (symmetry shield), a phase with polar bonding (nitrile) was also characterized.

For each of the 15 stationary phases, Fig. 4 depicts six parameters calculated by multivariate regression from the retention data measured under the given eluent conditions. The error bars represent the confidence ranges for a 95% significance level. It will be demonstrated in the following that the discussion of these parameters gives an insight into the prevailing retention mechanisms on the different columns.

Table 1. Test substances and related solvatochromic descriptors.

Compound	R_2	$\pi_2^{\ast}$	$\Sigma \alpha_2^H$	$\Sigma \beta_2^H$	V_x
Ethylbenzene	0.613	0.50	0	0.15	0.9982
Toluene	0.613	0.52	0	0.14	0.8573
Bromobenzene	0.882	0.73	0	0.09	0.8914
Chlorobenzene	0.718	0.65	0	0.07	0.8388
Caffeine	1.500	1.60	0	1.33	1.3632
Dimethyl phthalate	0.780	1.41	0	0.88	1.4288
Pyridine	0.631	0.84	0	0.52	0.6753
Acetophenone	0.818	1.01	0	0.48	1.0139
Methyl benzoate	0.733	0.85	0	0.46	1.0726
Ethyl benzoate	0.689	0.85	0	0.46	1.2135
Benzyl cyanide	0.751	1.15	0	0.45	1.012
N,N-Dimethylaniline	0.957	0.84	0	0.42	1.098
Anisole	0.708	0.75	0	0.29	0.916
o-Nitrotoluene	0.866	1.11	0	0.27	1.032
Nitrobenzene	0.871	1.11	0	0.26	0.891
Hydroquinone	1.000	1.000	1.16	0.60	0.834
p-Nitrophenol	1.070	1.72	0.82	0.26	0.9493
Methylparaben	0.90	1.37	0.69	0.45	1.1313
Ethylparaben	0.86	1.35	0.69	0.45	1.2722
Propylparaben	0.86	1.35	0.69	0.45	1.4131
Butylparaben	0.86	1.35	0.69	0.45	1.554
β-Naphthol	1.520	1.08	0.61	0.45	1.144
α-Naphthol	1.520	1.05	0.60	0.37	1.144
Phenol	0.805	0.89	0.60	0.30	0.7751
p-Cresol	0.820	0.87	0.57	0.32	0.916
o-Cresol	0.840	0.86	0.52	0.31	0.916
p-Ethylphenol	0.800	0.90	0.55	0.36	1.0569
3,5-Dimethylphenol	0.82	0.84	0.57	0.36	1.0569
2,6-Dimethylphenol	0.860	0.79	0.39	0.39	1.0569
p-Nitroaniline	1.220	1.91	0.42	0.38	0.991
Benzyl alcohol	0.803	0.87	0.33	0.56	0.916
Aniline	0.955	0.96	0.26	0.50	0.8162
o-Toluidine	0.966	0.92	0.23	0.59	0.9571
α-Naphthylamine	1.670	1.26	0.20	0.57	1.185

Table 2. Investigated stationary phases and their related parameters.

Stationary phase	Manufacturer	Ligand type	Particle size (µm)	Pore size (nm)	Surface area (m² g⁻¹)	%C approx.	Abbre- viation
Lichrospher 100 RP-18e	Merck	C18	5.0	10	350	21.6	M-C18e
Lichrospher 100 RP-18	Merck	C18	5.0	10	350	21.0	M-C18
Purospher 100 RP-18	Merck	C18	5.0	12	350	18.0	M-PURe
Purospher	Merck	C18	5.0	8	500	18.5	M-PUR
Lichrospher PAH	Merck	C18	5.0	15	200	20.0	M-PAH
Symmetry Shield RP-C18	Waters	C18	5.0	10	340	n.a.	SYM-C18
Symmetry Shield RP-C8	Waters	C8	5.0	10	340	15.0	SYM-C8
LiChrosorb RP-selectB	Merck	C8	5.0	6	300	11.4	M-RP-B
LiChrospher 100 RP-8	Merck	C8	5.0	10	350	12.5	M-C8
Aquapore OD-300	Applied Biosystems	C18	7.0	30	100	n.a.	A-C18
Synchropak RP-C18	Synchrom	C18	6.5	30	n.a.	7.5	S-C18
Aquapore Butyl	Applied Biosystems	C4	7.0	30	100	n.a.	A-C4
Synchropak RP-C4	Synchrom	C4	6.5	30	n.a.	7.5	S-C4
Zorbax SB 300 C8	Rockland Technologies	C8	5.0	30	45	1.5	Z-C8
Zorbax SB 300 CN	Rockland Technologies	CN	5.0	30	45	1.2	Z-CN

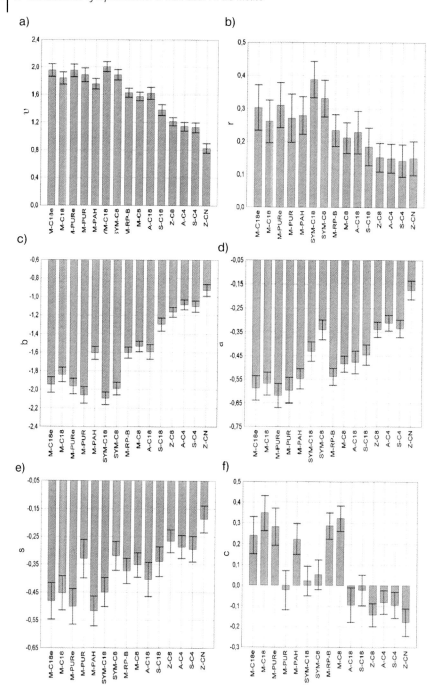

Fig. 4. Phase parameters (regression coefficients) for hydrophobicity (υ), n- and π-electron activity (r), H-bond donor acidity (b), H-bond acceptor basicity (a), polarizability (s), and intercept (c) calculated using Eq. (5) from the retention data of 34 model compounds on 15 different stationary phases. Error bars indicate the confidence range at a 95% significance level (from [12]).

The υ parameter (see Fig. 4a) is positive for all of the columns, which is indicative of a higher hydrophobic interaction in the stationary phase than in the mobile one. Neglecting the nitrile phase, υ is the most prominent parameter for all of the given columns, which confirms the expected dominance of hydrophobic effects in reversed-phase HPLC. Phases with high carbon content and high bonding density generally exhibit high hydrophobicity parameters. Moreover, C_{18} phases show higher hydrophobicity (average 1.9) than C_8 phases (average 1.7), and narrow pore materials lower energy for cavity formation than the wide pore phases (average 1.2). These insights express the differing hydrophobicities of the individual phases, which does not result from differing phase ratios (see Eq. 1) as is often invoked to explain the various retentive properties of these different types of reversed phases. The υ parameter of the polymer-bonded PAH column is rather low considering its high bonding density. This can be explained by the increased pore diameter (15 nm) or the more rigid arrangement of the alkyl chains at the surface, which serve to enhance the so-called shape selectivity for PAHs; this should also hamper cavity formation for hydrophobic solutes.

Besides the υ parameter, only the r parameter (see Fig. 4b) shows positive values. Hence, the interactions through n- or π-electrons are stronger in the stationary phase than in the eluent, which is not as obvious as the results on hydrophobicity. All r values are small, which reflects the minor difference between the stationary and mobile phases with respect to this interaction. The r values of the polar embedded phases (symmetry shield) are clearly higher than those of all the other columns, confirming the n-/π-electron activity of the incorporated carbamate moieties. The nitrile phase also shows a relatively prominent r parameter, which is easily traceable from the electronic structure of the surface groups. Whilst all other parameters of the cyano phase resemble those of the C_4 and C_8 phases, the r parameter indicates a higher contribution of electron interactions than with any other of the tested columns. Nevertheless, the nitrile phase r parameter is the smallest in its parameter set, whereas its υ parameter is 5.5 times the size. It can be concluded that the Zorbax SB 300 CN column mainly behaves as a reversed phase in the given system.

The b parameter (Fig. 4c) measures the difference in proton-donor activity between the two phases of the chromatographic system (complementary to the H-bond basicity of the solute) and opens the list of parameters with negative values. This results from the higher proton activity in the water-rich mobile phase, whilst only adsorbed water or residual silanol groups can act as donor moieties in the stationary phase. The negative sign further implies (considering other parameters as constant) a *decrease* in retention with an increase in the contribution of this parameter. Comparing the data depicted in Fig. 4c with those in Fig. 4a, it becomes obvious that the proton-donor parameter is as equally important in reversed-phase HPLC as the hydrophobic parameter. The influence of a specific interaction on the overall retention depends on the size of the system parameter as well as on the value of the related solute descriptor. Considering the H-bond acceptor descriptors $\Sigma \beta_2^H$ in Table 1, it becomes obvious that there is no strict correlation between this descriptor and the Brønsted basicities of the solutes.

Brønsted bases such as aniline, toluidine, and naphthylamine exhibit only medium values, whilst caffeine and dimethyl phthalate have the highest acceptor descriptors, despite not being Brønsted bases. Looking, for example, at the values of the main system parameters v and b of the Symmetry Shield C_{18} column as well as the related descriptors ($\Sigma\,\beta_2^H$, V_x) for the solute caffeine, a negative value of log k can be deduced, which was confirmed by a measured retention factor of less than unity (0.65). When comparing all of the b parameters of the 15 investigated columns, the strongly negative values of both of the Symmetry Shield phases are noticeable. This implies that the H-bond acidity of the stationary phase is markedly weaker than that of the eluent and it can be considered a confirmation of the carbamate group's shielding effect against residual silanol groups. Moreover, the lower b parameter of the phase Lichrospher RP-18e relative to that of Lichrospher RP-18 demonstrates the efficacy of the end-capping, in other words a reduction in the number of acidic silanol groups.

The "a" parameter (shown in Fig. 4d) represents differences in the H-bond acceptor capability between the two phases of the chromatographic system. This interaction of stationary phase and eluent with donor-active solutes is more pronounced in the mobile phase, as can be concluded from the negative sign of "a" for all of the columns. In contrast to the results for the donor activity difference of the chromatographic phases (high negative b values), the small sizes of the "a" parameter indicate the similarity of the mobile and stationary phases with respect to their acceptor activities, and thus a poor contribution to the overall retention. Additionally, the descriptor $\Sigma\,\beta_2^H$ is equal to zero for all solutes without any heteroatom-bound proton. For a reversed phase without incorporated polar groups, an acceptor activity is not chemically obvious. Although not a Brønsted base, the acceptor role is much more plausible for a carbamate moiety. This is the reason why both Symmetry Shield phases exhibit such small "a" values, since their acceptor activity is closer to that of the mobile phase than the basicity of the other columns in the set. The smaller "a" parameter of Symmetry Shield C_8 is due to the weaker influence of the shorter alkyl chain on the solute–carbamate interactions. The "a" parameter of the cyano phase shows the most marked distinction from those of the other phases among all of the regression parameters, which can be explained from the chemical structure.

The dipolarity/polarizability parameter s closes the list of the interaction parameters from Eq. (5). Again, the data show a negative sign, indicating more pronounced dipole or induced-dipole activity in the eluent than in the stationary phase. Principally, phases with higher carbon load show weaker dipolar interactions and thus more strongly negative s parameters. As expected, the polar embedded phases and the cyano phase exhibit smaller values. The difference between Symmetry Shield C_{18} and C_8 reflects the differences in the steric effects of the alkyl chains.

The intercept values of the multivariate regression (c in Eq. 5) are shown in Fig. 4f. They incorporate the phase ratio of the columns, as well as all other properties that do not depend on solute properties, or are not represented by the selected solute descriptors. Due to these complex contributions, the values cannot easily be discussed. It is obvious that all wide-pore materials, with their smaller specific surface area and thus smaller phase ratios, show negative intercepts, whereas the values for the high surface area phases are positive. The pronounced error bars for the c parameters indicate that the differences cannot be regarded as highly significant. Therefore, they are not further interpreted. On the other hand, the most important parameters, v and b, have the smallest confidence ranges and can be determined with the best precision. Generally, the precision of the system parameters is the most critical factor in all approaches for describing HPLC retention with the help of the solvation equation and given solute descriptors. This matter will be addressed in a later section.

2.1.5.4.2 The Influence of the Mobile Phase Expressed in LFER Parameters

In all of the foregoing discussions (related to [12]), the eluent conditions were kept constant during the determination of the parameters for all 15 columns. The influence of the eluent on the LFER parameters was studied by the same authors in a later publication [14]. For the combinations of five columns (LiChrospher 100 RP-18e, Purospher RP-18e, LiChrospher 100 RP-8, Symmetry Shield RP-C$_{18}$, and Symmetry Shield RP-C$_8$; see data in Table 2) and 12 different eluent conditions, the retention factors of 31 solutes were measured. One series with water/methanol, and another one with water/acetonitrile, both in 10% increments from 30% to 70% (v/v) without addition of a buffer, were carried out. Detailed results for all columns are not discussed here, as no specific additional information was obtained. To give an overview of the influence of the eluent on LFER parameters, Fig. 5 displays the different parameters averaged over all five columns as a function of the eluent acetonitrile (Fig. 5a) or methanol content (Fig. 5b). This plot shows the differences between methanol and acetonitrile as well as the role of the different parameters at one glance, but ignores the different characteristics of the individual columns. With both methanol and acetonitrile, the magnitudes of the positive and negative parameters decrease when the organic content is increased, since the mobile phase properties become increasingly adapted to those of the stationary phase. As the protic solvent methanol is more akin to water, this effect is more pronounced with acetonitrile. The predominance of the hydrophobicity parameter (v) and the H-bond donor parameter (b) is obvious, but the relative contribution of these mechanism decreases with increasing proportion of organic modifier. At 70% water content the parameters v and b describe 90% of the total variance, at 30% water content the contribution decreases to below 80%. The gain in importance of the s and r parameters indicates a shift in the direction of a normal-phase retention mechanism. To give a theory-based interpretation of the differences between the parameters in the methanol and acetonitrile eluents, the cohesion energy densities and proton-donor abilities of both organic solvents need to be considered. Methanol exhibits a 50% higher cohesion energy density, resulting in higher v parameters, and a markedly higher donor ability, and hence elevated b parameters when compared to acetonitrile. The two effects are compensatory, leading to similar overall elution strengths for both modifiers, with acetonitrile being slightly the stronger.

a)

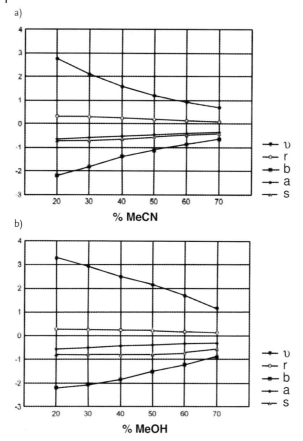

b)

Fig. 5. Regression coefficients (Eq. 5) obtained for the stationary phase M-C$_{18}$e as a function of percent of organic modifier (v/v) (from [14]).

Most of the observed parameter effects under variation of the mobile phase can be explained on a physico-chemical basis, but the information gained about different stationary phases as a result of variation of eluent conditions was found to be rather limited. Thus, a variation of the eluent is not mandatory for meaningful phase characterization with parameters from the solvation equation. However, it appears favorable to make such measurements at higher water content, whereby the individual characteristics of different phases prove to be more pronounced, but still under conditions where the alkyl chains are solvated (wetted) (~30% organic content).

2.1.5.4.3 The Prediction of Chromatographic Selectivity from LFER Data

In liquid chromatography, selectivity is the crucial optimization parameter in method development, whereby the prediction of the relative retention for a given pair of solutes with a specific eluent/column combination is a real challenge.

a)

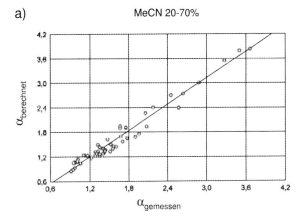

b)

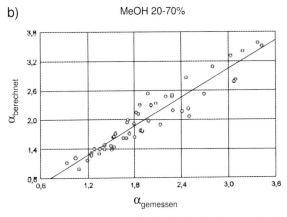

Fig. 6. Correlation of measured and calculated selectivities for five different eluent compositions and nine substance pairs on a Lichrospher 100 RP-18e column (from [14]).

Provided that the system (regression) parameters for a chromatographic method are known and the molecular descriptors for a "new" analyte are available, retention and thus selectivity should be predictable with the help of the LFER approach. Correlations of the calculated and the measured selectivities for nine pairs of solutes under six eluent compositions on a Merck Lichrospher 100 RP-18e column are shown in Fig. 6a for water/acetonitrile eluents and in Fig. 6b for water/methanol eluents (taken from [14]). This plot, however, does not illustrate a real prediction of solute retention, since the depicted measured retention data served at the same time as a basis for the calculation of the "theoretical" selectivities. Even under these less critical conditions, relative deviations (expressed as the distances from the best linear fit in the x-coordinate) reach values of up to 20%, due to the significant residues from the multivariate regression (see confidence ranges in Fig. 4). To discuss the impact of these uncertainties on the quality of a possible prediction, a real example is discussed here. Starting from a selectivity of

$\alpha = 1.30$ (theoretical value), an optimal chromatographic resolution of $R = 1.5$ can be expected at standard values for retention and column plate number. However, if the actual selectivity drops by just 10% to a value of $\alpha = 1.17$, the resolution will be as low as $R = 0.95$ and the method would be completely inadequate for practical use. Poor precision in the prediction of retention from LFER data, due to the significant uncertainty in regression parameters, is common to all approaches using given analyte descriptors, e.g., with the solvation equation (Eq. 5). The obvious reason for this is a limited relevance, accuracy, orthogonality, and completeness of the considered descriptors, which are usually determined by methods that are run under conditions far removed from those of liquid chromatography, where the complex phenomena of surface interactions come into play. The data presented in the foregoing sections can certainly extend the fundamental knowledge of retention in HPLC and serve as a basis to classify stationary phases according to different parameters. To enhance the success and speed of method development, they unfortunately proved to lack the required precision. Moreover, they are only applicable if appropriate solute descriptors are available, which is certainly not the case, e.g., for a newly synthesized pharmaceutical compound.

2.1.5.5 An Empirical Approach to the Determination of LFER Solute Parameters (Descriptors) from HPLC Data

2.1.5.5.1 How does this Strategy Differ from the Use of Predetermined Solute Descriptors

In this section, a more recently published variant of a quantitative relationship akin to an LFER, but without the use of given solute descriptors and based on modified system parameters, is introduced [16]. All calculated data are derived solely from HPLC measurements, which constitutes the most important difference compared to the use of the solvation equation (Eq. 5). With this method, no *a priori* assumptions on the role of a certain parameter and its relation to a specific retention mechanism are made. The assignment to chromatographic principles is developed in the framework of the interpretation of each determined parameter, considering the chemical properties of the contributing test analytes. It should be mentioned that the senior authors of the relevant publications, John Dolan and Lloyd Snyder, have also contributed the following chapter to this book (Chapter 2.1.6), in which the application of their strategy to a wide variety of columns is presented. In this chapter, the applied procedure is explained and the performance of this approach critically assessed.

For the sake of clarity, the chromatographic meaning of each parameter (assigned during the course of their determination) will be introduced together with the equation, in order to compare the parameters with those of Eq. (5). In this new approach, less relevant parameters, such as s and r from Eq. (5) have been omitted, but two additional parameters are incorporated. The formation of surface charges due to dissociation of residual silanol groups implies the possibility of cation-exchange retention. This contribution has not been considered in the strategy described above, although the related interaction is energetically important.

Moreover, the chromatographic retention and selectivity is always affected to a certain extent by the interplay of the molecular configuration of analyte and stationary phase. This so-called molecular or steric recognition is one of the main reasons for the excellent selectivities that can be achieved in liquid chromatography. These parameters form the basis of the following equation (Eq. 6) to describe the retention relative to that of a reference solute. Therefore, Eq. (6) does not include an intercept value, which usually reflects, among other things, the phase ratio of the column.

$$\log(k/k_{ref}) = \log \alpha = H\,\eta' + S\,\sigma' + A\,\beta' + B\,\alpha' + C\,\kappa' \tag{6}$$

The solute "descriptors" (η', σ', β', α', κ') refer to the properties of the selected test analytes and are not known prior to the experiments. The authors do not refer to them as descriptors, but as solute parameters, since they represent all properties complementary to those of the chromatographic system in defining the chromatographic retention of the given analyte.

The chromatographic interpretation of the system descriptors, which are mostly determined prior to the determination of the solute parameters, is as follows:

H hydrophobic interaction parameter
S steric selectivity parameter
A H-bond donor (acidity) parameter
B H-bond acceptor (basicity) parameter
C cation-exchange parameter (at dissociated silanols)

2.1.5.5.2 The Experimental Plan

The determination of both the system and solute parameters is based on a greater experimental and mathematical effort. The authors used a large set of test analytes comprising between 60 and 90 widely differing solutes. Besides an extensive selection of neutral substances to cover a wide range of polarities, weakly basic compounds such as aniline and pyridine derivatives, as well as strong bases such as tricyclic antidepressants and other secondary and tertiary amines, were included. In later publications, solutes with a permanent positive charge, e.g., berberine, were added to the set. Among the acidic components, only weak acids such as benzoic acid derivatives or pharmaceutical compounds with carboxylic groups (e.g., ketoprofen) were selected.

In the first publication [16], the experimental strategy was applied to just ten different stationary phases, in part different modifications of the same silica-based material. Phases with incorporated polar functions were not included. The eluent conditions were kept constant. The chromatographic experiments with all of the neutral substances were carried out with a H_2O/CH_3CN (1 : 1, v/v) eluent with buffer additives, while basic and acidic solutes were eluted with a 31.2 mM phosphate buffer at pH 2.8. The selected pH value does not imply very critical conditions, as it suppresses dissociation of most residual silanol groups and of weakly acidic solutes. A variation of the eluent conditions will be discussed in a later section.

2.1.5.5.3 Determination of the Five LFER Parameters – A Procedure in Eight Steps

The mathematical plan for determining the parameters from the chromatographic data is divided into eight steps, as discussed in the following. In a **first step**, the retention of all i selected test solutes is related to that of ethylbenzene (ideally hydrophobic substance by definition) according to Eq. (7). This results in a data point $\log \alpha_{i,P}$ for each column P.

$$\log \alpha_{i,P} = \log \frac{k_{i,P}}{k_{Et,P}} \tag{7}$$

In the **second step**, a reference phase (RP) is selected, which, in the given case, is HP Zorbax SB C_{18} (SB 100) with a maximum degree of alkyl derivatization. The correlation of all of the $\log \alpha_{i,P}$ values with the $\log \alpha_{i,RP}$ values of the reference column has been checked for all of the substances (see Fig. 6). The correlation of the all $\log \alpha_{i,RP}$-values with the $\log \alpha_{i,RP}$-values of the reference column was checked for all substances. When the reference phase is compared to a very similar stationary phase (e.g., HP Zorbax SB C_{18} with only 90% alkyl derivatization) the correlation of the data adopts high values $(r = 0.9996)$ as expected. However, when a phase like Inertsil ODS-3 is compared, the standard error of the correlation s_y was more than 7 times higher $(r = 0.9822)$. Especially the strongly basic substances (amitriptyline, nortriptyline, diphenhydramine and propranolol) and N,N'-dimethyl acetamide scatter markedly from the linear correlation. These first correlation experiments enable the identification of all ideally hydrophobic compounds within the pool of test solutes. The selection criterion is a maximum deviation from the linear fit of 0.01 $\log \alpha$ units. This can be fulfilled for 24 solutes from the complete set of 67 test substances. For these 24 compounds, a pure hydrophobic retention mechanism is assumed. Hence, the relative retention of these compounds can be described by a phase parameter $\boldsymbol{H}$ (interpreted as hydrophobicity) and a related solute parameter, resulting in Eq. (8) for each solute i on a specific column P:

$$\log \alpha_{i,P} \approx \eta_i' \cdot H_P \tag{8}$$

In the **third step**, the H parameter is calculated for each of the P stationary phases. Assuming an H value of 1.000 for the reference phase (H_{BP}), this can be done from the data of the 24 ideal substances by applying Eq. (9):

$$\log \alpha_{i,P} = \frac{H_P}{H_{BP}} \cdot \log \alpha_{i,BP} \tag{9}$$

In the **fourth step**, an approximate value for the solute hydrophobicity parameters $\eta_{i,P}''$ of all 67 test substances is calculated according to Eq. (10), using the hydrophobicity parameter of the given phase and the measured retention (selectivity) of the solute:

$$\log \alpha_{i,P} = \eta_{i,P}'' \cdot H_P \tag{10}$$

For each test compound, an average value $\eta_i''(\text{avg})$ is calculated over all stationary phases together with the related standard deviation s. From an elevated value of s, an increased deviation of the related solute from pure hydrophobic retention can be deduced. Such solutes should exhibit significant contributions of the other retention parameters $S\sigma'$, $A\beta'$, $B\alpha'$, and $C\kappa'$ and serve as a basis for their determination. With the criterion $s > 0.017$, 35 "non-ideal" substances can be selected from the list of 67 compounds.

In the **fifth step**, the logarithmic deviation of the relative retention from an ideally hydrophobic relative retention, calculated according to Eq. (10) using an averaged solute hydrophobicity parameter $\eta_i''(\text{avg})$, is determined and specified as $\Delta_{i,P}$ (Eq. 11).

$$\Delta_{i,P} = \log\alpha_{i,P} - \eta_i''(\text{avg}) \cdot H_P \tag{11}$$

In Fig. 7, the deviation values $\Delta_{i,P}$ for one non-ideal substance (propranolol, #48) are correlated with those of four other non-ideal solutes over all ten columns. In the case of diphenhydramine (#47), this correlation is excellent (Fig. 7a), whilst poor correlations were found for propranolol with prednisone (#39), N,N-dimethylacetamide (#45), and 2-nitrobenzoic acid (#64), as depicted in Figs. 7b–d. This correlation experiment helps to classify the test solutes. It follows that from the

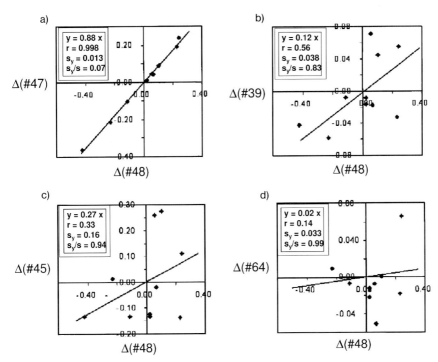

Fig. 7. Correlation of the deviation measure Δ (test compound #) from ideal hydrophobic behavior (see text) according to Eq. (11). A good correlation (a) is indicative of similar chromatographic behavior of two compounds and allows them to be grouped together for the calculation of a distinct phase parameter. (b–d) show compounds with poor correlation and hence significantly different chromatographic behaviors. Test compounds: propranolol (#48), diphenhydramine (#47), prednisone (#39), N,N-dimethylacetamide (#45), which could not be grouped with any other test compound, and 2-nitrobenzoic acid (#64) (from [16]).

quotient of the related standard error of correlation s_y (a measure of the similarity in the chromatographic behavior of two solutes) and the standard deviation of $\Delta_{i,P}$ (Eq. 11) over all ten columns (a measure of non-ideal retention behavior of the given solute i), suitable substance groups for the calculation of further parameters can be defined. The criteria are $s_y/s < 0.5$ and a high value of the correlation coefficient r.

In **step 6**, the identified solute groups from step 5 are assigned to a specific contribution to the chromatographic process on the basis of the chemical structure of group members. It should be emphasized again that no *a priori* assumptions about the role of each parameter in the chromatographic process have to be made with this strategy, but the assignment is made during an interpretation based on the chemical structure of the test solute. To obtain approximate values for all parameters except H, the $\Delta_{i,P}$ of the following substances were considered as a measure of their deviation from ideal hydrophobic behavior:

- $\Delta_{i,P}$ of *N,N*-dimethylacetamide (#45) as parameter *A* (H-donor acidity of stationary phase). No other solute was found to group with #45, which was the only cyclic carbonic acid amide in the list.

- The average of $\Delta_{i,P}$ of amitriptyline (#46), diphenhydramine (#47), propranolol (#48), nortriptyline (#49), and prolintane (#50) as parameter *C* (cation exchange at dissociated silanol groups). These substances are secondary or tertiary amines and thus strong Brønsted bases.

- The average of $\Delta_{i,P}$ of diclophenac acid (#56), mefenamic acid (#57), ketoprofen (#58), 4-*n*-butylbenzoic acid (#60), 4-*n*-pentylbenzoic acid (#61), 4-*n*-hexylbenzoic acid (#62), 3-cyanobenzoic acid (#63), 2-nitrobenzoic acid (#64), and 3-nitrobenzoic acid (#65) as parameter *B* (H-acceptor basicity of stationary phase). These substances are weak acids and thus potent H-donors.

- The average of $\Delta_{i,P}$ of benzophenone (#32), *cis*-chalcone (#33), *trans*-chalcone (#34), *cis*-4-nitrochalcone (#35), *trans*-4-nitrochalcone (#36), *cis*-4-methoxychalcone (#37), *trans*-4-methoxychalcone (#38), prednisone (#39), hydrocortisone (#40), flunitrazepam (#43), and 5,5-diphenylhydantoin (#44) as parameter *S* (steric hindrance to analyte insertion into the stationary phase). These substances differ from ethylbenzene (#3) mainly through their molecular size.

In this way, a preliminary set of parameters for a complete description of the chromatographic system is obtained. It should be mentioned that the parameters do not represent only the properties of the stationary phase, but more exactly the differences between the mobile and stationary phases in relation to a specific parameter.

In **step 7**, a multivariate regression using the parameters obtained in steps 3 and 6 (*H*, *S*, *A*, *B*, and *C*) and the log α values measured on all ten columns is carried out to calculate all solute parameters (similar to the descriptors) by Eq. (6). The standard deviation of these parameters in relation to log α values did not exceed 0.005.

In the last step (**step 8**), Eq. (6) is again used for a multivariate regression. This time, the solute parameters or descriptors (η', σ', β', α', κ') are kept constant in order to calculate the final set of phase parameters (**H**, **S**, **A**, **B**, and **C**). The standard deviation in these important parameters was around 0.004 (in logarithmic units). The relative error in the selectivity resulting from these very precise parameters is calculated to lie between 1 and 2%. This excellent precision must be regarded as the main success of such a very elaborate LFER approach.

The individual values of the parameters of the ten stationary phases should not be discussed here. The highest and always positive values are contributed by the hydrophobicity parameter (**H**), whilst the values for H-donor (**A**), H-acceptor (**B**), cation-exchange activity (**C**), and steric discrimination (**S**) vary considerably between the different phases and adopt both positive and negative values. The smallest contribution comes from the parameters **B** and **S**. For further information, the reader should refer to the chapter by Snyder and Dolan (Chapter 2.1.6).

2.1.5.5.4 Variation of the Eluent Conditions

An LFER system that enables the determination of HPLC system parameters with very high precision should also be amenable to monitoring the behavior of specific stationary phases under different chromatographic conditions. Based on the method described above, the same authors characterized three phases (two Zorbax C_{18} with different coverage and one Symmetry C_{18}) with different organic eluent components (acetonitrile, methanol, and tetrahydrofuran), different organic contents, at different pH values, and at different temperatures [17]. The individual impact on solute, on the one hand, and system parameters on the other, was in some respects surprising. The main contribution to changes in retention under varying eluent conditions results from changes in solute parameters rather than parameters of the chromatographic system. The contribution of the column parameters to retention changes is found to be at most 25%. This demonstrates that solute parameters with this LFER method are no longer constant for given test substances, but are strongly system-dependent variables. This finding explains at once why the column parameters are not at all in accordance with the data found by applying a classical LFER approach with given (and thus invariant) solute descriptors. When chromatographic systems are characterized by the solvation equation (Eq. 5), the column parameters describe the effect on the system of eluent variations alone. Nevertheless, the results are in good accordance with the solvophobic theory. It is obvious that changes in the eluting solvent properties strongly influence the energy balance when a specific analyte leaves the closing mobile phase cavity to enter the freshly created stationary phase cavity due to the altering solvation energy. This process is clearly analyte-dependent. From this result, it can be deduced that the eluent conditions are not crucial when stationary phases are compared using this empirical LFER strategy. However, one important exception from this conclusion must be noted. When the pH of the mobile phase is altered, the **C**-parameter of the stationary phase can change considerably. This finding is not surprising, since the pH controls the degree of silanol dissociation and thus the cation-exchange capacity of the stationary phase.

The C-parameter is calculated from the retention data of strong bases such as amitriptyline (#46) or nortriptyline (#49), as well as substances with permanent positive charges such as berberine (#91) or bicuculline methiodide (#92). In Fig. 8, the influence of pH on the retention of these solutes on three markedly different stationary phases is depicted. The pH was adjusted with a 60 mM citrate buffer, as citric acid provides a more continuous buffer capacity over this pH range due to its closer pK_a values as compared to, e.g., phosphoric acid. A pronounced decrease in retention with increasing pH was observed with Inertsil, which is at variance with the theoretical behavior of the silanols. It must be taken into consideration, however, that for both Inertsil and Symmetry columns the retention is very weak ($k \ll 1$), which might have biased the accuracy of the data. Moreover, the authors mention that this trend for Inertsil was reversed when a phosphate buffer was used to adjust pH 3 and pH 7 instead of citrate. This confirms the general observation that retention mechanisms in reversed-phase chromatography become very complex when strong ionic interactions are involved.

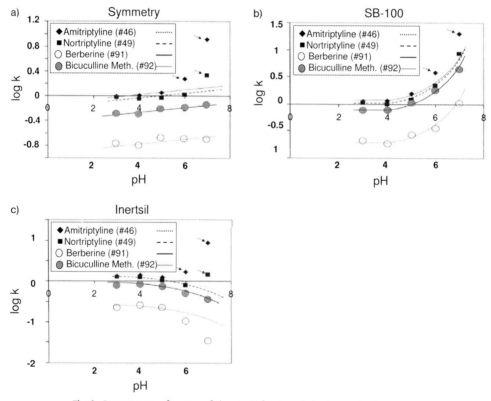

Fig. 8. Retention as a function of eluent pH for strongly basic or cationic substances on three different stationary phases. Conditions: acetonitrile/60 mM citrate buffer (1 : 1, v/v), 35 °C (from [17]).

Table 3. Values of C-parameter for 7 distinct phases determined at pH 2.8 and 7.0.

Phase	$C_{2.8}$	$C_{7.0}$	$C_{7.0} - C_{2.8}$
Inertsil	−0.35	−0.32	0.03
Symmetry	−0.21	−0.05	0.16
SB-100	0.09	0.85	0.76
SB-90	0.05	0.78	0.73
Eclipse	0.04	0.17	0.13
YMC 15	−0.10	−0.15	−0.05
YMC 16	0.01	0.02	0.01

Table 3 lists the calculated *C*-parameters for eight different phases at both "extreme" pH values 2.8 and 7.0. It is obvious from the mathematical procedure that the relative change in berberine retention (normalized to toluene retention) directly reflects the change in the *C*-parameter of the stationary phase. Assuming a linear relationship, the *C*-parameter for any pH can be calculated with the help of Eq. (12) from the berberine retention at this pH. The measurement of berberine retention was carried out at each pH value by adjusting the eluent with 30 mM phosphate.

$$C\,(\text{pH} = x) = C\,(\text{pH} = 2.8) + \log(k_x/k_{2.8}) \tag{12}$$

The third column in Table 3 reflects the increase in the *C*-parameter when switching from pH 2.8 to 7.0, thus increasing the ion-exchange capacity of the stationary phase. The two non-endcapped Zorbax phases SB-100 and SB-90 show an extraordinary behavior, which must be related to a greater presence of acidic silanol groups.

Provided that the effects on the *C*-parameter can be neglected, it can be concluded that a meaningful phase characterization using the empirical LFER approach can be carried out under one single set of eluent conditions.

2.1.5.5.5 Stationary Phase Characterization with Empirical LFER Parameters

The high precision of the chromatographic parameters determined on the basis of the empirical LFER prompts application of this methodology for a systematic characterization of stationary phases for HPLC to describe their individual selectivities. Moreover, these data should extend the basic knowledge on retention mechanisms in liquid chromatography. In a continuing paper, the authors discussed the physico-chemical basis of selectivity in HPLC with the help of the determined LFER parameters. Thus, the correlation of the parameters with properties of the stationary phases such as alkyl chain length, bonding density, average pore diameter, presence of endcapping, metal content or the incorporation of polar groups (polar endcappings, polar embedded phases) into the bonding were considered.

In a first follow-up publication [18], general investigations into this subject are described. In three later publications, 87 stationary phases based on ultra-pure (type B) silica were studied with regard to these correlations [19], then a further 87 phases based on a metal-rich (type A) silica [20], and finally 21 reversed phases with embedded polar groups or polar endcapping [21].

The hydrophobicity parameter H, which is correlated with the methylene group selectivity and the solvation parameter v (other model), increases with increasing alkyl chain length, increasing bonding density, and decreasing pore diameter. For phases with additional polar groups, H is markedly smaller, even when alkyl chain coverage is similar. This is also the case for type A materials as compared to the type B ones. These results may sound trivial, but one should not forget that H provides a precise and statistically evaluated quantitative measure of hydrophobic retention, which did not exist in this manner before.

The steric parameter S reflects the resistance of the stationary phase to undergo insertion of solute molecules with a higher "thickness". It exhibits very similar correlations with stationary phase properties such as the hydrophobicity parameter H. However, the S parameter cannot be related to the so-called shape-selectivity parameter defined by Sander and Wise [22–24], which describes the ability of phases to differentiate between planar and twisted polyaromatic hydrocarbons (PAH). These two parameters do not correlate at all and obviously characterize very different properties.

The parameter A, describing the hydrogen-bonding acidity, is significantly higher with type A phases than with the type B ones, which are based on ultra-pure silicas. Embedded polar groups decrease the parameter A, whilst a polar end-capping increases it somewhat. Moreover, this parameter sharply decreases when a phase is subjected to standard endcapping to reduce the number of residual silanol groups. Basically, these findings confirm the general assumption that the A parameter can be related to undissociated silanol groups. Applying this model, the A parameter should decrease with increasing pH (especially with type A materials), since the essential proton should be lost concomitantly as the cation-exchange capacity is created (rise of the C parameter). This dependence, however, could not be detected and no consistent explanation was provided by the authors. This can be taken as a further indication that a fundamental understanding of the role of silanol groups in reversed-phase chromatography is still very incomplete.

The most difficult parameter for a chemical interpretation in the case of a simple silica-based reversed-phase without additional polar groups is the B parameter, which characterizes the stationary phase H-acceptor basicity. With type B silicas, it is mainly related to adsorbed water at the stationary phase surface. With some of the type A phases, a correlation with the metal content may be seen, but this is not a general rule. Some polar modified phases exhibit pronounced B parameters, but a universal dependence on the chemical identity of the polar group cannot be found.

The last parameter to discuss is the pH-dependent C-parameter, which describes the cation-exchange activity of the reversed phase. As might be expected, the size of C as well as the magnitude of its pH dependence is more pronounced with type A phases and is clearly reduced through endcapping. The introduction of a polar group into the alkyl chain (shielding group) leads to increased **C** values. This could not be consistently observed for phases with polar endcapping.

The authors further defined a sequence of importance of each of the five parameters on the individual selectivity of reversed phases for the solution of given separation problems. Although the retention is mainly governed by the hydrophobic mechanism (**H** parameter), this parameter contributes least to the individual selectivities of the various reversed phases. It is by no means a new perception, but the comparison of the parameter lists for different phases confirms that selectivity in RP-HPLC is mainly based on the so-called secondary interactions. According to the empirical LFER approach, the importance of the phase parameters for selectivity control increases in the following sequence:

$$H < B < S < A \ll C \tag{13}$$

2.1.5.6   Concluding Remarks on LFER Applications in HPLC

Linear free energy relationships enable the characterization of chromatographic systems on a sound physico-chemical basis. They can provide detailed and precise information on stationary phase properties, thus leading to new principles of classification. The practical uses for the chromatographer and the developer of stationary phases are manifold. Stationary phases for use during method development can be selected in a more measured way and less empirically. The transfer of an existing HPLC method to another stationary phase should be facilitated. It gives users and manufacturers precise tools to compare different batches of stationary phases of the same brand and to obtain reliable quantitative information with regard to the property in which the phases may differ.

In recent papers [19–21], a correlation between LFER parameters and stationary phase properties such as average pore width, alkyl chain length, bonding density, polar modification, etc., has been described. Such interrelations should extend the knowledge on retention mechanisms in liquid chromatography and hence serve to create alternative strategies for the design of new improved stationary phases.

An efficient support from LFERs to speed-up method development and optimization is not yet conceivable. Besides the huge effort needed to determine solute parameters of "new" compounds, even the most powerful LFER strategies do not yet offer the required accuracy in the prediction of selectivities. This vision, however, has been a focus of interest of HPLC experts for a long time. The future will show whether an LFER strategy is capable of bringing HPLC closer to the goal of computer-assisted precise prediction of retention on a physico-chemical basis.

References

1 M. H. Abraham, J. C. McGowan, *Chromatographia* 23 (1987), 243.

2 M. J. Kamlet, R. W. Taft, *J. Am. Chem. Soc.* 98 (1976), 377.

3 M. J. Kamlet, R. W. Taft, *J. Am. Chem. Soc.* 98 (1976), 2886.

4 M. J. Kamlet, J. L. M. Abboud, R. W. Taft, *Prog. Phys. Org. Chem.* 99 (1977), 6027.

5 M. J. Kamlet, J. L. M. Abboud, R. W. Taft, *Prog. Phys. Org. Chem.* 13 (1981), 485.

6 M. H. Abraham, *Chem. Soc. Rev.* 22 (1993), 73.

7 M. H. Abraham, *J. Phys. Org. Chem.* 7 (1994), 672.

8 M. H. Abraham, *Pure Appl. Chem.* 65 (1993), 2503.

9 M. H. Abraham, J. Andonian-Haftvan, G. S. Whiting, A. Leo, *J. Chem. Soc., Perkin Trans.* II (1994), 1777.

10 P. C. Sadek, P. W. Carr, R. M. Doherty, M. J. Kamlet, R. W. Taft, M. H. Abraham, *Anal. Chem.* 57 (1985), 2971.

11 M. H. Abraham, *Chem. Soc. Rev.* 23 (1993), 660.

12 A. Sandi, L. Szepesy, *J. Chromatogr. A* 818 (1998), 1.

13 A. Sandi, L. Szepesy, *J. Chromatogr. A* 818 (1998), 19.

14 A. Sandi, M. Nagy, L. Szepesy, *J. Chromatogr. A* 893 (2000), 215.

15 H. Engelhardt, H. Löw, W. Götzinger, *J. Chromatogr. A* 544 (1991), 371.

16 N. S. Wilson, M. D. Nelson, J. W. Dolan, L. R. Snyder, R. G. Wolcott, P. W. Carr, *J. Chromatogr. A* 961 (2002), 171.

17 N. S. Wilson, M. D. Nelson, J. W. Dolan, L. R. Snyder, P. W. Carr, *J. Chromatogr. A* 961 (2002), 195.

18 N. S. Wilson, M. D. Nelson, J. W. Dolan, L. R. Snyder, P. W. Carr, L. C. Sander, *J. Chromatogr. A* 961 (2002), 217.

19 J. J. Gilroy, J. W. Dolan, L. R. Snyder, *J. Chromatogr. A* 1000 (2003), 757.

20 J. J. Gilroy, J. W. Dolan, P. W. Carr, L. R. Snyder, *J. Chromatogr. A* 1026 (2004), 77.

21 N. S. Wilson, J. J. Gilroy, J. W. Dolan, L. R. Snyder, *J. Chromatogr. A* 1026 (2004), 91.

22 L. C. Sander, S. A. Wise, *Anal. Chem.* 59 (1987), 2309.

23 L. C. Sander, S. A. Wise, *Anal. Chem.* 61 (1989), 1749.

24 L. C. Sander, S. A. Wise, *Anal. Chem.* 67 (1995), 3284.

2.1.6
Column Selectivity in Reversed-Phase Liquid Chromatography

Lloyd R. Snyder and John W. Dolan

The selectivity of reversed-phase liquid chromatography (RP-LC) columns is known to vary, even columns with the same ligand (e.g., C_{18}). Column selectivity can also vary from batch to batch for columns claimed to be equivalent by the manufacturer. For different reasons, it is sometimes necessary to locate a replacement column for a given assay that will provide the same separation as the previous column. In other cases, as in HPLC method development, a column of very different selectivity may be needed – in order to separate peaks that overlap on the original column. For each of these situations, means for measuring and comparing column selectivity are required. Until recently, no such characterization of column selectivity was able to guarantee that two different columns can provide equivalent separation for any sample or separation conditions.

The present chapter describes the "hydrophobic-subtraction" model of RP-LC column selectivity, for characterizing columns in terms of five fundamental column properties (**H**, **S***, **A**, **B**, **C**). Because the hydrophobic-subtraction model can predict retention (values of k) within a few percent, columns with sufficiently similar values of **H**, **S***, etc. should provide equivalent separation for any sample. Similarly, columns with very different values of **H**, **S***, etc. should provide quite different separation. Several examples of the use of values of **H**, **S***, etc. for the selection of similar columns have been reported, some of which are discussed below. The present chapter also compares the selectivity of different column types (e.g., alkyl-silica columns, phenyl columns, cyano columns, etc.) and summarizes values of **H**, **S***, etc. for several commercial C_{18} columns.

2.1.6.1 Introduction

By the early 1990s, a general picture had emerged of reversed-phase liquid chromatography (RP LC) on alkyl-silica columns. Retention was attributed primarily to *solvophobic* or *hydrophobic interaction* [1, 2], but with contributions from other solute–column interactions [3–5]. Thus, ionized silanols (–SiO⁻) can retain protonated bases by cation-exchange, and neutral silanols (–SiOH) can interact with proton-acceptor solutes through hydrogen bonding. The shape of the solute molecule further affects sample retention, leading to so-called column shape selectivity.

Some of the above (and other) contributions to column selectivity (for neutral solute molecules) have been incorporated into the *solvation equation* model [6] for RP LC retention:

$$\log k = \underset{(i)}{C_1} + \underset{(ii)}{r\,R_2} + \underset{(iii)}{s\,\pi_2^H} + \underset{(iv)}{a\sum\alpha_2^H} + \underset{(v)}{b\sum\beta_2} + \underset{(vi)}{v\,V_x} \qquad (1)$$

Here, C_1 is a solute-independent constant, and the quantities r, s, a, b, and v characterize column selectivity for a given set of separation conditions (for a definition of the terms in Eq. (1), see the list of abbreviations at the end of this chapter). Thus, terms (ii), (iii), and (vi) together determine the hydrophobic interaction between solute and column, term (iv) describes the effects of hydrogen bonding between acidic (donor) solutes and basic (acceptor) groups in the column, and term (v) represents the contribution of hydrogen bonding between basic solutes and acidic column groups. Eq. (1) ignores contributions to retention from shape selectivity, cation-exchange, and some other less well documented solute–column interactions; it is also not accurate enough (±10–15% in k, 1 standard deviation) for quantitative column characterization.

An alternative, more widely used approach for specifying column selectivity is based on retention measurements of suitable test solutes which purport to measure the above solute–column interactions: hydrophobic interaction, shape selectivity, hydrogen bonding of (a) basic solutes by acidic column groups or (b) acidic solutes by basic column groups, and cation exchange with ionized silanol groups (see Fig. 1). The measurement of the selectivity of 135 different RP-LC columns in this way has recently been described [7]. However, the reliability of this and other past approaches for characterizing column selectivity has been questioned [8].

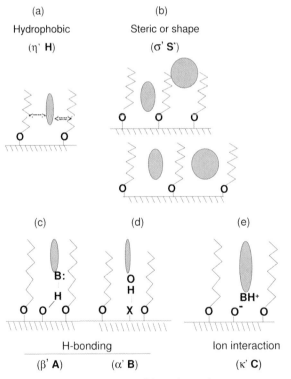

Fig. 1. Pictorial representation of five solute–column interactions of Eq. (2). "X" for α' **B** interaction varies with column type.

Over the past few years, an alternative method for specifying column selectivity has been proposed [9–19], which will form the basis for the following discussion. It should be noted that there is no "best" column selectivity, because separation depends on both the column and the sample; a column which is "best" for one sample will not be "best" for all other samples.

2.1.6.2 The "Subtraction" Model of Reversed-Phase Column Selectivity

We have noted above that hydrophobic interaction of the solute and column represents the main contribution to RP LC retention and column selectivity. The "subtraction model" assumes that we first subtract the major contribution of hydrophobicity to RP LC retention, in order to better see contributions to retention from remaining solute–column interactions. The further analysis of these minor contributions then leads to a general equation for RP LC retention and column selectivity [9, 13]:

$$\log \alpha = \log(k / k_{ref}) = \underset{(i)}{\eta' \, \mathbf{H}} - \underset{(ii)}{\sigma' \, \mathbf{S}^*} + \underset{(iii)}{\beta' \, \mathbf{A}} + \underset{(iv)}{\alpha' \, \mathbf{B}} + \underset{(v)}{\kappa' \, \mathbf{C}} \qquad (2)$$

Here, k is the retention factor of a given solute, k_{ref} is the value of k for a nonpolar reference solute (ethylbenzene in the present treatment), and the remaining selectivity-related symbols represent either empirical, eluent-dependent properties of the solute (η', σ', β', α', κ') or eluent-independent properties of the column ($\mathbf{H}$, $\mathbf{S}^*$, $\mathbf{A}$, $\mathbf{B}$, $\mathbf{C}$). Terms (i)–(v) of Eq. (1) correspond to various solute–column interactions as in Fig. 1. Thus, the various column parameters measure the following column properties: $\mathbf{H}$, *Hydrophobicity*; $\mathbf{S}^*$, *Steric* resistance to insertion of bulky solute molecules into the stationary phase (similar to, but not the same as "shape selectivity" [4]); $\mathbf{A}$, column hydrogen-bond *Acidity*, mainly attributable to non-ionized silanols; $\mathbf{B}$, column hydrogen-bond *Basicity*, presently believed to result (for some columns) from absorbed water in the stationary phase; and $\mathbf{C}$, column *Cation*-exchange activity, due to ionized silanols. The parameters η', σ', etc. denote complementary properties of the solute (see the nomenclature list). The five solute–column interactions of Eq. (2) are further illustrated in Fig. 1.

Equation (2) has been evaluated for about 150 different solutes and a similar number of modern alkyl-silica columns made from high purity (type B) silica; the average accuracy of Eq. (2) for these solutes and columns was ±1% in α [12]. Because of the intentionally wide variety of solute molecular structures studied, and the use of columns of varying ligand density, pore diameter, and ligand length (C_1–C_{30}, but mainly C_8 and C_{18}), it is believed that Eq. (2) has captured all significant solute–column interactions. This claim cannot be made for any previous procedure for characterizing column selectivity. Table 1 presents values of $\mathbf{H}$, $\mathbf{S}^*$, etc. for several common C_{18} columns made from type B silica.

Equation (2) has also been extended to other commonly used kinds of RP LC columns: (a) older alkyl-silica columns made from impure (type A) silica, (b) columns with polar groups such as amide or carbamate, which are either embedded into the column ligands or used to end-cap unreacted silanols after a conventional bonding, (c) cyano columns, (d) phenyl columns, (e) columns with

Table 1. Values of **H**, **S**✶, etc. for some common C_{18} columns made from type B silica (data from [12]).

Column[a]	H	S✶	A	B	C(2.8)	C(7.0)
Agilent						
Zorbax Eclipse XDB-C_{18}	1.077	0.024	−0.064	−0.033	0.054	0.088
Zorbax Extend C_{18}	1.098	0.05	0.012	−0.041	0.030	0.016
Zorbax Rx-18	1.076	0.04	0.307	−0.039	0.096	0.414
Akzo Nobel						
Kromasil 100-5C_{18}	1.051	0.035	−0.069	−0.022	0.038	−0.057
Alltech						
Alltima C_{18}	0.993	−0.014	0.037	−0.013	0.093	0.391
Beckman						
Ultrasphere ODS	1.085	0.014	0.174	0.068	0.279	0.382
Dionex						
Acclaim C_{18}	1.033	0.017	−0.142	−0.026	0.086	−0.003
ES Industries						
Chromegabond WR C_{18}	0.979	0.026	−0.159	−0.003	0.320	0.283
GL Science						
Inertsil ODS-2	0.994	0.032	−0.045	−0.005	−0.116	0.773
Inertsil ODS-3	0.991	0.021	−0.142	−0.021	−0.473	−0.333
Hamilton						
HxSil C_{18}	0.847	0.073	0.302	0.014	0.230	1.055
HiChrom/ACT						
ACE5 C_{18}	1.000	0.026	−0.095	−0.006	0.143	0.096
Higgins Analytical						
Targa C_{18}	0.977	−0.019	−0.070	0.000	0.013	0.175
Jones						
Genesis C_{18} 120A	1.005	0.003	−0.068	−0.006	0.139	0.124
Macherey Nagel						
Nucleodur 100-C_{18} Gravity	0.868	0.032	−0.240	0.000	−0.158	0.631
Nucleodur C_{18} Gravity	1.056	0.041	−0.097	−0.025	−0.080	0.316
MacMod/Higgins						
PRECISION C_{18}	1.003	0.003	−0.041	−0.009	0.079	0.341

[a] Source or supplier shown first, then individual columns from that source.

Table 1. (continued)

Column[a]	H	S*	A	B	C(2.8)	C(7.0)
Merck						
Chromolith RP18e	1.003	0.028	0.009	−0.014	0.103	0.187
LiChrospher 60 RP-Select B	0.747	−0.060	−0.042	0.006	0.108	1.773
Purospher RP-18	0.585	0.254	−0.560	−1.309	−1.934	1.109
Purospher STAR RP18e	1.003	0.013	−0.069	−0.035	0.018	0.044
Superspher 100 RP-18e	1.030	0.025	−0.028	−0.011	0.352	0.266
Chromolith RP18e	1.003	0.028	0.009	−0.014	0.103	0.187
Nacalai Tesque						
COSMOSIL AR-II	1.017	0.011	0.128	−0.028	0.116	0.494
COSMOSIL MS-II	1.032	0.041	−0.129	−0.012	−0.117	−0.027
Nomura						
Develosil ODS-MG-5	0.964	−0.039	−0.163	−0.002	−0.012	0.051
Develosil ODS-UG-5	0.997	0.025	−0.145	−0.003	0.150	0.155
Phenomenex						
Luna C_{18}(2)	1.003	0.023	−0.121	−0.006	−0.269	−0.173
Prodigy ODS (3)	1.023	0.024	−0.129	−0.011	−0.195	−0.133
SynergiMAX RP	0.989	0.028	−0.008	−0.013	−0.133	−0.034
Restek						
Allure C_{18}18	1.115	0.043	0.112	−0.045	−0.048	0.066
Restek Ultra C_{18}	1.055	0.030	−0.068	−0.021	0.008	−0.066
SGE						
Wakosil II $5C_{18}$AR	0.998	0.075	−0.055	−0.034	0.070	0.010
Supelco						
Discovery C_{18}	0.985	0.026	−0.126	0.005	0.176	0.154
Thermo/Hypersil						
Hypersil Beta Basic-18	0.993	0.032	−0.097	0.003	0.163	0.126
Hypersil BetamaxNeutral	1.098	0.036	0.067	−0.031	−0.039	0.011
HyPURITY C_{18}	0.981	0.025	−0.089	0.004	0.192	0.168
Varian						
OmniSpher 5 C_{18}	1.055	0.050	−0.033	−0.029	0.121	0.057
Waters						
Atlantis dC_{18} b	0.918	−0.032	−0.191	0.003	0.036	0.087
DeltaPak C_{18} 100A	1.028	0.018	−0.017	−0.010	−0.051	0.024
J'Sphere H80	1.132	0.060	−0.023	−0.067	−0.242	−0.161
J'Sphere L80	0.763	−0.039	−0.214	0.000	−0.399	0.346
J'Sphere M80	0.927	−0.027	−0.121	−0.003	−0.293	0.140
Symmetry C_{18}	1.053	0.062	0.020	−0.020	−0.302	0.124
XTerra MS C_{18}	0.985	0.012	−0.141	−0.014	0.133	0.051
YMC Basic	0.821	−0.006	−0.235	0.028	0.070	0.093
YMC Pro C_8	0.890	0.014	−0.214	0.007	−0.322	0.020

Table 2. Comparison of the selectivities of different column types. Average values from [12–16].

Column	H^b	S^*	A	B	C(2.8)	C(7.0)
Type B C$_8$	0.83	−0.01	−0.16	0.02	0.02	0.31
Type B C$_{18}$	1.00	0.01	−0.07	−0.01	0.05	0.17
Type A C$_{18}$	0.84	−0.06	0.12	0.05	0.78	1.13
Polar group embedded [25]	0.68	0.00	−0.54	0.17	−0.65	0.13
Polar group endcapped [25]	0.94	−0.02	−0.01	0.01	−0.14	0.27
Cyano [26]	0.41	−0.11	−0.58	−0.01	0.07	0.67
Phenyl [27]	0.60	−0.16	−0.23	0.02	0.16	0.74
Fluoroalkyl [27]	0.7	−0.03	0.1	0.04	1.03	1.42
Fluorophenyl [27]	0.63	0.14	−0.26	0.01	0.55	1.1
Bonded zirconia [11]	1.03	−0.01	−0.43	0.05	2.08	1.98

perfluoroalkyl or perfluorophenyl groups, and (f) bonded zirconia columns. Table 2 summarizes average values of **H**, **S***, etc. for different kinds of RP LC columns. From Table 2, we can see that average values of the column parameters vary widely: **H**, 0.41 to 1.03; **S***, −0.16 to 0.14; **A**, −0.58 to 0.12; **B**, −0.01 to 0.17; **C**(pH 2.8), −0.65 to 2.08; **C**(pH 7.0), 0.13 to 1.98.

So far, values of **H**, **S***, etc. that characterize column selectivity have been reported for about 200 columns [12–16]. Values for another hundred columns are included in a commercial database (Column Match®; Rheodyne LLC, Rohnert Park, CA). A procedure for measuring values of **H**, **S***, etc. has been reported and shown to give comparable values for the same columns in inter-laboratory testing [17]. Table 2 summarizes average values of **H**, **S***, etc. for several different types of columns. Within a given column type, further changes in these column parameters are typical, as illustrated in Table 1 for some type B C$_{18}$ columns.

2.1.6.3 Applications
Given a way to characterize the relative selectivity of different columns (values of **H**, **S***, etc.), several applications of these column-selectivity data are possible.

2.1.6.3.1 Selecting "Equivalent" Columns [18]
Routine RP LC assay procedures are often carried out over periods of months or years, as well as in different laboratories and different parts of the world. During the application of such a procedure over time, many columns may be required. For various reasons, it may prove difficult or impossible to obtain a replacement column with sufficiently similar selectivity from the original source. In such cases, it is necessary to locate an equivalent replacement column from a different source – or a column of different part number from the same source. During the

development of an RP LC procedure, it is also recommended that one or more back-up columns of equivalent selectivity (from different sources) be specified. The need to find a column that is of similar selectivity as an original column is therefore fairly common.

A column-comparison function based on values of **H**, **S***, etc. for columns 1 and 2 has been derived [12]:

$$F_s = \{[12.5\,(\mathbf{H}_2 - \mathbf{H}_1)]^2 + [100\,(\mathbf{S}_2^* - \mathbf{S}_1^*)]^2 + [30\,(\mathbf{A}_2 - \mathbf{A}_1)]^2$$
$$+ [143\,(\mathbf{B}_2 - \mathbf{B}_1)]^2 + [83\,(\mathbf{C}_2 - \mathbf{C}_1)]^2\}^{1/2} \tag{3}$$

Here, $\mathbf{H}_1$ and $\mathbf{H}_2$ refer to values of **H** for columns 1 and 2, respectively (and similarly for values of $\mathbf{S}_1^*$ and $\mathbf{S}_2^*$, etc.). F_s can be regarded as the distance between two columns whose values of **H**, **S***, etc. are plotted in five-dimensional space, with weighting factors (12.5, 100, etc.) added for a sample of "average" composition. It was found [12] that if $F_s \leq 3$ for two columns 1 and 2, variations in α should be $\leq 3\%$, so that the two columns are likely to provide equivalent selectivity and separation for different samples and conditions.

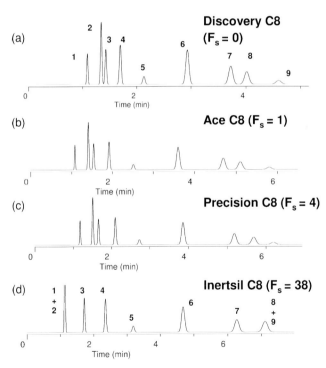

Fig. 2. Comparisons of column selectivity for a given sample and RP-LC procedure. Sample: (1) N,N-diethylacetamide; (2) nortriptyline; (3) 5,5-diphenylhydantoin; (4) benzonitrile; (5) anisole; (6) toluene; (7) cis-chalcone; (8) trans-chalcone; (9) mefenamic acid. Columns (15 × 0.46 cm, 5 μm particles) identified in the figure. Experimental conditions: 50% acetonitrile/pH 2.8 buffer, 35 °C, 2.0 mL min^{-1}. (Reprinted from [18]).

An example of the application of Eq. (3) is shown in Fig. 2, for the separation of a mixture of neutral, basic, and acidic compounds on four different columns. Values of F_s from Eq. (3) are shown for the three columns of Fig. 3b–d, each of which is compared with the Discovery C_8 column of Fig. 3a. The values of F_s for the ACE C_8 (b) and Precision C_8 (c) columns are relatively small ($F_s \leq 4$), and separation on these columns is therefore expected to be (and is) quite similar to that on the Discovery C_8 column. For the Inertsil C_8 column, $F_s = 38$, indicating that this column has a selectivity that is very different from that of the Discovery C_8 column, which indeed it does; note, for example, the co-elution of bands nos. 1/2 and 8/9 in Fig. 3d.

The sample of Fig. 2 was chosen from the test solutes used to measure values of **H**, **S***, etc. according to Eq. (3). A better test of Eq. (3) and values of **H**, **S***, etc. for measuring column selectivity has recently been reported [18]. Twelve different routine RP LC separations were selected from three different pharmaceutical laboratories, whereupon (successful) attempts were made to select equivalent replacement columns for each separation on the basis of Eq. (3). One of these separations is illustrated in Fig. 3, where the original separation on an ACE C_8 column (a) is compared with that on three other columns (b–d) with $1.3 \leq F_s \leq 248$. The separation with the Discovery C_8 column (b) is seen to be reasonably similar to that in (a), as expected from its value of $F_s = 1.3$. However, separation on the

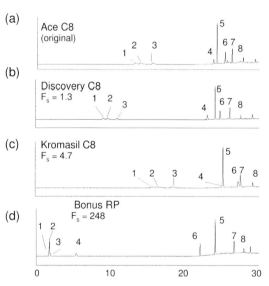

Fig. 3. Comparative separations of a pharmaceutical sample on an original column (a) and three possible replacement columns (b–d). The sample contained strong bases and carboxylic acids. Columns (15 × 0.46 cm, 5 μm particles) identified in the figure. Conditions: gradient separation with solvents (A) and (B); (A) is pH 2.7 buffer; (B) is acetonitrile; the gradient was 10/10/22/88/88% (B) in 0/5/15/25/27 min; 1.0 mL min^{-1}. See Ref. [18] and text for other details. Sample composition and detailed experimental conditions are proprietary. (Reprinted from [18]).

Kromasil C$_8$ and Bonus RP columns, (c) and (d), is rather different, as anticipated from their values of $F_s \geq 4.7$.

Note that the separations of Fig. 3 involve conditions different from those used to measure values of **H**, **S***, etc. However, values of **H**, **S***, etc. are little affected by experimental conditions other than mobile phase pH. This is particularly true for columns that are more similar (smaller values of F_s^*), and for conditions that are not far removed from those used to measure values of **H**, **S***, etc. (50% acetonitrile/buffer; 35 °C). **C** varies with mobile phase pH, and a value of **C** at a given pH can be estimated by linear interpolation between values at pH 2.8 (**C**[2.8]) and pH 7.0 (**C**[7.0]).

Modification of Equation (3) as a Function of the Sample

In Eq. (3), it is assumed that every possible compound may be present in the sample. When this is the case, each of the column parameters (**H**, **S***, etc.) can be important in affecting separation selectivity for that sample. If one or more of these column parameters has little or no effect on the separation, however, it is useful to reduce its contribution to the final value of F_s, because the resulting lower values of F_s make it more likely that a suitable column (i.e., one with $F_s \leq 3$) can be found. Specifically, the presence or absence of certain compound types leads to a re-weighting of the last two terms of Eq. (1):

$$F_s^* = \{[12.5\,(\mathbf{H}_2 - \mathbf{H}_1)]^2 + [100\,(\mathbf{S}_2^* - \mathbf{S}_1^*)]^2 + [30\,(\mathbf{A}_2 - \mathbf{A}_1)]^2$$
$$+ [143\,x_B(\mathbf{B}_2 - \mathbf{B}_1)]^2 + [83\,x_C(\mathbf{C}_2 - \mathbf{C}_1)]^2\}^{1/2} \tag{4}$$

Here, x_B and x_C (with values between 0 and 1) represent possible correction factors that depend on sample composition. For example, if bases are absent from the sample, the term $x_C \approx 0$, because values of **C** mainly affect the retention of ionized basic solutes. For similar reasons, if carboxylic acids are absent from the sample, $x_B = 0$. Note that if x_B and x_C are assumed to be equal to one (equivalent to setting $F_s^* = F_s$), maximum values of F_s^* result, with a decrease in the number of possible replacement columns with $F_s^* \leq 3$. When a basic compound is partly ionized, there is a reduced contribution of **C** to the separation. As a rough rule, weak bases such as anilines or pyridines have $x_C \approx 0.1$ for a mobile phase with pH < 6 and $x_C \approx 0$ for pH ≥ 6. Similarly, strong bases (aminoalkyl derivatives) have $x_C \approx 0.1$ for pH ≥ 7 and $x_C \approx 1$ for pH < 6. See Ref. [18] for details.

Figure 4 shows an example from a study [18] that illustrates the use of Eq. (4); it shows the gradient separation of a complex mixture containing carboxylic acids but no bases. Eleven components of the sample are of interest (indicated by "*"), and it was necessary to separate each of these compounds from adjacent peaks with baseline resolution ($R_s \geq 1.5$). The original separation on a Luna C$_{18}$(2) column meets this requirement (Fig. 4a). Three other columns with $0.8 \leq F_s^* \leq 10.1$ were selected as possible replacements for the original column; the separation with each of these columns is shown in Fig. 4b–d. It can be seen that virtually the same separation was obtained with the Prodigy ODS (3) (b) and Inertsil ODS-3 (c) columns, as expected from their low values of F_s^* (0.8 and 2.3, respectively).

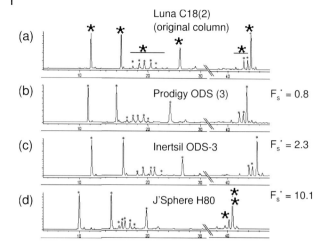

Fig. 4. Comparative separations of a pharmaceutical sample on an original column (a) and three possible replacement columns (b–d). The sample contained carboxylic acids and no bases. Columns (15 × 0.2 cm, 5 μm particles) identified in the figure. Conditions: gradient separation with solvents (A) and (B); (A) was 5% acetonitrile/pH 6.8 buffer; (B) was 95% acetonitrile/buffer; the gradient was 0/19/25/50/100% (B) in 0/5/28/40/ 60 min; 0.2 mL min^{-1}. See Ref. [18] and text for other details. Sample composition and detailed experimental conditions are proprietary. (Reprinted from [18]).

However, the J'Sphere H80 column with $F_s^* = 10.1$ gives a poor separation (complete overlap of the last two bands). In most of the other ten separations reported in [18], one or more successful replacement columns could be identified in the same way as in Figs. 3 and 4. Where attempts at replacing the original column were unsuccessful, this could be anticipated from the values of F_s^* of the different columns compared with that of the original column.

Columns with values of F_s or F_s^* moderately larger than 3 are often suitable replacements, meaning that columns with $F_s^* < 6$ can be considered as *possible* replacement columns. When the critical resolution (value of R_s for the least resolved pair of bands) is > 2 for the original separation, larger changes in α are usually allowable for the replacement column. As a rough rule, this means that allowable values of F_s^* for equivalent columns can be as large as 1.5 times the critical resolution. For example, with a critical resolution $R_s = 4$, columns with $F_s^* \leq 6$ are likely to be suitable replacements for the original column. Similarly, columns with $F_s^* \leq 12$ are *possible* candidates for a separation with a critical resolution of 4.

2.1.6.3.2 Selecting Columns of Very Different Selectivity

During RP LC method development, a need for a change in separation selectivity is often encountered [20]. Similarly, in some cases there is a need for *orthogonal separation*, where a very different column selectivity can be used to check for the

Table 3. A comparison of average column selectivities by column type.
Values of F_s at pH 2.8 for the various column types of Table 2 (Eq. 15).

	C_{18} B[a]	C_8 B[a]	C_{18} A[b]	EPG[c]	PE[d]	CN[e]	Phenyl[f]	F-alk[g]	F-phen[h]
C_8 B[a]	6								
C_{18} A[b]	62	64							
EPG[c]	65	61	122						
PE[d]	16	14	77	51					
CN[e]	21	18	64	66	27				
Phenyl[f]	21	19	54	73	30	15			
F-alk[g]	82	85	21	143	98	83	74		
F-phen[h]	44	47	31	104	60	48	44	45	
Zr[i]	170	172	110	228	186	168	161	89	129

[a] Type B C_{18} or C_8 column; [b] type A C_{18} column; [c] embedded polar group column; [d] polar end-capped column; [e] cyano column; [f] phenyl column; [g] perfluoroalkyl column; [h] perfluorophenyl column; [i] bonded zirconia column.

presence of sample compounds which were not present during method development; see also Chapter 1.1. In either case, a major change in column selectivity is one option; values of F_s or $F_s^* \gg 3$ can be used to select a column of different selectivity. The examples of Figs. 2 to 4 are suggestive of the latter possibility.

The average values of **H**, **S***, etc. in Table 2 can be used to compare the selectivity of different RP LC column types in terms of their F_s values (Eq. 3), as summarized in Table 3. For example, type B C_8 and C_{18} columns are rather similar *on average*, with $F_s = 6$ for two such columns. This does not mean that every C_8 and C_{18} column has a similar selectivity (e.g., examples in Table 1 of type B C_{18}), just that on average there is not much difference in selectivity between a C_8 and a C_{18} column. Similarly, polar-endcapped columns ("PE" in Table 3) are relatively similar to type B C_8 ($F_s = 14$) or C_{18} ($F_s = 16$) columns. For the largest *average* change in selectivity, assuming a type B C_{18} column as the original column, a bonded zirconia column (Zr, $F_s = 170$) or a fluoroalkyl column (F-alk, $F_s = 82$) is most likely to result in a large change in selectivity. Since the F_s values in Table 3 will change with mobile phase pH or sample composition (presence of acids or bases), these comparisons of column selectivity are mainly useful as qualitative guides. Actual values of F_s^* for two columns are much more reliable predictors of relative column selectivity or orthogonality.

Once a primary, RP LC assay procedure has been selected, an orthogonal separation can be readily ceveloped [21]. A column of very different selectivity is selected, based on a large value of F_s compared to the column used in the primary assay procedure. The use of a different column plus a change in mobile-phase organic solvent (e.g., methanol replacing acetonitrile) generally provides a large enough change in separation selectivity to resolve overlapping peaks from the primary separation. This procedure was used to develop orthogonal assays for nine primary assays [21]; for eight of these assays, the change in separation selectivity was judged adequate. A further increase in separation selectivity can be achieved by a further change in the column or mobile phase pH.

2.1.6.4 Conclusions

An understanding of column selectivity is important for various reasons, but mainly when it is necessary to replace one column by another with either similar or very different selectivity. Column selectivity can be accurately characterized by five measurable properties of the column: **H**, *Hydrophobicity*; **S***, *Steric* resistance to insertion of bulky solute molecules into the stationary phase; **A**, column hydrogen-bond *Acidity*; **B**, column hydrogen-bond *Basicity*; **C**, column *Cation-exchange* activity. Values of these column parameters (**H**, **S***, etc.) have now been measured for a large number of different RP LC columns. Given values of **H**, **S***, etc., it is possible to compare two columns in terms of selectivity by means of a column-comparison function F_s. Commercial software is available (Column Match®; Rheodyne LLC, Rohnert Park, CA, USA) that facilitates the convenient comparison of any two columns in terms of selectivity; the software also includes values of **H**, **S***, etc. for about 300 different columns.

Nomenclature
(only for special terms that relate to the present chapter)

a	stationary phase hydrogen-bond basicity (Eq. 1)
A	relative column hydrogen-bond acidity, related to number and accessibility of silanol groups in the stationary phase (Eq. 2)
b	stationary phase hydrogen-bond acidity (Eq. 1)
B	relative column hydrogen-bond basicity (Eq. 2)
C	relative column cation-exchange activity, related to number and accessibility of ionized silanol groups in the stationary phase (Eq. 2)
C(2.8)	value of **C** at pH 2.8
C(7.0)	value of **C** at pH 7.0
F_s	column matching function (Eq. 3)
F_s^{*}	value of F_s corrected for absence of acids or bases (Eq. 4)
H	relative column hydrophobicity (Eq. 2)
RP LC	reversed-phase liquid chromatography
s	dipolarity/polarizability parameter for the stationary phase (Eq. 1)
S*	relative steric resistance to insertion of bulky solute molecules into the stationary phase; as **S*** increases, bulky solute molecules experience resistance (Eq. 2)
x_B, x_C	correction factors in Eq. 4)
α	separation factor for two solutes; also, k/k_{ref} (Eq. 2)
α'	relative solute hydrogen-bond acidity (Eq. 2)
α_2^{H}	solute hydrogen-bond acidity in solution (Eq. 1)

β'	relative solute hydrogen-bond basicity (Eq. 2)
β_2	solute hydrogen-bond basicity in solution (Eq. 1)
η'	relative solute hydrophobicity (Eq. 2)
κ'	relative charge on solute molecule (positive for cations, negative for anions) (Eq. 2)
π_2^H	dipolarity/polarizability parameter for solute (Eq. 1)
σ'	relative steric resistance of solute molecule to penetration into stationary phase (σ' is larger for more bulky molecules) (Eq. 2)
ν	free energy associated with creating a cavity in the stationary phase (Eq. 1)

References

1 W. R. Melander, Cs. Horvath, *High-Performance Liquid Chromatography. Advances and Perspectives*, Vol. 2 (Ed.: Cs. Horvath), Academic Press, New York, 1980, p. 113.

2 P. W. Carr, D. M. Martire, L. R. Snyder (Eds.), *Retention in Reversed-Phase Liquid Chromatography, J. Chromatogr.* 656 (1993).

3 K. Kimata, K. Iwaguchi, S. Onishi, K. Jinno, R. Eksteen, K. Hosoya, M. Araki, N. Tanaka, *J. Chromatogr. Sci.* 27 (1989), 721.

4 L. C. Sander, S. A. Wise, *J. Chromatogr.* 656 (1993), 335.

5 E. Cruz, M. R. Euerby, C. M. Johnson, C. A. Hackett, *Chromatographia* 44 (1997), 151.

6 C. F. Poole, S. K. Poole, *J. Chromatogr. A* 965 (2002), 263.

7 M. R. Euerby, P. Petersson, *J. Chromatogr. A* 994 (2003), 13.

8 H. A. Claessens, *Trends Anal. Chem.* 20 (2001), 563.

9 N. S. Wilson, M. D. Nelson, J. W. Dolan, L. R. Snyder, R. G. Wolcott, P. W. Carr, *J. Chromatogr. A* 961 (2002), 171.

10 N. S. Wilson, M. D. Nelson, J. W. Dolan, L. R. Snyder, P. W. Carr, *J. Chromatogr. A* 961 (2002), 195.

11 N. S. Wilson, M. D. Nelson, J. W. Dolan, L. R. Snyder, P. W. Carr, L. C. Sander, *J. Chromatogr. A* 961 (2002), 217.

12 J. J. Gilroy, J. W. Dolan, L. R. Snyder, *J. Chromatogr. A* 1000 (2003), 757.

13 J. J. Gilroy, J. W. Dolan, L. R. Snyder, *J. Chromatogr. A* 1026 (2004), 77.

14 N. S. Wilson, J. Gilroy, J. W. Dolan, L. R. Snyder, *J. Chromatogr. A* 1026 (2004), 91.

15 D. H. Marchand, K. Croes, J. W. Dolan, L. R. Snyder, *J. Chromatogr. A*, 1062 (2005), 57.

16 D. H. Marchand, K. Croes, J. W. Dolan, L. R. Snyder, R. A. Henry, K. M. R. Kallury, S. Waite, P. W. Carr, *J. Chromatogr. A*, 1062 (2005), 65.

17 L. R. Snyder, A. Maule, A. Heebsch, R. Cuellar, S. Paulson, J. Carrano, L. Wrisley, C. C. Chan, N. Pearson, J. W. Dolan, J. J. Gilroy *J. Chromatogr. A*, 1057 (2004), 49.

18 J. W. Dolan, A. Maule, D. Bingley, L. Wrisley, C. C. Chan, M. Angod, C. Lunte, R. Krisko, J. Winston, B. Homeier, D. M. McCalley, L. R. Snyder, *J. Chromatogr. A*, 1057 (2004), 59.

19 L. R. Snyder, J. W. Dolan, P. W. Carr, *J. Chromatogr. A*, 160 (2004), 77.

20 L. R. Snyder, J. J. Kirkland, J. L. Glajch, *Practical HPLC Method Development*, 2nd ed., Wiley-Interscience, New York, 1997.

21 J. Pellett, P. Lukulay, Y. Mao, W. Bowen, R. Reed, M. Ma, R. C. Munger, J. W. Dolan, L. Wrisley, K. Medwid, N. P. Toltl, C. C. Chan, M. Skibic, K. Biswas, K. A. Wells, L. R. Snyder, *J. Chromatogr. A*, 1101 (2006), 122.

2.1.7
Understanding Selectivity by the Use of Suspended-State High-Resolution Magic-Angle Spinning NMR Spectroscopy

Urban Skogsberg, Heidi Händel, Norbert Welsch, and Klaus Albert

2.1.7.1 **Introduction**

It is of essential importance to have an idea about the retention mechanism that is operative between an analyte and a chromatographic sorbent in order to predict the retention behavior for a series of analytes eluted from a chromatographic sorbent. To probe the interactions between analytes and chromatographic sorbents, tedious examinations are often carried out by injecting analytes onto a packed column and then analyzing the chromatographic peaks in terms of the retention factors. Even if the test analytes are carefully chosen, one obtains at best an educated guess of the interactions taking place between the analyte and the stationary phase rather than direct proof. Of course, the prediction of retention order becomes even more difficult when the stationary phase has the capacity to interact with an analyte through more than one type of interaction.

The retention of an analyte is dependent on the fraction of analyte present in the stationary phase compared to the fraction of analyte that remains in the mobile phase, i.e., a high equilibrium constant K between the analyte and the sorbent will result in a long retention time. Hence, the observed macroscopic chromatographic peak is the sum of all contributing interactions that take place between the analyte and the chromatographic sorbent. This means that not only the interactions that contribute to the selectivity, the selective ones, but also the ones that reduce the selectivity, the non-selective ones, contribute to the retention.

Since K is proportional to the retention factor k, the sum of all the possible interactions between the analyte and the sorbent can be summarized according to Eq. (1), where the indices 1, 2, 3, 4, 5, etc. correspond to the type of interaction that exists between the analyte and the chromatographic sorbent.

$$k = k_1 + k_2 + k_3 + k_4 + k_5 + k_{etc} \tag{1}$$

Hence, the outcome of a chromatogram, in terms of retention factors and selectivity factors, for a series of analytes eluted from different stationary phases is dependent on which of the interactions between the sorbents and the analytes is dominant. This is exemplified by observing the elution orders of a series of vitamin A metabolites eluted from four different stationary phases (see Fig. 1) [1].

In chromatogram A, all of the analytes are eluted from the C_{18} sorbent according to their different hydrophobic interactions. However, when the phase is changed to a C_{18} sorbent with embedded polar groups (B), a calix[4]arene-based sorbent (C), or a C_{30} sorbent (D), significant changes in selectivity are observed, even though the mobile phase composition is the same in all experiments. Here, it can be seen that when the potential for hydrogen-bonding interactions is introduced (Fig. 1b), a reversal in the relative positions of some peaks is observed. Also, when

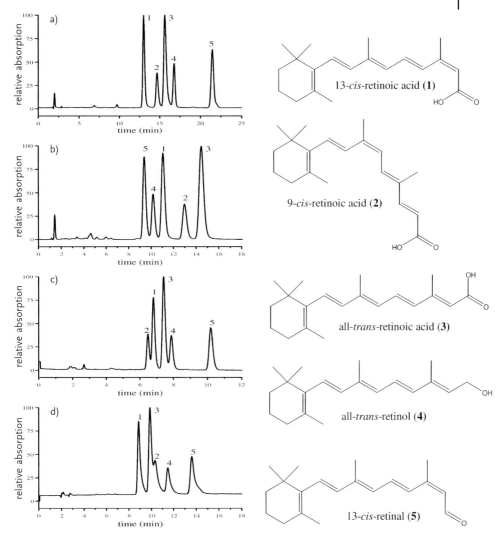

Fig. 1. Chromatograms showing the separations of analytes **1–5** obtained from the phases **I–IV**. (a) Phase **I**; (b) Phase **II**; (c) Phase **III**; (d) Phase **IV**. Conditions: mobile phase, acetonitrile (80%) in water containing 0.1% trifluoroacetic acid (TFA).

inclusion complexes are formed, the retention mechanism is changed for the shape-constrained analytes (Fig. 1c).

If a sorbent has the potential to interact with analytes through more than one retention mechanism, it should be possible to adjust the chromatographic conditions in order to strengthen just one of the interactions and thereby obtain different selectivity between the analytes. This is illustrated here for the case of the elution of vitamin A metabolites **1–5** from a stationary phase with embedded polar groups at different mobile phase compositions (see Fig. 2).

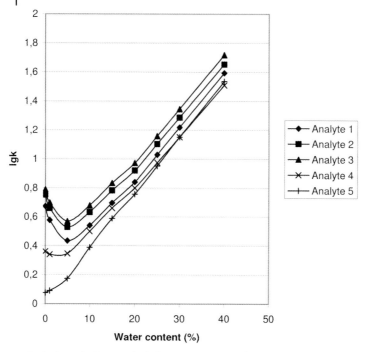

Fig. 2. A graph showing the lg *k* vs. water content relationship when the analytes **1–5** were eluted from a phase with embedded polar groups using different concentrations of acetonitrile in the mobile phase.

It was shown that by choosing a sorbent that separated the analytes through a retention mechanism based on both hydrogen-bonding and hydrophobic inter-actions instead of hydrophobic interactions alone, it was possible to reduce the analysis time from 22 min to only 7 min.

In order to observe intramolecular as well as intermolecular interactions between molecules, NMR is undoubtedly the method of choice [2]. Several reports have shown that solution-state ^{1}H NMR spectroscopy can be used to map proximities between a stationary phase and an analyte and thereby extract valuable information concerning the retention behavior of the analyte [3]. However, when solution-state NMR is used, both the selector and the analyte are dissolved during the measurements, precluding direct comparison to a chromatographic system. In other words, solution-state NMR investigations do not fully consider the complexity and the impact of the actual sorbents, for example the problems associated with the immobilization technique and the interactions with the silanol groups of the silica support. Also, the solid-state MAS NMR technique cannot alone give a complete picture of the retention behavior due to the absence of a mobile phase during the experiments.

Therefore, the method of choice when NMR is used to extract information about chromatographic behavior is to use suspended-state high-resolution magic-angle spinning (HR/MAS) NMR spectroscopy; see experimental set-up in Fig. 3 [4].

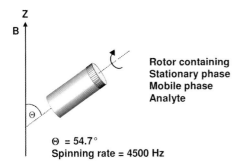

Θ = 54.7°
Spinning rate = 4500 Hz

Fig. 3. Experimental set-up of the suspended-state HR/MAS NMR rotor.

In this experiment, a chromatographic sorbent is suspended in a deuterated solvent in the presence of an analyte. The rotor is then spun at the magic angle of Θ = 54.7° to the magnetic field (B).

The fast rotation of the rotor (4500 Hz) at the "magic angle" will reduce chemical shift anisotropies as well as reduce the dipole-dipole interactions within the solid structure and make it possible to record ^{13}C spectra with good resolution. However, in order to reduce the dipolar interaction between protons and to enable the recording of a solid-state ^1H NMR spectrum of a solid sample, significantly higher spinning rates must be used. In suspended-state NMR experiments, the addition of a solvent to the sorbent increases the mobility of the surface structures, or ligands, and thereby minimizes susceptibility distortions and residual dipolar couplings between protons, leading to better resolution of the ^1H NMR spectrum (see Fig. 4) [5].

Hence, this NMR technique allows us to record the NMR spectrum of a dissolved analyte in the presence of a chromatographic sorbent, and therefore mimic a static chromatographic situation. In the suspended-state HR/MAS NMR experiment, the resolution of the signals of the dissolved analyte will be comparable to that in a solution-state NMR experiment. HR/MAS NMR spectroscopy also allows the acquisition of ^1H NMR spectra of stationary phases bonded to silica supports with high resolution compared to the corresponding solid-state NMR spectra. The HR/MAS ^1H NMR spectrum of a (*tert*-butylcarbamoyl)quinine selector covalently bonded to silica, suspended in [D_4]methanol, shows that a full structure elucidation of the selector is possible (see Fig. 4).

2.1.7.2 Is the Comparison Between NMR and HPLC Valid?

Although a mathematical treatment of chemical shift data is beyond the scope of this chapter, some basics are presented here; for the interested reader, there is plenty of published literature [6, 7]. It is important to first point out that the mathematical treatment of NMR data presented here is only valid when a 1 : 1 complex is formed between a selector and an analyte.

A signal observed in an NMR spectrum will have a different position on the NMR chemical shift scale if the nucleus has a different electronic environment

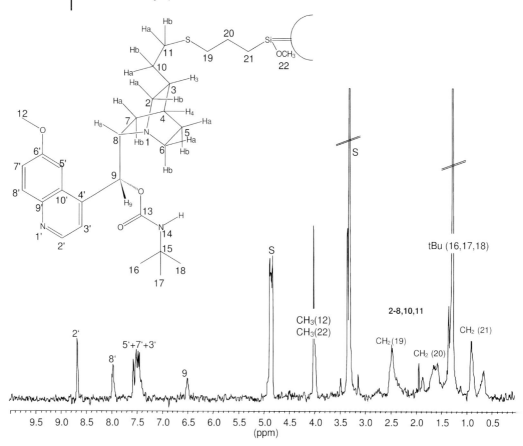

Fig. 4. Suspended-state HR/MAS NMR spectrum of a chromatographic sorbent suspended in [D$_4$]methanol.

according to Eq. (2), where the B$_0$ is the external magnetic field, σ is the electronic density, and γ is the gyromagnetic constant [8].

$$v_1 = (\gamma/2\,\pi)\,(1 - \sigma)\,B_0 \qquad (2)$$

It is obvious, therefore, that if the electronic density (σ) around a nucleus in an analyte is influenced by some kind of perturbation, the position of its NMR signal will be changed. Therefore, if a proton in an analyte interacts with some part of a chromatographic sorbent added to the suspended-state rotor, its signal will shift from its original chemical shift position.

In chromatography, as well as in NMR, the formation of different selector–analyte complexes between the stationary phase (S) and the analytes (A$_1$, A$_2$, etc.) takes place, where the equilibrium for the formation of these complexes is given by Eq. (3).

$$S + A \underset{}{\overset{K}{\rightleftharpoons}} SA \tag{3}$$

Hence, the chromatographic selectivity obtained from a chromatographic sorbent for a series of analytes is dependent on the different strengths of the complexes formed between the sorbent and the respective analytes. In order to calculate the equilibrium constant K for complex formation between a sorbent and an analyte, the concentration of the free and complexed analyte as well as the amount of selector immobilized on the support must be known (see Eq. 4).

$$K = [SA]/([S][A]) \tag{4}$$

However, two cases must be considered in the NMR experiment, i.e., one needs to distinguish whether the exchange between the selector and analyte is fast or slow on the NMR time scale. If the reaction rate is slow on the NMR time scale, two individual signals should appear on the chemical shift scale, one corresponding to the free analyte and one corresponding to the complexed form. Since the amount of selector is known, it is no problem to calculate K from the measured integrals of the NMR signals of the free and complexed analyte.

A more complex situation arises when the free and complexed analyte undergo fast exchange on the NMR time scale. It is then impossible to directly integrate the signals of the free and complexed analyte since a time-averaged spectrum of the free and complexed analyte will be obtained. However, the total concentration of A can be expressed as the mole fraction of the free and complexed analyte according to Eq. (5).

$$X_A + X_{SA} = 1 \tag{5}$$

Therefore, the observed NMR chemical shift of A (δ_A^{obs}) can be expressed as an average of the mole fractions of the free (δ_A) and complexed analyte (δ_{SA}) (see Eq. 6).

$$\delta_A^{obs} = X_A \delta_A + X_{SA} \delta_{SA} \tag{6}$$

When two enantiomers (A_1 and A_2) are present with a chiral stationary phase (S), a chemical shift difference may be observed under special conditions see Eq. (7).

$$\Delta\delta = X_{SA1}(\delta_{SA1} - \delta_0) - X_{SA2}(\delta_{SA2} - \delta_0) \tag{7}$$

In Fig. 5, a chemical shift difference between the methine protons in a mixture of (±)-O,O'-dibenzoyl-tartaric acid (DBTA) in the presence of the chromatographic sorbent Kromasil-TBB (TBB = N,N'-diallyl-L-tartardiamide bis(4-tert-butylbenzo-ate)) can be observed. The splitting of the signals is due to simultaneous formation of two different diastereomeric complexes, one between (+)-DBTA and Kromasil-TBB and the other between (–)-DBTA and Kromasil-TBB. The magnitude of this splitting is dependent on the amount of sorbent, i.e., the selector-to-analyte ratio, as well as on the solvent used. This proves that there is an equilibrium situation between the sorbent and the analytes. The experiment gives direct evidence of the chiral discriminating ability of the chromatographic sorbent.

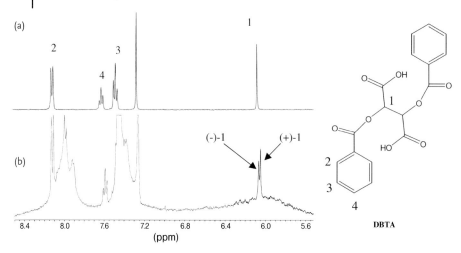

Fig. 5. Suspended-state HR/MAS NMR spectra of (a) (±)-DBTA dissolved in CDCl₃; (b) (±)-DBTA dissolved in CDCl₃; in the presence of Kromasil-TBB.

However, the only chemical shift that is known under conditions of rapid exchange is δ_A; neither δ_{AS} nor X_{AS} is known. Therefore, the observed chemical shift displacement of the analyte peak is used to calculate K through gradually increasing the amount of the selector relative to a fixed concentration of analyte. Hence, the interactions that take place between analytes and a stationary phase both in an HPLC column and in a suspended-state NMR rotor can be described by the equilibrium constant K. Therefore, a comparison of NMR data and retention factors k is possible.

2.1.7.3 The Transferred Nuclear Overhauser Effect (trNOE)

How do we know that the analyte and stationary phase interact in the NMR rotor if we do not observe any chemical shift displacements? The NOESY experiment is nowadays a well-established NMR experiment for observing through-space proximities within a molecule as well as between molecules. With the use of NOE, it is possible to distinguish how far different spin systems are from each other, since the magnitude of the NOE is proportional to the distance between the spin systems (r^{-6}). It is known that the sign of the NOE, either positive or negative, is dependent on the size of the molecule due to the different relaxation mechanisms that are operative in a large (zero quantum) or small (double quantum) molecule, respectively [2]. This is a result of the fact that a small molecule has a short correlation time (τ_c), i.e., the molecule rotates rapidly, while a large molecule has a long τ_c. These relaxation mechanisms change over at around a molecular size of 1000 Da. Thus, molecules with a molecular weight of over 1000 Da will have negative NOE cross-peaks, whereas molecules with a weight under 1000 Da will have positive NOE cross-peaks. However, if a low molecular weight analyte interacts with a large selector molecule, i.e., a chromatographic sorbent, the NOE from the

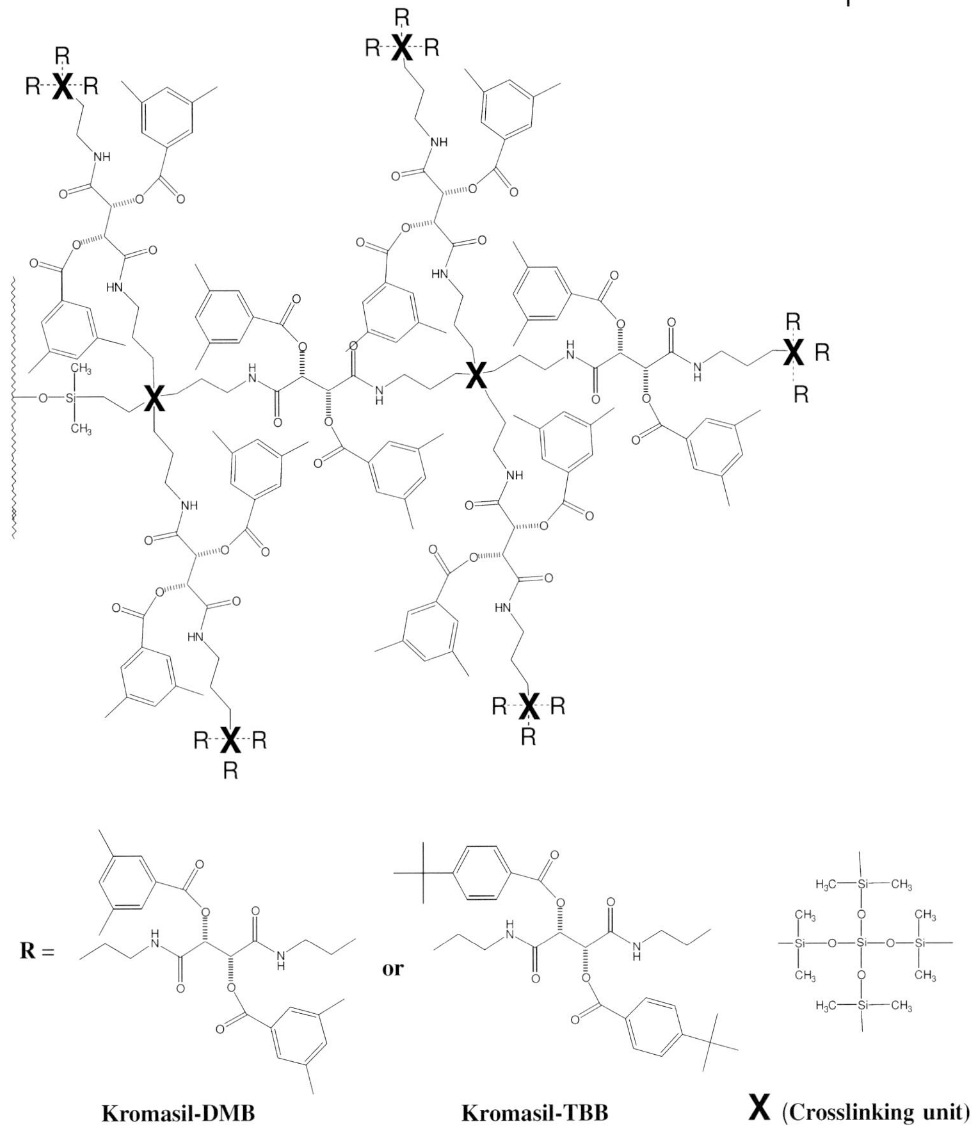

Fig. 6. Schematic representation of the Kromasil-DMB and Kromasil-TBB sorbents.

large selector molecule can be transferred to the small analyte. Hence, the low molecular analyte will, for a short time (ca. 100 ms), show a negative NOE cross-peak in the recorded spectra [4, 5, 9]. This is because, as long as they interact, the small molecule will have the same τ_c as the large selector molecule. It is important to realize that no such trNOE effect can be recorded if the interactions are too weak, i.e., if too few molecules have the memory effect, or too strong, i.e., if there are no free molecules from which to measure the memory effect.

By comparing the magnitudes of the trNOEs for a given cross-peak in an analyte, a quantitative estimation of the interaction strengths with different sorbents can be obtained. Hence, the correlation with chromatography is proven, for example, by comparing the trNOE intensities obtained for norethisterone acetate (**6**) and chlormadinone acetate (**7**) in the presence of the sorbents Caltrex AIII (phase **A**) and Caltrex BIII (phase **B**) with the corresponding retention factors obtained from HPLC (see Tables 1 and 2). From Table 1, it can clearly be seen that analyte **6** shows the same trNOE cross-peak magnitude in the presence of both phases **A** and **B**, while analyte **7** shows a stronger trNOE cross-peak in the presence of phase **A** than with **B**. This means that analyte **6** interacts equally strongly with the phases **A** and **B**, while analyte **7** interacts more strongly with phase **A** than with phase **B**. This is corroborated by the chromatographic data presented in Table 2 [10]. It should be noted that the comparison between the NMR data and chromatography is only valid when the same solvent composition is used in the NMR experiments and as mobile phase in the chromatographic experiment.

However, no comparison of the interaction strengths can be made between analytes **6** and **7** since the trNOEs were measured between different atoms. The intramolecular distance (r) is proportional to the NOE with an r^{-6} dependence, and therefore different magnitudes of NOE will be observed if r differs between the molecules. For a valid comparison, it is important to know that the intramolecular distances between the groups giving rise to the trNOE cross-peak are the same. Therefore, this technique is well-suited for chiral discrimination studies,

Table 1. Comparison of trNOE cross-peaks obtained from suspended-state HR/MAS NMR experiments on analytes 6 and 7 in the presence phases A and B suspended in [D$_4$]methanol (80%) and D$_2$O.

Analytes (chosen cross-peak)	Cross-peak intensity in the presence of phase A	Cross-peak intensity in the presence of phase B
6 [H-4-H-2]	−0.01	−0.01
7 [H-4-H-7]	−0.20	−0.10 −0.12[a]

[a] Duplicate measurement.

Table 2. *k* values of analytes **6** and **7** eluted from phases **A** and **B** using different concentrations of methanol and water in the mobile phases.

Phase	Analyte	MeOH 80% k	MeOH 75% k	MeOH 70% k	MeOH 65% k	MeOH 60% k
A	6	1.61	2.64	4.47	8.37	14.21
	7	1.78	2.88	4.84	9.13	15.40
B	7	1.39	2.41	4.17	7.64	14.55
	6	1.61	2.83	4.91	9.18	17.76

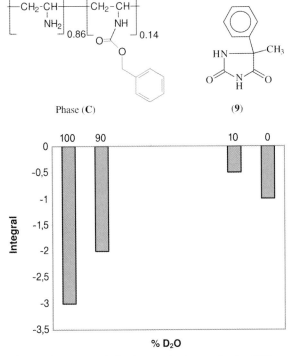

Phase (**C**) (**9**)

Fig. 7. Relative integrals for the negative intramolecular NOE cross-peak between the phenyl and methyl groups of methylphenylhydantoin (**9**) in a suspension of **9** and phase **C** in $D_2O/[D_3]$acetonitrile solvent mixtures.

where different enantiomers interact with a chiral stationary phase. Otherwise, the same molecule can only be compared in different environments, as shown in Fig. 7. Different magnitudes of trNOE were recorded when methylphenyl-hydantoin (**9**) was allowed to interact with a silica gel coated with a polyvinylamine (phase **C**) suspended in different mixtures of water/acetonitrile [4].

It can easily be seen that the NOE cross-peak magnitude will reflect the retention order when a mobile phase composed of water/acetonitrile is used. An interesting increase in interaction strength is observed when the water concentration is reduced from 10% to 0%; this is believed to reflect a change in retention mechanism from a hydrophobic dominated one, at water contents above 10%, to a hydrogen-bonding dominated one, at water contents below 10%.

2.1.7.4 Suspended-State ¹H HR/MAS T_1 Relaxation Measurements

The T_1 relaxation time for protons, measured using an inversion-recovery pulse sequence, is known to be shorter when the number of relaxation pathways is increased. This means that if an analyte proton is in closer proximity to one selector compared to another, this should be reflected in a shorter T_1 relaxation time. Therefore, suspended-state ¹H HR/MAS T_1 relaxation measurements can be used to map differences in proximities between analytes and chromatographic sorbents.

Table 3. Suspended-state HR/MAS-NMR T_1 measurements on (R)-**8** and (S)-**8** in the presence of Kromasil-DMB using [D_8]propan-2-ol (5%) in C_6D_{12} as solvent.

Analyte signal	T_1 values for (R)-**8** (s)	T_1 values for (S)-**8** (s)
H-1	1.7	2.3
H-2	1.4	1.9
H-3	1.4	1.9
H-4	1.5	1.9
H-6	1.6	2.2

However, if the intramolecular distances in two molecules are the same, for example in enantiomers, their T_1 times can be compared and used to predict retention order. From a comparison of the T_1 values obtained for (R)- and (S)-1,1'-binaphthyl-2,2'-diol (**8**) in the presence of Kromasil-DMB (DMB = N,N'-diallyl-L-tartardiamide bis(3,5-dimethylbenzoate)), as shown in Table 3, it can be seen that significantly shorter values were observed for (R)-**8** than for (S)-**8** for each of the respective signals [9]. This indicates closer proximity of (R)-**8** to Kromasil-DMB, and hence stronger interactions. This was confirmed by the retention order when a mixture of (R)- and (S)-**8** was eluted from a column packed with the Kromasil-DMB sorbent using the same mobile phase mixture as in the NMR experiment (see Fig. 8).

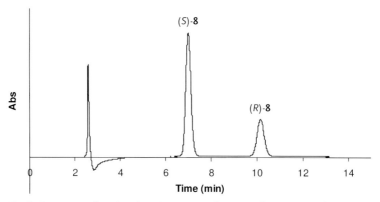

1,1'-Binaphthyl-2,2'-diol (**8**)

Fig. 8. Separation of (R)-**8** and (S)-**8** on Kromasil-DMB as the stationary phase with propan-2-ol (5%) in cyclohexane as mobile phase; flow rate 1 mL min^{-1}.

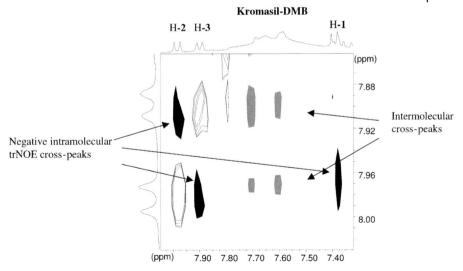

Fig. 9. Suspended-state HR/MAS trNOESY spectrum of (*S*)-**8** dissolved in CDCl₃ in the presence of Kromasil-DMB showing both intramolecular cross-peaks (black) and intermolecular cross-peaks (grey).

2.1.7.5   Where do the Interactions Take Place?

Several different suspended-state HR/MAS NMR experiments can be performed in order to obtain information aboutly exact where the interactions take place between an analyte and a sorbent. In Fig. 9, the suspended-state HR/MAS trNOESY spectrum of (*S*)-**8** dissolved in CDCl₃ in the presence of Kromasil-DMB, shown in Fig. 5, shows both negative intramolecular trNOE cross-peaks (black), confirming the interaction with the sorbent, and intermolecular (grey) cross-peaks between the aromatic residue in the selector and the aromatic ring in the analyte [9].

2.1.7.6   Hydrogen Bonding

Suspended-state HR/MAS ¹H NMR spectroscopy can also be used to detect hydrogen bonding between analytes and stationary phases. Functional groups such as alcohols, amines, etc., that have the capacity for hydrogen bonding can show a chemical shift displacement or a signal broadening. In Fig. 10, it can clearly be seen that the signal corresponding to the hydroxyl group present in (*S*)-**8** is significantly broadened when Kromasil-DMB is added to the rotor [9].

The magnitude of the broadening or the chemical shift displacement is of course dependent on the S/A ratio used in the experiment.

2.1.7.7   Some Practical Considerations

Since not all solvents used as mobile phases can be obtained in deuterated form, or are prohibitively expensive, the chromatographic conditions must be tailored to a mobile phase that can be substituted by a more NMR-compatible solvent

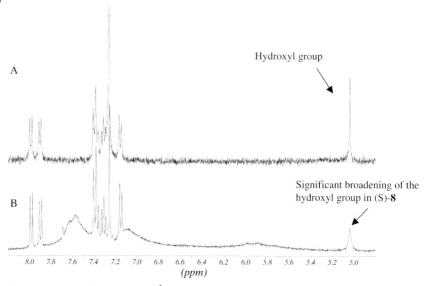

Fig. 10. Suspended-state HR/MAS ^{1}H NMR spectra of (A) free (S)-**8** dissolved in CDCl$_3$, and (B) (S)-**8** dissolved in CDCl$_3$ in the presence of Kromasil-DMB.

mixture. This is generally not a problem for reversed-phase chromatography since water, methanol, and acetonitrile are available in deuterated form. However, under normal-phase chromatographic conditions, it can be a good idea to change from *n*-hexane to cyclohexane or to use chloroform where possible; these changes will, of course, change retention and selectivity in the chromatograms. The choice of solvent will also affect the type of interaction, i.e. more hydrogen bonding when a nonpolar solvent is used. Therefore, if solvent mixtures are used in the experiments, for example [D$_8$]propan-2-ol and [D$_{12}$]cyclohexane, the amount of the polar component will determine the strength of the interaction. Since NMR detection is less sensitive than the UV detection used in HPLC, larger amounts of analyte must be used in the NMR experiment. Therefore, the problem of low solubility of the analyte must be considered.

Another feature that needs to be considered is the fact that the NMR experiment is a static experiment and hence no consecutive bond making and bond breaking will take place as in a flow situation. Therefore, is it important to have a correct selector-to-analyte ratio (S/A value) during the NMR measurements. Both the choice of solvent and the S/A ratio will affect the amount of complexed analyte, due to the equilibrium situation between the sorbent and the analyte. Therefore, care must be taken that there is no unwanted signal overlap of the analyte and selector signals. Since an equilibrium situation is involved, it is important to use pure solvents in order to prevent competing interactions with possible impurities present in the solvent. It is also important that the interaction with the sorbent does not result in a conformational change of the analyte and hence result in different intramolecular distances.

2.1.7.8 **Future Aspects**

The use of suspended-state HR/MAS NMR spectroscopy to study the interactions between chromatographic sorbents and analytes of interest is still in its infancy. In principle, all of the sophisticated NMR experiments developed for protein ligand exchange studies might be used to investigate the sorbents. Therefore, experiments such as saturation transfer difference (STD) NMR, whereby the part of the analyte that interacts with the sorbent can be revealed, will give valuable information on retention mechanisms.

References

1 U. Skogsberg, D. Zeeb, K. Albert, *Chromatographia* 59 (2004), 1.

2 D. Neuhaus, P. Williamson, *The Nuclear Overhauser Effect in Structural and Conformational Analysis*, VCH, Weinheim, 2000.

3 C. Yamamoto, E. Yashima, Y. Okamoto, *J. Am. Chem. Soc.* 124 (2002), 12583.

4 H. Händel, E. Gesele, K. Gottschall, K. Albert, *Angew. Chem.* 115 (2003), 454; *Angew. Chem. Int. Ed.* 42 (2003), 438.

5 C. Hellriegel, U. Skogsberg, K. Albert, M. Lämmerhofer, N. M. Maier, W. Lindner, *J. Am. Chem. Soc.* 126 (2004), 3809–3816.

6 I. Fielding, *Tetrahedron* 56 (2000), 6151.

7 K. A. Connors, Binding Constants, John Wiley & Sons, USA, 1987.

8 T. D. W. Claridge, High-Resolution NMR Techniques in Organic Chemistry, Pergamon, The Netherlands, 1999.

9 U. Skogsberg, H. Händel, D. Sanchez, K. Albert, *J. Chromatogr. A* 1023 (2004), 215–223.

10 U. Skogsberg, H. Händel, E. Gesele, T. Sokoließ, U. Menyes, T. Jira, U. Roth, K. Albert, *J. Sep. Sci.* 26 (2003), 1–6.

2.2
Optimization in Normal-Phase HPLC

Veronika R. Meyer

2.2.1
Introduction

The normal-phase (NP) mode is not often used in high-performance liquid chromatography, probably because many analysts suppose that it is too complicated or even unreliable. This is by no means true; rather, it is a valuable tool for the separation of a large group of analytes. Silica, as the prototype of an NP stationary phase, is superior for the separation of isomers (*cis/trans*, positional isomers, and diastereomers). An example is shown in Fig. 1. The silica surface is rigid (in contrast to the flexible hydrocarbon chains of the usual reversed-phase materials) and the analytes interact in a sterically defined manner with the polar groups that

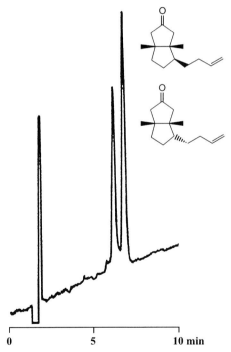

Fig. 1. Separation of two diastereomers (bicyclic ketones synthesized by M. Thommen, University of Bern, Department of Organic Chemistry, 1994). Column: 25 cm × 3.2 mm; stationary phase: LiChrosorb SI 60 5 μm; mobile phase: hexane/isopropanol, 99 : 1; 1 mL min^{-1}; temperature: ambient; detection: refractive index.

HPLC Made to Measure: A Practical Handbook for Optimization. Edited by Stavros Kromidas
Copyright © 2006 WILEY-VCH Verlag GmbH & Co. KGaA, Weinheim
ISBN: 3-527-31377-X

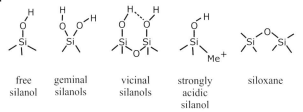

| free silanol | geminal silanols | vicinal silanols | strongly acidic silanol | siloxane |

Fig. 2. The various polar groups on the surface of silica.

are present there (Fig. 2), namely free, geminal, vicinal (with hydrogen bridge), and strongly acidic silanols with a cation such as Fe^{3+} nearby, plus the siloxane groups. Geminal and vicinal silanols are neutral, but the free silanols are weakly acidic. The acidic groups can act as cation exchangers and therefore they may give rise to problems with basic analytes. Highly pure silicas are almost free of cationic impurities and the strongly acidic silanol groups are absent.

In normal-phase liquid chromatography, optimization is based on the properties of the solvents that can be used as eluents, namely localization and basicity. A localizing solvent interacts significantly with the adsorptive centers of the stationary phase. A basic solvent acts as a proton acceptor in hydrogen bonding. Usually, a weak and therefore non-localizing and non-basic solvent such as hexane is part of the mixture. The second solvent can be dichloromethane (non-basic, non-localizing), acetonitrile or ethyl acetate (non-basic, localizing) or methyl *tert*-butyl ether (basic, localizing).

2.2.2
Mobile Phases in NP-HPLC

In NP-LC, the user has a much larger choice of solvents with different properties than in the reversed-phase mode. Table 1 gives an overview and lists the following data:

- Strength $\varepsilon°$, a parameter based on the adsorption energy of the solvent.
- Viscosity; it is an advantage of NP eluents that their viscosity is low (dioxane and isopropanol being exceptions), which results in low back-pressure and high theoretical plate numbers.
- UV cut-off.
- Localization, the capacity to interact strongly with the adsorptive centers of the stationary phase.
- Basicity, the property to act as a proton acceptor in hydrogen bonds. This parameter is irrelevant for non-localizing solvents, and alcohols have different selectivity properties.

The solvents can be presented in a selectivity triangle according to their acidic, basic, and dipole properties (α, β, and π^*), as shown in Fig. 3. It is obvious that more pronounced changes in selectivity can be expected if the test separations are tried with solvents that are far apart from each other in the triangle.

Table 1. Solvent properties.

Solvent	Strength $\varepsilon°$	Viscosity [mPa s]	UV cut-off [nm]	Locali- zation	Basicity
Hexane	0	0.33	190	–	
Heptane	0	0.41	190	–	
Diisopropyl ether	0.22	0.37	220	–	
Diethyl ether	0.29	0.24	205	+	+
Dichloromethane	0.30	0.44	230	–	
Chloroform	0.31	0.57	245	–	
1,2-Dichloroethane	0.38	0.79	230	–	
Triethylamine	0.42	0.38	230	+	+
Acetone	0.43	0.32	330	+	–
Dioxane	0.43	1.54	220	+	+
THF	0.48	0.46	220	+	+
MTBE	0.48	0.35	220	+	+
Ethyl acetate	0.48	0.45	260	+	–
Acetonitrile	0.50	0.37	190	+	–
Isopropanol	0.60	2.3	210	+	
Ethanol	0.68	1.2	210	+	
Methanol	0.73	0.6	205	+	
Acetic acid	large	1.26	260	+	

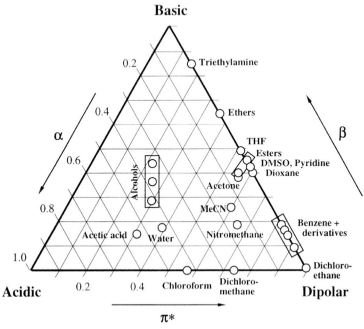

Fig. 3. The selectivity triangle of organic solvents:
α: acidic properties; β: basic properties; π*: dipole properties.

Solvents of equal strength will yield similar k values of the analytes, but in practice the observed retention times and resolutions can differ due to their selectivity properties. As an example, *tert*-butyl methyl ether and acetonitrile have similar strengths, but the former is basic whereas the latter is not. The same can be true for solvent mixtures. The separation of the two diastereomers shown in Fig. 1 was tried with several mixtures (on LiChrosorb SI 60) and the following results were obtained:

- Hexane/isopropanol, 99 : 1, $k_1 = 3.4$, $\alpha = 1.14$; this is the separation presented.
- Hexane/methyl *tert*-butyl ether, 80 : 20, $k_1 = 4.2$, $\alpha = 1.12$.
- Hexane/ethyl acetate, 90 : 10, $k_1 = 7.2$, $\alpha = 1.11$.
- Hexane/tetrahydrofuran, 90 : 10, $k_1 = 2.5$, $\alpha = 1.07$.

The separation was not possible in the reversed-phase mode (on LiChrospher 100 RP-18 with water/methanol, 30 : 70, which gave $k = 3.4$).

A rational approach for method development and optimization is the use of a weak solvent denoted as A and a stronger solvent denoted as B. Usually, A is hexane or heptane. Three different types of solvent B are tried as the second component of a binary mixture as follows:

a) A non-localizing solvent, e.g. dichloromethane.
b) A non-basic, localizing solvent, e.g. acetonitrile or ethyl acetate. Acetonitrile is poorly miscible with alkanes and a small amount of dichloromethane is probably necessary to obtain a useful eluent. Note the high UV cut-off of ethyl acetate, which limits the amount that can be added to the solvent A if UV detection is to be performed.
c) A basic, localizing solvent, e.g. methyl *tert*-butyl ether.

Figure 4 presents binary mixtures of equal strengths. As an example, mixtures with 40% dichloromethane, 2% methyl *tert*-butyl ether, 4% tetrahydrofuran, 2% ethyl acetate or less than 0.5% isopropanol in hexane all have a similar elution strength of approximately $\varepsilon° = 0.22$. If a higher strength is necessary (because the analytes are rather polar), solvents other than hexane can be used as the component A. For example, if methyl *tert*-butyl ether is the solvent B, 60% in hexane gives a similar strength as 20% in dichloromethane ($\varepsilon° = 0.4$), but the selectivity may differ.

Instead of mixing all these solvents for use in HPLC trials, it can be much simpler (and cheaper) to perform the first experiments on small thin-layer chromatography plates coated with a silica phase. An eluent is only useful if the resulting R_f values are between 0.1 and 0.5. An R_f value of 0.3 can be expected to give a k value of approximately 2 on the column.

Note the mixing properties of the various solvents as shown in Fig. 5. Only a few solvents are fully miscible with all others from hexane to water, namely acetone, glacial acetic acid, dioxane, absolute ethanol, isopropanol, and tetrahydrofuran.

It may be necessary to add a small amount of triethylamine to the mobile phase if basic analytes need to be separated. Similarly, a trace of acetic, trifluoro-acetic or formic acid can be useful for acidic analytes. However, additives should only be used when necessary (simple eluents are less prone to handling errors).

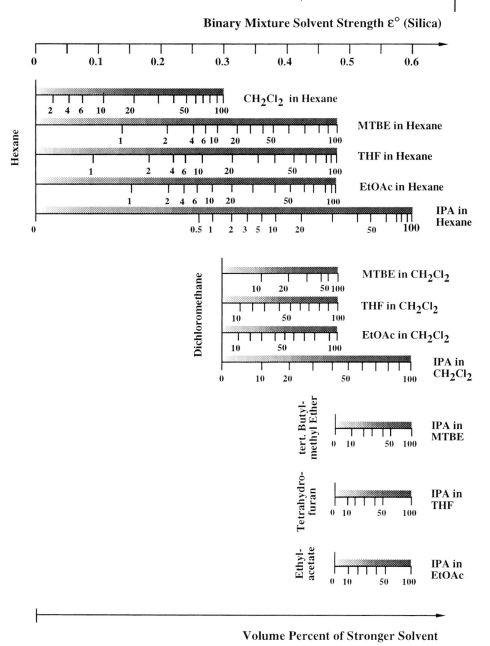

Fig. 4. Elution strengths of binary mixtures. If a vertical line is drawn through the nomogram, all of the corresponding mixtures will have the same strength. MTBE = methyl *tert*-butyl ether.

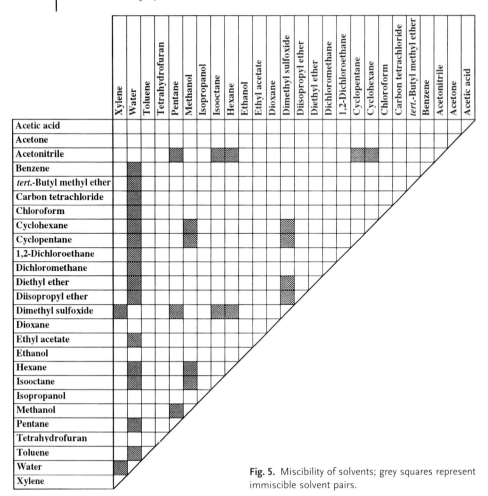

Fig. 5. Miscibility of solvents; grey squares represent immiscible solvent pairs.

2.2.3
Stationary Phases in NP-HPLC

Silicas can be basic, neutral or acidic, so it might be advisable to try different brands. A basic stationary phase is not suitable for acidic analytes, and vice versa. Moreover, the various commercially available silicas differ in purity with regard to their content of heavy metal ions (mainly iron), giving rise to different selectivities. Last but not least, the greater the specific area of the stationary phase, varying in the range 400–900 $m^2 g^{-1}$ depending on the pore size, the larger will be the k values when using the same mobile phase.

It is possible to apply gradients to silica gel; see Fig. 6 for an example. There is no problem with re-equilibration if the solvent B is non-localizing, e.g. dichloromethane. Other solvent systems have also been successfully used, although a

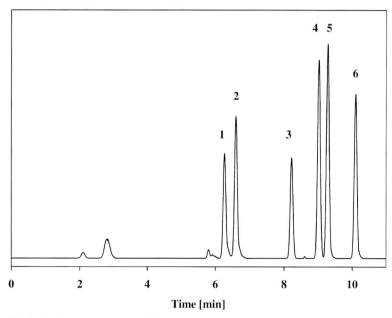

Fig. 6. Gradient separation on silica.
Column: 25 cm × 3.2 mm; stationary phase: LiChrosorb SI 60 5 μm;
eluent A: *n*-hexane; eluent B: dichloromethane;
gradient: 0–2 min 100% A, 2–12 min 0–80% B; flow: 1 mL min^{-1};
temperature: ambient; detector: UV 254 nm.
Compounds: (1) 2-phenylethylbromide, (2) 1,4-diphenylbutane, (3) phenetole,
(4) nitrobenzene, (5) *trans*-chlorostilbene oxide, 6 = Sudan red 7B.

strict temperature control may be necessary. It can be assumed (but has not yet been investigated) that silicas with lower metal content are more convenient for gradient separations, and that a used column (with some dirt collected on the surface of its packing) is more rugged than a new one.

Other polar phases, i.e., silicas with covalently bonded nitrile, nitro, amine, or diol groups, are valuable alternatives to silica. They are less polar (hence the mobile phase can be weaker) and have different selectivities. However, in most cases, they are used in the reversed-phase mode with aqueous mobile phases. Normal-phase applications of some importance are the separation of aromatics on nitro-derivatized phases, and of non-ionic surfactants on diol-derivatized phases. The theoretical plate numbers are usually lower than with non-derivatized silica. Activity control (see Section 2.2.4 on troubleshooting of NP separations) is not necessary. The stability of the chemical bonding can be limited.

A rarely used polar stationary phase is alumina (aluminum oxide; alox). It has a different selectivity to silica, and its packings show higher theoretical plate heights (i.e., the number of theoretical plates per unit length is lower). In contrast to silica, irreversible adsorption of analytes can be a problem.

2.2.4
Troubleshooting in Normal-Phase HPLC

One of the reasons why silica is not used by many analysts, even for separation problems that would best be solved with this stationary phase, are the decades-old reports of the kind:

- equilibration times are long after a solvent change,
- retention times are poorly reproducible,
- the linearity of the adsorption isotherm is poor and therefore so is the useful range of sample sizes that can be applied.

The reason for such problems lies in the heterogeneity of the silica surface with regard to its adsorptive sites (the silanol groups). The effects increase with decreasing polarity of the eluent. New studies on this topic are lacking. It can be assumed that these problems have been over-estimated in the past. In order to overcome them, it was proposed that a moderator should be added to the mobile phase, i.e., a small amount of a solvent with high localization. One strategy is to add 0.05% acetonitrile to the solvent A (hexane). Note that even such a small amount of acetonitrile is not freely miscible with alkanes, and that careful mixing is necessary until the droplets of the additive disappear. Another approach is to use eluents that are half-saturated with water. This is best achieved by preparing a bottle of water-saturated eluent by adding enough water to obtain a two-phase system, and then mixing it with dry eluent in a 1 : 1 ratio.

It can be assumed that new columns are more prone to such problems than used ones; in the latter, the most active adsorption centers will be blocked with contaminations from preceding injections.

Whereas it seems that a minor degree of column contamination could be advantageous, nobody wants to really spoil the stationary phase. A simple thin-layer chromatography test will reveal whether strongly retained compounds are present in a sample. If this is the case, it is recommended that a pre-column is used. However, it can even be advisable to first clean the sample by passing it though a solid-phase extraction cartridge with silica packing or though a self-made short glass column with coarse silica.

Solvents with low boiling point, such as pentane or diethyl ether, are less convenient. Usually, it is no problem to find substitutes with higher boiling point, namely hexane or methyl *tert*-butyl ether in these cases. If a low-boiling eluent needs to be used, it is advisable to install a pressure restrictor at the detector outlet. This can be a Teflon tube of length 10 m and inner diameter 0.25 mm.

In order to avoid problems with water contamination, it is best to use a dedicated HPLC instrument for normal-phase separations. If the same instrument is used for both reversed-phase and normal-phase separations, it will be necessary to rinse it exhaustively with tetrahydrofuran or isopropanol before the new eluent is pumped through. When the instrument is prepared for a normal-phase separation with a high proportion of a nonpolar solvent (i.e., with a low proportion of the more polar solvent B), it is also necessary to rinse it for a sufficiently long time with the new eluent.

Although the "adventure" of chromatography began over 100 years ago with normal-phase separations, this mode is underestimated today and is used less frequently than it deserves. It is an elegant technique for a large number of analytes that are soluble in non-aqueous solvents.

References/Further Reading

H. Engelhardt: The role of moderators in liquid-solid chromatography, *J. Chromatogr. Sci.* 15 (1977), 380–384.

L. R. Snyder, J. J. Kirkland: Introduction to Modern Liquid Chromatography, Chapter 9: Liquid-solid chromatography, Wiley-Interscience, New York, 2nd ed., 1979, p. 349–409.

K. Ballschmiter, M. Wössner: Recent developments in adsorption liquid chromatography (NP-HPLC), *Fresenius J. Anal. Chem.* 361 (1998), 743–755.

F. Geiss: Fundamentals of Thin-Layer Chromatography, Chapter X: Transfer of TLC Separations to Columns, Hüthig, Heidelberg, 1987, p. 398–419.

P. Jandera: Gradient elution in normal-phase high-performance liquid chromatographic systems, *J. Chromatogr. A* 965 (2002), 239–261.

P. Jandera, Gradient elution in liquid column chromatography, in: Advances in Chromatography, P. R. Brown, E. Grushka, S. Lunte (Eds.), Dekker, New York, Vol. 43 (2005), 1–108.

J. L. Glajch, J. J. Kirkland, L. R. Snyder: Practical optimization of solvent selectivity in liquid-solid chromatography using a mixture-design statistical technique, *J. Chromatogr.* 238 (1982), 269–280.

V. R. Meyer: An example of gradient elution in normal-phase liquid chromatography, *J. Chromatogr. A* 768 (1997), 315–319.

M. D. Palamareva, V. R. Meyer: New graph of binary mixture solvent strength in adsorption liquid chromatography, *J. Chromatogr.* 641 (1993), 391–395.

L. R. Snyder: Principles of Adsorption Chromatography, Marcel Dekker, New York, 1968.

L. R. Snyder, P. W. Carr, S. C. Rutan: Solvatochromically based solvent-selectivity triangle, *J. Chromatogr. A* 656 (1993), 537–547.

L. R. Snyder, J. J. Kirkland, J. L. Glajch: Practical HPLC Method Development. Chapter 6.6: Retention in normal-phase chromatography, p. 268–282; Chapter 6.7: Optimizing the separation of non-ionic samples in normal-phase chromatography, p. 282–288, Wiley-Interscience, New York, 2nd ed., 1997.

2.3
Optimization of GPC/SEC Separations by Appropriate Selection of the Stationary Phase and Detection Mode

Peter Kilz

2.3.1
Introduction

Gel-permeation chromatography (GPC, also known as size-exclusion chromatography (SEC)) is the established method for determining the properties of synthetic and natural macromolecules and biopolymers in solution. The vast majority of macromolecular samples are heterogeneous with regard to molar mass, chemical composition, end-group functionality, etc. This is illustrated in Fig. 1. GPC is the only technique by which property distributions for a wide range of applications can be reliably and efficiently measured. Therefore, GPC is superior to other techniques such as viscometry, osmometry or light scattering, which only allow the determination of property averages. The accurate and complete measurement of property distributions is of the utmost importance because the polydisperse nature of the macromolecules determines both the desired and unfavorable macroscopic product properties. In general, only individual property distributions

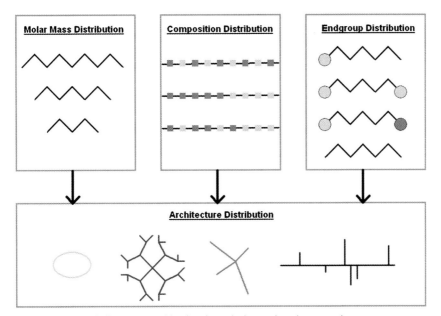

Fig. 1. Analytical challenges caused by the physical, chemical, and structural complexity of macromolecular samples.

HPLC Made to Measure: A Practical Handbook for Optimization. Edited by Stavros Kromidas
Copyright © 2006 WILEY-VCH Verlag GmbH & Co. KGaA, Weinheim
ISBN: 3-527-31377-X

Table 1. Practical differences between HPLC and GPC.

	HPLC	*GPC*
Sample preparation	Fast dissolution, sample degradation unlikely	Slow dissolution avoid shear, microwave, etc.
Chromatogram appearance	Many narrow peaks	(one) broad peak (plus trash peaks in RI)
Required result	a) Qualitative analysis b) Quantitative analysis	a) Molar mass averages b) Molar mass distribution
Information derived from	a) Peak sequence b) Peak area	a) Absolute peak position b) Peak shape
Calibration	Detector response → amount	Retention → molar mass
Detection	Single detector (UV, DAD)	Multiple detectors (UV, RI, LS, viscometry, FTIR, etc.)

can be accurately determined by a single separation method. In this chapter, GPC-related techniques are discussed, which allow the measurement of molar mass distributions (MMD) and averages thereof, and chemical composition distributions (CCD) and averages thereof, as well as ways to determine functionalities. Important parameters and useful detection techniques (hyphenation) for optimal GPC work are also discussed.

Since most readers will be experienced in general HPLC applications, major differences in the practice of GPC are outlined in Table 1. The separation in GPC is governed by the well-known size-exclusion mechanism; its principles can be found in the relevant literature [1]. In the ideal case, only the conformational entropy change causes retention. Secondary enthalpic interactions should be avoided by appropriate selection of the phase system. Since GPC does not measure molar masses directly, the retention axis has to be calibrated with so-called narrow polymer standards in order to transform peak position to molar mass. The absolute position of the peak(s) and its shape(s) are evaluated to determine the molar mass average and molar mass distribution, respectively.

2.3.2
Fundamentals of GPC Separations

GPC separations require an interaction-free phase system. The separation efficiency in GPC is proportional to the pore volume, which limits the decrease in column dimensions. In many cases, GPC analyses are performed on serial combinations of columns. In most applications, silica and surface-modified silica columns do not meet the requirements for GPC columns and polymeric packings are preferred. Important characteristics for GPC column packings are the inertness of the sorbent material (chemistry), the accessibility of the pore structure, a high pore volume, and fast diffusion (optimal mass transfer) [1a].

Table 2. Different requirements for column packings for HPLC and GPC applications.

Property of sorbent	Requirements for	
	HPLC columns	GPC columns
Pore size distribution	Narrow	Medium
Pore size	Less important	Small to large
Pore volume	Less important	High
Particle size distribution	Narrow	Narrow
Surface area	Large	Less important
Surface chemistry	Homogeneous	Inert
Pore architecture	High surface area	High accessibility
Mass transfer	High	High
Axial dispersion	Low	Low

Only a small number of column and sorbent manufacturers develop column packing materials for GPC work. Specialized GPC columns should preferably be used for GPC analyses because they are designed to meet the GPC column criteria. Some design differences between GPC and HPLC columns are listed in Table 2.

Table 3 gives an overview of available packings and their major uses (this table is not comprehensive and is intended to show only the major features and uses in GPC applications) [2].

Table 3. Synopsis of major packings for GPC applications.

	Polymer packings			Inorganic packings		
Polarity	Nonpolar	Medium	Polar	Nonpolar	Medium	Polar
Chemistry	Styrene-DVB	Acrylates	OH-acrylic (ionic)	SiO_2-C_{18}	SiO_2-diol	SiO_2
Solvents	Nonpolar medium	All	Aqueous	Selective only		
Producers	Few	Three	Few	Many		
Examples	PSS SDV PSS Polefin PL Gel Styragel TSK-H Shodex A,K	PSS HEMA PSS GRAM PSS Novema Shodex ToSo	PSS HEMA Bio PSS Suprema PL Aquagel UltraHydrogel TSK-PW Shodex OH (Ionic: SO_3H, amides)	Very many		
Use	Lipophilic samples	Universal	OH: universal SO_3H: saccharides	Limited	Some proteins	Very limited

2.3.2.1 Chromatographic Modes of Column Separation

All GPC separations require an interaction-free diffusion of the sample molecule in solution into and out of the pore structure of the column packing material. In general, this goal is easier to achieve in organic media than in aqueous eluents. The reason for this is that aqueous mobile phases have many more parameters that need to be correctly adjusted (e.g., type of salt, salt concentration, pH, addition of organic modifier, concentration of co-solvent). Additionally, water-soluble macromolecules have more ways in which to interfere with the stationary phase, i.e. through their charged functional groups, hydrophobic and/or hydrophilic regions, etc. These types of parameters have to be balanced in a valid GPC experiment. In order to obtain a "pure" GPC separation, the polarity of stationary phase, the polarity of eluent, and the polarity of sample have to be matched. This type of "magic triangle" is shown in Fig. 2; the dominance of size separation is only maintained in the center of the triangle, where the phase system is balanced. Otherwise, specific interactions will occur, which will be superimposed upon the normal GPC elution behavior.

The basic principle of chromatographic separations can be described in terms of thermodynamics by means of the distribution coefficient K:

$$K = a_s / a_m = \exp(\Delta G / RT) \tag{1}$$

a is the activity (concentration) of the molecule in the stationary phase (index s) and the mobile phase (index m);

ΔG is the change in free energy between the species in the stationary phase and the mobile phase.

In GPC separations, the enthalpic contribution to the free energy term vanishes when we assume no energetic interaction between the analyte and sorbent:

$$K_{SEC} = \exp(\Delta S / R), \quad 0 < K_{SEC} \leq 1, \quad \Delta H = 0 \tag{2}$$

ΔS is the entropy loss when a molecule enters a pore of the stationary phase.

In the case of no steric exclusion of the molecule from parts of the stationary phase, the retention can be described by the enthalpic term alone:

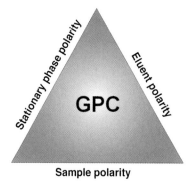

Sample polarity

Fig. 2. Balancing polarities of the phase system in GPC applications for interaction-free separations.

$$K_{\mathrm{HPLC}} = \exp(\Delta H / RT), \quad K_{\mathrm{HPLC}} \geq 1, \quad \Delta S \approx 0 \tag{3}$$

ΔH is the enthalpy change when a molecule is adsorbed by the stationary phase.

The above expressions describe the two ideal extremes of chromatography (GPC and HPLC), when there is no change in entropy or enthalpy, respectively. Another mode of chromatographic behavior is encountered when the enthalpic and entropic contributions balance, i.e. when the change in free energy disappears ($\Delta G = 0$). This mode is called liquid adsorption chromatography at the critical adsorption point (LACCC). The polymeric nature of the sample (that is, the repeating units) does not contribute to the retention of the species. Only defects (such as end groups, comonomers, branching points, etc.) contribute to the separation of the molecules. Figure 3 illustrates this behavior and shows the dependence of retention volume on the molar mass for the different modes of chromatography.

Using this basic theory of separation [1d], we can modify our experimental conditions to shift the separation into the desired chromatographic mode. In many cases, this can be done without buying new columns, by simply adjusting the polarity of the mobile phase.

Let us consider the separation of poly(methyl methacrylate) (PMMA) on a non-modified silica column as an example: In THF (medium-polarity eluent), the PMMA elutes in size-exclusion mode, because the dipoles of the methyl methacrylate (MMA) repeating units are masked by the dipoles of the THF. Using non-polar toluene as eluent on the same column, the separation is governed by adsorption, because the dipoles of the carbonyl group of the PMMA will interact with the dipoles on the surface of the stationary phase. The separation of PMMA in critical adsorption mode can be achieved by selecting an appropriate THF/toluene mixture as eluent. In this case, all PMMA samples will elute at the same time, regardless of their different molar masses. PMMA samples with different end groups will be separated with high selectivity, with no overlapping of size separation effects.

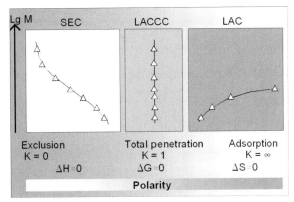

Fig. 3. Different modes of chromatography as seen in the elution order of samples with different molar masses.

2.3.2.2 **GPC Column Selection Criteria and Optimization of GPC Separations**

The major parameter for column selection is the intended application [1, 2]. A balance of mobile phase polarity in relation to the polarity of the stationary phase and sample polarity is important for pure GPC separations. In general, users tend to select their columns according to the mobile phase that they need to use.

Another important parameter for column selection is the correct choice of sorbent porosity. The pore size of the sorbent determines the fractionation range of the column. The best way of making this choice is by looking at the calibration curves of the columns, which are published by the column vendor.

Table 4 summarizes various ways to optimize GPC separations by proper selection and combination of GPC columns.

Table 4. Optimization of GPC separations.

Task	Optimized by
Better peak separation	Addition of similar columns; use of 3 μm columns, if $M < 100$ kg mol^{-1}; use of 2D chromatography (see Chapter 3.2)
Better separation of high molar mass fractions	Addition of column(s) with large pore width
Better separation of small molecules, additives, solvents, eluent, etc.	Addition of column(s) with small pore width
Avoiding exclusion peaks	Addition of column with large pore width
Not all compounds detected	Change phase system (column, eluent) use universal detector (RI)
Faster analyses	Use of HighSpeed columns
Faster calibrations	Use of calibration cocktails (ReadyCal)
Fast screening of unknown samples	Use of (less) linear mixed-bed columns
Better reproducibility	Use of internal standard correction (flow marker)

2.3.2.2.1 **Selection of Pore Size and Separation Range**

The molar mass range of the samples to be investigated determines the column porosity that should be selected in order to achieve the most efficient separation. The larger the pores in the column packing, the higher the molar mass of the samples that can be characterized. Unfortunately, there is no general nomenclature to allow easy selection of column pore sizes. Each manufacturer has its own system for pore size designation. The simplest way of ascertaining which columns will be useful for a selected task is to consult the calibration curve that every manufacturer shows in their literature. Figure 4 shows the calibration curves (on the left) and the recommended molar mass separation ranges (on the right) for PSS SDV columns.

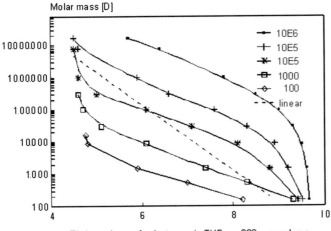

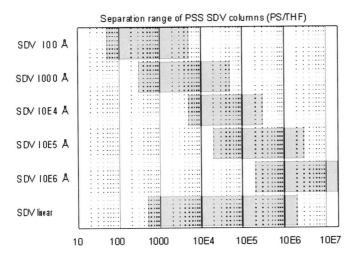

Fig. 4. Separation characteristics of PSS SDV GPC columns for organic eluents.

The highest efficiency for a separation is determined by the lowest slope of the calibration curve.

2.3.2.2.2 Advantages and Disadvantages of Linear or Mixed-Bed Columns

Discussing pore size selection of columns leads directly to the issue of whether to use single-porosity columns or so-called linear or mixed-bed columns, which contain mixtures of different pore sizes in a single column. Both types of columns have advantages and disadvantages [2], as shown in Table 5.

An example may show how the different concepts come into effect in a real-life laboratory environment. This example is based on column selections many laboratories use for ordinary, general purpose work.

Table 5. Comparison of single porosity-type and linear mixed-bed GPC columns.

Pore type	Advantages	Disadvantages
Single porosity	Efficient Optimized Flexible Low cost Good for QC	Viscous fingering
Mixed-bed or linear	Fast (screening work) "universal" Inject band dilution	Low efficiency Column combination

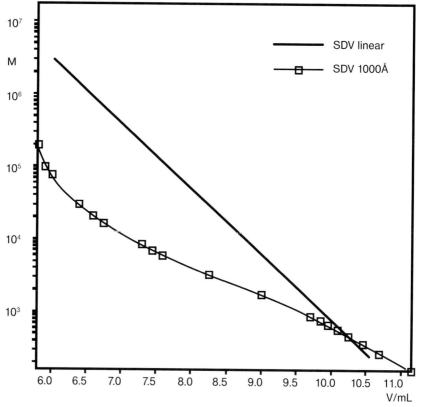

Fig. 5. Comparison of molar mass separation range and separation efficiency of a linear mixed-bed column and a single pore size column.

(A) *Conventional column set consisting of a series of columns with single pore sizes:*

1. "Universal" column set:

   PSS SDV 5 μm 10^6, 10^5, 1000 Å   Separation range: 10^7 to 200 D

   Resolution: $R_{sp} \approx 15$,

   Plate count: $N_{th} \approx 50000 \text{ m}^{-1}$

2. Optimized oligomer column set:

   PSS SDV 3 μm 1000, 100, 50 Å   Separation range: 30000 to 50 D

   Resolution: $R_{sp} \approx 25$,

   Plate count: $N_{th} \approx 85000 \text{ m}^{-1}$

(B) *Mixed-bed column combination with:*

   PSS SDV 5 μm linear (3×)   Separation range: 3×10^6 to 500 D

   Resolution: $R_{sp} \approx 15$,

   Plate count: $N_{th} \approx 50000 \text{ m}^{-1}$

The performances of column sets (A1) and (B) are similar; column set (B) is, however, more expensive. Using a specially selected column set for oligomer separation (as in column set A2), the user obtains much improved resolution and a separation range that corresponds directly to the samples under investigation. In the case of mixed-bed columns, the separation range is determined by the column and not by the application or the user, and therefore flexibility is sacrificed. This can easily be seen in Fig. 5.

The linear column (PSS SDV 5 μm linear) has a wider molar mass fractionation range while the analysis time remains roughly the same. Therefore, the slope of the calibration curve is much steeper and the resolution will be poorer in this case. The second column, with a single pore size (PSS SDV 5 μm 1000 Å) separates only below 50 000 D, but does this very efficiently in the same time.

2.3.2.3 HighSpeed GPC Separations

The pore volume of the column packing has been shown to be one of the major factors influencing peak resolution in GPC [1, 2]. True HighSpeed separations with good resolution require special HighSpeed columns, which allow fast flow rates, possess high separation volumes, and allow solutes to access the pores easily [3]. PSS is currently the only vendor of HighSpeed columns for GPC. Their HighSpeed columns replace conventional columns on a one-to-one basis, which allows for trouble-free method transfer from an existing conventional application to a HighSpeed application. HighSpeed GPC can be accomplished in about 1 min (cutting analysis time to about 10% of conventional) while maintaining similar resolution on existing instrumentation [3b]. Figure 6 shows a comparison of a GPC separation of polystyrene standards in THF on a conventional column and on a HighSpeed column, as performed on the same instrument.

The precision and accuracy of HighSpeed separations have been investigated in relation to various applications. Both the accuracy of molar mass results and the reproducibility have been comparable to results obtained with conventional columns [3b].

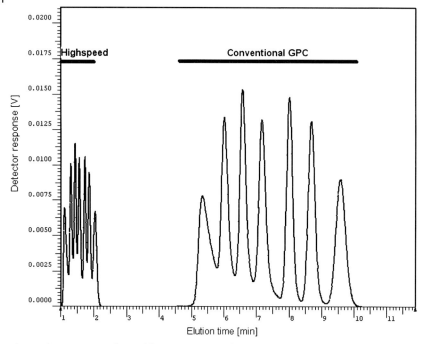

Fig. 6. Chromatogram obtained from a conventional GPC column (right) compared to that obtained from a HighSpeed GPC column (left); tests on identical instrument with polystyrene standards in THF.

Figure 7 shows the superposition of 10 out of 60 repeated runs of a commercial polycarbonate sample run in THF. They overlap almost perfectly. Each run took about 2.5 min; the total run time for 60 repeats was about 2 h.

The overall time savings can even be greater when taking the complete analytical process into account. The total run time of an instrument consists of the preparation and equilibration time, the time needed for running the calibration standards, and the run time for the unknown samples. If ten individual standards are used for calibration and ten samples are run, the total run time on a conventional system will be about two days and a night. On a HighSpeed system, the same work will require only about 3 h and so can easily be done in a single day [3b].

The cost-saving aspects of high-throughput GPC techniques can be substantial and have been evaluated for different scenarios.

Polyalkenes, other synthetic polymers, and water-soluble macromolecules have been investigated using HighSpeed GPC systems. HighSpeed GPC can be a big time-saver in two-dimensional chromatography applications, which require about 10 h of analysis time for cross-fractionation. This can be reduced by a factor of ten to about 1 h, which makes it much more appealing for many laboratories. Details of these and additional HighSpeed applications can be found in Refs. [3b, 4].

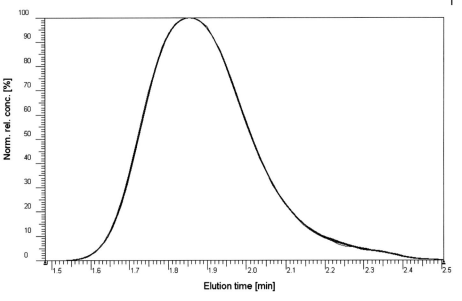

Fig. 7. Overlay of 10 out of 60 repeats of a commercial polycarbonate run in THF on PSS SDV 5 μm HighSpeed, 10^3, 10^5 Å column; measured Mw = (29610 ± 150) g mol^{-1} (nominal sample molar mass quoted by producer: 30000 g mol^{-1}).

2.3.3
The Role of Comprehensive Detection in the Investigation of Macromolecular Materials

Since most recent polymers are complex mixtures, in which the composition and molar mass depend in multiple ways on the polymerization kinetics, the polymerization technique, and the process conditions, the polymer processing parameters must be carefully controlled and monitored to obtain polymeric materials with the desired properties. It is essential to understand the influence of molecular parameters on polymer properties and end-use performance. Molar mass distribution and average chemical composition may no longer provide sufficient information for process and quality control nor define structure–property relationships. Modern characterization methods now require multidimensional analytical approaches rather than average properties of the whole sample [5].

The analytical requirements for complex macromolecules are fundamentally different from those of low molar mass organic samples, where single molecules are to be determined. The major analytical task in relation to polymers is the comprehensive determination of property distributions. The molecular heterogeneity of a certain complex polymer can be represented either by a three-dimensional diagram (Fig. 8a) or by a so-called "contour plot" (Fig. 8b). Using appropriate analytical methods, the type and concentration of the different functionality fractions must be determined and, within each functionality, the

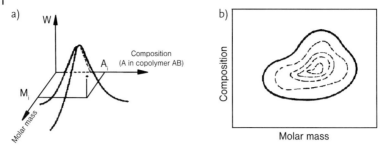

Fig. 8. Representations of property distributions of polydisperse macromolecules; (a) three-dimensional map; (b) contour map (view from top of (a)).

molar mass distribution has to be obtained. To do this, two different methods must be combined, each of which is preferably selective towards one type of heterogeneity. For example, a chromatographic method separating solely with respect to functionality could be combined with a molar mass selective method. Another approach would be the separation of the sample into different molar mass fractions, which are then analyzed with respect to functionality [6].

Numerous methods have been established for determining the chemical composition drift in relation to molar mass distribution [6, 7]. These include

Table 6. Optimization of detection methods in GPC applications [12].

Task	Optimized by
Not all compounds detected	Use of a non-specific detector (RI) Use of a different eluent
Investigation of chemical composition	Use of two concentration detectors (see Section 2.3.3) Use of information-rich detectors (see Section 2.3.3.1)
Identification of compounds	Use of FTIR detection (see Section 2.3.3.3) Use of MS or NMR detection (if molar masses are low)
Determination of additives, pigments, plasticizers	Use of RI/DAD and FTIR detection (see Section 2.3.3.3)
Determination of end-groups and/or functionalities	Use of both a universal *and* a specific detector (e.g., RI and UV/DAD)
Molar mass measurement without standards	Use of light-scattering detection (see Section 2.3.3.4.1) Use of viscosity detection (see Section 2.3.3.4.2)
Determination of absolute molar masses of copolymers and blends	Use of viscosity detection (see Section 2.3.3.4.2)
Investigation of (hyper-) branched samples	Use of light-scattering detection (see Section 2.3.3.4.1) Use of viscosity detection (see Section 2.3.3.4.2)
Investigation of complex structures and architectures	Simultaneous use of light-scattering and viscosity detection (see Section 2.3.3.4.2)

precipitation, partition, and cross-fractionation techniques (see Chapter 3.2 on two-dimensional chromatography in this book). The aim of these efforts is to obtain fractions of narrow composition and/or molar mass distribution which can be subsequently analyzed by information-rich detectors or detector combinations. Table 6 gives an overview of the application of detection methods for the investigation of macromolecular materials in addressing various GPC tasks.

2.3.3.1 Coupling of Liquid Chromatography with Information-Rich Detectors

Liquid chromatography of polymers is often understood as being synonymous with size-exclusion chromatography. GPC separates polymers according to the size of the macromolecules by entropic interactions and enables the molar mass distribution of a sample to be evaluated. However, in addition to size-exclusion phenomena, other types of interaction between the macromolecules and the stationary and mobile phases of a chromatographic system can be used for separation. Liquid adsorption chromatography (LAC) uses enthalpic interactions to separate, for example, copolymers according to chemical composition. Finally, liquid chromatography at the critical point of adsorption (LC-CC) can be used for functionality-type separation by balancing entropic and enthalpic interactions in the chromatographic system [1b].

Size-exclusion chromatography is the premier polymer characterization method for determining molar mass distributions. For homopolymers, condensation polymers, and strictly alternating copolymers, there is an unequivocal relationship between elution volume and molar mass. Thus, chemically similar polymer standards of known molar mass can be used for calibration. However, for GPC of random and block copolymers and branched polymers no simple correlation exists between elution volume and molar mass because of possible compositional heterogeneity of these materials. The dimensional distribution of macromolecules can, in general, only be unambiguously correlated with MMD within one heterogeneity type. For samples consisting of molecules of different chemical composition, the distribution obtained represents an average of dimensional distributions of these molecules and, therefore, cannot be attributed to a certain type of macromolecule [5].

The limitation of using GPC without taking into account the determination of the MMD of polymer blends or copolymers results from the following consideration. For a linear homopolymer distributed only in molar mass, fractionation by GPC results in one molar mass being present in each retention volume. The polymer at each retention volume is monodisperse. If a blend of two linear homopolymers is separated, then two different molar masses can be present in one analytical fraction. If a copolymer is analyzed by GPC, then a multitude of different combinations of molar mass, composition, and sequence length can be combined to give the same hydrodynamic volume. In this case, fractionation with respect to molecular size is completely ineffective in assisting the analysis of composition or MMD.

Three on-line detection methods are used to try to characterize copolymers with respect to MMD and/or composition by GPC:

1. Conventional GPC utilizing multiple concentration detection.
2. On-line analysis of GPC fractions with a light-scattering detector.
3. On-line viscometry.

Combinations of different separation techniques have also been effectively used for the comprehensive analysis of polydisperse complex macromolecules. A specific chapter of this book is dedicated to this methodology (see Chapter 3.2).

2.3.3.2 Copolymer GPC Analysis by Multiple Detection

Conventional GPC data processing is unable to determine other important polymer properties such as copolymer composition or copolymer molar mass. The reason for this is that GPC separation is based on hydrodynamic volume rather than the molar mass of the polymer and that molar mass calibration data are only valid for polymers of identical structure. This means that polymer topology (e.g., linear, star-shaped, comb, ring or branched polymers), copolymer composition, and chain conformation (isomerization, tacticity, etc.) determine the *apparent* molecular weight. The main problem in copolymer analysis is the calibration of the GPC instrument for copolymers with varying comonomer compositions. However, even if the bulk composition is constant, second-order chemical heterogeneity has to be taken into account, i.e. composition will generally vary for a given chain length.

Several attempts have been made to solve the calibration dilemma. Some are based on the universal calibration concept, which has been extended for copolymers; another approach to copolymer calibration is multiple detection. The advantages of multiple detection lie in its flexibility and the fact that it yields the composition distribution as well as molar masses for the copolymer under investigation [7]. This method requires molar mass calibration and an additional detector response calibration to determine the chemical composition at each point of the elution profile. No other kind of information, parameters or special equipment are necessary to carry out this kind of analysis and to calculate compositional drift, bulk composition, and copolymer molar mass [7a].

In order to characterize the composition of a copolymer of k comonomers, the same number of independent detector signals d is necessary in the GPC experiment; for example, in the case of a binary copolymer, two independent detectors (e.g., UV and RI) are required to calculate the composition distribution $w_k(M)$ and the overall (bulk) composition $\bar{w}_k$. The detector output U_d of each detector d is the superposition of all individual responses from all comonomers present in the detector cell at a given elution volume V. Therefore,

$$U_d(V) = \sum_d f_{dk} \cdot c_k(V) \qquad (4)$$

where f_{dk} is the response factor of comonomer k in detector d, and c_k is the true concentration of comonomer k in the detector cell at elution volume V. The detector response factors are determined in the usual way by injecting homopolymers for each comonomer of known concentration and correlating this concentration with the area of the corresponding peak. Where no homopolymers have been available, model compounds have been used to estimate the detector response factors.

In the case of a binary copolymer, the weight fraction, w_A, of comonomer A is then given by:

$$W_A(V) = \left\{ 1 + \frac{\left[U_1(V) - \dfrac{f_{1B}}{f_{2B}} \cdot U_2(V) \right]\left[f_{1A} - \dfrac{f_{1B}}{f_{2B}} \cdot f_{2A} \right]}{\left[U_1(V) - \dfrac{f_{1A}}{f_{2A}} \cdot U_2(V) \right]\left[f_{1B} - \dfrac{f_{1A}}{f_{2A}} \cdot f_{2B} \right]} \right\}^{-1} \tag{5}$$

Obviously, the sum of all comonomer weight fractions is unity. The accurate copolymer concentration and the distribution of the comonomers across the chromatogram can be calculated from the apparent chromatogram and the individual comonomer concentrations.

The accuracy of the compositional information is not affected by the polymer architecture. Deviations from the true comonomer ratios are only possible if the detected property is dependent on the local environment. This is the case if neighboring-group effects are operative. The probability of electronic interactions causing such deviations is very low, because there are too many chemical bonds between the two different monomer units. Other types of interactions, especially those which proceed across space (e.g., charge-transfer interactions), may influence composition accuracy.

The major difficulty in the determination of the copolymer molar mass distribution is the fact that the GPC separation is based on the molecular size of the copolymer chain. Its hydrodynamic radius, however, is dependent on the type of comonomers incorporated into the macromolecule and their placement (sequence distribution). Consequently, there can be a coelution of species having different chain lengths *and* different chemical compositions. The influence on the chain size of different comonomers copolymerized into the macromolecule can be measured by GPC elution of homopolymer standards of this comonomer. Unfortunately, the influence of the comonomer sequence distribution on the hydrodynamic radius cannot be described explicitly by any theory at present. However, there are limiting cases that can be discussed to evaluate the influence of the comonomer placement in a macromolecular chain.

From a GPC point of view, the simplest copolymer is an alternating copolymer $(AB)_n$, which can be treated exactly like a homopolymer with a repeating unit (AB). The next simplest copolymer architecture is an AB block copolymer, in which a sequence of comonomer A is followed by a block of B units. The only hetero contact in this chain is the A–B link, which can influence the size of the macromolecule. Hydrodynamically, the A segment and the B segment of the AB block copolymer will behave like a pure homopolymer of the same chain length. In the case of long A and B segments in the AB block copolymer, the isolated A–B link acts as a defect position and will not change the overall hydrodynamic behavior of the AB block copolymer chain. Consequently, the molar mass of the copolymer chain can be approximated by the molar masses of the respective segments. Similar considerations apply for ABA, ABC, and other types of block structures, as well as for comb-shaped copolymers with low side-chain densities.

In such cases, the copolymer molar mass M_c can be determined from the interpolation of homopolymer calibration curves $M_k(V)$ and the weight fractions w_k of the comonomers k according to:

$$\lg M_c(V) = \sum_k w_k(V) \cdot \lg M_k(V) \qquad (6)$$

Copolymer molar mass averages $M_{n,c}$, $M_{w,c}$, etc. and copolymer polydispersity D_c are calculated as in conventional GPC calculations using the copolymer molar mass calculated from Eq. (6).

In cases where the number of hetero contacts can no longer be neglected, this simplified reasoning breaks down and copolymer molar masses cannot be measured accurately by GPC alone. This is the case with statistical copolymers, polymers with only short comonomer sequences and high side chain densities. In such cases, more powerful and universal methods have to be employed, e.g. 2D separations (see Chapter 3.2).

Block copolymers are an important class of polymers used in many applications from thermoplastic elastomers to polymer blend stabilizers. They are most often synthesized by ionic polymerization, which is both costly and sometimes difficult to control. However, block copolymer properties strongly depend, for example, on the exact chemical composition, the block molar mass, and the block yield. These parameters can be evaluated in a single experiment using copolymer GPC with multiple detection.

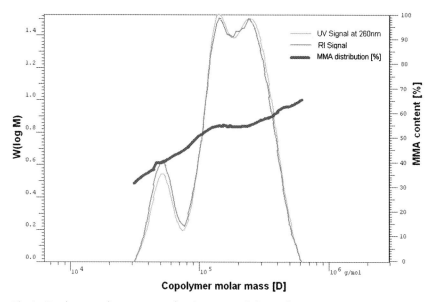

Fig. 9. Simultaneous determination of molar mass and chemical composition distributions of a complex styrene/MMA block copolymer using GPC in combination with RI and UV detection.

Figure 9 shows the molar mass distribution of a styrene/MMA block copolymer as measured using RI and UV detection. The RI responds to the styrene and MMA units, whereas the UV tuned to 260 nm predominantly picks up the presence of styrene in the copolymer. After detector calibration, the styrene and MMA content in each fraction can be measured. The MMA content distribution (black solid line) is superimposed on the MWD of the product in Fig. 9. It is obvious that the MMA content is not constant throughout the MWD, but continuously increases with the molar mass. The trimodal MWD itself shows only the presence of three different species. The MMA content information clearly reveals that the copolymerization process did not produce the block structure, but that the MMA was added to styrene chains of different molar mass.

2.3.3.3 Simultaneous Separation and Identification by GPC-FTIR

As shown above, GPC is a powerful separation tool for fractionating macromolecules according to size. However, it is limited in its ability to identify sample constituents such as additives, processing agents, residual monomers and solvents, etc. Infrared spectroscopy is a perfect companion as it is strong in the determination of different structural elements (functional groups, conformations, etc.), as long as they are present in a pure state. The combination of GPC as a separation technique with FTIR detection as an identification method allows on-line deformulation of complex compounds and mixtures. Unfortunately, most GPC solvents show strong IR absorption bands. Consequently, useful IR spectra from flow-through cells are best obtained in favorable GPC solvents such as dichloromethane, chloroform, and tetrachloromethane.

More universal FTIR detection can be achieved when the mobile phase is removed from the sample prior to spectral analysis. The sample fractions are then measured in a pure state without interference from solvents [6b, 6c, 8]. A widely used commercial LC-FTIR interface is manufactured by LabConnection. The LC-Transform strips volatile mobile phases by nebulizing the HPLC eluent and spraying it onto a rotating germanium (IR-transparent) disk forming a solid, time-resolved deposit. In an off-line second step, the disk is placed in the sample compartment of a standard FTIR spectrometer and IR spectra of KBr quality can

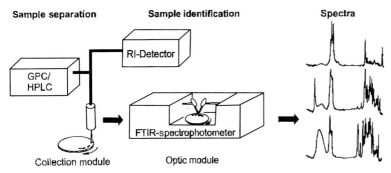

Fig. 10. Experimental set-up of HPLC-FTIR detection using on-line sample collection on a Ge disk and off-line spectroscopic analysis.

be recorded at each position in the reflection mode. The design of the interface is shown in Fig. 10.

Figure 11 shows the deformulation of a blister foil used in food packaging. From a single GPC-FTIR run, the following information could be obtained:

- the type and nature of the polymer used (peak A: PVC)
- the molar masses and molar mass distribution of the polymer (peak A)
- identification of the additives (peaks B–E)
- quantification of all additives in the packaging foil
- identification and quantification of the processing agent (peak F)

The application of coupled SEC-FTIR becomes inevitable when blends comprising copolymers have to be analyzed. Very frequently, components of similar molar masses are used in polymer blends. In these cases, the resolution of GPC is not sufficient to resolve all component peaks.

Many research groups have applied this synergistic combination of GPC separation and IR detection for various samples in many application areas; details can be found in Refs. [6b, 6c]. Willis and Wheeler demonstrated the determination of the vinyl acetate distribution in ethylene/vinyl acetate copolymers, the analysis of branching in high-density polyethylene, and the analysis of the chemical composition of a jet oil lubricant. Provder et al. showed that all of the additives in powder coatings could be positively identified by SEC-FTIR through comparison with known spectra. Even biocides in commercial household paints could be analyzed.

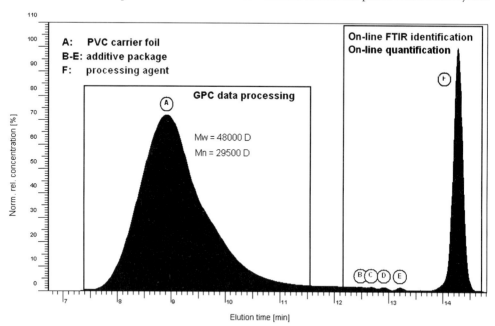

Fig. 11. Simultaneous determination of molar mass, compound identification, and quantification of a PVC food packaging foil by coupling of a separation method (GPC) with spectroscopic detection.

Comparison of a PS-PMMA blend with a corresponding copolymer gave information on the chemical drift. In the analysis of a competitive modified vinyl polymer sample by SEC/FTIR, some of the components of the binder could be readily identified (vinyl chloride, ethyl methacrylate, acrylonitrile), and an epoxidized drying oil additive was also detected. An analysis of styrene-butadiene copolymers, including a determination of the styrene/butadiene ratio and of the microstructure of the butadiene units (*cis/trans*, 1,2-/1,4-units), was performed by Pasch et al.

2.3.3.4 Application of Molar Mass-Sensitive Detectors in GPC

To overcome the problems associated with classical GPC of complex polymers, molar mass-sensitive detectors have been introduced into GPC instruments. Since the response of such detectors depends on both concentration and molar mass, they have to be combined with a concentration-sensitive detector. This set-up allows the direct measurement of molar mass in each analytical fraction and so there is no longer reliance on a calibration curve generated from reference polymer standards. Molar mass-sensitive detectors based on Rayleigh light scattering or intrinsic viscosity measurements can be used for this purpose [9].

The following types of molar mass-sensitive detectors are in common use:
- low-angle laser light scattering detector (LALLS)
- multi-angle laser light scattering detector (MALLS)
- differential viscometer

2.3.3.4.1 Light-Scattering Detection

On-line light-scattering measurements overcome the calibration dilemma in GPC analyses by direct determination of molar masses independently of the nature of the sample or its architecture. In such set-ups, there is no longer a need to use reference polymer standards as calibrants to relate molecular size to molar mass.

A light-scattering detector measures the scattered light of a laser beam passing through the detector cell at various observation angles. The (excess) intensity $R(q)$ of the scattered light at an angle q is related to the weight-average of molar mass M_w according to:

$$K^* c / R(q) = [1/M_w + P(q)] + 2\,A_2 c \qquad (7)$$

where c is the concentration of the polymer, A_2 is the second virial coefficient, and $P(q)$ describes the angular dependence of the scattered light. K^* is the optical constant containing Avogadro's number N_A, the wavelength λ_0, the refractive index n_0 of the solvent, and the refractive index increment dn/dc of the sample:

$$K^* = 4\,B^2 n_0^2 \,(dn/dc)^2 /(\lambda_0^4 / N_A) \qquad (8)$$

In a plot of $K^* c/R(q)$ *vs.* $\sin^2(q/2)$, M_w can be obtained from the intercept and the radius of gyration, R_g, from the slope. A multi-angle measurement provides additional information on molecular size, structure, conformation, aggregation state, etc.

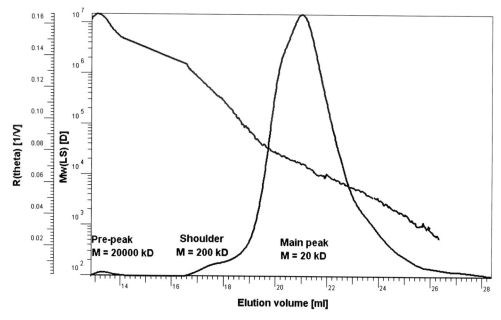

Fig. 12. Direct measurement of molar mass (diagonal line) of a PVB sample by GPC-MALLS irrespective of the structural differences in the sample.

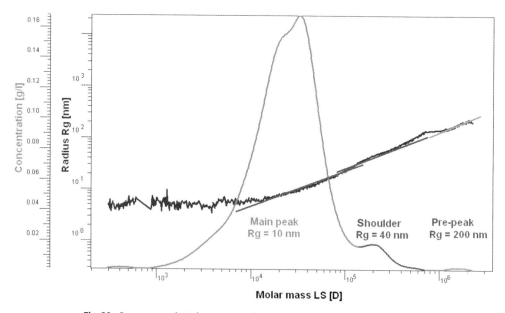

Fig. 13. Structure analysis by GPC coupled with multi-angle light scattering by direct measurement of molar mass and size (R_g).

Figure 12 illustrates an example of direct molar mass measurement by light-scattering in a GPC experiment for a poly(vinyl butyral) sample. For each analytical fraction (elution volume slice), the LS detector determines the correct M_w and subsequently the molar mass distribution can be determined without any further assumptions. Obviously, this sample contains species of very different molar mass *and* structure, as can be further elucidated by molecular size measurement, which is possible from the multi-angle light-scattering data analysis (cf. Fig. 13).

Various kinds of light-scattering instruments have been used for such analyses. However, the most useful devices are modern multi-angle systems (MALLS) because of their versatility and precision. Light-scattering instruments with only a single angle are less universally applicable and are limited to certain molar mass ranges or sample types.

2.3.3.4.2 Viscometry Detection

Another very useful approach for obtaining molar mass information of complex polymers is the coupling of GPC to a viscosity detector. The viscosity of a polymer solution is closely related to the molar mass (and architecture) of the polymer molecules. The product of polymer intrinsic viscosity $[\eta]$ and molar mass M is proportional to the size of the polymer molecule (the hydrodynamic volume). Based on the Einstein–Stokes theory, the molar mass, M, of an unknown sample can be determined directly in a GPC experiment from the known molar mass, M_{std}, of a polymer standard and its intrinsic viscosity, $[\eta]_{std}$, irrespective of sample type or architecture.

$$M = [\eta]_{std} \, M_{std} \, /[\eta] \tag{9}$$

This behavior is generally referred to as "universal calibration" [10]. In conventional calibrations, the logarithm of molar mass is plotted versus elution volume; different samples yield different calibration curves. In a universal calibration, the logarithm of the product of molar mass and intrinsic viscosity is plotted versus elution volume. In this plot, the calibration curves of very different samples all fall on a single line (the so-called universal calibration curve, see Fig. 14).

Viscosity measurements in GPC can be performed by measuring the pressure drop across a capillary, which is proportional to the viscosity η of the flowing liquid (the viscosity of the pure mobile phase is denoted as η_0). The relevant parameter, intrinsic viscosity $[\eta]$, is defined as the limiting value of the ratio of specific viscosity ($\eta_{sp} = (\eta - \eta_0)/\eta_0$) to vanishing concentration c:

$$[\eta] = \lim(\eta - \eta_0)/\eta_0 c = \lim \eta_{sp} /c \quad \text{for } c \to 0 \tag{10}$$

Thus, the concept of universal calibration also provides an appropriate calibration for polymers for which no calibration standards exist.

Various experimental designs for on-line viscometers have been investigated since the late 1960s. The most useful viscometry detection technique is based on a balanced or non-balanced four-capillary bridge design. Signal artefacts and sometimes strong dependence on flow rate changes make other experimental configurations less attractive for general use.

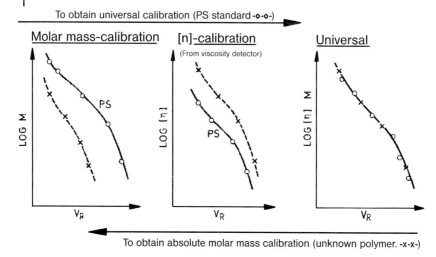

To obtain universal calibration (PS standard -o-o-)

To obtain absolute molar mass calibration (unknown polymer. -x-x-)

Fig. 14. Comparison of conventional and universal calibration curves and how molar mass and intrinsic viscosities depend on molecular structure.

Due to the problems encountered with SEC-LS and SEC-viscometry, a combination of these techniques has been developed, whereby three on-line detectors are incorporated into a single GPC system. In addition to the concentration detector, an on-line viscometer and an LS instrument are used in the GPC (so-called triple detection). This allows absolute molar mass determination for polymers that are very different in chemical composition and molecular conformation. The usefulness of this approach has been demonstrated in a number of applications [11].

2.3.4
Summary

Advanced GPC techniques are well-established and allow the comprehensive characterization of complex macromolecules. Modern detector technologies open up new ways of investigating various properties with high sensitivity even in the low concentration ranges used in chromatography. Hyphenated chromatographic methodologies, in their various forms, are currently among the most promising and powerful methods for the fractionation and characterization of complex sample mixtures in different property coordinates. This chapter offers just a glimpse of the possibilities, and newly emerging techniques such as LC-NMR and various LC-MS(n) combinations offer even wider capabilities (see Chapters 3.3 and 3.4).

References

1 (a) W. W. Yau, J. J. Kirkland, D. D. Bly, Modern Size-Exclusion Liquid Chromatography, Wiley, New York, 1979; (b) G. Glöckner, Liquid Chromatography of Polymers, Hüthig, Heidelberg, 1982; (c) S. Mori, H. G. Barth, Size-Exclusion Chromatography, Springer, Berlin, 1999; (d) H. Pasch, B. Trathnigg, HPLC of Polymers, Springer, Berlin, 1997.

2 P. Kilz, Design, Properties and Testing of Columns and Optimization of SEC Separations, in *Column Handbook for Size-Exclusion Chromatography* (Ed.: C.-s. Wu), Academic Press, San Diego, 1999.

3 (a) P. Kilz, HighSpeed GPC Methodologies, in *Encyclopedia of Chromatography* (Ed.: J. Cazes), Marcel Dekker, New York, 2002 (online edition); (b) P. Kilz, Methods and Columns for HighSpeed SEC Separations, in *Handbook for Size-Exclusion Chromatography and Related Techniques* (Ed.: C.-s. Wu), Marcel Dekker, New York, 2003.

4 (a) H. Pasch, P. Kilz, Fast liquid chromatography for high-throughput screening of polymers, *Macromol. Rapid Commun.* **24**, 104 (2003); (b) J. C. Meredith, A current perspective on high-throughput polymer science, *J. Mat. Sci.* **38**, 4427 (2003).

5 H. G. Barth, "Hyphenated Polymer Separation Techniques. Present and Future Role", in *Chromatographic Characterization of Polymers. Hyphenated and Multidimensional Techniques* (Eds.: T. Provder, H. G. Barth, M. W. Urban), Chapter 1, *Adv. Chem. Ser.* 247, American Chemical Society, Washington, DC, 1995.

6 (a) G. Glöckner, *Gradient HPLC of Copolymers and Chromatographic Cross-Fractionation*, Springer, Berlin, 1991; (b) H. Pasch, *Adv. Polym. Sci.* **150**, 1 (2000); (c) P. Kilz, H. Pasch, *Coupled LC Techniques in Molecular Characterization*, in Encyclopedia of Analytical Chemistry (Ed.: R. A. Myers), Wiley, Chichester, 2000; (d) P. Kilz, *Chromatographia* **59**, 3 (2004).

7 (a) F. Gores, P. Kilz, *Copolymer Characterization Using Conventional SEC and Molar Mass-Sensitive Detectors*, in Chromatography of Polymers (Ed.: T. Provder), Chapter 10, ACS Symp. Ser. 521, American Chemical Society, Washington, DC, 1993; (b) P. Kilz, Copolymer Analysis by LC Methods including 2D, in *Encyclopedia of Chromatography* (Ed.: J. Cazes), Marcel Dekker, New York, 2001, pp. 195; (c) H. Schlaad, P. Kilz, Determination of MWD of Diblock Copolymers with Conventional SEC, *Anal. Chem.* **75**, 1548 (2003).

8 (a) J. N. Willis, L. Wheeler, in *Chromatographic Characterization of Polymers. Hyphenated and Multidimensional Techniques* (Eds.: T. Provder, H. G. Barth, M. W. Urban), Chapter 17, *Adv. Chem. Ser.* 247, American Chemical Society, Washington, DC, 1995; (b) P. C. Cheung, S. T. Balke, T. C. Schunk, in *Chromatographic Characterization of Polymers. Hyphenated and Multidimensional Techniques* (Eds.: T. Provder, H. G. Barth, M. W. Urban), Chapter 19, *Adv. Chem. Ser.* 247, American Chemical Society, Washington, DC, 1995.

9 (a) C. Jackson, H. G. Barth, in *Molecular Weight Sensitive Detectors for Size-Exclusion Chromatography* (Ed.: C.-s. Wu), Chapter 4, Marcel Dekker, New York, 1995; (b) H. G. Barth, *Hyphenated Polymer Separation Techniques. Present and Future Role*, in Chromatographic Characterization of Polymers. Hyphenated and Multidimensional Techniques (Eds.: T. Provder, H. G. Barth, M. W. Urban), Chapter 1, *Adv. Chem. Ser.* 247, American Chemical Society, Washington, DC, 1995.

10 H. Benoit, P. Rempp, Z. Grubisic, *J. Polym. Sci.* **B5**, 753 (1967).

11 W. W. Yau, *Chemtracts – Macromol. Chem.* **1**, 1 (1990).

12 More applications and examples at www.polymer.de/applications

2.4
Gel Filtration/Size-Exclusion Chromatography (SEC) of Biopolymers – Optimization Strategies and Troubleshooting

Milena Quaglia, Egidijus Machtejevas, Tom Hennessy, and Klaus K. Unger

Gel filtration or Size Exclusion Chromatography (SEC) is still a method widely practised for the separation of biopolymers, although its application areas changed little since its introduction in 1960. As an introduction, the specific aspects of SEC of biopolymers in comparison to other HLC variants are outlined in Sections 2.4.2 and 2.4.3. The main focus of the chapter lies on the optimization of the chromatographic resolution: selection of column, optimum flow rate, eluent composition, injection conditions, detection methods (Section 2.4.4). Information on troubleshooting is also provided. Section 2.4.5 reviews the contemporary application areas of modern SEC of biopolymers: determination of molecular weight, SEC as a tool to monitor conformational changes of proteins, SEC in downstream processing and SEC columns based on the restricted access principle for the automated integrated sample clean up in proteomics.

2.4.1
Where Are We Now and Where Are We Going?

The evolution of gel filtration as a method for sizing biopolymers according to their hydrodynamic volume and molecular weight started in 1959, when Porath and Flodin developed cross-linked dextrans as SEC packings and advocated the novel technology [1]. The product was soon marketed as Sephadex (Separation Pharmacia Dextran) by Pharmacia AB, Uppsala, Sweden. Sephadex is the classical packing in gel filtration. The follow-up products differ from the original ones in that they have a smaller particle size, a higher degree of cross-linking, and thus an enhanced mechanical stability [2].

Later, other suitable cross-linked polymers with electroneutral hydrophilic surface groups were developed, as well as porous silicas with graduated pore diameters and chemically bonded hydrophilic groups [3–7]. The market leaders in this field are Amersham Bioscience at Uppsala and Tosoh Biosciences.

SEC has its place in the purification and isolation of biopolymers, e.g. insulin, and other therapeutic drugs as a module in downstream processing [8, 9]. Mass spectrometry has, however, comprehensively displaced SEC for molecular weight determination. A current field of application is in the monitoring of conformational changes and aggregation of proteins. Micro-SEC with capillary columns has found application in the SEC of synthetic polymers in process analysis.

Recently, packings have been developed that combine SEC properties with retentive features. The porous particles bear dual surface functionalities: hydrophilic neutral groups on the outer surface and hydrophobic or ion-exchange groups on the inner surface. Proteins with a molecular weight beyond the exclusion limit

HPLC Made to Measure: A Practical Handbook for Optimization. Edited by Stavros Kromidas
Copyright © 2006 WILEY-VCH Verlag GmbH & Co. KGaA, Weinheim
ISBN: 3-527-31377-X

are excluded from the pores and washed out with the dead volume of the column, while smaller proteins and peptides are adsorbed at the inner surface. After washing the column, the adsorbed compounds are preferably eluted in the back-flush mode by applying a strong solvent and transferred to the next separation column. The process of loading, washing, elution, and regeneration can be accomplished automatically by column switching devices.

Columns based on the restricted access principle introduced by Pinkerton in 1983 [10] and improved by Boos [11] for low molecular weight compounds are now successfully applied to the integrated sample clean-up and multi-dimensional LC/MS peptide mapping of biofluids [12].

2.4.2
Theory in Brief

Gel filtration is a size-exclusion chromatography mode, whereby proteins and peptides are resolved according to their hydrodynamic volumes. The most simple and widely accepted approach to describe the retention in size-exclusion chromatography (SEC) is by considering a selective permeation of macromolecules into the pores of the packing on the basis of their molecular size (more precisely, hydrodynamic volume) (see Fig. 1). It is possible to distinguish three cases:

Case A: large molecules cannot penetrate into the packing pores and will be eluted with an elution volume (V_e) corresponding to the interparticle volume of the column (V_0);

Case B: small molecules freely enter the pores of the packing by penetrating the pore volume (V_i) and will be eluted with an elution volume corresponding to the sum of V_0 and V_i, equal to the total accessible volume of the column (V_t);

$$V_t = V_0 + V_P \tag{1}$$

Case C: intermediate sized molecules can partially penetrate the V_i and will be eluted with an elution volume between V_0 and V_t. The elution volume increases with decreasing hydrodynamic volume of the biopolymer.

The V_e of a solute in SEC can therefore be described as:

$$V_e = V_0 + K_{SEC} V_i \tag{2}$$

where K_{SEC} is the distribution coefficient of the biopolymer and expresses the fraction of pore volume accessible to the solute. The distribution coefficient will be zero for a totally excluded molecule and unity for a molecule capable of penetrating all the pores. Under ideal conditions, no molecule can be eluted with a K_{SEC} greater than 1 or less than 0. If K_{SEC} is greater than 1, non-specific interactions between the solute and the support are involved; if it is less than 0, channels are likely to be present in the chromatographic bed and the column has to be repacked. Between K_{SEC} values of approximately 0.15 and 0.80, there is a linear relationship between K_{SEC} and V_e, as illustrated in Fig. 1. The linear portion of the curve corresponds to the working molecular size range of the column and,

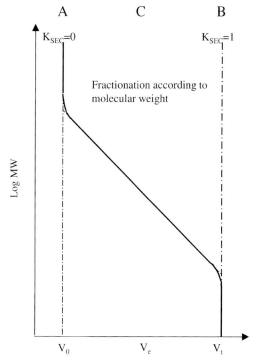

Fig. 1. Calibration curve for the mechanism of size-exclusion chromatography under ideal conditions;
(a) corresponds to the exclusion limit, (b) to total permeation, and
(c) to the region in which the molecules are fractionated according to their MW.

in general, the smaller the slope of the curve, the better is the resolution between proteins of different molecular weights. Proteins of about 1.5–2 decades of magnitude of molecular weight can be eluted from an SEC column across the linear range. In order to cover a molecular range between 1000 and 10 million, two or three columns with an appropriate fractionation range are therefore required. For optimum separation, it has been found best to couple columns that differ in pore diameter by a factor of 10 to 20 [13].

It is important to underline that the calibration curve is valid only for proteins of the same shape, e.g. globular proteins. It has in fact been proven that the radius of gyration of a spherical polymer is less than that of a linear polymer of the same molecular weight.

Appropriate markers for V_0 of SEC columns are apoferritin (467 kDa), thyroglobulin (670 kDa), and glutamic dehydrogenase (998 kDa). As V_t markers, sodium azide and DNP-alanine are usually preferred [14, 15]. The selection of the low molecular weight probe is a problem, because V_e is unlikely to be truly independent of solute size and can be affected by solute surface interactions. Recently, studies on the size dependence of the retention of non-interacting low molecular weight solutes have been carried out [16].

By using calibration curves (log MW against V_e) for standard proteins, it is possible to estimate the molecular weight of an unknown compound from its elution volume. One important reminder is that the approach to molecular weight determination is based on the assumption that the protein has a globular and symmetrical shape. It should be emphasized in this context that a given protein may change its conformation as a function of pH and ionic strength of the eluent.

Another important factor that readily influences K_{SEC} is the degree of hydration of the molecules. The polar nature of proteins is responsible for multiple associations between the molecules and water, which can be controlled by adjusting the pH and the ionic strength of the solution. The degree of hydration of proteins is normally between 0.3 and 1.0 and causes the radius to be larger than it would be in a non-hydrated or dry state and therefore the K_{SEC} becomes smaller.

In SEC, the capacity factor (k) is normally expressed as the exclusion chromatography ratio (k^*), calculated from the following equation:

$$k^* = (V_e - V_0)/V_0 \tag{3}$$

The main difference between the two parameters is that k has an infinitive maximum value and k^* has a maximum value given by the following ratio:

$$k^*_{max} = V_i / V_0 \tag{4}$$

and therefore k^* is dependent on the pore volume and on the interstitial volume of the column.

The aim of any chromatographic process is the resolution of the sample mixture. The simplest equation expressing the chromatographic resolution (R_s) between two peaks is:

$$R_s = 2\,(V_{e2} - V_{e1})/w_2 + w_1 \tag{5}$$

where V_{e2} and V_{e1} are the elution volumes of two peaks and w_2 and w_1 are the corresponding widths at 10% of the peak height. For two solutes of similar peak widths and height, R_s can be defined as:

$$R_s = (V_{e2} - V_{e1})/2\,(\sigma_2 + \sigma_1) \tag{6}$$

where σ_2 and σ_1 are the standard deviations of peaks 1 and 2.

Keeping in mind that:

$$M_w = D_1 e^{-D_2\,V_e} \tag{7}$$

where D_1 is the intercept of the calibration curve and D_2 is the slope of the linear semi-log calibration. By combining Eqs. (5) and (6), we obtain:

$$R_s = \ln(M_{w2}/M_{w1})/2\,D_2(\sigma_2 + \sigma_1) \tag{8}$$

where M_{w1} and M_{w2} are the molecular weights of two standards. D_2 depends on the appropriate choice of column packing material and is proportional to the reciprocal of the column length, L; σ is proportional to the square root of the column length [17]. High resolution will therefore be achieved on a column with a small value of $D_2(\sigma_2 + \sigma_1)$ and hence increasing the column length will increase the resolution.

In gel filtration, as in other chromatographic techniques, many factors influence the final resolution: column dimension, particle size and distribution, pore size, sample volume, and flow rate. The molecular weight range over which a gel filtration column can separate molecules is referred to as the column selectivity or the column efficiency and it is influenced by the conditions used to produce narrow peaks. The ability to counteract the band-broadening effects in size-exclusion chromatography processes is particularly important, since selectivity is fixed once the packing is chosen. Important contributions to band-broadening effects are related to mass transfer, to diffusion along the column, and to eddy diffusion.

In order to move the sample zone down the column one has to induce a deviation from equilibrium. If the zone moves too fast, some of the molecules at its leading edge will not have enough time to permeate the packing, while some at the rear of the sample zone will not have enough time to leave and will lag behind. The consequence of these two effects is a substantial broadening caused by incomplete mass transfer, which can be minimized by a slow flow rate. High molecular weight solutes are more influenced by these effects, due to their low diffusion rate.

Despite the opposite effect of the flow rate on the diffusion along the column, diffusion within the column acts, of course, to broaden the sample zone, and it is higher at low flow rates. Small analytes diffuse more rapidly than large ones and are therefore more influenced by the diffusion phenomena.

A divergent dependence of the mass transfer and of the diffusion along the column on the band-broadening in relation to the flow rate of the analysis is obvious. A compromise must be obtained, since different effects are observed for small and large molecules. The optimal flow rate is low for large proteins and high for small proteins.

Eddy diffusion depends entirely on the characteristics of the packing and column bed quality, and it can be minimized by using uniform, spherical particles and a good packing technique.

2.4.3
SEC vs. HPLC Variants

In ideal SEC, no interactions between the sample and the support should occur; only the pore dimension and the pore size distribution and therefore the volumes related to the porosity and to the interparticle volume can influence the separation. This makes SEC an easy and quick method to handle and to optimize in comparison with other chromatographic techniques such as ion-exchange chromatography or reversed-phase chromatography. Analyses of proteins in SEC are in fact carried out in the isocratic mode and the choice of the mobile phase is only directed to minimize the interactions between the sample and the support surface; it does not influence the retention of the analyte when ideal conditions are achieved. Due to the absence of interactions between the support and the analyte, the loadability of the column depends simply on the pore volume and not on the surface area. The additional limits regarding viscosity, solubility, and volume of the sample further decrease the mass loadability of the column in comparison with standard chromatographic techniques, where the loadability of the column can be modulated even by changing the mobile phase. One way to overcome the low loadability is to use long columns or to couple more columns with the same SEC characteristics. This is in contrast to the standard HPLC modes, where short and fast columns are preferred. The elution window in SEC is also simply dependent on the difference between the accessible and interstitial volumes of the column. The choice of the column will therefore mainly depend on the mass range of the sample and on the pore size of the packing, not on the modification of the surface of the column. This produces a narrow detection window, which can only be amplified by coupling columns characterized by different pore size ranges. Even the elution pattern is exclusively determined by the M_w of the sample and it is not influenced, under ideal conditions, by changing the mobile phase or the surface modification.

Gel filtration may therefore appear to be a very easy protein analysis technique, when no broad elution window or high loadability are required. However, the operation conditions need to be carefully optimized. This is easy in the analysis of synthetic polymers, but the analysis of proteins with complex structures often becomes a difficult issue. The need for stationary phases that do not interact with the sample, and the choice of a mobile phase that minimizes potential interactions, together with the need to preserve the biological activity of the sample, are frequently not fulfilled and non-ideal conditions may prevail.

2.4.4
Optimization Aspects in SEC of Biopolymers

Gel-filtration experiments can be successfully carried out by choosing the right column and the correct operation conditions. It is first necessary to select a column designed for SEC of biopolymers with fractionating properties correlated to the sample that has to be analyzed. Most important is the selection of an appropriate mobile phase with regard to the buffer, the pH, and the ionic strength. The correct sample mass and volume should then be assessed in order to obtain acceptable SEC conditions.

2.4.4.1 Column Selection and Optimal Flow Rate
The gel-filtration medium is normally packed into a stainless steel column characterized by an *i.d.* between 4 and 25 mm and a length between 300 and 600 mm. Four important factors in characterizing SEC packings are the average particle size, the particle size distribution, the average pore diameter, and the pore size distribution. The particle size and the particle size distribution have a significant influence on the column efficiency, which is most decisive in SEC, because the biopolymers are eluted isocratically in a small elution window with

limited selectivity according to their molecular weight range. The particles normally used have an average diameter of 5–15 µm, mostly 10 µm, and a narrow size distribution of about 10–20% standard deviation from the mean.

The pore size of the support in SEC is decisive for the fractionation range of the column. Proteins, which are more frequently investigated by gel filtration, have a molecular weight range between 10 and 500 kDa, and the particles normally used in these cases have pore diameters between 10 nm and 50 nm. Peptides or small molecules with M_w lower than 20 kDa are generally analyzed on supports with pores of about 5 nm. Materials with higher pore sizes, such as 100 nm, can be used for proteins of mass 1,000 kDa. Some limitations arise when using 5 nm pore size supports, which usually exhibit high specific surface areas: in this case, solute–surface adsorption phenomena of the proteins can become notable and have to be suppressed by choosing appropriate mobile phases. When supports with large pores are employed (pore size of 100 nm), the low mechanical stability of the packing may become a problem.

To extend the separation range, SEC columns with two different pore sizes, e.g. 5–30 nm and 100 nm, are combined. The columns should be coupled in order of decreasing molecular weight exclusion limit.

At optimal packing conditions, the interstitial volume should be between 35 and 45% of the total volume of the column. Interstitial volumes lower than this are normally due to fine particles or mechanical instability of the packing. An important parameter characterizing a SEC column is the ratio V_i/V_0, called the "phase ratio". By assuming a constant interstitial volume for all spherical supports and a stationary phase volume corresponding to 15–40% of the empty column, the phase ratio will vary between 0.5 and 1.5 [7]. In order to improve the phase ratio, the intraparticle volume (pore volume) should be increased by using longer columns or coupling several columns of the same type. By doubling the length of the column, the slope of the calibration curve (D_2) will be reduced by a factor of two and the resolution will improve as expected from Eq. (7), while the separation range will remain the same.

By coupling more columns or using very long columns, it is possible to improve the resolution and the fractionation window. However, the back-pressure and the analysis time will also increase. A reasonable compromise should therefore be sought.

As discussed above, interactions between the biopolymeric solutes and the surface of the SEC packing should be minimized. Since proteins have electrostatic and hydrophobic regions in their structures, the surface of the support should be neutral or hydrophilic. Packings based on dextrans, agarose or polyacrylamide are normally employed in classical SEC, but their swelling properties and their low mechanical stability do not allow their application in HPSEC. For the latter, commonly used packings are silica-based materials and controlled-porosity organic polymers. The silica-based materials are generally available in a pore diameter range between 2 and 1000 nm. They show a number of significant advantages, such as pressure stability and very well defined porosity. The porosity of the material can, in fact, be controlled by the method of synthesis and by post-treatments.

Silica-based supports cannot be employed as native materials due to the presence of silanol groups, which cause adsorption and denaturation of proteins. Furthermore, the silanol groups act as weak acidic groups, and therefore as cation exchangers in the pH range 4–9. A chemical modification of the surface of the silica particles is therefore essential. The surface should be hydrophilic, free from charged sites, and chemically stable toward the eluents. The functional groups mainly utilized are hydroxyl, ether, and amide, which are anchored via a hydrocarbon spacer to the surface of the matrix. The compositions of commercial columns are not normally specified. The first successful surface-modified inorganic SEC support for proteins was synthesized by Regnier et al. [18]. Tosoh Bioscience has carried out continuous research in this field and offers a set of columns TSK-GEL SW and SWXL designed for the gel filtration of proteins and peptides. TSK-GEL SW and SWXL columns are packed with silica particles of carefully controlled pore sizes for HPSEC. These silica-based columns are bonded with hydrophilic compounds to reduce the adsorption of proteins. The highly efficient 5 μm TSK-GEL SWXL columns provide high resolution and accurate molecular weight determinations. A new set of columns called TSK-GEL SuperHZ, packed with silica-based particles of diameter 3 μm, has recently been introduced onto the market. These columns should significantly reduce analysis times and improve resolution.

Alternative supports for gel filtration are based on highly cross-linked organic polymers, where the pore size is controlled by the degree of cross-linking. The surface modification is in this case not an issue, because it is normally easy to obtain polymers with a hydrophilic and neutral surface. Other limitations, such as mechanical stability, are, however, associated with the use of organic cross-linked polymers. The rigidity of the polymer can be improved by increasing the degree of cross-linking, but this also adds hydrophobicity and decreases the pore volume. Amersham Bioscience produces the most widely used columns based on organic polymers, which are diverse types of polymer-based columns designated for specific applications (Superdex, Sephacryl, Superose, Sephadex). The matrix of the Superdex columns, for example, was designed for short run times and high recovery and is based on highly cross-linked porous agarose particles modified with dextran. Sephacryl is an alternative to Superdex for large-scale applications. It is composed of a medium prepared by covalently cross-linking allyl dextran with N,N'-methylene bisacrylamide to form a hydrophilic matrix of high mechanical strength. The porosity was matched in such a way as to develop columns with five different selectivities. Superose columns are characterized by a very broad fractionation range, but they are not suitable for industrial-scale separations. The matrix is basically a highly cross-linked agarose. Sephadex columns are more designed for rapid group separation of high and low molecular weight substances, such as desalting, buffer-exchange, and sample clean-up purposes. Sephadex is prepared by cross-linking dextran with epichlorohydrin, and the different types of Sephadex vary in their degree of cross-linking and hence in their degree of swelling and selectivity for specific molecular sizes. Toyo Soda also produces some organic-based matrices for gel filtration called TSK PW gels.

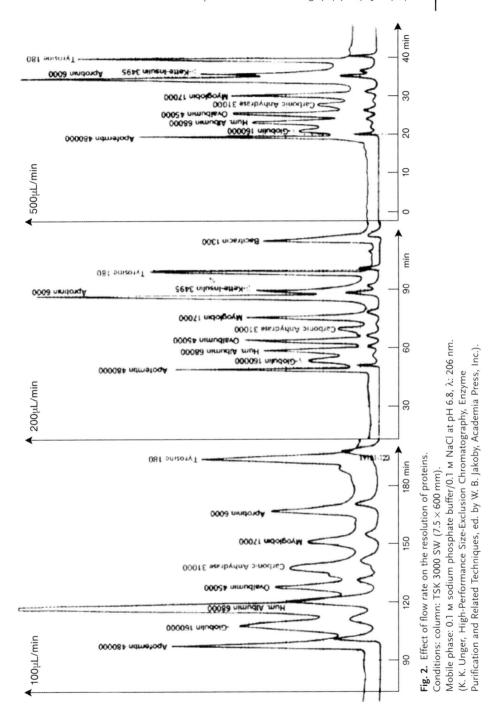

Fig. 2. Effect of flow rate on the resolution of proteins.
Conditions: column: TSK 3000 SW (7.5 × 600 mm).
Mobile phase: 0.1 M sodium phosphate buffer/0.1 M NaCl at pH 6.8, λ: 206 nm.
(K. K. Unger, High-Performance Size-Exclusion Chromatography, Enzyme
Purification and Related Techniques, ed. by W. B. Jakoby, Academia Press, Inc.).

The choice of the column is therefore mainly determined by the sample properties and by the application. In HPSEC, columns with micro-particulate packings are normally used because short analysis times and high efficiencies are required. On the other hand, columns with coarse particles are selected for preparative applications.

As discussed above, the flow rate should be low to provide sufficient time for equilibration without incurring a decrease in performance due to longitudinal diffusion of the biopolymeric solutes along the column (Fig. 2). The operating flow rates normally used are lower than the flow rates used in HPLC. A flow rate of 0.5–1 mL min^{-1} is generally used for columns with an *i.d.* of 7.5 mm, which is far from the flow rates used in HPLC for columns of this type.

Comments Concerning Column Handling

As a rule, one should carefully follow the recommendations of the column supplier and the classical maintenance rules applicable to HPLC. For example, if the column is not employed overnight, a low flow rate of the eluent should be maintained. All columns should be stored, after washing with a diluted buffer, in a solution containing 0.05% *w/w* sodium azide in water. The working conditions of SEC columns with regard to flow rate and column back-pressure should be respected. Particular attention should be paid to columns with swelling packing properties. They should be operated under the conditions quoted by the supplier with appropriate maximum flow rates.

As frequently pointed out in the above, the flow rate for SEC columns is much lower than in other HPLC variants. It is therefore advisable to check the influence of the flow rate on the elution volume and on the column performance with standard proteins. However, low flow rates result in long analysis times. Thus, a compromise should be made. As a rule of thumb, the optimum flow rate increases proportionally with the average particle diameter of the packing at constant *i.d.* of the column.

2.4.4.2 Optimization of the Mobile Phase

The primary roles of the mobile phase in gel filtration are to solubilize all of the components of the sample and to minimize the interactions between the sample and the stationary phase. It is obvious that if the sample is not completely solubilized, the fractionation of the components will not be exclusively by size. The choice of the mobile phase regarding the suppression of the analyte-surface interactions is more delicate. A mobile phase for gel filtration is normally composed of a buffer and salt, and, where necessary, a stabilizer is added in order to suppress the analyte–support interactions. The two principal types of interactions involved in SEC are electrostatic and hydrophobic interactions.

Electrostatic interactions can be minimized by reducing the pH or by increasing the salt concentration (Fig. 3). A reduction of the pH will reduce the electrostatic interactions on HPSEC columns because their residual surface charges tend to become less negative (silanol groups of silica-based supports or carboxylic acid groups on some polymeric supports). The pH, however, has to be adjusted within

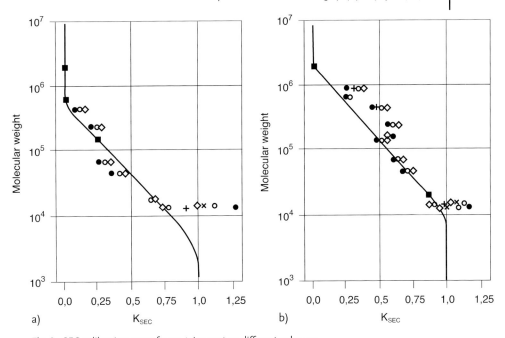

Fig. 3. SEC calibration curve for proteins on two different columns
(9.4×250 mm) at 308 K:
(a) Zorbax Bio series GF 250, (b) Zorbax Bio series GF 450.
Conditions: mobile phase 50 mM phosphate buffer at pH 7 with
increasing salt content; sodium chloride concentrations as follows:
($\square$) 0 mM, ($\bullet$) 100 mM, (+) 200 mM, ($\bigcirc$) 300 mM, ($\times$) 400 mM, ($\Diamond$) 500 mM.

a certain range, because of the limited stability of the biological macromolecules
and the limited chemical stability of the SEC packing. An increase in the ionic
strength can also decrease the electrostatic interactions, but it should be emphasized that an increase in the ionic strength will promote hydrophobic interactions,
which can only be reduced by decreasing the ionic strength. A good compromise
for protein analysis was found to be a buffer at near-neutral pH (5–8) with a salt
concentration between 0.05 and 0.2 M. The buffers normally used are phosphate,
acetate, citrate, Tris, and acetate. The salts employed to assert the ionic strength
are generally sodium chloride, sodium sulfate, ammonium acetate, and ammonium formate.

The use of a denaturant or non-denaturing detergents can sometimes be useful
in order to thoroughly investigate the structure of a given protein or to additionally
suppress hydrophobic interactions.

Strong denaturants such as sodium dodecyl sulfate (SDS) and guanidine are
responsible for the conversion of native protein structures into random coil
conformations. In many cases, this induces a uniform protein conformation,
which increases the resolution and leads to an improvement in the accuracy of
the protein molecular weight determination. It has also been observed that the

addition of a denaturant strongly influences the hydrodynamic volumes of proteins. Calibration curves calculated by using the same proteins with or without denaturants are different.

In many cases, the use of a denaturant is not necessary, but it becomes essential when, even after adjusting the pH and the ionic strength of the buffer, hydrophobic protein–support or protein–protein interactions are still present. This behavior is frequently observed with proteins displaying large hydrophobicity values and which are also relatively insoluble in buffers and therefore tend to aggregate. The hydrophobic protein–protein interactions can often be reduced or completely eliminated by using detergents in the buffer.

SEC in the presence of SDS appears to offer more versatility for the determination of M_w, but guanidine hydrochloride is more suitable for proteins of less than about 40 kDa [19].

The most widely used detergent in SEC is sodium deoxycholate, particularly because of its low UV absorbance at wavelengths at which proteins are normally detected and because of the knowledge regarding its interactions with proteins [20]. Other detergents include Triton X-100 and Nonidet P40. The limitation of these detergents is their strong UV absorbance in the range in which proteins absorb and the formation of unacceptably large micelles under common chromatographic conditions.

Several organic solvents have also been employed in SEC of proteins under denaturing conditions. Acetonitrile is most favored due to its volatile nature and UV-transparency, but at contents above 30% it leads to poor solubility of proteins in solution. It is therefore preferred for proteins with low molecular weight and peptides.

The use of denaturants is very useful when the compositions of more complex structures such as cell organelles, viruses or multimeric enzymes have to be investigated. Denaturants destroy the non-covalent protein–protein bindings.

The analysis of peptides by gel filtration is more complicated due to their heterogeneity (hydrophobic or cationic properties), their poorly defined conformations (which vary between linear and globular), and also the paucity of columns that are suitable for biomolecules with molecular weights of less than 10 kDa.

To sum up, much care has to be taken in selecting the mobile phase composition in order to obtain reproducible results in SEC. The primary rule is to always use freshly prepared eluents. The pH and the ionic strength should be optimized for the given sample. SEC of strongly hydrophobic proteins such as membrane proteins requires extensive experience and is not suited as a task for beginners.

2.4.4.3 Sample Preparation

The composition of the solvent used to solubilize the sample does not significantly influence the resolution, as long as it does not influence the sample stability. Proteins are generally characterized by a high degree of tertiary structure stabilized by van der Waals forces, electrostatic and hydrophobic interactions, as well as hydrogen bonding. By attenuating these forces, the risk of denaturing and/or precipitation is maximized and this should be taken into consideration during sample preparation. However, as a primary precaution, each sample should pass through filters with a pore diameter of 1 μm before being injected in order to eliminate any particulate matter, which could block the column or simply give spurious results.

The sample preparation of small molecules for gel filtration is normally unproblematic, but when handling proteins the effect of solvents on protein conformation and association should be considered.

In some applications, such as protein purification, preservation of the biological activity of the proteins is required. If this is the case, the proteins should be solubilized in a minimum volume of either sodium phosphate/sodium chloride buffer or phosphate buffer containing sodium deoxycholate.

In other applications, such as molecular weight determination or analysis of the protein structures, denaturation of the proteins is required and SDS or guanidine should be added to the buffer solution to disrupt hydrogen bonds, electrostatic interactions, and/or hydrophobic interactions. However, many proteins also contain disulfide linkages, which can compromise the complete dissociation of multimeric proteins. To obtain a complete and uniform denaturation, sulfhydryl reagents, such as 2-mercaptoethanol, should be used in order to break the S–S bonds by reduction.

After use, samples should be stored in a refrigerator in closed containers.

2.4.4.4 Sample Viscosity and Sample Volume – Two Critical Parameters at Injection

Due to the generally high molecular weights of the samples applied in SEC, the viscosity could emerge as an issue when working at high concentrations. A useful piece of advice is that in SEC the relative viscosity of the sample should be at most half of that of the mobile phase, which corresponds to a protein concentration of about 70 mg mL^{-1} in dilute buffer [21]. The sample viscosity is therefore only a problem when stabilizing agents are used or the loadability has to be improved. In order to increase the loadability of a column, without influencing the viscosity of the sample, a larger volume of a diluted sample could be injected. A limit of this operation mode is, however, the volume loadability of the column. The maximum volume load is achieved when the plate number is decreased by 20%. This value corresponds to approximately 1–5% of the total column volume. Smaller injection volumes do not necessarily improve the resolution. To obtain a high resolution, a sample volume of 0.5–4% of the total column volume is generally recommended. For most applications, the sample volume should not exceed 2% of the total column volume to achieve maximum resolution. Analytical separations and separations of complex samples should not exceed a sample volume of 0.5%.

Only in certain cases, such as in desalting processes, a high resolution is achieved under analytical conditions and the column can be considerably overloaded up to 30% of the total column volume [2]. Also, for group separations, sample volumes of up to 30% of the total column volume can be applied.

2.4.4.5 Detection Methods

Detectors such as those based on on-line light-scattering, UV absorbance, or refractive index measurements are employed to characterize the molecular weights of polypeptides, simple proteins, and glycoproteins or to determine the stoichiometry of protein complexes [22].

Insoluble peptides produced from tryptic or cyanogen bromide-digested proteins may be effectively analyzed by SEC with SDS in the mobile phase in order to separate families of peptides that may be subsequently analyzed by a different technique (e.g., mass spectrometry). Hyphenation of these techniques is particularly valuable, considering the fact that size-exclusion chromatography does not accurately determine molecular mass but hydrodynamic volume [23]. However, the most commonly used SEC protocols for the analysis of proteins are not compatible with mass spectrometric detection. The protocols that are compatible with MS detection use volatile mobile phases, but do not always give adequate results with respect to the chromatography. To find a suitable mobile phase, several parameters have to be considered to achieve a size-dependent separation of proteins and to avoid non-specific interactions with the stationary phase. Lecchi and Abramson [24] tested different chromatographic conditions, taking into consideration ionic strength, protein solubility, and volatility of the solvent. The recommended mobile phase consisted of 1 M triethylammonium formate in water/THF (2 : 1, *v*/*v*) at pH 3. This mobile phase dissolved and eluted all of the standard proteins that were tested, with results comparable to those obtained using conventional involatile modifiers in the mobile phase (e.g., 50 mM K_2HPO_4 and

Table 1. Troubleshooting.

Observation	Reasons for	Action advised
Poor resolution	1. Unsuitable column material	1. Choose a support with a narrower fractionation range or a smaller particle size
	2. Column dimensions not optimal	2. Use longer columns or couple two columns
	3. Non-optimal flow rate	3. Reduce the flow rate
	4. Sample size	4. Reduce the sample concentration and/or the sample volume
High back-pressure	Precipitation of proteins during the analysis	Filter the eluent and add a denaturing agent
Skewed peaks	1. Adsorption of proteins on the column	1. Add salt to the buffer
	2. Air bubbles in the column	2. Repack the column and sonicate the buffer
	3. Unsuitable buffer	3. Choose a different mobile phase
Loss of protein activity	1. Adsorption of protein on the column	1. Add salt
	2. Separation of subunits	2. Change the mobile phase

200 mM NaCl at pH 6.5). Many proteins were successfully analyzed, including thyroglobulin, catalase, collagen, transferin, albumin, carbonic anhydrase, lysozyme, and insulin. For most SEC columns, a volatile buffer solution containing the salt ammonium acetate proved to be suitable for the analysis of polysaccharides (dextrans, heparin) or nucleic acid polymers (RNA, oligonucleotides) [24].

Since the mechanism behind SEC is entropy-driven, while enthalpic interactions are suppressed, non-covalent complexes are not dissociated in the separation process. Furthermore, SEC is a very efficient method for removing involatile salts and buffers prior to mass spectrometric analysis and allows direct and sensitive analysis of the complexes present in a biological matrix with high salt concentrations, which are often necessary for complex formation and stability.

Consequently, the possibility of introducing complex biological mixtures directly into the mass spectrometer without any additional steps is an enormous advantage. A further benefit can be realized if we ensure that any chromatography step employed performs critical but separate functions. First, it should remove involatile salt components from the mixture, replacing them with an appropriate amount of volatile salt just before analysis by MS. Exchange of alkali metal cations with volatile cations (NH_4^+) results in improvements in the sensitivity of the subsequent mass spectra. Large biomolecules are usually highly charged and may form adducts with inorganic ions. Therefore, the chromatographic procedure that separates the components of the mixture should not adversely affect the complexes or at worst should only slightly perturb the non-covalent interactions. It is worth emphasizing that the role of the SEC portion of this methodology is not to determine mass; that is the role of the mass spectrometer.

The SEC step is also used for separation based on gentle gel–pore interactions that conserve non-covalent complexes. The bonus of using SEC for this is that it is an excellent method for mildly removing involatile salts and buffers prior to mass spectrometric analysis.

Table 1 gives an overview over troubleshooting.

2.4.5
Applications

2.4.5.1 High-Performance SEC

Procedures using microparticulate SEC columns have been optimized for the preparative separation and quantitative determination of the relative proportions of major size classes of polypeptides of rice proteins from several varieties [25]. The advantage of applying SEC is that chromatographic resolution is not disturbed when using a strong dissociation-supporting solvent containing both a detergent and a chaotropic agent to isolate and fractionate the rice proteins (0.1 M acetic acid, 3 M urea, 0.01 M cetyltrimethylammonium bromide). Hamada [25] reported that the protein load and the flow rate significantly affected V_e. Increasing the flow rate enhanced the retention of the peaks by 9%. Increasing the injection load increased the V_e values of the protein. Decreasing the concentration of injected protein from 175 to 1.75 mg mL^{-1} resulted in a shift of M_r of the peak from 220 to 130 kDa.

2.4.5.2 Determination of Molecular Weight

SEC is a simple and fast method for estimating the molecular weight of a protein in its native form based on its elution position. However, there are several problems associated with this conventional SEC approach. One is that the elution position depends not only on the molecular weight of the protein, but also on its shape. Another problem is that the elution position will change if the protein has any tendency to interact with the column matrix. In addition, when a protein or a protein-protein complex bears carbohydrate moieties, these usually have a disproportionately large effect on the elution position, so SEC may not be able to accurately determine the molecular weight of the polypeptide. Thus, the obtained data should be treated with extreme care, especially when drawing conclusions or speculating about possible mechanisms. For example, the total molecular weight estimated by SEC for stern cell factor from Chinese hamster ovary is 113,000, whereas the true value is 53,100 [22]. This error by more than a factor of two reflects the fact that the analyzed biomolecule is highly asymmetric and, therefore, behaves hydrodynamically as though it were larger when subjected to SEC. This example illustrates the frailty or possible danger of using the elution position alone to estimate the molecular weight of a protein.

2.4.5.3 Gel Filtration as a Tool to Study Conformational Changes of Proteins

SEC has proven to be a useful technique in the observation and control of protein folding and aggregation. Its mild conditions and short analysis times, along with compatibility with many different eluent compositions, maximize its utility. Specific detection methods, such as UV absorption spectroscopy or laser light scattering with low-angle or multi-angle methods can allow the identification of aggregates or even conformational changes. SEC has been used to perform a buffer exchange from a solubilizing to a refolding solution [26]. During the process, unfolded lysozyme was gradually refolded as it traversed the column and its mobile phase environment changed to the refolding solution. This gradual solvent-exchange process minimized protein aggregation.

A common characteristic of protein-folding reactions is a dramatic decrease in the size of the conformational ensembles. The physical characterization of native, intermediate, and unfolded conformers of tryptophan synthase is based on apparent radius, which has been measured by several techniques [27]. Comparing SEC, analytical ultra-centrifugation, dynamic light-scattering, and small-angle X-ray scattering, SEC showed the smallest data deviations (4.8%, 9.2%, 3.4%, and 17.3% for the native form; 8.1%, 9.7%, 13.1%, and 30.8% for the unfolded form, respectively).

A SEC refolding process has been successfully used to fully renature lysozyme at high concentrations. The process was based on the different hydrodynamic characteristics of folded and unfolded proteins and their interaction with the SEC column packing. Changes in the Stokes' radius, the hydrodynamic volume, and, consequently, the partition coefficient, occur when lysozyme is refolded from urea in an SEC column. This method utilizes SEC columns to perform controlled buffer exchange from solubilizing to refolding solutions, aggregate removal, and protein folding. Physical strategies for enhancing protein refolding rely on separating refolding molecules until they are in a stable, non-aggregating conformation. The reduced diffusion of proteins in SEC packings has been shown to suppress the non-specific interactions of partially folded molecules and, therefore, to reduce aggregation. Hen egg-white lysozyme and bovine carbonic anhydrase have been successfully refolded from initial protein concentrations of up to 80 mg mL^{-1} using a Sephacryl S-100 column [26].

Initially, the denatured unfolded protein has a random coil configuration and, thus, a large hydrodynamic radius. The protein is therefore excluded from the majority of the pores of the packing. When the eluent (refolding buffer) is applied, the local solvent environment becomes less denaturing, and the protein starts to refold. This results in a smaller hydrodynamic radius as the protein develops a more compact and native-like structure. At this stage, the protein can move further into the pore space in its partially folded form. It should be noted that the partition coefficient of the protein between the eluent and the packing increases over time as the protein folds and becomes more compact. Because of the porous nature of the packing, the mass transport of the protein into the pores is diffusion-limited. This assists the refolding by reducing intermolecular aggregation. The refolding reaction is completed within the pores and there is minimal likelihood of protein aggregation. The protein is finally eluted from the column in its active native form. The size-exclusion characteristic of this process ensures that any aggregates that may have formed are removed from the column first, and the chaotropic species, which have low molecular weights in comparison to the protein are eluted after the refolded protein.

A partially folded protein penetrates the SEC packing to a greater extent with respect to available pore space or surface area in comparison to the unfolded species, thus minimizing protein–protein aggregation and simultaneously increasing the refolding yield. The protein species are further kept apart during elution as the folded state requires a greater volume to elute than does the unfolded protein. The success of this process will depend on the degree of resolution inherent to the size-exclusion packing.

The size-exclusion protein refolding of lysozyme is relatively unaffected by temperature. Increasing the column temperature from 20 °C to 50 °C results in a decrease in the regained specific activity of 5%, and a decrease in the elution volume of 8 mL (2.3%) [26].

2.4.5.4 Gel Filtration in Preparative and Process Separations (Downstream Processing)

SEC is also useful for peptide separations. However, the resolving power of available size-exclusion columns for peptide mixtures is not comparable to that attainable by reversed-phase HPLC (applying gradient elution).

Few SEC columns are available for the preparative chromatography of biopolymers with molecular weights above one million. Josić et al. [28] have performed such SEC separations with high molecular weight plasma proteins. The SEC separation of biopolymers with molecular weights over 300,000 requires time-consuming optimizations.

Biotechnologically relevant products such as recombinant proteins, plasmids, etc., require extensive processing for purification and isolation. In such cases, downstream processes are applied, composed of sequences of successive chromatographic and non-chromatographic steps to selectively remove impurities and to enrich the target compounds. The preferred techniques are SEC, ion-exchange chromatography, and affinity chromatography [9]. There are a number of special features to be considered for work on a process scale: large volumes have to be handled, which requires large-bore columns with coarse packing materials generating a low column back-pressure.

Gel filtration in downstream processing is often employed for desalting, rebuffering, and group fractionation of biopolymers.

The injection volume amounts to ca. 5% of the column volume for fractionation and 15–30% for a desalting step. The column length varies between 60 and 180 cm at *i.d.* of 5 to 180 cm [9].

2.4.5.5 SEC Columns Based on the Principle of Restricted Access and Their Use in Proteome Analysis

The principle of the restricted access support is based upon the presence of two chemically different surface properties of porous silica particles (Fig. 4). The outer surface of the particles (25–40 mm o.d.) is highly biocompatible, it possesses diol modification, and hence is hydrophilic, while the pore surface chemistry is tailored as a hydrophobic dispersion phase with C_{18}-functionalities [11] or as a strong cation-exchanger with SO_3H functionalities. An advantage of these adsorbents stems from the simultaneous occurrence of two chromatographic separation mechanisms: selective reversed-phase interaction or ion-exchange chromato-

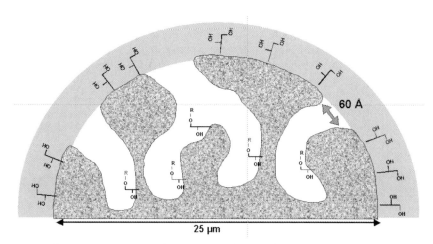

Fig. 4. Schematic representation of an SCX-RAM silica particle.
The external surface is coated with hydrophilic, electroneutral diol groups for the exclusion of high molecular weight components (> 15 kDa); the internal surface is functionalized with ion-exchange groups accessible to low molecular weight components, which may be trapped by electrostatic interactions.

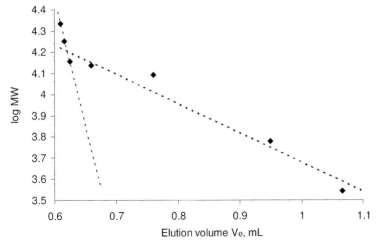

Fig. 5. Determinations of the SCX-RAM exclusion limit using standard proteins (α-chymotrypsin, myoglobin, lysozyme, ribonuclease, cytochrome *c*, insulin (subunit) and insulin B chain). In order to suppress the ion-exchange properties of the support, SEC was performed with 1 M NaCl with the other parameters set to pH 3, in 20 mM phosphate buffer with 10% acetonitrile.

graphy of lower molecular mass analytes, and size-exclusion chromatography of the macromolecular sample constituents. By regulating the pore size of the particles, a physical restriction barrier is adjusted to regulate the size range of molecules that may penetrate and, in the case of penetration, may be trapped in the functionalized pore structure. A pore size of 6 nm allows access to the pores only for analytes with a molecular mass below 15 kDa (Fig. 5). Proteins (> 15 kDa) can thus be eluted with the void volume directly into the waste. Smaller analytes, however, such as drugs and metabolites from body fluids, or pesticides or hormone residues from milk or animal tissue samples, may enter the pores and interact with the *n*-alkyl chains or ion-exchange groups bound to the inside of the pores. When dealing with complex samples, e.g., human biofluids, sample clean-up and fractionation into matrix and target analytes can be achieved. Depending on the ligands present in the pores, small molecules with hydrophobic or ionic properties can be selectively enriched.

The features described above, when elegantly combined with column switching, become a powerful tool for direct analysis of biofluids in a fully automated, multidimensional way. This selective stationary phase has found application in a number of areas, such as bioanalytical quantification of drugs [11], analysis of endogenous peptides [29], and immunoassay techniques [29, 30].

When operating restricted access columns, some basic rules should be followed. It is strongly recommended that the analysis be performed within the linear range of the adsorption isotherm of the sorbent in question in order to avoid the over-

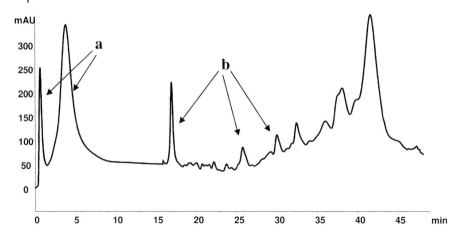

Fig. 6. Typical SCX-RAM column separation profile:
Peaks (a) correspond to physical exclusion by pore size.
Trapped retained biomolecules are separated by applying a gradient in the second step (b).
Conditions: column: LiChrospher 60 SO$_3$/Diol, 25 × 4 mm *i.d.*,
flow rate: 0.5 mL min^{-1}, gradient from 0 to 1 M NaCl in 20 mM KH$_2$PO$_4$, pH 2.5,
containing 5% MeCN in 30 min.
Sample: 100 µL human hemofiltrate (3.7 mg mL^{-1}), UV detection at 214 nm.

loading effect (the number of interaction sites, especially for bigger molecules, is limited), such that effective and reproducible enrichment of the analytes of interest can be achieved. One should keep in mind that the operational flow rate has an enormous impact on the molecular size distribution when employing a RAM column. Higher flow rates can shift the molecular range of the trapped molecules to lower values as smaller molecules need less time to penetrate the pores. Also, higher flow rates could alter the hydrodynamic volume of the biomolecules. Higher molecular mass molecules will be enriched by operating at lower flow rates. Column temperature affects the viscosity of the mobile phase and, consequently, the diffusion ratio, and influences the speed of mass transfer. Carefully performed optimizations of the chromatographic parameters ensure the success of the analysis.

The analysis of peptides of human hemofiltrate on a SCX-RAM column under optimized conditions allows the exclusion of unwanted large molecules (and those that are not cationic) (Fig. 6, a) and, consequently, these are directed to waste. After appropriate washing, a salt gradient elution allows analysis of the components of interest (Fig. 6, b), which may be directed to other chromatographic dimensions to gain additional resolution or to perform desalting/mobile phase exchange and subsequent analysis employing mass spectrometry.

References

1 J. Ch. Janson, *Chromatographia* 1987, 23, 361.
2 Gel Filtration: Principles and Methods, Amersham Biosciences AB, Uppsala, 2002.
3 W. W. Yau, J. J. Kirkland, D. D. Bly, Modern Size-Exclusion Chromatography, John Wiley and Sons, New York, 1979.
4 K. K. Unger, High-Performance Size-Exclusion Chromatography, in Enzyme Purification and Related Techniques (Ed.: W. B. Jakoby), Academic Press, Orlando, 1984.
5 P. L. Dubin, Aqueous Size-Exclusion Chromatography, Elsevier, Amsterdam, 1988.
6 V. Dawkings, Packings in Size-Exclusion Chromatography, in Packings and Stationary Phases in Chromatographic Techniques (Ed.: K. K. Unger), Marcel Dekker, New York, 1990.
7 K. M. Gooding, F. E. Regnier, Size-Exclusion Chromatography, in HPLC of Biological Macromolecules (Eds.: K. M. Gooding, F. E. Regnier), Marcel Dekker, New York, 2002.
8 J. Ch. Janson, P. Hedman, Large Scale Chromatography in Advances in Biochemical Engineering, Volume 25 (Ed.: A. Fiechtler), Springer Verlag, Heidelberg, 1982.
9 G. Jagschies, Process-Scale Chromatography, in Ullmann's Encyclopedia of Industrial Chemistry, Vol. B, VCH, Weinheim, 1988.
10 T. C. Pinkerton, *J. Chromatogr.* 1991, 544, 13.
11 K. S. Boos, A. Rudolphi, *LC&GC International* 1997, 15, 602.
12 K. Wagner, T. Miliotis, G. Marko-Varga, R. Bischoff, K. K. Unger, *Anal. Chem.* 2002, 74, 809.

13 W.W. Yau, J. J. Kirkland, D. D. Bly, Modern Size-Exclusion Chromatography, John Wiley and Sons, New York, 1979.
14 M. E. Himmel, P. G. Squire, *Int. J. Pept. Protein Res.* 1981, 17, 365.
15 E. Pfankoch, K. C. Lu, F. E. Regnier, H. G. Barth, *J. Chromatogr. Sci.* 1980, 18, 430.
16 H. Dai, P. L. Dublin, T. Andersson, *Anal. Chem.* 1998, 70, 1576.
17 K. K. Unger, Enzyme Purification and Related Techniques, 104, Academic Press, Inc., 1984, p. 7.
18 F. F. Regnier, R. Noel, *J. Chromatogr.* 1978, 156, 239.
19 R. Montecarlo, *Aqueous Size-Exclusion Chromatography* 1988, 40, 10.
20 C. Tanford, *Adv. Protein Chem.* 1968, 23, 121.
21 H. G. Barth, *J. Chromatogr. Sci.* 1980, 18, 409.
22 J. Wen, T. Arakawa, J. S. Philo, *Anal. Biochem.* 1996, 240, 155.
23 J. Cavanagh, R. Thompson, B. Bobay, L. M. Benson, S. Naylor, *Biochemistry* 2002, 41, 25, 7859.
24 P. Lecchi, F. P. Abramson, *Anal. Chem.* 1999, 71, 2951.
25 J. S. Hamada, *J. Chromatogr. A* 1996, 734, 195.
26 B. Batas, H. R. Jones, J. B. Chaudhuri, *J. Chromatogr. A* 1997, 766, 109.
27 P. J. Gualfetti, M. Iwakura, J. C. Lee, H. Hihara, O. Bilsel, J. A. Zitzewitz, C. R. Matthews, *Biochemistry* 1999, 38, 13367.
28 D. Josić, H. Horn, P. Schulz, H. Schwinn, L. Britsch, *J. Chromatogr. A* 1998, 796, 289.
29 A. Oosterkamp, H. Irth, L. Heintz, G. Marko-Varga, U. R. Tjaden, J. van der Greef, *Anal. Chem.* 1996, 68, 4101.
30 P. Onnerfjord, S. Eremin, J. Emneus, G. Marko-Varga, *J. Chromatogr. A* 1998, 800, 219.

2.5
Optimization in Affinity Chromatography

Egbert Müller

2.5.1
Introduction to Resin Design and Method Development in Affinity Chromatography

The application and preparation of affinity resins was first described in a paper by Cuatrecasas in 1968 [1]. In this work, nearly all features of this technique were explored. Affinity chromatography is the youngest of the four major purification methods used in biochromatography. In 1968, ion exchange (IEX), hydrophobic interaction (HIC), and gel filtration (GF) were already well established. What new opportunities in the purification of biomolecules were then added by affinity chromatography?

A major target in downstream processing is the isolation of a specific biomolecule from a solution that may contain several thousand substances. The ideal situation is a single-step purification procedure, but the statistical probability of isolating a pure molecule from such a mixture by one chromatography step is determined by a Poisson distribution and falls exponentially with the number of compounds [2]. The only way to improve this situation is to increase the affinity of the resin for the target molecule. The affinity constant is defined as [3]:

$$K_{Aff} = \frac{[L \cdot P]}{[P] \cdot [L]} \tag{1}$$

where
[P]   is the equilibrium concentration of the target protein
[L]   is the equilibrium concentration of the ligand
[L · P] is the equilibrium concentration of the resin-bound protein

The inverse value is the dissociation constant (K_D). In ion-exchange chromatography, K_D varies in the range 10–0.1 μM [4]. Specific binding capacity is acceptable in this range for large-scale applications. Protein impurities also carry charges and are thus difficult to separate from the target protein. The purification result remains relatively poor and this is the same for other non-affinity techniques. There was thus a need to develop resins with ×100 to ×1000 lower dissociation constants [5]. The objective was to achieve specificity of interaction so high that, in an ideal case, only the target compound would be bound. Antigen–antibody interaction is one example which comes close to this objective. This was the main aim in the early days of affinity chromatography. In reality, this is not achievable because non-specific interactions invariably prevent the isolation of the pure product.

There is also the problem of elution of tightly bound biomacromolecules from a resin, so it is a good compromise to use affinity ligands with dissociation

HPLC Made to Measure: A Practical Handbook for Optimization. Edited by Stavros Kromidas
Copyright © 2006 WILEY-VCH Verlag GmbH & Co. KGaA, Weinheim
ISBN: 3-527-31377-X

constants somewhere between those of ion-exchange resins and highly biospecific resins. This is the case with the class of group-specific ligands [6].

Affinity resins can be divided into those employing low molecular weight and those with high molecular weight ligands. Frequently used low molecular weight ligands include [7]:

- *p*-Aminobenzamidine for the purification of serine proteases [8]; phenylboronate for the purification of glycoproteins [9]; Cibacron Blue F3GA (reactive dye) for the isolation of phosphatases, kinases, albumin, and interferon [10]; imino-diacetic acid and tris(carboxy)ethylenediamine (chelates) for immobilized metal chelate affinity chromatography (IMAC) (purification of His-tagged fusion proteins) [11]; thiopyridine and ligands for thiophilic adsorption (a sulfo group in proximity to a thioether) [12]; amino acids such as lysine for the purification of plasminogen, and arginine [13], biotin, small peptides, nucleotides, and carbohydrates [14]. Virtually any molecule with reactive groups can be immobilized on a resin.

High molecular weight ligands offer even more possibilities. They may be grouped into carbohydrates, polynucleotides, and proteins. The most prominent examples are:

- Heparin and dextran sulfate for the purification of plasma proteins and DNA-binding proteins [15].

- Streptavidin [16]; proteases (pepsin, papain, staphylococcus aureus V8 protease) [17]; antibodies and antibody-binding proteins such as Protein A, Protein G or Protein M [18]; lectins such as concanavalin and jacalin [19].

The number of applications is enormous. Nowadays, many new ligands are prepared by combinatorial chemistry [20, 21]. These ligands include carbohydrates, peptides, and polynucleotides. Molecules are first produced, screened, and then synthesized. A central problem in affinity chromatography is often the synthesis of resins, which is both time-consuming and expensive. Most such resins are prepared by highly qualified experts. It is essential for commercially used resins to exhibit caustic stability during use. Only a few resins are used on a commercial scale. These include Protein A for antibody purification, heparin for plasma fractionation, immobilized dyes (especially Reactive Blue), and iminodiacetic acid and nitrilotriacetic acid for the purification of His-tagged fusion proteins [22]. Protein A resin applications have increased vastly in recent years and account for about 80% of all affinity applications. One reason for this has been the successful use of antibodies as therapeutics against cancer [23].

Despite affinity methods having been used for more than 20 years, there are still serious problems that have not yet been resolved. A major problem is the bleeding or leaching phenomenon, which is a slow but persistent release of bound ligand. Bleeding occurs with high molecular weight ligands such as Protein A as well as with low molecular weight ligands such as immobilized dyes [24, 25]. This may be caused by protease activity on protein ligands or non-specific binding of molecules for the lower molecular weight ligands.

Nevertheless, affinity chromatography is used today both for production and analysis, and the application areas are still increasing.

Since the 1980s, the concept of high-performance affinity chromatography has been developed [26]. High resolution in chromatography is a function of particle size. With small particles in the region of 20 μm, high separation efficiency is possible. High resolution is not required for affinity resins with biospecific ligands because the selectivity is high enough to separate the target molecule from contaminants in a single-step elution and a gradient is not required as with reversed-phase or ion-exchange separations.

The situation is different for some resins with group-specific ligands, where salt, pH or imidazole gradients can be used to obtain high resolution [27].

In downstream processing, the optimal particle size is closely related to the purpose of the process step. Downstream chromatography consists of three general stages [28]:

1. Capture (removal of water, quantitative adsorption, and purification with moderate to low selectivity).
2. Intermediate purification (medium selectivity).
3. Final polishing (high selectivity to obtain the specified purity).

For each step, a preferred particle size can be used:

During capture the particle diameter is often around 100 μm, for intermediate purification it is in the region of 60 μm, and for polishing it is generally smaller than 30 μm [30]. For cost reasons, the objective is to reduce the number of purification operations and minimize intermediate steps. One way is to increase selectivity in the capture step. This concept has been used for a long time in antibody purification. A protein resin is used for capture in a primary capture step, which often guarantees a purity of greater than 90%. Particle size is one of several parameters that can be refined. A commercially useful affinity resin is a careful balance of base matrix, surface chemistry, and ligand. Two types of affinity resin are currently available: (1) ready to use, and (2) activated resins. The purification result is determined by the application. An immunoaffinity resin used for recombinant Factor VIII purification will not be the same as that supplied for the purification of plasma-derived Factor VIII.

There is presently no such thing as a universal affinity resin, and the general complexity of interactions suggests that this may never be possible to achieve. It is thus necessary to design a resin for each individual application by empirical methods. The procedure may be speeded up by the use of mathematical methods and this will be described later. The influence of the base material will first be discussed in the following section.

2.5.2
Base Matrix

Many ready-for-use affinity materials are currently available on the market. In the absence of a common ligand, one is faced with a preparation problem. One must decide which base resin should be used and which immobilization method is most suitable.

The selection of the base matrix is very important. Support materials for biochromatography are prepared from either natural or synthetic polymers. For protein purification, most gels are designed for low- or medium-pressure application. Sometimes, silica gel or controlled pore glass is used. Some examples are shown in Table 1.

These materials are used for the different modes of biochromatographic separation, ion exchange, hydrophobic interaction, gel filtration, and also affinity chromatography. Some materials are underivatized but not commercially available. In general, resins suitable for gel filtration of proteins are good candidates for affinity chromatography.

There are some basic requirements for affinity resins [31]. The resin should contain derivatizable groups (e.g., -OH groups). The matrix must be chemically and mechanically stable. This means that it must remain stable in solvents or under acidic or basic conditions. The packed resin must resist compression under flow. Above all, the material must be hydrophilic. Interestingly, mechanical stability and hydrophilicity are almost mutually exclusive properties. Mechanical stability in synthetic polymers is connected to the presence of aromatic rings and a high degree of cross-linking. Aromatic rings also increase non-specific interactions and higher degrees of cross-linking decrease the number of available -OH groups [32].

Sepharose resins are hydrophilic, but they are only stable up to flow rates of about 500 cm h^{-1} [33]. Polystyrene resins are more mechanically stable, but unmodified they are too hydrophobic for protein chromatography and the surface must be rendered hydrophilic by costly modification procedures [34]. Methacrylate resins are an ideal compromise. They are not too hydrophobic and they offer excellent mechanical stability.

Table 1. Support materials for biochromatography.

Supplier	Trade name	Support material
Amersham Bioscience	Sepharose®	Agarose
Amersham Bioscience	Sephadex®	Dextran
Amersham Bioscience	Resource®/Source®	Polystyrene
Biorad	Macro-Prep®	Methacrylate
Tosoh Bioscience	TSK-Gel, Toyopearl®	Methacrylate
Millipore	Cellufine®	Cellulose
Ciphergen	Trisacryl®	Acrylamide
Röhm	Eupergit C	Methacrylate

Mean pore size is equally as important as particle size. Affinity support matrices are all porous materials. This both permits and restricts mass transfer within the particle [35]. Typically, diffusion, film resistance, and (in affinity chromatography) slow reaction kinetics all restrict the transport of large proteins within the pores. The low diffusion coefficients of large proteins is a major problem. This can be overcome by increasing the pore size to values above 100 nm. However, large pores reduce the accessible surface area and so decrease binding capacity. A more attractive solution is the application of polymeric surface modification procedures, which will be described later [36].

Such technical solutions must be robust for industrial applications, and long-term product availability and cost efficiency become critical issues.

2.5.3
Immobilization Methods

Many ready-to-use resins are available commercially. Newer proprietary ligands resulting from the use of combinatorial library sources pose their own problems. The immobilization method influences the properties of the final resin significantly and the production of ligand may require a high level of technical competence. Thus, a basic knowledge of immobilization chemistry is required by the customer.

2.5.4
Activation Methods

Activation of the resin requires that a bifunctional reagent is attached to the resin surface. One group attaches to the resin and the other group is used for subsequent immobilization of the ligand:

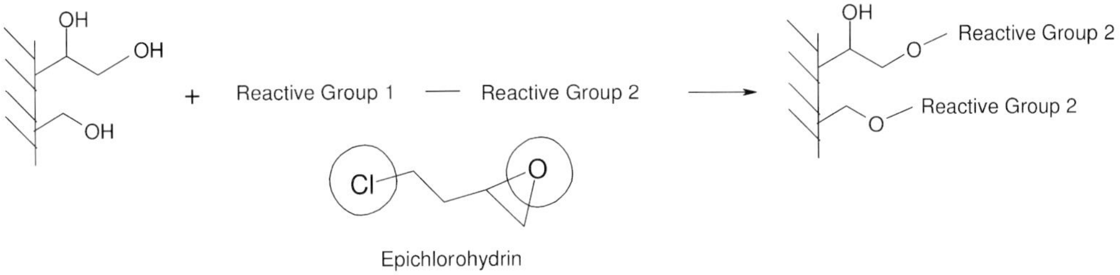

Epichlorohydrin

Fig. 1. Preparation of an activated resin (example: epichlorohydrin).

Reactions and reaction products on surfaces are difficult to follow and to detect due to the lack of suitable analytical methods.

At the surface of base resins, single -OH groups and diol groups are often present, which cause the immobilization to be somewhat unpredictable. This has been described for the classical cyanogen bromide activation [37]. Diol groups occur in large numbers on carbohydrate-based resins and also in synthetic resins, where glycidylmethacrylate is the source. Some activation methods, such as oxidation with periodate, only work with diol group containing resins [36].

Some selected activation methods are described here:

1. Cyanogen Bromide Activation [38]

Cyanogen bromide activation is one of the oldest activation methods. It is versatile and easy, and can be applied with good success to a wide range of ligands. The toxicity of cyanogen bromide is so high, however, that applications outside of the research laboratory are questionable. The activated resin is also sensitive to moisture, which makes it hard to achieve reproducible results. The activated resin contains a mixed population of activated groups, which results in differently bound protein ligands.

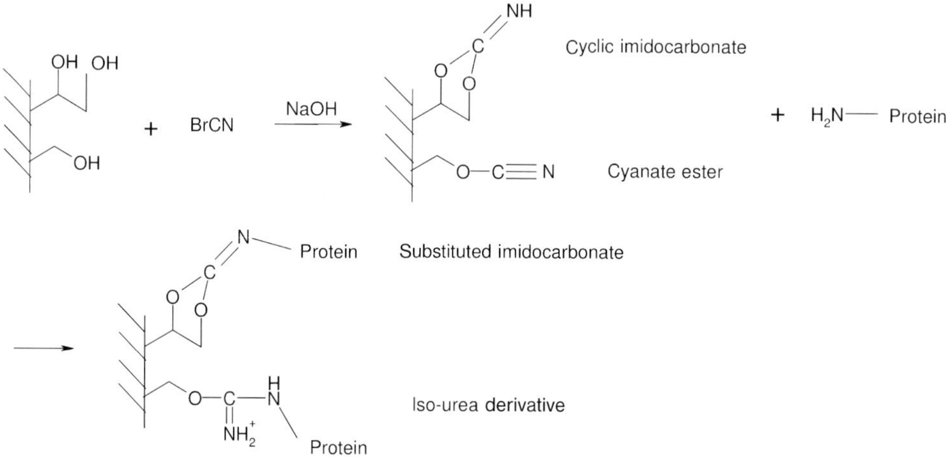

Fig. 2. Cyanogen bromide activation.

2. *N*-Hydroxysuccinimide (NHS) Activation (Fig. 3) [39]

In contrast to cyanogen bromide activation, where the resulting bond to the coupled protein is a hydrolytically labile isourea derivative, NHS activation uses a more stable amide bond.

Hydroxy group containing matrices can be directly activated using *N,N'*-disuccinylcarbonate [40].

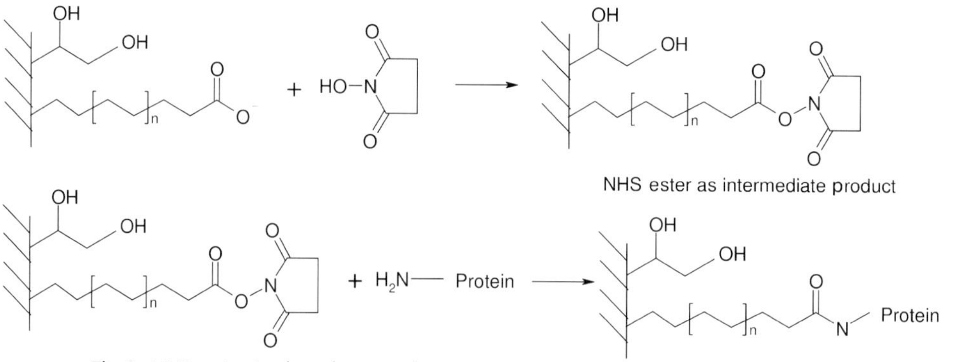

Fig. 3. NHS-activation based on a carboxy group containing resin.

3. Tresyl activation (Fig. 4) [41]

Tresyl activation results in a connection to a protein via a secondary amine bond. This could cause trouble with non-specific interactions, but this is suppressed by the addition of a salt to the working buffer. Results of NMR investigations by Jennissen [42] suggest that the final bond is in fact a sulfonamide bond rather than a secondary amine bond.

The immobilization of proteins by all three methods can be accomplished at physiological pH values. This limits the denaturation of the proteins.

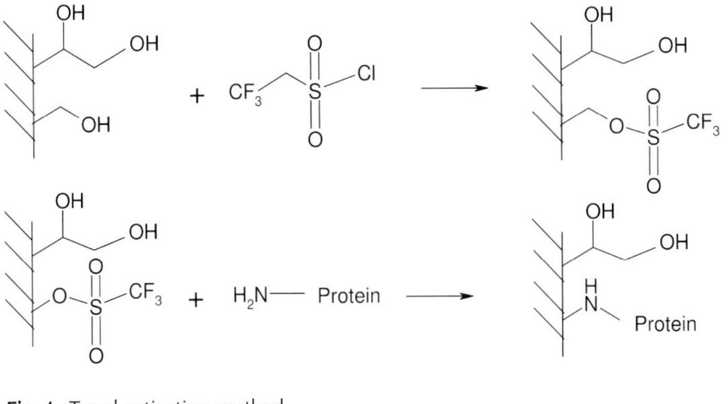

Fig. 4. Tresyl activation method.

Epoxy activation methods with 1,4-butanediol diglycidyl ether and epichlorohydrin are recommended for the immobilization of low molecular weight compounds and stable proteins (Fig. 5) [43].

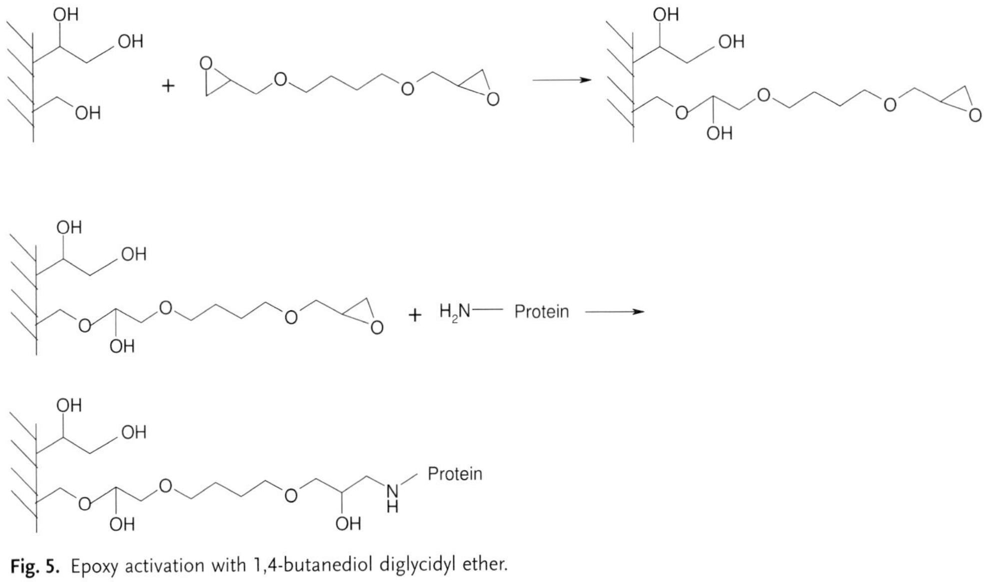

Fig. 5. Epoxy activation with 1,4-butanediol diglycidyl ether.

A free hydroxy group is always formed besides the secondary amine group. This increases the hydrophilic property of the resin. The general drawback is that a high pH is required for immobilization. High concentrations and temperature (40 °C) can increase the yield. Some resins with group-specific ligands (such as iminodiacetic acid) are produced by this method. Other popular immobilization methods include:

1. Hydrazide   immobilization of carbohydrates and glycoproteins [44].
2. Formyl   immobilization of amino group containing compounds [45] and subsequent reduction with borohydride.
3. Cyanuric chloride   immobilization of reactive dyes [46].
4. Divinylsulfone   similar to epoxy activation [47].
5. Carbonyl diimidazole   similar to NHS activation [48].
6. Azlactone   smooth ring-opening reaction [49].
7. Carbodiimide   water-soluble carbodiimide for the coupling of proteins [50].

There are many more protocols described throughout the literature. The major problem is identification of the best solution for a particular application. It is advantageous to use a commercially available resin first (cyanogen bromide, *N*-hydroxysuccinimide, tresyl or epoxy), for which an established immobilization protocol is available. A simple immobilization buffer is recommended for method screening experiments.

The following buffer systems can be used: For sensitive proteins a simple phosphate buffer (50–100 mM) with pH ≈ 7 is recommended. For immobilizations above pH 9, a borate, carbonate or hydrogencarbonate may be used. The addition of a certain amount of salt, 1–3 M sodium chloride, 1 M phosphate buffer or 0.5 M sodium sulfate can enhance the binding by a supporting hydrophobic interaction [51]. Careful washing of the resin is then recommended to remove non-covalently bound proteins.

2.5.5
Spacer

The spacer issue is often discussed in affinity chromatography. Sometimes a spacer is useful, but sometimes not [52]. A spacer should be designed to improve the accessibility of ligands to proteins. The benefit of a spacer differs between low molecular weight ligands and the larger ligands, or proteins. With smaller ligands an advantage can be achieved with spacer technology; it is more questionable with larger ligands. In general, there are three different ways to connect a ligand to a resin surface [36]:

1. Direct ligand connection to the resin surface (no spacer).
2. Spacer with a single ligand attached.
3. Polymeric spacer with multiple ligand attachment.

The architecture of each arrangement is shown in Fig. 6.

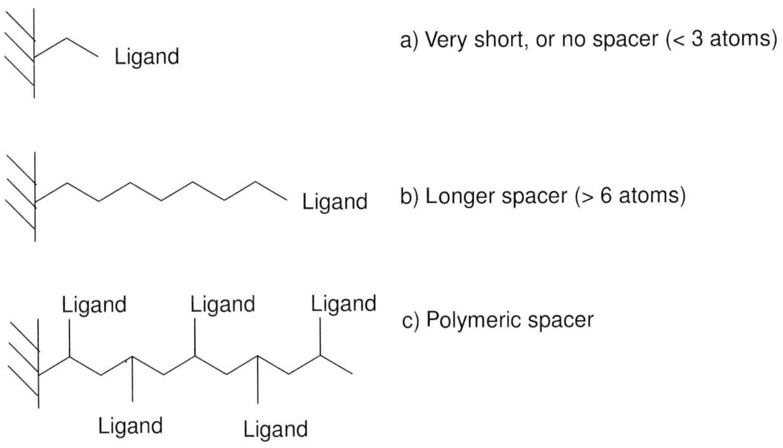

a) Very short, or no spacer (< 3 atoms)

b) Longer spacer (> 6 atoms)

c) Polymeric spacer

Fig. 6. Ligand arrangements at the particle surface.

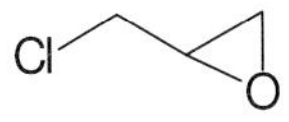

Fig. 7. Epichlorohydrin.

The properties of these ligand architectures can be described using the example of epoxy activation. Some resin suppliers use glycidylmethacrylate or allyl glycidyl ether as the constituent monomer for resin preparation. After preparation, the resin contains a large number of epoxy groups. Examples include Toyopearl Epoxy HW70 EC 5 mmol g^{-1} (Tosoh), Eupergit C ca. 2 mmol g^{-1} (Röhm Darmstadt), and SepaBeads FP-HP 3 mmol g^{-1} (Mitsubishi Chemicals). The epoxy content was measured by non-aqueous titration with perchloric acid by the method of Pribl [53]. The problem with such material is the inaccessibility of most of the epoxy groups. These resins are cheap and are used for enzymatic catalysis. However, they are not suitable for the preparation of resins with group-specific ligands. Some suppliers immobilize via coupled epichlorohydrin, but the three-carbon spacer is not long enough compared with the more normal six-carbon spacer.

Longer spacer molecules include 6-aminocaproic acid, 1,6-diaminohexane, diaminodipropylamine, 1,4-butanediol diglycidyl ether, and polyethylene glycol. The highest possible ligand density that can be obtained using such a spacer is lower compared to that in resins in which the monomers already bear the ligands. Most of the commercially available affinity resins are prepared with a spacer. The ligand density is sufficient for most applications.

A frequently used spacer is 1,4-butanediol diglycidyl ether [43]. The molecule is more hydrophilic than a six-carbon spacer molecule. A disadvantage is the high pH value required for ligand coupling.

A third alternative is offered by polymeric spacers. These can be prepared by a wide variety of methods. Examples are dextran-coated beads [54] or polyvinyl-

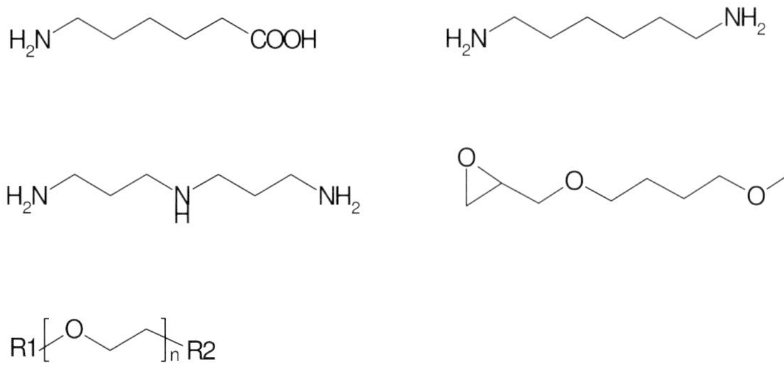

Fig. 8. Spacer molecules.

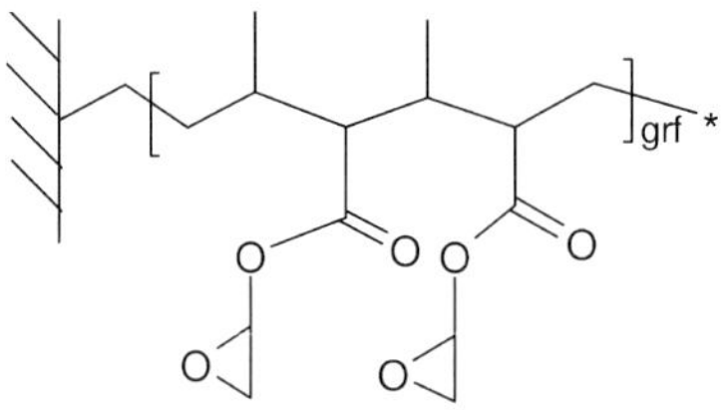

Fig. 9. Grafted glycidylmethacrylate.

alcohol-modified particles [55]. Covalently bound polymers can also be prepared by surface-initiated radical graft polymerization with suitable monomers. By using glycidylmethacrylate, the grafted polyglycidylmethacrylate has the following structure [56].

The possible ligand density with grafted glycidylmethacrylate is about $1-2$ mmol g^{-1}. The resultant ligands are all very accessible. This material is commercially available.

Epoxy resins with alternative spacer group chemistries have been used for the preparation of resins for immobilized metal-chelate affinity chromatography (IMAC). The following materials have been synthesized [56]:

1. Reaction of Toyopearl 650M with epichlorohydrin.
2. Reaction of Toyopearl 650M with 1,4-butanediol diglycidyl ether.
3. Ce^{4+}-initiated graft polymerization on Toyopearl 650M with glycidyl methacrylate.

Resins for IMAC were prepared by subsequent reaction with iminodiacetic acid in alkaline solution. The reactions were performed under optimized conditions to guarantee maximum ligand density. The results are shown in Table 2.

The ligand utilization of the metal chelate resin can be characterized by the ratio of ligand density to lysozyme binding capacity. This is shown in Fig. 10.

The longer spacer provides an enormous improvement in binding capacity compared to the small spacer. The polymeric coating is the best of all these methods. For smaller ligands, a spacer is always beneficial. Similar results have also been described by R. Hahn for Factor VIII-specific peptides [57].

Table 2. Comparison of different epoxy-activated resins.

Resin	Epoxy group density [mmol g^{-1}]	Ligand density IDA [mg mL^{-1}]	Lysozyme binding capacity Cu^{2+}-gel [mg mL^{-1}]
Epichlorohydrin (3-C-atom spacer)	0.9	4.4	5
1,4-Butanediol diglycidyl ether (12-atom spacer)	0.5	5.4	35
Glycidyl methacrylate (polymeric spacer)	1.3	5.2	50

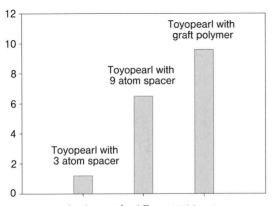

Lysozyme binding capacity/ligand density

Fig. 10. Ligand utilization for different IMAC resins.

2.5.6
Site-Directed Immobilization

Site-directed immobilization is the connection of ligand at a defined point to orientate the ligand optimally for specific interaction. This is shown schematically in Fig. 11 for an artificial peptide.

The peptide should contain lysine, the terminal amino group, and a cysteine. The recognition site for the target molecule is in the amino group containing region. Immobilization should not take place here. The trick is to immobilize the peptide only via the -SH group. This can be achieved using an epoxy-activated resin. To prevent co-immobilization onto the amino groups, the reaction is controlled at pH values below 8 and over a short duration (~1 h). This limits immobilization to the -SH groups as these react much faster than amino groups. In the absence of cysteine, the amino groups must be selectively protected, which increases the cost of the resin and makes the preparation more involved.

Site-directed immobilization on proteins is more complicated as more anchor groups are available. An elegant example is the immobilization of glycoproteins

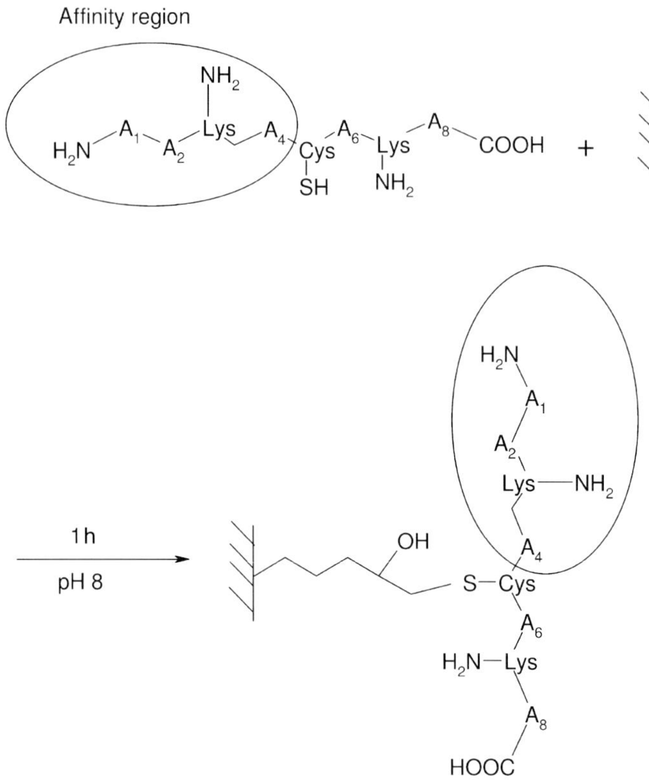

Fig. 11. Site-directed immobilization of a synthetic peptide.

through oxidation of the carbohydrate residues and subsequent immobilization by reductive amination [58].

Site-directed immobilization of proteins is controversial and much discussed. Wilchek and Taron showed that the binding of proteins to activated resins is mediated by only one or two surface-located residues, which suggests that site-directed immobilization is possible [59].

2.5.7
Non-Particulate Affinity Matrices

Affinity materials are not exclusively prepared from particles. Other porous materials can be used in a similar way. These include monoliths made from polymers [62], silica gel [61] or membranes [60]. Here, transport within the pores is driven by convection and is not limited by the slow diffusion processes as in particles. The dynamic protein-binding capacity is not dependent upon linear flow rate and high-volume throughput becomes possible. These materials will become important in the future, but at the present time most applications are based on particles in a packed column.

2.5.8
Affinity Purification

The primary objective of a chromatographic screening purification is to obtain sufficient purified protein for preliminary experiments. Affinity chromatography is the method of choice in this type of research, avoiding complex method development of multi-stage purification protocols.

After sample preparation, such as cell disruption, centrifugation, and/or filtration, a small amount of solution (a few mL) is applied to a small column.

It is advisable to have some ready-to-use commercially available affinity columns to hand, such as IMAC resins, *p*-aminobenzamidine, Reactive Blue, heparin, Protein A for antibody purification, as well as some activated resins such as tresyl, epoxy or *N*-hydroxysuccinimide.

Most resins can be stored for more than two years, but activated resins are hydrolytically labile and should be used immediately upon opening the container [63].

Usually, an affinity purification consists of the following steps:

1. Equilibration
2. Sample application
3. Wash
4. Elution
5. Regeneration

The scenario is shown schematically in Fig. 12.

The above procedure is described in detail in comprehensive textbooks and will not be covered here [9]. It is, however, important to establish optimal binding and elution conditions. The binding condition should be both selective and strong.

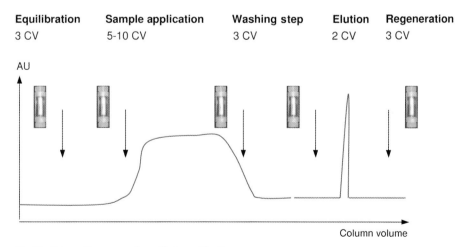

| Equilibration | Sample application | Washing step | Elution | Regeneration |
| 3 CV | 5-10 CV | 3 CV | 2 CV | 3 CV |

Fig. 12. Schematic course of an affinity purification.

This requires careful adjustment of ionic strength and pH, and often other reagents such as protease inhibitors, detergents or solvents must be added. The sample should be dissolved in loading buffer. The loading volume should be low (about 1 CV) for weakly binding compounds, but can be larger for strongly binding molecules. Low flow rates (< 30 cm h^{-1}) can be advantageous with low kinetic time constants for macromolecule interaction. In affinity chromatography, in contrast to other chromatography modes, selective or non-selective elution is possible. Selective elution conditions may be achieved by adding enzyme substrates or nucleotides (for the elution of alcohol dehydrogenase, the addition of NADP$^+$ is required [64]), or carbohydrates can be used for the elution of bound glyco-proteins from a lectin column [65]. Under such elution conditions, the substrates or co-factors compete with the proteins for attached ligands. Non-selective elution conditions include pH, salt, and temperature steps. In general, affinity methods present more parameters to adjust than other chromatography modes. This can increase method optimization time.

One way to minimize method development time is the use of small screening column sets. Such columns are simple and the user is not faced with technical problems such as column packing or clogged frits and can start immediately. Some resin suppliers provide sets of small columns (about 1 mL) filled with a range of materials including affinity resins. These are low-budget versions made from plastics and would not be appropriate for process development or up-scaling. They are more suitable for preliminary resin or mode screening. Such sets are available from Amersham Pharmacia (HITRAP) and from Tosoh (ToyoScreen). ToyoScreen columns are shown in Fig. 13.

An alternative way of optimizing affinity chromatography is the use of mathematical methods. A simple approach is described in the next paragraph.

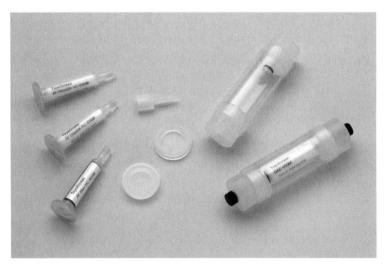

Fig. 13. Some 1 mL screening columns with cartridge (ToyoScreen, Tosoh).

2.5.9
Factorial Design for the Preparation of Affinity Resins

The large number of variables to be optimized necessitates the application of a mathematical method. The literature describes the preparation of a vast number of affinity resins. The resin should by well-suited for the intended application. But how good is good? Should it be possible to improve the purification by modifying the immobilization conditions (salt type and concentration, ligand density or pH)?

Often, these problems are solved by trial and error, changing one at a time. However, this is not efficient with such a large number of variables.

A more rigorous and effective approach is to employ factorial design methods [66]. Among the simplest of these are the Box–Hunter designs. Here, every variable can be adjusted to two levels; a lower level and a higher level (e.g., 6 or 8 for pH, or 1 mg mL^{-1} and 5 mg mL^{-1} for ligand concentration in the solution). The values can be transformed to +1 and –1, which makes the factorial design easier to work through. The equations for transformation are:

$$z_{pH} = \left(pH - \frac{pH_{high} + pH_{low}}{2}\right) \cdot \frac{1}{(pH_{high} - pH_{low})/2} \tag{2}$$

$$z_C = \left(C - \frac{C_{high} + C_{low}}{2}\right) \cdot \frac{1}{(C_{high} - C_{low})/2} \tag{3}$$

A similar equation can be derived for each separate parameter. The resulting data can be described in a scheme (Table 3) showing the binding capacity of an affinity material as a function of pH and ligand concentration. For example, in the first experiment both variables are adjusted to the low level so that immobilization proceeds at pH 6 with a ligand concentration of 1 mg mL^{-1}.

The effect of pH and ligand concentration can then be calculated using the following equations:

$$pH\ effect = \frac{1}{2} \cdot [(Y_3 - Y_1) + (Y_4 - Y_2)] \tag{4}$$

$$Ligand\ concentration\ effect = \frac{1}{2} \cdot [(Y_2 - Y_1) + (Y_4 - Y_3)] \tag{5}$$

Table 3. Box–Hunter design for two parameters.

pH	Ligand concentration C [mg mL^{-1}]	Interaction, ligand concentration, pH	Binding capacity Q [mg mL^{-1}]
–	–	+	Y_1
–	+	–	Y_2
+	–	–	Y_3
+	+	+	Y_4

In Table 3, the interaction between pH and ligand concentration is also considered. A low or high pH value may decrease the concentration of a peptide or protein ligand by hydrolysis.

$$\text{Interaction pH--ligand concentration} = \frac{1}{2} \cdot [(Y_1 - Y_2) + (Y_4 - Y_3)] \tag{6}$$

The signs of Y_i can be taken from Table 3. An arbitrary binding capacity value can be calculated from the summary equation:

$$Q = \frac{Y_1 + Y_2 + Y_3 + Y_4}{2} + \text{pH effect} \cdot z_{pH} + \text{ligand concentration effect} \tag{7}$$
$$\cdot z_C + \text{pH effect} \cdot \text{ligand concentration effect} \cdot z_{pH} \cdot z_C$$

The above example is a full factorial design, in which 2^2 experiments are required. With increasing number of parameters, the number of experiments increases accordingly; e.g., for six parameters, 2^6 experiments must be performed. If interactions are counted as a parameter, this is known as a fractional factorial design and the number of experiments is reduced. The following example illustrates how the factorial design method is employed for the development of an immobilization method.

Immobilization of soya bean trypsin inhibitor (STI) on Toyopearl AF Tresyl 650M

The capacities of prepared resins were determined by measurement of the binding capacity for porcine trypsin.

Immobilization conditions

STI (10 or 80 mg) was dissolved in a solution of 0.1 M phosphate buffer with 0.5 M sodium chloride for experiments at pH 6, and in 0.1 M carbonate buffer with 0.5 M sodium chloride for experiments at pH 8, and ca. 0.4 g of dried Toyopearl AF Tresyl resin was added. Unreacted tresyl groups were deactivated by reaction with 0.1 M Tris-HCl solution (10 mL, pH 8.5, 1 h).

Capacity determination

The resins were packed into a glass column (1 mL) and the capacity was determined by loading trypsin (porcine) (10 mg) in 50 mM Tris-HCl buffer, 0.5 M sodium chloride, and 20 mM $CaCl_2$ at pH 8.5. Elution was performed with 0.1 M acetic acid in 0.5 M sodium chloride at pH 3, and the capacity was determined by UV absorption at 280 nm.

The ligand concentration, pH value, temperature, and reaction time were varied. This model can be expanded to include more parameters. This might include qualitative elements such as two separate resin types.

Table 4 gives the upper and lower values chosen for the above parameters.

Table 4. Upper and lower parameter level.

	Concentration [mg mL^{-1}]	pH	Temperature [°C]	Reaction Time [h]
Upper value	20	8	25	16
Lower value	2.5	6	4	0.5

The choice of level depends on the experience of the user and is not decided by the program. In general, levels should not be chosen to be wide apart (e.g., pH 1 and pH 13) since the linearity of factorial design is limited to a narrow range.

The factorial method shown in Table 5 is a fractional factorial design. The unlikely four-component interaction vector (concentration, pH, temperature, reaction time) was covered by the pH vector, which reduces the number of experiments from 16 to 8. For further details of set-up and applications of these designs, the relevant technical literature should be consulted [67]. Many programs are available to help the user to develop a factorial design.

The effect of each variable can also be calculated (Table 6).

Table 5. Experimental matrix for the immobilization of soyabean trypsin inhibitor on Toyopearl AF Tresyl 650M.

Trial	Concentration [mg mL^{-1}]	Temperature [°C]	Reaction time [h]	pH	Capacity [mg mL^{-1}]
1	2.5	4	0.5	6	0.1
2	20	4	0.5	8	0.5
3	2.5	25	0.5	8	2.1
4	20	25	0.5	6	0.3
5	2.5	4	16	8	2.1
6	20	4	16	6	0.4
7	2.5	25	16	6	0.5
8	20	25	16	8	5.4

Table 6. Influence of the reaction conditions on binding capacity.

Variable	Effect
Concentration	0.45
Temperature	1.3
Reaction time	1.35
pH	1.95

Discussion

A change in concentration from 2.5 mg mL^{-1} to 20 mg mL^{-1} alters the binding capacity by only 0.45 mg mL^{-1}. This is a surprising and valuable result as it shows that it is possible to effectively immobilize the protein at low concentration. The pH has the greatest impact on the immobilization result. Changing the pH from 6 to 8 increases the binding capacity by 2 mg mL^{-1}.

The binding capacity can be described by a linear equation as a function of the above parameters as follows:

$$\text{Binding capacity} = 0.0514\,[\text{Conc.}] + 0.1238\,[\text{Temp.}] + 0.1742\,[\text{Time}] + 1.95\,[\text{pH}] - 16.0356$$

It is impossible to display the binding capacity as a function of all of the parameters, but it can be visualized as a function of two while the others remain constant. In Fig. 14, it is displayed as a function of reaction time and pH:

The advantage of this factorial design technique is the ability to examine the influence of all immobilization parameters independently as they are orthogonal. Figure 14 indicates that half of the maximum binding capacity is achieved after 30 min, which means that the Tresyl activation is a fast reaction.

Extrapolation from the linear model is dangerous. One might expect that capacity will continue to increase beyond pH 8, but in fact the binding capacity decreases above pH 9. This is not predicted by the available data.

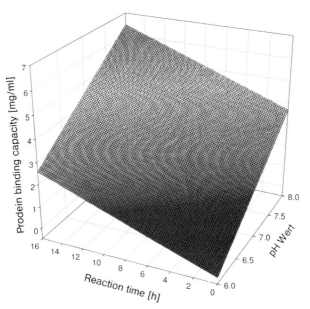

Fig. 14. Trypsin binding capacity of a trypsin inhibitor resin prepared with Toyopearl AF Tresyl as a function of pH and reaction time (temp. 25 °C, concentration: 20 mg mL^{-1}).

It is possible to obtain fast and reliable results with this factorial design procedure. An important condition for the successful application of factorial design methods is that the experimental plan must be rigorously adhered to, which requires repeated measurements. It cannot be denied that the linear designs are frequently too simple and that quadratic plans with more than two levels are required. Nevertheless, the factorial design technique is a valuable tool for minimizing the development time of affinity resins. The following (selected) experimental programs can be recommended: Cornerstone™ (www.brooks-pri.com), Design-Expert® (www.statease.com), Minitab™ (www.minitab.com), Statistica™ (www.statsoft.de), Systat™ (www.spssscience.com), Stavex™ (www.aicos.com).

2.5.10
Summary of Immobilization

1. Literature – and Database Search. If the same or a similar substance has already been immobilized on any support material, it may be successfully used for experiments.

2. Immobilization Method. For the immobilization of proteins, a selection of amino, hydroxy, mercapto, and carboxy group specific methods should be used. For example, cyanogen bromide, tresyl, *N*-hydroxysuccinimide, epoxy, divinylsulfone, cyanuric chloride, and carbodiimide. The most promising method can then be further optimized.

3. Buffer. If possible, use simple buffers; phosphate for pH values around 7 and carbonate or hydrogencarbonate for pH > 8. The stability of the ligand should be tested in the absence of the resin.

4. Wash Procedure. The resins should be carefully washed after immobilization. The wash procedure should include a sudden change in pH, e.g. from acidic to slightly alkaline (4 to 8) and a cleaning step with 1 M sodium chloride.

5. Removal of Residual Groups. Unreacted groups can be removed by reaction of the resins with 1 M Tris buffer, ethanolamine, or mercaptoethanol.

6. If capacity is too low and cannot be increased, try a resin with a spacer.

7. If all immobilization attempts have failed, contact a resin supplier to see if there are any other resin types available.

References

1 P. Cuatrecasas, M. Wilchek, C. B. Anfinsen, Selective enzyme purification by affinity chromatography, *Proc. Natl. Acad. Sci. USA* 61 (1968), 636–643.

2 In: J. C. Giddings, Unified Separation Science, Wiley, New York, 1992, pp. 131.

3 G. Skidmore, B. J. Horstmann, H. A. Chase, Modelling single-component protein adsorption to the cation exchanger Sepharose® FF, *J. Chromatogr. A*, 498 (1990), 113–128.

4 In: J.-C. Janson, L. Ryden, Protein Purification, VCH, Weinheim, 1989, pp. 283.

5 W. H. Scouten, Affinity Chromatography, Wiley, 1981, pp 141.

6 X. Wu, K. Haupt, M. A. Vijayalakshmi,
Separation of Immunoglobulin G by
high-performance pseudoaffinity
chromatography with immobilized
histidine, *J. Chromatogr.* 584 (1992), 35.

7 C. R. Lowe, P. D. G. Dean, Affinity
Chromatography, Wiley, New York, 1974.

8 P. D. G. Dean, F. A. Middle,
C. Longstaff, A. Bannister,
J. J. Dembinski, in Affinity
Chromatography and Biological
Recognition (Eds.: I. M. Chaiken,
M. Wilchek, I. Parikh), Academic Press,
Orlando, 1983, pp 433–443.

9 Affinity Chromatography: Principles
and Methods; Handbook published
by Amersham Bioscience, Uppsala,
Sweden.

10 M. D. Scawen, Dye Affinity Chromatography, *Anal. Proc.* 28 (1991), 143.

11 J. Porath, IMAC-immobilized metal
affinity based chromatography,
Trends Anal. Chem. 7 (1988), 254.

12 A. Liehme, P. Heegard, Thiophilic
adsorption chromatography:
the separation of serum proteins,
Anal. Biochem. 192(1) (1991), 64.

13 I. Blomquist, Characterization of
ECH-Lysine Sepharose FF – A New
Affinity Support, Poster at the Interlaken
Conference Series, 2003.

14 J. H. Pazur, Affinity chromatography for
the purification of lectins (a review),
J. Chromatogr. 215 (1981), 361.

15 Heparin-Sepharose® CL-6B for Affinity
Chromatography; Amersham
Bioscience, Uppsala, Sweden.

16 E. A. Bayer, M. Wilchek, *Meth. Biochem.
Anal.* 26 (1980), 1, Wiley, New York.

17 S. J. Gutcho, "Immobilized Enzymes:
Preparation and Engineering
Techniques", Noyes Data Corporation,
Park Ridge, New Jersey, 1974.

18 J. J. Langone, Protein A of
Staphylococcus aureus and related
immunoglobulin receptors produced
by streptococci and pneumonococcoi,
Adv. Immunol. 32 (1982), 157–252.

19 P. J. Hogg, D. J. Winzor, Studies on
lectin-carbohydrate interactions by
quantitative affinity chromatography
systems with galactose and ovalbumin

as saccharidic ligand, *Anal. Biochem.* 163
(1987), 331–338.

20 Z. Zhang, W. Zhu, T. Kodadek, Selection
and application of peptide-binding peptides, *Nature Biotechnology* 18 (2000), 71–74.

21 L. B. McGown, M. J. Joseph, J. B. Pitner,
G. P. Vonk, C. P. Linn, The nucleic acid
ligand. A new tool for molecular
recognition, *Anal. Chem.* 67 (1995),
663A–668A.

22 F. H. Arnold, Metal-affinity separations:
A new dimension in protein processing,
Bio/Technol. 9 (1991), 152.

23 Drug and Market Development
Application, Antibody Therapeutics,
Production, Clinical Trials and Strategic
Issues, 2001, November.

24 J. Lasch, F. Janowski, Leakage stability
of ligand-support conjugates under
operational conditions, *Enzyme Microb.
Technol.* 10 (1988), 312–314.

25 P. Santambien, P. Girot, I. Hulak,
E. Boschetti, Immunoenzymatic assay
of dyes currently used in affinity
chromatography for protein purification,
J. Chromatogr. 597 (1992), 315–322.

26 E.-K. Cabrera, M. Wilchek, Silica
containing primary hydroxyl groups for
high-performance affinity chromatography, *Anal. Biochem.* 159(2) (1986),
267–272.

27 M. D. Barcolod, R. el Rassi, High-
performance metal chelate interaction
chromatography of proteins with silica-
bound ethylenediamine-*N,N'*-diacetic
acid, *J. Chromatogr.* 512 (1990), 237–247.

28 In G. Sofer, L. Hagel, Handbook of
Process Chromatography, Academic
Press, 1997, pp. 53.

29 In "Seamless Scale-up in Process
Chromatography", Tosoh Bioscience
GmbH, 1996.

30 R. I. Fahrner, G. S. Blank, G. A. Zapata,
Expanded-bed Protein A chromatography of r-hMAb in comparison to
packed-bed chromatography,
J. Biotechnol. 75 (1999), 273–280.

31 F. Ahmed, K. D. Cole, Affinity liquid
chromatographic systems, *Sep. and
Purif. Meth.* 29(1) (2000), 1–25.

32 In B. Tieke, "Makromolekulare Chemie",
VCH, 1997, pp 272.

33 In "Gel Filtration", Amersham
Bioscience, Uppsala.

34 N. B. Afeyan, Flow through for the high-performance liquid chromatographic separation of biomolecules: perfusion chromatography, *J. Chromatogr. A.* 519 (1990), 1.

35 F. H. Arnold, H. W. Blansch, C. R. Wilke, Analysis of affinity separations, *Chem. Eng. J.* 30 (1985), B9–B23.

36 E. Müller, "Coupling Reactions" in "Protein Liquid Chromatography", *J. Chromatogr.*, Library vol. 61 (Ed.: M. Kastner), Elsevier, 2000.

37 R. Axen, S. Ernback, *Eur. J. Biochem.*, Chemical fixation of enzymes to cyanogen halide activated polysaccharide carriers, 18 (1971), 351–360.

38 R. Axen, J. Porath, S. Ernback, Chemical coupling of peptides and proteins to polysaccharides by means of cyanogen bromide, *Nature* 214 (1967), 1302–1303.

39 R. H. Allen, P. W. Majerus, Isolation of vitamin B12 binding proteins using affinity chromatography, *J. Biol. Chem.* 247 (1972), 7709.

40 M. Wilchek, T. Miron, Activation of Sepharose with *N,N*-disuccinimidyl carbonate, *Appl. Biochem. Biotechnol.* 11 (1985), 191.

41 K. Nilsson, K. Mosbach, Tresyl chloride-activated supports for enzyme immo-bilization, *Methods Enzymol.* 135 (1987), 65–78.

42 T. Zumbrink, A. Demiroglou, H. P. Jennissen, Analysis of affinity supports by ^{13}C CP/MAS NMR spectroscopy: application to carbonyl-diimidazole- and novel tresyl chloride-synthesized agarose and silica gels. *J. Mol. Rec.* 8 (1995), 363–373.

43 L. Sundberg, J. Porath, Preparation of adsorbents for biospecific affinity chromatography; I. Attachment of group-containing ligands to insoluble polymers by means of bifunctional oxiranes, *J. Chromatogr.* 90 (1974), 87.

44 M. Wilchek, R. Laurel, In Methods in Enzymology (Eds.: W. B. Jakoby, M. Wilchek), Academic Press, New York, 34 (1974), 475.

45 P. O. Larsson, M. Glad, L. Hansson, M. O. Mansson, S. Ohlson, K. Mosbach, High performance liquid chromato-graphy, *Adv. Chromatogr.* 22 (1983), 41–85.

46 F. Quadri, The reactive triazine dyes: their usefulness and limitations in protein purification, *Trends Biotechnol.* 3 (1985), 7.

47 J. Porath in Methods in Enzymology (Eds.: B. Jakoby, M. Wilchek), Academic Press, New York, 34 (1974), 13–30.

48 M. T. W. Hearn, In Methods in Enzymology (Ed.: K. Mosbach), Academic Press, New York, part B, 135 (1987), 102–117.

49 P. L. Colemann, M. M. Walker, D. S. Milbraith, D. M. Stauffer, J. K. Rassmussen, L. R. Krepski, S. M. Heilmann, Immobilization of Protein A at high density on azlactone-functional polymeric beads and their use in affinity chromatography, *J. Chromatogr.* 512 (1990), 345.

50 A. Tengblad, A comparative study of the binding of cartilage link protein and the hyaluronate-binding region of the cartilage proteoglycan to hyaluronate-substituted Sepharose gel, *Biochem. J.* 199 (1981), 297.

51 J. B. Wheatley, D. E. Schmidt Jr., Salt-induced immobilization of affinity ligands onto epoxide-activated supports, *J. Chromatogr. A* 849 (1999), 1–12.

52 P. O'Carra, S. Barry, T. Griffin, Spacer arms in affinity chromatography – the need of more rigorous approach, *Biochem. Soc. Trans.* 1 (1973), 289.

53 M. Pribl, Bestimmung von Epoxyendgruppen in modifizierten chromatographischen Sorbentien und Gelen, *Fresenius Z. Anal. Chem.* 303 (1980), 113–116.

54 A. L. Dawidowicz, D. Wasilewsak, S. Radkiewicz, *Chromatographie* 42 (1992), 49.

55 I. Y. Galev, B. Mattiasson, Shielding affinity chromatography, *Bio/Technology* 12 (1994), 1086.

56 E. Müller, Habilitationsschrift, Polymere Oberflächenbeschichtungen – eine Methode zur Herstellung von Träger-materialien für die Biochromatographie, Magdeburg, 2003.

57 R. Hahn, K. Amatschek, R. Necina, D. Josić, A. Jungbauer, Performance of affinity chromatography with peptide ligands: influence of spacer, matrix composition, and immobilization chemistry, *Int. J. Biochrom.* 5 (2000), 175–185.

58 D. J. O'Shannessy, W. L. Hoffmann, Site-directed immobilization of gluco-proteins on hydrazide-containing solid supports, *Biotechnol. Appl. Biochem.* 9 (1987), 488.

59 M. Wilchek, T. Miron, Oriented versus random protein immobilization, *J. Biochem. Biophys. Methods* 55 (2003), 67–70.

60 C. Charcosset, Protein A immun-oaffinity hollow fibre membranes for immunoglobulin G purification: experimental characterization, *Biotech. Bioeng.* 48 (1995), 415–427.

61 N. Ishizuka, Designing monolithic double-pore silica for high-speed liquid chromatography, *J. Chromatogr. A* 797 (1998), 133–137.

62 R. Hahn, K. Pfleger, E. Berger, A. Jungbauer, Direct immobilization of peptide ligands to accessible pore sites by conjugation with a placeholder molecule, *Anal. Chem.* 75(3) (2003), 543–548.

63 Homepage of Tosoh Bioscience, www.tosohbioscience.com

64 J. Aradi, A. Zsindely, A. Kiss, M. Szaboles, M. Schablik, Separation of Neurospora crassa myo-inositol-1-phosphate synthase from glucose-6-phosphate dehydrogenase by affinity chromatography, *Prep. Biochem.* 12(2) (1982), 137–151.

65 R. L. Esterday, I. M. Esterday, in Immobilized Biochemicals and Affinity Chromatography (Ed.: R. B. Dunlap), Plenum Press, New York, 1974, pp. 123–133.

66 P. D. Haaland, Experimental Design in Biotechnology, Marcel Dekker, New York, 1989.

67 G. Retzlaff, G. Rust, J. Waibel, Statistische Versuchsplanung, Weinheim, 1978.

2.6
Optimization of Enantiomer Separations in HPLC

Markus Juza

2.6.1
Introduction

In the previous chapters, the methods of RP and NP HPLC have been discussed in detail. These methods are ideally suited for qualitative and quantitative analytical investigations of complex mixtures of substances. However, they do not allow any statement about the chiral composition of the analyte. Therefore, another liquid chromatographic method is used for the determination of the enantiomeric composition of samples. Enantioselective HPLC, as the method is called, can be viewed as a complementary measurement for the characterization of analytes. This is not only true for samples of pharmaceutical origin, for which monitoring of achiral and enantiomeric purity are mandatory [1], but also for agrochemical products, flavors, and fragrances.

Enantioselective HPLC wrongly has the reputation of being a difficult to develop method that is very unreliable and mastered only by a few specialists. However, the multitude of commercially available stationary phases (CSPs) [2] with significantly different retention mechanisms makes an overview of the different factors contributing to a successful separation almost impossible.

Therefore, in the following paragraphs first the basic principles of enantio-selective HPLC are introduced, which show many similarities to those known from RP and NP chromatographies. After this introduction, the CSPs most frequently applied in industrial practice are briefly described. Deliberately, *not all* commercially available stationary phases are listed and phases that are only available in academia are not described. In the limited space of this chapter, no review of the historical developments that have led to today's CSPs is given.

2.6.2
Basic Principles of Enantioselective HPLC

HPLC is one of the most universal methods for determining the enantiomeric composition of substances and mixtures in a short time frame. Its application is not restricted to molecules in which chirality is based on a quaternary carbon atom with four different substituents; it can also be employed for compounds containing a chiral silicon, nitrogen, sulfur, or phosphorus atom. Likewise, asymmetric sulfoxides or aziridines, the chirality of which is based on a lone electron pair, can be separated. Chirality can also be traced back to helical structures, as in the case of polymers and proteins; to the existence of atropiso-merism, the hindered rotation about a single bond, as observed, for example, in the case of binaphthol, or to spiro compounds.

The fundamental principle of enantioselective HPLC is based on the formation of labile diastereomeric complexes of the two enantiomers with the chiral selector of the stationary phase [3]. The enantiomer that forms the less stable complex will be eluted earlier, while the enantiomer that forms the more stable complex will be eluted later. The ratio of the two retention factors k determines the separation factor for the enantioselectivity [4] α (Eq. 1) of a stationary phase for two enantiomers at a certain temperature and for a defined solvent composition.

$$\alpha = \frac{k_2}{k_1} \tag{1}$$

The determination of retention factors has already been discussed in Chapter 1.1. In enantioselective HPLC, the void time ("dead time") is measured by the injection of a non-adsorbed compound, e.g., tri-*tert*-butylbenzene or a low molecular weight alcohol. In practice, separation factors between 1.5 and 2.5 are commonly found, yet in some cases values of $\alpha > 20$ have been reported [5, 6]. A selectivity of $\alpha = 1$ shows that no enantiomer separation is possible under the conditions used; values below 1 are not defined since the separation factor is always referred to the second eluting enantiomer. It is not possible to determine the α value of a racemate when using a temperature or solvent gradient.

Very often, the aim is not to determine the selectivity of the column, but the enantiomeric purity of the sample under investigation. The enantiomeric purity, often wrongly called in laboratory jargon "chiral purity", is generally quoted as enantiomeric excess (% *ee*) (Eq. (2)).

$$\%ee = \frac{\text{main enantiomer} - \text{lesser enantiomer}}{\text{main enantiomer} + \text{lesser enantiomer}} \cdot 100 \tag{2}$$

Since the two enantiomers have identical physical properties, it is possible to calculate the *ee* directly from the relative area percentages or the peak areas (see Fig. 1). The determination of the *ee* requires a baseline resolution of the two enantiomers, and it is given in %.

The resolution of an enantiomer separation (R_s) is dominated by three different factors, the number of plates (N), the enantioselectivity α, and the retention (cf. Equation for R in Section 1.1.3). For practical purposes, Eq. (3) is often used [7], in which $w_{1/2}$ is the width of the peak at half height.

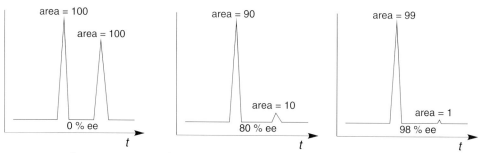

Fig. 1. Determination of *ee* with sufficient baseline separation.

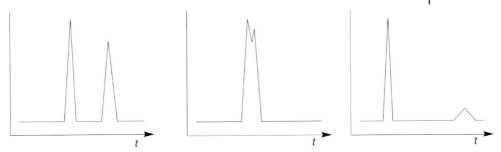

Fig. 2. Resolution of a separation. Left: baseline separation ($R_s > 1.5$) allows a precise calculation of the *ee*; middle: $R_s \ll 1.5$ (peak overlap) does not allow for a reliable calculation of the *ee*; right: $R_s \gg 1.5$ can complicate the calculation of *ee*, but might be useful for preparative purposes.

$$R_S = 1.18 \cdot \frac{(t_{R2} - t_{R1})}{(w_{1/2w_1} + w_{1/2_2})} \tag{3}$$

A precise calculation of the *ee* is only possible if $R_s > 1.5$. If R_s is much greater than 1.5, an inaccurate integration of the second peak can cause problems with the precision of the determined *ee* value. If R_s is much lower than 1.5, the peaks will be overlapped and a reliable analysis will no longer be possible. If only one injection has been made, it is generally not meaningful to quote more than one decimal place, e.g. **98.9%** *ee*.

2.6.2.1 Thermodynamic Fundamentals of Enantioselective HPLC

As discussed above, chiral recognition in HPLC (and all other kinds of enantioselective chromatography) relies on the formation of intermediate diastereomeric complexes. For a racemic sample (identical amounts of (*R*)- and (*S*)-enantiomers), the two association constants (K_S and K_R) have to be different in order to observe an enantiomer separation (see Fig. 3) [3]. Both enantiomers are in solution and can interact with the selector, which is attached to the silica. If, for example, the complexation constant is higher for the (*S*)-enantiomer, it is bound more strongly to the chiral stationary phase and is eluted later.

Since the capacity factor of a compound is connected via the phase ratio Φ with the association constant (Eqs. 4 and 5), the ratio of the retention factors *k* of the two enantiomers can be used to determine the ratio of the complexation constants (Eq. 6).

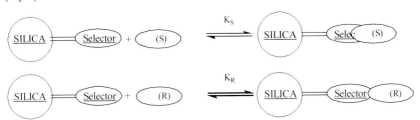

Fig. 3. Formation of intermediate diastereomeric complexes, $K_S > K_R$, in schematic view.

$$k_S = K_S \cdot \Phi \tag{4}$$

$$k_R = K_R \cdot \Phi \tag{5}$$

$$\frac{k_S}{k_R} = \frac{K_S}{K_R} = \alpha \tag{6}$$

This ratio allows estimation of the difference in free energy of the enantiomer separation, $-\Delta\Delta G_{S,R}$, and provides an insight into the order of magnitude of the interaction energies (see Fig. 4).

$$-\Delta\Delta G_{S,R} = -\Delta G_S - (-\Delta G_R) = RT \ln \frac{K_S}{K_R} = RT \ln \alpha \tag{7}$$

Whenever using Eq. (7), one should be aware that various different interactions between CSP and analyte can contribute to retention, which is caused partially by achiral reciprocation. An overview of the different forces involved in chiral recognition and their magnitudes is given in Table 1.

Table 1. Intermolecular interaction forces and their relative strengths (according to [8]).

Interaction forces	Lipophilicity	Relative strength [kJ mol⁻¹]
Ionic	Polarity	40
Via H-bonds		40
Without H-bonds		20
Ion-dipole		4–17
H-bonds		4–17
Van der Waals		4–17
Induced dipole		2–4
Aryl–aryl charge transfer		4–17
Hydrophobic	Hydrophobicity	4

2.6.2.2 Adsorption and Chiral Recognition

The interaction of a chiral analyte with a selector requires a temporary adsorption of the enantiomers on the surface of the stationary phase (see Fig. 4). Different models have been proposed to explain the reversible formation of energetically different quasi-diastereomeric molecular associates; however, none of them can explain all of the observed retention mechanisms. Association and dissociation constants for enantiomers interacting with a CSP can be different, since the rates at which the diastereomeric associates are formed can differ. Analogously, the dissociation rates can be different.

If the interaction of one of the enantiomers is only small, it will not be adsorbed and, in extreme cases, can be eluted within the void time. However, in most cases both enantiomers show some retention, requiring an equilibrium of adsorption and desorption. In some cases, one might observe that one or both of the enantiomers are strongly adsorbed and are not eluted during the run time of the

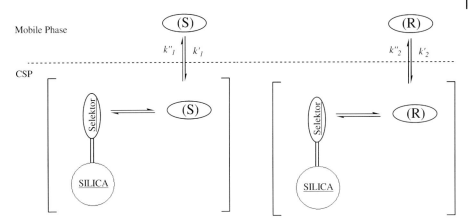

Fig. 4. The enantiomers form diastereomeric molecular associates with the CSP. Association constants k'_1 and k'_2 and dissociation constants k''_1 and k''_2 need not be identical.

chromatogram. Such behavior often necessitates a change of eluent system to purge the column of adsorbed components.

Almost all chiral stationary phases (see Section 2.6.3) have a heterogeneous surface. Besides the chiral selector molecule, the support (often based on silica) can also interact with the analyte and contribute to retention. In order to describe the adsorption, one often uses a modified Langmuir adsorption isotherm (Eq. 8) [9].

$$n_i = \lambda \cdot c_i + \frac{\bar{N}_i \cdot K_i \cdot c_i}{1 + \sum_{k=1}^{2} K_k \cdot c_k} \tag{8}$$

In this equation, n_i and c_i are the adsorbed and the mobile phase concentrations, λ is a dimensionless coefficient, and K_i is the equilibrium constant; the upper limit of n_i is given by $\bar{N}_i$. The linear term describes the interaction of achiral carrier particles with the analyte, while the Langmuir term describes the saturation behavior of the chiral selector. This behavior can be understood by inspection of Fig. 5, which shows in an idealized form the adsorption isotherms of two enantiomers.

The Henry constants H_i give the slope of the adsorption isotherms under dilute conditions, i.e. under analytical conditions (Eq. 9), for the two enantiomers.

$$n_i = H_i \cdot c_i \tag{9}$$

The ratio of the Henry constants is equivalent to the separation factor α (cf. Eq. 1).

Adsorption isotherms and Henry constants are only really of significance in preparative enantioselective HPLC and simulated moving bed (SMB) applications.

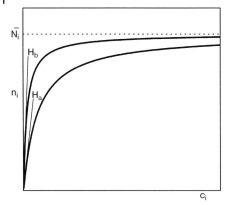

Fig. 5. Competitive adsorption isotherms, Henry constants, and saturation limit.

However, one should be aware that at high sample concentrations a competition of the two enantiomers for the chiral selector takes place and that CSPs are more easily overloaded than RP or NP phases since in almost all chiral stationary phases no more than 20% of the phase contributes to chiral recognition.

Overloading a chiral stationary phase leads to a fronting of the peaks, as the adsorption sites necessary for chiral recognition are occupied and the interaction between the analyte and CSP is not sufficient (see Fig. 6). In Fig. 6, typical retention behavior for a racemate at increasing sample volumes is shown. Both R_s and α become smaller at higher injection volumes and ultimately no longer allow for baseline resolution. The retention times become shorter.

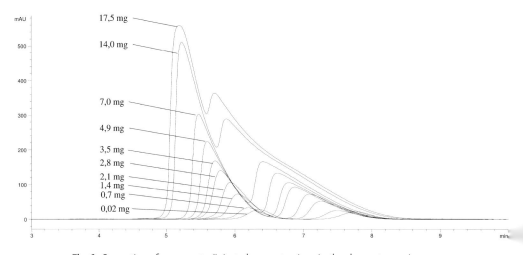

Fig. 6. Separation of a racemate (injected amounts given in the chromatogram) on Chiralcel OJ (250 × 4.6 mm); eluent: *n*-heptane/2-propanol (98 : 2, *v*/*v*); flow 1.0 mL min^{-1}; detection at 254 nm.

2.6.2.3 Differences to Reversed-Phase and Normal-Phase HPLC

The differences between achiral and enantioselective HPLC are smaller than is generally assumed. Enantioselective HPLC does not require special units or detectors, and all physico-chemical fundamentals of the chromatography are identical. However, it should be noted that many of the described enantioselective HPLC separations are performed under normal-phase conditions. Applications of the popular water/methanol or water/acetonitrile gradients are in the minority, since many CSPs are not compatible with these gradient systems or do not show any retention.

2.6.2.4 Principles for Optimization of Enantioselective HPLC Separations

The fundamentals introduced in the previous sections allow some conclusions regarding optimization of enantioselective HPLC separations.

Eq. (1) implies that in order to obtain a sufficient separation factor α, a sufficient difference in the retention factors must be obtained. Therefore, gradients are rarely applied in enantioselective HPLC, which would speed up isocratic separations.

The equation for R_S (see Section 1.1.3) shows that for the optimization of an enantioselective separation, besides the separation factor α and the retention factor k, a sufficient number of plates N must also be available if baseline resolution is required. For this reason, for some years there has been a trend in enantioselective HPLC towards smaller particle sizes. Today, many CSPs are available as 5 μm materials.

Eq. (7) allows a statement to be made about the temperature dependence of α (Eq. 10) by applying the Gibbs–Helmholz equation ($G = H - TS$) [3].

$$\ln \alpha = -\frac{\Delta \Delta H}{RT} + \frac{\Delta \Delta S}{R} \tag{10}$$

Almost all enantiomer separations show a lower separation factor at higher temperatures; when the temperature is increased further, enantiomer separation can be completely suppressed. This means that the entropy term in the above equation approaches the value of the enthalpy term with rising temperature, and that at a certain temperature entropy and enthalpy cancel one another out ($\ln \alpha = 0$). For this reason, temperature increase and gradients can be neglected for the optimization of enantioselective HPLC.

2.6.3
Selectors and Stationary Phases

Chromatographic enantiomer separation can be performed in two different ways. The so-called direct method is based, as described above (Section 2.6.2.4), on the formation of a diastereomeric transient complex between the chiral selector and the analyte. The selector can be present in various forms in the column, either linked directly to the silica particle, or as a coating on the carrier particles. However, it can also be dissolved in the mobile phase as an additive. In contrast, the indirect

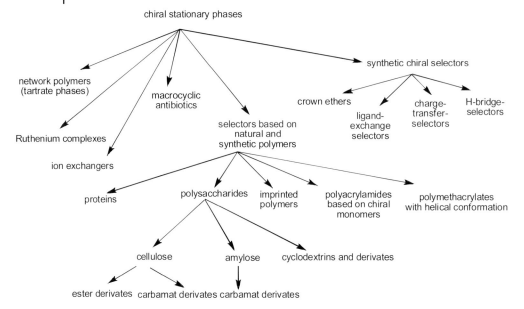

Fig. 7. Overview of the most commonly used chiral stationary phases.

method is based on chemical reaction of the enantiomers with an enantiomerically pure derivatizing reagent. After derivatization, two diastereomeric covalent compounds are formed, which can be resolved on an achiral stationary phase. The advantages and disadvantages of the indirect method are discussed in Section 2.6.7.

Figure 7 gives an overview of the most important classes of chiral stationary phases that can be employed for direct enantiomer separation. It should be noted that not all of these phases are currently commercially available.

According to their chemical structures, CSPs can be divided into three different groups. A multitude of chiral stationary phases is derived from (modified) natural or synthetic polymers, e.g., the polysaccharides, proteins or polyacrylamides. A second type of selectors is based on large chiral ring systems, such as cyclodextrins, macrocyclic antibiotics, and crown ethers. The last group comprises molecules of small and medium size, such as amino acids and their derivatives, alkaloids, and fully synthetic selectors.

Chiral stationary phases can exist in different forms [10] (see Fig. 8). Some selectors can be used as particulate phase materials, such as polymeric cellulose triacetate. Polymeric cellulose and amylose derivatives are often coated onto silica carrier particles so that only 20% of the CSP consists of the chiral selector. This combination of stationary phase and chiral polymer combines good chromatographic properties (due to the homogeneous particle size distribution) with a high density of chiral adsorption sites in the polysaccharide derivatives. Another approach is selected for the so-called brush-type CSPs. In these, the chiral selector is covalently bound to the surface of the silica particles. These phases show high chemical inertness and allow the use of a multitude of different mobile phases.

Table 2. Structures and properties of some commercially available chiral stationary phases.

Structure of the CSP	Chiral selector	Compatible solvents[a]	Trade name
(polysaccharide structure) R = CH$_3$	Microcrystalline cellulosetriacetate	Ethanol	MCTA/CTA-I, Chiralcel CA-1
R = (phenyl)	Cellulosetribenzoate	Ethanol	Chiralcel OB
R = (NH–phenyl)	Cellulose tris(phenyl-carbamate)	Alkane/alcohols (100/0–0/100), Alcohols, Acetonitrile	Chiralcel OJ
R = (NH–3,5-dimethylphenyl)	Cellulose tris(3,5-dimethyl-phenylcarbamate)	Alkane/alcohols (100/0–0/100), Alcohols, Acetonitrile	Chiralcel OD, Chiralpak IB
(amylose structure) R = (NH–3,5-dimethylphenyl)	Amylose tris(3,5-dimethyl-phenylcarbamate)	Alkane/alcohols (100/0–0/100), Alcohols, Acetonitrile	Chiralpak AD, Chiralpak IA
R = (CH$_3$/NH–phenyl, (S)-methylbenzyl)	Amylose tris[(S)-methyl-benzylcarbamate]		Chiralpak AS-V

a) Examples, not exhaustive.

Table 2. (continued)

Structure of the CSP	Chiral selector	Compatible solvents [a]	Trade name
R = C$_4$H$_9$ R =	O,O'-bis(4-tert-butylbenzoyl)-N,N'-diallyl-L-tartardiamide	Alkane/alcohols Alkane/tetrahydrofuran Alkane/dioxan Chlorinated solvents	Kromasil CHI-TBB
	O,O'-bis(3,5-dimethylbenzoyl)-N,N'-diallyl-L-tartardiamide		Kromasil CHI-DMB
	(S,S)-4-(3,5-dinitrobenzamido)-tetrahydrophenanthrene	Alkane/alcohols Alkane/chlorinated solvents Acetonitrile/ethyl acetate	(S,S)-Whelk-O-2 (analogous: O-1)
	(R,R)-N-(3,5-dinitrobenzoyl)-diphenyl-ethylenediamine		Ulmo

Table 2. (continued)

Structure of the CSP	Chiral selector	Compatible solvents[a]	Trade name
	Vancomycin	Alcohols/acetic Acid/triethylamine Tetrahydrofuran/water Alkane/alcohols	Chirobiotic V (analog: Chirobiotic R, T)

[a] Examples, not exhaustive.

Table 2. (continued)

Structure of the CSP	Chiral selector	Compatible solvents[a]	Trade name
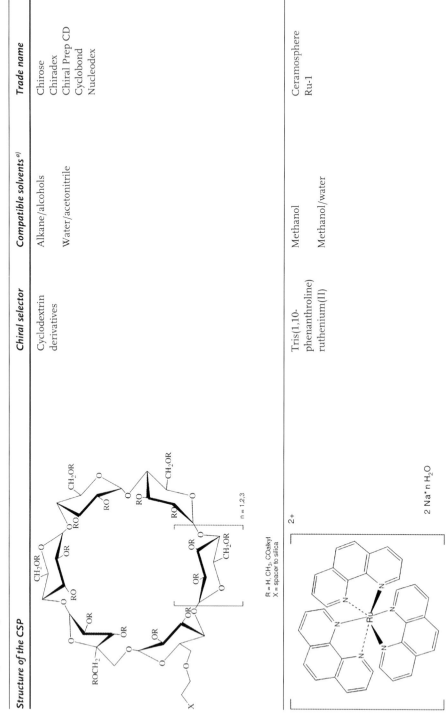	Cyclodextrin derivatives	Alkane/alcohols Water/acetonitrile	Chirose Chiradex Chiral Prep CD Cyclobond Nucleodex
	Tris(1,10-phenanthroline) ruthenium(II)	Methanol Methanol/water	Ceramosphere Ru-1

R = H, CH₃, COalkyl
X = spacer to silica

n = 1,2,3

2+

2 Na⁺ n H₂O

Table 2. (continued)

Structure of the CSP	Chiral selector	Compatible solvents[a]	Trade name
	Poly[(S)-N-acryloylphenyl-alanine ethyl ester]	Alkane/alcohols Alkane/cyclic ether Dichloromethane Toluene/ethyl acetate	Chiraspher
	ʟ-proline bound to polyacrylamide	Copper(II) acetate buffer/ acetic acid	Chirosolve-pro
	Chinin- and chinidin carbamate	Methanol/ammonium acetate Water/organic solvents	Chiralpak QD-AX Chiralpak QN-AX

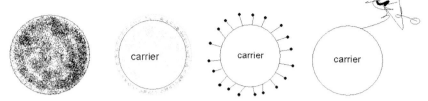

Fig. 8. From left to right: schematic representation of pure and coated polysaccharide phases, brush-type phase, protein phase.

In a similar way, proteins can also be bound to silica; however, the low amount of chiral selector significantly restricts the loading capacity of these phases.

For chiral stationary phases, spherical silica particles of high quality are used almost exclusively, which, in chromatographic terms, exhibit far superior performance to irregular silica.

In Table 2, the structures and properties of some frequently employed CSPs are summarized. It should be noted that this table represents only a very small part of more than 1300 known chiral stationary phases. Nevertheless, these phases have facilitated more than 95% of all known racemate resolutions by HPLC.

The selection of the mobile phase plays a significant role in enantioselective HPLC. Four different types of eluents can be distinguished. Besides the well-known reversed-phase and normal-phase modes, for enantiomer separations so-called polar organic and polar ionic eluents are also used (see Table 3).

Table 3. Mobile phase modes in enantioselective HPLC.

Mode	*Normal phase*	*RP*	*Polar organic*	*Polar ionic*
Solvents	Organic solvents	Water and organic modifiers	Polar alcohols	Polar alcohols and volatile buffers (acids and bases)
Compatible CSPs	Daicel, Whelk	Daicel, Whelk and others	Daicel cyclodextrin	Chirobiotic

In the following sections, method selection and optimization of separations for some of the CSPs described in Table 2 are discussed.

2.6.4
Method Selection and Optimization

Method selection and optimization of separations might require different conditions depending on the selector and the stationary phase. The following paragraphs give an overview of the main features of commonly employed CSPs. It is recommended that before using a new CSP the information supplied by the manufacturer should always be studied carefully in order to avoid inadvertent destruction of the CSP by selecting an unsuitable eluent.

2.6.4.1 Cellulose and Amylose Derivatives

Natural products such as cellulose and amylose have proved to be of great versatility for enantiomer separations. Though poor chiral selectors in their native state, they become highly effective when their hydroxyl functions are derivatized, particularly with aromatic moieties through ester or carbamate linkages, and they are suitably immobilized on silica supports [11].

These materials are composed of small chiral units that are regularly repeated along the polymeric chain, hence the density of active sites capable of chiral recognition is very high, and this results in a high loading capacity. Microcrystalline cellulose triacetate and cellulose tribenzoate have, in some instances, been used for chiral separations as neat, non-coated particles. However, they are inconvenient to pack, are of limited chemical stability, and show low efficiency.

To overcome these limitations, the Japanese company Daicel markets a family of derivatized cellulose and derivatized amylose CSPs. These chiral polymers are not covalently attached to the support, but are coated on a silanized, wide-pore silica gel. Consequently, some caution has to be exercised when choosing mobile phase solvents for use with such CSPs. The stationary phases derived from cellulose are named "Chiralcel", and those derived from amylose are termed "Chiralpak". Table 2 gives an overview of the ester and carbamate derivatives, which are the most versatile phases. These CSPs are most often used in the normal-phase mode (i.e., with solvents such as hexane and ethanol). Utilizing polar solvents such as alcohols or acetonitrile as eluent systems for enantioselective separations, rather than traditional alcohol/alkane mixtures, is advantageous for preparative separations since many pharmaceutical compounds are more soluble in the former systems. This increased solubility in polar solvents allows for higher mass loads and higher productivity. Moreover, shorter retention times are frequently observed when employing polar solvents. Furthermore, phase variants are available, which can be used under aqueous and gradient conditions. The specific advantages of these stationary phases lie in their loading capacity and their ability to resolve a great range of racemates, e.g., alcohols, carbonyl compounds, lactones, aromatic compounds, and many more.

Table 4. Systematic method development protocol for Daicel stationary phases.

Eluent or eluent mixture (composition in vol%)	Chiralpak AD	Chiralpak AS	Chiralcel OD	Chiralcel OJ
n-heptane/ethanol (90 : 10)	x	x	x	x
n-heptane/ethanol (80 : 20)	x	x	x	x
n-heptane/2-propanol (90 : 10)	x	x	x	x
n-heptane/2-propanol (80 : 20)	x	x	x	x
Acetonitrile/2-propanol (90 : 10)	x	x		
Acetonitrile	x	x	x	x
Methanol	x	x	x	x
Ethanol/methanol (50 : 50)	x	x	x	x
Ethanol	x	x	x	x

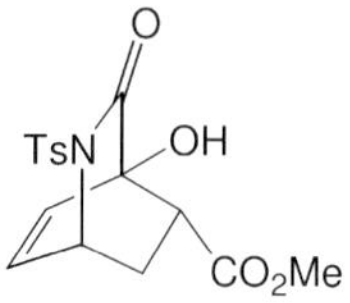

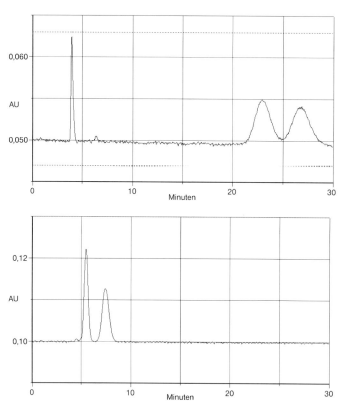

Fig. 9. Racemic bicyclo[2.2.2]octene derivative (**1**).

Basic compounds, such as primary and secondary amines, and acidic compounds, such as acids or phenols, have a tendency to show a tailing of peaks. This tailing can be efficiently suppressed by adding a basic (DEA, TEA) or acidic (TFA, acetic acid) additive (0.1–0.5%) to the eluent. Daicel stationary phases are available as 5, 10, and 20 μm materials. The 20 μm particle size is most often used for preparative separations (see Section 2.6.6).

In industrial practice a systematic method development protocol using different stationary and mobile phases is often used [12, 13]. An example of such a protocol is outlined in Table 4.

The protocol summarized in Table 4 allows the separation of 70–80% of all racemates by using four different columns and nine eluent combinations.

Fig. 10. Enantiomer separation of racemate **1** on Chiralpak AS.
Top: eluent: isohexane/ethanol (85 : 15, *v/v*); bottom: eluent: ethanol.
Column dimensions 250 mm × 4.6 mm *i.d.*; flow: 1.00 mL min^{-1}; detection: 254 nm.

An example is given in Fig. 10. The racemic [2.2.2]octene derivative (**1**, Fig. 9) can be separated using Chiralpak AS as stationary phase and isohexane/ethanol (85 : 15, *v/v*) as mobile phase (see Fig. 10, left). A systematic screening of stationary and mobile phase showed, however, that by using pure ethanol as eluent the retention times can be reduced significantly (Fig. 10, right). The optimization of the separation conditions thus led to a threefold higher sample throughput.

2.6.4.2 Immobilized Cellulose and Amylose Derivatives

The main disadvantage of coated polysaccharide CSPs is their very limited stability against chlorinated solvents, ethers and various other eluents. This limitation was overcome by Francotte and co-workers[1] a decade ago and has resulted in the recently introduced immobilized versions of Chiralpak AD and Chiralcel OD, which have been named Chiralpak IA and IB (produced by Daicel). They contain the same selectors (amylose and cellulose 3,5-dimethylphenyl carbamate), which are no longer coated, but connected to the silica particle via a proprietary immobilisation technology. The columns can be used with all types of miscible organic solvents, progressing from the standard mobile phases compatible with the coated-type polysaccharide-derived CSPs (cf. Table 4) to mobile phases containing chloroform ($CHCl_3$), ethyl acetate (EtOAc), tetrahydrofuran (THF), tert-butylmethyl ether (TBME) and toluene, among others.

The option to use the "non-standard" solvents in the mobile phase opens up new possibilities for unique selectivities. There are also no limitations on the sample injection solvent with the immobilized phases. Solvents such as dichloromethane, acetone, THF, dimethylformamide (DMF) or even dimethyl sulfoxide (DMSO) can be safely and effectively used as sample diluents. This is highly beneficial for the automation of injections for samples coming directly from various synthetic media, *e.g.* from lab-automation experiments. On the other side the immobilisation changes the selectivity of the stationary phase and compounds not resolved on the coated CSPs can be resolved by the immobilized CSP.

Two groups of solvents can be identified in regard to immobilized polysaccharide CSPs. Group A includes the solvents commonly used for coated polysaccharide phase, such as alkanes (*n*-hexane, *iso*-hexane, *n*-heptane), the lower alcohols and acetonitrile. Group B contains solvents employed in classical normal phase chromatography and which allow adjusting the elution strength and polarity of the eluent. In Table 5 some typical solvent combinations from groups A and B are summarized that allow for an efficient method development.

In order to modify retention times and to improve peak shapes the proportion of alkane can be increased or decreased. In many cases it is useful to add a small amount of alcohol to enhance selectivity and/or reduce retention times and/or improve the method. Addition of alcohols to certain mixtures, such as alkane/dichloromethane will strongly modify separations. In the case of TBME (last column in Table 5) EtOH can be substitued by another alcohol or THF.

---

[1] E. Francotte, in: Preparative Enantioselective Chromatography, Ed. G. Cox, Blackwell Publishing, Oxford, 2005, p. 69–70.

Table 5 Typical mobile phases for immobilized polysaccharide phases

Group A	Alkane	Alkane	Alkane	Alkane	Alkane	Alkane	Ethanol
Group B	CHCl$_3$	EtOAc	THF	CH$_2$Cl$_2$	toluene	acetone	TBME
Typical starting conditions (v : v)	50 : 50	60 : 40	70 : 30	60 : 40	30 : 70	75 : 25	2 : 98
Optimization range	75 : 25 to 0 : 100	80 : 20 to 0 : 100	90 : 10 to 50 : 50	75 : 25 to 0 : 100	70 : 30 to 0 : 100	90 : 10 to 60 : 40	20 : 80 to 0 : 100

Using the following four solvent mixtures a straightforward approach for the screening of these phases can be made:

a) Alkane/THF 70 : 30
b) MtBE/EtOH 98 : 2
c) Alkane/Chloroform 50 : 50
d) Alkane/Ethyl acetate 60 : 40

With these four solvent combinations, chances are good to identify good separations, without investing too much time in looking for other possibilities. The immobilized CSPs offer of course much more possibilities, but in view of time and budget constraints frequently encountered in an industrial environment it is essential to make a compromise between the desire to find the "optimal conditions" and the objective to deliver the HPLC separation when it is needed.

2.6.4.3 Stationary Phases Derived from Tartaric Acid

Immobilized network polymers based on *O,O'*-diaroyl derivatives of (+)-(2*R*,3*R*)-*N,N'*-diallyl tartrate diamide (see Table 3) represent another class of synthetic CSPs recently obtained from a readily available C$_2$-symmetric starting material from the chiral pool. The covalent bonding to functionalized silica imparts these tartrate phases with high stability towards mobile phase additives such as alcohols, ethers, and chlorinated solvents [12]. Immobilization and cross-linking combine the efficiency and loading capacity of a "brush-type" structure with the recognition of chiral polymers. In aprotic solvents, chiral recognition is achieved mainly through hydrogen bonds and π-π interactions of the aromatic esters. In practical applications, these phases show limited potential since by using polar solvents the formation of hydrogen bonds is hindered. In the optimization of a separation, a non-polar "base solvent" such as heptane or *tert*-butyl methyl ether (TBME) is combined with a modifier. The nature of the modifier is oriented towards polarity of the analyte (see Table 6). Acidic and basic additives, such as acetic acid, formic acid or triethylamine (0.01–0.1%), improve peak form and enantioselectivity.

2.6.4.4 π-Acidic and π-Basic Stationary Phases

The best known CSPs of the π-acidic and π-basic types are the "Pirkle phases", which are classified into π-acceptor and π-donor phases. The most frequently

Table 6. Systematic method development protocol for chiral tartrate phases.

Base solvent	Modifiers		
	Very polar and basic analytes	Semi-polar analytes	Hydrophobic analytes
Heptane or TBME			
	2-propanol	Ethyl acetate	
	Ethyl acetate	Tetrahydrofuran	TBME
		Dioxane	Toluene
		Acetone	

used π-acceptor phases are derived from the amino acids phenylglycine (DNBPG) or leucine (DNBLeu), covalently or ionically bonded to 3-aminopropyl silica gel. The π-donor phases (e.g., with naphthylalanine as the chiral moiety) allow the separation of π-acceptor analytes by virtue of the "reciprocity concept" [13].

Usually, these CSPs are used under normal-phase conditions, requiring a non-polar solvent such as hexane to enable π-electron sharing or anchoring of the analyte to the CSP. In the reversed-phase mode, anchoring involves either strong hydrogen bonding or ion exchange. Pirkle phases exhibit high efficiency and a significant loadability, but are intrinsically limited to π-interactions for chiral resolution.

A significant advantage in comparison with selectors derived from the chiral pool is the option of obtaining both enantiomers of the selector through stereo-selective synthesis. When one changes the chirality of a selector, e.g. from (R,R) to (S,S), the elution order of the enantiomers of the analyte is changed, which can sometimes be advantageous in the determination of *ee*.

A multitude of racemate resolutions can be achieved on the Whelk-O 1 phase, which represents a combination of π-electron donor and π-electron acceptor (see

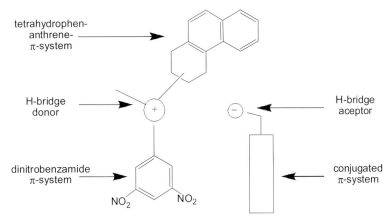

tetrahydrophen-
anthrene-
π-system

H-bridge
donor

H-bridge
aceptor

dinitrobenzamide
π-system

conjugated
π-system

NO₂   NO₂

Fig. 11. Systematic representation of the functional groups of the Whelk-O 1 phase and an analyte. The stereoselective interaction may rely on π-π interactions and/or hydrogen bonds and dipole-dipole interactions.

Table 2 and Fig. 11). In addition, due to the covalent nature of Whelk-O 1, this CSP is compatible with all commonly used mobile phases, including aqueous systems.

The so-called ULMO CSP is based on a 3,5-dinitrobenzoyl derivative of diphenylethylenediamine covalently bound to silica [14].

The optimization of enantiomer separations follows a similar approach as shown in Table 6. Under normal-phase conditions, Pirkle-type phases require an apolar "base solvent", which is modified in terms of its elution strength by adding one or more additives. For example, if at the selected solvent composition the analyte elutes within the dead time, the amount of polar additive must be reduced, e.g. from 50% to 25%. If under these conditions no partial separation can be observed, it is recommended to change to another Pirkle phase.

2.6.4.5 Macrocyclic Selectors, Cyclodextrins, and Antibiotics

A broad range of macrocyclic compounds can be used as chiral selectors for enantioselective HPLC. Besides synthetic crown ethers, derivatized cyclodextrins and cyclic antibiotics are also used as chiral stationary phases. Enantiomer separation employing these compounds is often based on host–guest interactions [15], whereby the cyclic molecules form an inclusion compound or an association complex with the analyte.

Chirally substituted crown ethers (e.g., Crownpack CR$^+$ produced by Daicel) are useful for enantiomer separations of primary amines and amino acids under reversed-phase conditions.

Cyclic oligomers of α-(1,4)-linked D-glucose molecules (6, 7, 8, etc.) are a homologous series of macromolecules of different size, known as cyclodextrins. Depending on ring size they are referred to as α, β, γ, etc. As with the polysaccharides, underivatized cyclodextrins are not useful in HPLC; however, by conversion of the hydroxyl functions to ethers, esters or carbamates, chemically stable stationary phases can be obtained [16]. By this methodology, cyclodextrins can also be covalently bound to silica. The inner surface of the conical rings is relatively hydrophobic and allows the separation of apolar analytes. Phases based on cyclodextrins are marketed under the names Chirose by Chiralsep, Nucleodex by Macherey & Nagel, and ChiralPrep by YMC (see Table 2). Almost all commercial products are based on β-cyclodextrin, which contains seven glucose subunits.

Cyclodextrin phases can be operated in RP and normal-phase modes, as well as in the polar ionic mode (see Table 3). Aqueous eluent systems are better suited for the formation of inclusion complexes of apolar or aromatic moieties in the inner cavity of the macrocycle. The selectivity of the cyclodextrins depends on ring size, pH, ionic strength, the nature and concentration of modifier, and temperature.

When cyclodextrins are used in normal-phase mode, chiral recognition is based on the interaction of the outer functional groups through hydrogen bonds, dipole-dipole interactions, and π-π interactions. The normal-phase mode is used for analytes that show a low solubility in aqueous eluents; here, mostly alkane/alcohol mixtures are used.

In the polar mode, the mobile phase is very often acetonitrile containing some protic additives, such as methanol, acetic acid or triethylamine. The formation of inclusion complexes is suppressed by acetonitrile, while simultaneously the formation of hydrogen bonds to the secondary hydroxyl functions at the outer rim of the macrocycle is enhanced. Most compounds separated in the polar organic mode bear amino functions, or in a few cases carboxylic or phenolic groups. Separations can be optimized by the addition of acetic acid or triethylamine, ranging from 0.002% to 2.5%, while the eluent consists of 85–100% acetonitrile and 15–0% methanol. An increase in the percentage of methanol will lead to a decrease in retention times for strongly adsorbed analytes.

Macrocyclic glycopeptide antibiotics such as Vancomycin, Teichoplanin, and Ristocecin A can be covalently bound via hydroxy or amino groups to silica for use as CSPs [17]. All glycopeptides contain an aglycone moiety that is able to form a cavity, at the rim of which carbohydrates are bound. Furthermore, this group of chiral selectors possesses a large number of stereogenic centers and functional groups allowing for ionic and other interactions between the phase and analyte. The phases are sold under the trade names Chirobiotic V, R, and T (by Astec). These CSPs are most often used in polar ionic mode (see Table 3), for which methanol/acetic acid/triethylamine mixtures (100 : 0.1 : 0.1, $v/v/v$) are often successful. For optimization, the ratio between acid and base (or their volatile salts, NH_4OAc, NH_4OCF_3) is varied from 4 : 1 to 1 : 4.

Figure 12 shows the optimization sequence for an enantioselective separation of a racemic ester and its free acid formed by hydrolysis. The structure of the substance cannot be presented due to pending patent applications.

The enantiomer separation of the ester can be performed on the Daicel stationary phase Chiralpak AD using acetonitrile/isopropanol/diethylamine (95 : 5 : 0.2, $v/v/v$) as eluent; however, the free acid (racemic) co-elutes without even partial separation along with the second enantiomer of the ester (see Fig. 12a). A preliminary screening of the Chirobiotic phases in the polar ionic mode (methanol/acetic acid/triethylamine, 100 : 0.1 : 0.2, $v/v/v$) showed a separation of the racemic ester on Chirobiotic V (see Fig. 12b), while Chirobiotic R and T could not resolve the ester. The racemate of the free acid partially co-eluted with the ester (see Fig. 12c). By increasing the ratio of acetic acid/triethylamine from 1 : 2 to 1 : 8 and decreasing the operation temperature to 10 °C, the enantiomer separation of the ester was improved (see Fig. 12d). A further increase of the acid/base ratio as well as a decrease in temperature did not result in an improved enantiomer separation (see Fig. 12e). By coupling two Chirobiotic V columns and adding tetrahydrofuran to the eluent (see Fig. 12f) the separation of the free acid from the main product was enhanced, allowing quantitative determination of the enantiomers of the free acid besides the racemic ester.

Special advantages of the Chirobiotic phases are their capacity to separate salts of analytes directly into enantiomers and their suitability for LC-MS hyphenation. The phases can also be used under RP conditions, e.g. with methanol/water/triethylamine mixtures (25 : 75 : 0.1, $v/v/v$) at pH 6.0 or in methanol/water gradients. For normal-phase conditions, a mixture of hexane/ethanol (40 : 60, v/v) has proven to be useful.

448 2 *Characteristics of Optimization in Individual HPLC Modes*

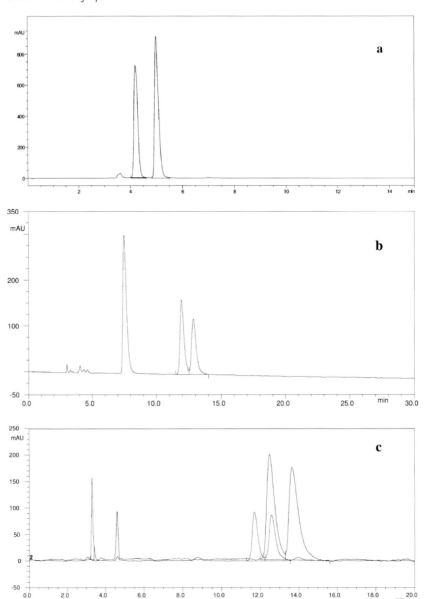

Fig. 12. (a–c) Separation of a racemic ester and the free racemic acid
in various compositions; flow: 1.0 mL min⁻¹; detection: 270 nm;
column dimensions: 250 mm × 4.6 mm *i.d.*;

(a) Chiralpak AD 10 μm, acetonitrile/isopropanol/diethylamine
(95 : 5 : 0.2, *v/v/v*), *T* = 23 °C;

(b) Chirobiotic V, 5 μm, methanol/acetic acid/triethylamine
(100 : 0.1 : 0.2, *v/v/v*), *T* = 23 °C;

(c) Chirobiotic V, 5 μm, methanol/acetic acid/triethylamine
(100 : 0.1 : 0.2, *v/v/v*), *T* = 23 °C.

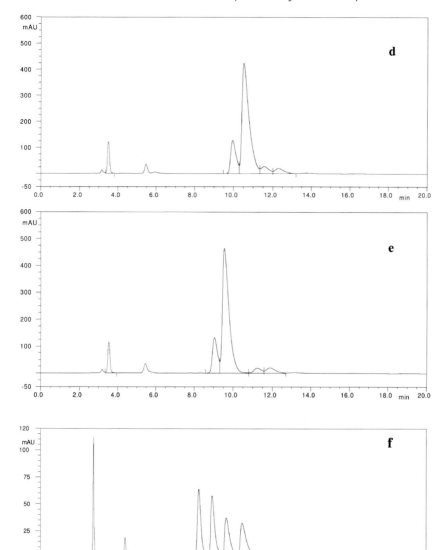

Fig. 12. (d–f) Separation of a racemic ester and the free racemic acid in various compositions; flow: 1.0 mL min^{-1}; detection: 270 nm; column dimensions: 250 mm × 4.6 mm *i.d.*;

(d) Chirobiotic V, 5 μm, methanol/acetic acid/triethylamine
     (100 : 0.4 : 0.1, *v/v/v*), *T* = 10 °C;

(e) Chirobiotic V, 5 μm, methanol/acetic acid/triethylamine
     (100 : 0.4 : 0.02, *v/v/v*), *T* = 5 °C;

(f) 2 × Chirobiotic V, 5 μm, methanol/tetrahydrofuran/acetic acid/triethylamine
     (98 : 2 : 0.1 : 0.05, *v/v/v/v*), *T* = 10 °C.

2.6.4.6 Proteins and Peptides

Proteins are high molecular weight biopolymers consisting of many small chiral subunits capable of forming a three-dimensional tertiary structure. When proteins are covalently bound to silica it is possible to use them as very effective and sufficiently stable CSPs. The adsorption of small molecules is very often stereospecific [18]. Immobilized protein phases are often used in combination with aqueous buffers, which are compatible with many analytes originating from biological or medical trials. Enantioselectivity can be significantly influenced by pH, type of buffer, and the concentration of modifiers. The choice of modifiers requires some care; depending on the selected protein, only a certain amount of organic solvents can be tolerated. If this amount is exceeded, the proteins are irrevocably destroyed. The low loading capacity of these phases renders them useless for preparative applications. Since protein phases can be employed under RP conditions, the use of (salt) gradients is possible. With such a gradient, the peak form of the second eluting (and more adsorbed) enantiomer can be improved. This second peak often exhibits a low number of plates, since kinetic effects hinder the equilibration between the different spatial orientations adopted by the protein and analyte during molecular recognition.

2.6.4.7 Ruthenium Complexes

A relatively new type of stationary phase is based on sodium magnesium silicates, in which metal atoms have been replaced by an optically active tris(1,10-phenanthroline)ruthenium(II) complex [19]. These stationary phases (Ceramosphere by Shiseido; see Table 2) have a high loading capacity due to their large specific surface area and are used at temperatures of 50 °C or higher in the polar normal-phase mode. For optimization of the separation the methanol content can be varied, and in order to shorten retention times the temperature can be increased.

2.6.4.8 Synthetic and Imprinted Polymers

Chiral polymers can be obtained in two different ways. The use of a chiral catalyst during polymerization can lead to helical structures, as observed in polysaccharides. The other synthesis path uses chiral monomers, which are polymerized to give a chiral polymer capable of folding to a supramolecular structure [20]. For application in HPLC, all of these polymers must be coated onto silica, since they are unable to withstand the high pressures encountered in HPLC. Currently, chiral stationary phases based on polyacrylates or polymethacrylates play only a minor role. Chirasphere (Merck) is derived from a silica material coated with poly(*N*-acryloyl-(*S*)-phenylalanine ethyl ester) and can be used for the separation of β-blockers in the normal-phase mode. The chiral polymethacrylates Chiralpak OP and Chiralpak OT (Daicel) are able to separate aromatic compounds into their enantiomers.

The so-called "imprinted" polymers are obtained by polymerization of monomers in the presence of chiral template molecules [21]. The templates are washed out of the stationary phase after completion of the polymerization. After this treatment, the polymer is supposed to have a chiral surface. Due to low selectivities, small pore volume, and slow mass transfer kinetics, these polymers have not yet been commercialized.

2.6.4.9 Metal Complexation and Ligand-Exchange Phases

Ligand-exchange phases (e.g. Chiralpak WH) are obtained by binding a chiral chelating agent to the carrier. In the presence of a suitable transition metal ion, e.g. copper(II), a molecular complex between the ligand of the stationary phase and the analyte is formed [22]. Compounds particularly well-suited for this type of chiral recognition are α-amino acids, hydroxy acids, and small peptides. In contrast to all of the enantioselective HPLC methods discussed thus far, ligand-exchange chromatography requires the presence of a chelatable metal ion in the mobile phase. Other ligands are necessary for the chelation, such as polar solvent molecules or anions.

For the optimization of the separations, the pH and type of buffer, the ionic strength, the nature and concentration of the ligands and modifiers, as well as the temperature must be considered. These numerous factors explain why ligand-exchange chromatography has decreased in importance for enantiomer separations over the last years; today it is mainly used for the separation of α-amino acids (e.g., Crownpak WH by Daicel).

2.6.4.10 Chiral Ion Exchangers

Quinine and quinidine carbamate phases have recently been commercialized by Daicel [23]. These selectors derived from natural alkaloids of the chinolin-type show great potential for enantiomer separations of polar analytes, such as N-derivatized amino acids and peptides, amido sulfonic acids, and non-derivatized acids.

The chiral selectors possess a tertiary amine in a binding cleft along with additional hydrogen bonding sites, allowing very strong interactions with chiral acids. The combination of strong interactions and the steric hindrance of the binding site cleft allows for high resolution separations of molecules containing acidic functional groups.

Since chinin and chinidine are pseudo enantiomers, the elution order of enantiomers can often be reversed by selection of the appropriate stationary phase. The phases can be used under RP conditions as well as in polar ionic mode. The nitrogen atom of the quinine ring can be protonated under acidic conditions (ammonium acetate buffer; pH 6). Most separations are accomplished in the polar organic and reversed phase modes at pH < 8, where the anion-exchange mechanism allows for ionic interactions. Three key parameters can be distinguished to optimize the separations, i.e. the type of organic solvent, the counter ion concentration and the pH. A typical suitable mobile phase to start a method development would be methanol/acetic acid/ammonium acetate (98 : 2 : 0.5, v : v : w), conditions used frequently in method development of the Chirobiotic phases. In order to modify the enantioselectivity various amounts of methanol and acetonitrile can be tested. In case the retention times are too long the counter ion concentration can be increased, which will decrease k values without affecting the α-value of the separation. Also the ratio of acidic and basic modifier can effect retention times and enantioselectivity. The less acidic the buffer becomes, the stronger its elution strength will become and therefore retention times will be decreased at higher apparent pH. Chiralpak QD-AX and QN-AX columns can also be used under reversed phase conditions, when the method development in the polar organic mode was not successful. Under these conditions aqueous buffers (acetate, formiate, phosphate, etc.) containing 90–70% of an organic modifier (methanol or acetonitrile) can be adjusted with TEA or aqueous ammonia solution to the desired pH.

2.6.5
Avoiding Errors and Troubleshooting

Besides the usual reasons for mistakes that are familiar from achiral HPLC [24], there are other sources of errors in enantioselective HPLC, which, in many cases, can be excluded by taking simple precautions. In view of the price of HPLC columns, some basic care can result in significant cost savings.

2.6.5.1 Equipment and Columns – Practical Tips
If an HPLC unit has previously been operated with buffers under RP conditions, it is necessary to ensure that no residue of aqueous eluent is left in the system. This can be achieved by sufficiently long flushing of the system first with water and then with ethanol before an enantioselective column intended for use under NP conditions is installed. Also, enantioselective HPLC columns have to be conditioned before a good separation performance can be expected. Furthermore, one should keep in mind that some eluents used under normal-phase conditions, such as heptane and methanol, are immiscible.

It is a laboratory myth that prolonged storage times destroy chiral selectors and render enantioselective columns unusable. Many suppliers recommend a (non-aqueous) solvent for storage, to which the column should be conditioned if one intends to store it for a long time. It should be kept in mind that many suppliers continually strive to optimize the production of stationary phases, and thus small differences in enantioselectivity between newer and older columns can sometimes be observed. When the history of a column is not known and it has been stored for a long time, it is recommended that it is first rinsed according to protocols given by the supplier and that a test separation is performed. Typical test substances are racemates of Tröger's base, binaphthol or *trans*-stilbene oxide, which can be separated by a multitude of CSPs.

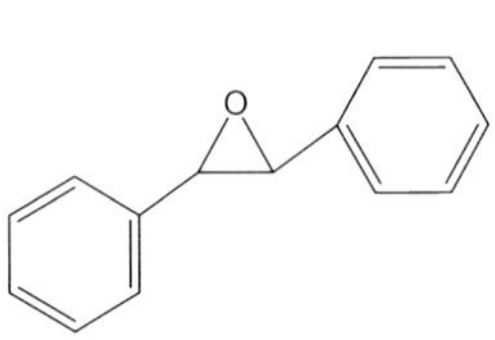

Fig. 13. *trans*-Stilbene oxide (**2**).

For polysaccharide phases, the history of the column can have a decisive effect. For example, if such a column was used with a basic eluent, there is a high probability that after flushing it with an acidic eluent only a poor separation will be observed. This behavior (memory effect [25]) can be attributed to an irreversible reorientation of the helical structures of cellulose and amylose phases. Strongly basic or acidic compounds can effect such changes. A simple solution to avoid these effects is to allocate separate sets of columns for neutral, basic, and acidic conditions (in order to avoid errors in laboratory practice, labeling of columns with colored adhesive tape has proved to be quite useful, e.g. using a red tape for columns run under acidic conditions).

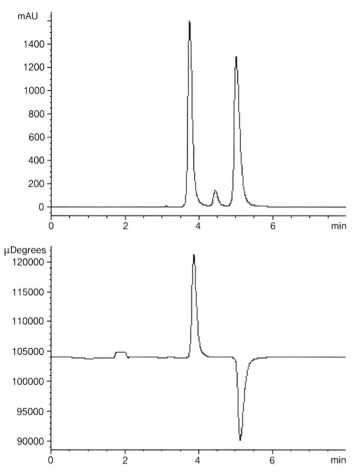

Fig. 14. Enantiomer separation of *trans*-stilbene oxide (**2**) on (*S*,*S*)-Whelk-O 1 [26] (column: 250 mm × 4.6 mm *i.d.*, d_p = 5 µm); eluent: *n*-hexane/isopropanol (60 : 40, *v*/*v*), injected amount: ~0.1 mg, flow: 1.00 mL min^{-1}, *T* = 35 °C; top: UV detection at 254 nm; bottom: on-line polarimetric detection, first eluting enantiomer: positive signal, second enantiomer: negative signal.

2.6.5.2 Detection

Besides UV detection for enantiomer separations, on-line polarimetric detection can also be used (see Fig. 14). This detection mode allows a distinction of the optical rotations of the two compounds during the separation and to make a statement about the enantiomeric composition of the sample even in the event of partial co-elution with other compounds.

Figure 14 shows the enantiomer separation of *trans*-stilbene oxide (**2**) (Fig. 13) on the Pirkle CSP Whelk-O 1. The two enantiomers induce opposite rotations of plane-polarized light; the achiral *meso* form of *trans*-stilbene oxide elutes between the two enantiomers and shows no signal under polarimetric detection.

2.6.5.3 Mistakes Originating from the Analyte

Whenever a racemate is baseline-separated in HPLC, integration must yield an enantiomeric ratio of 50 : 50. As soon as a deviation of > 1% is observed, a co-elution of one or more impurities with one of the enantiomers is evident. A determination of the *ee* is then not possible.

Another phenomenon that can only be observed in enantioselective HPLC is so-called enantiomerization, which is equivalent to a racemization of the separated enantiomers during chromatography [27]. Such enantiomerizations lead to characteristic chromatograms, as shown in Fig. 16 for compound **3** (see Fig. 15).

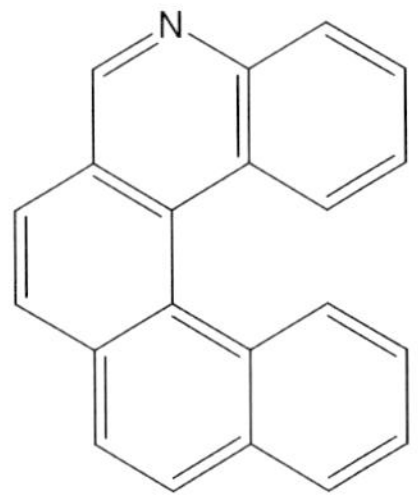

H-5

Fig. 15. 5-Aza[5]helicene (**3**).

The plateau between the two enantiomer peaks is no artefact, but the result of the interconversion of the separated enantiomers. In the extreme case (or when the temperature is increased) both enantiomers are eluted as one peak.

2.6.6
Preparative Enantioselective HPLC

The factors influencing preparative enantiomer separations are quite similar to those of analytical chromatography. Number of plates, retention factors, and selectivity all influence resolution (see Section 1.1.3).

In addition, particle size and column length influence the efficiency and speed of enantiomer separations. For preparative applications, it is recommended that one uses particle sizes of 16–20 μm. Spherical stationary phases of these particle

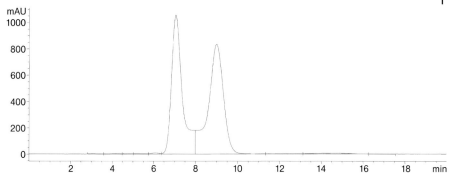

Fig. 16. Enantiomer separation of **3** on Chiralcel OD (250 mm × 4.6 mm i.d.),
eluent: methanol/ethanol/DEA (50 : 50 : 0.25, *v/v/v*), flow: 1.00 mL min^{-1},
detection at 254 nm, *T* = 23 °C. Enantiomerization of **3** during chromatographic
enantiomer separation.

sizes do not exhibit large plate numbers, but they allow work at high flow rates
with low system pressure, facilitating high throughput of samples.

Where possible, preparative enantiomer separations are performed under
overload of the column, i.e. no baseline separation is attempted [28]. In practice,
one often observes that the second eluting enantiomer is obtained with a lower *ee*
than the first eluting enantiomer. Therefore, it is advantageous if a separation
system can be found in which the target enantiomer elutes first.

2.6.6.1 Determination of the Loading Capacity

After the conditions for the enantiomer separation have been optimized, the
loading (injected amount) on the analytical column (16–20 μm particle size) is
increased in order to determine the maximum amount that will still guarantee
the desired purity and recovery. With the assumption that a pilot-scale column
shows an identical efficiency to an analytical column, the loading can be
determined according to Eq. (11):

$$M_P = M_A \cdot \frac{L_P}{L_A} \cdot \frac{d_P^2}{d_A^2} \qquad (11)$$

Here, M_P is the maximum amount that can be separated on a pilot scale and
M_A is the optimized amount obtained from the overloaded injections. L_P and L_A
are the lengths of the pilot and analytical columns, respectively; d_P and d_A represent
the inner diameters of the two columns. The maximum loading factor is constant.
In order to increase the amount that can be separated with a single injection, it is
useful to select a column with a larger inner diameter. The often (erroneously)
chosen increased bed length does indeed increase the maximum separable
amount, but it inevitably leads to an increased back-pressure, longer retention
times, and results in a lower productivity.

2.6.6.2 Determination of Elution Volumes and Flow Rates

To obtain identical retention times as under analytical conditions, the flow rate under preparative conditions has to be adjusted accordingly (see Eq. 12), assuming that the same particle size is used.

$$F_P = F_A \cdot \frac{L_P}{L_A} \cdot \frac{d_P^2}{d_A^2} \tag{12}$$

F_A and F_P are the volumetric flow rates on the analytical and pilot scales. The elution time on pilot scale (t_P) can be determined from Eq. (13):

$$t_P = t_A \cdot \frac{L_P}{L_A} \cdot \frac{d_P^2}{d_A^2} \cdot \frac{F_A}{F_P} \tag{13}$$

The volume to be collected (V_P) can be calculated according to Eq. (14):

$$V_P = V_A \cdot \frac{L_P}{L_A} \cdot \frac{d_P^2}{d_A^2} \tag{14}$$

The easy scalability of enantioselective HPLC can be demonstrated for the separation of omeprazole [29], one of the blockbuster drugs for the treatment of ulcers (see Fig. 17). The (S)-enantiomer, esomeprazole with the trade name Nexium™, is used as an improved active pharmaceutical ingredient against inflammation and ulcers of the oesophagus (reflux oesophagitis), heartburn/pyrosis (gastrooesophagal reflux), and ulcers of the duodenum after infection with *helicobacter pylori*.

After method optimization using the Daicel CSP Chiralpak AS, the enantiomers can be separated using alcohols as eluents (Fig. 18). Under analytical conditions, an enantiomer separation can be achieved within 12 min.

The enantiomer separation can be transferred to preparative scale using the same mobile and stationary phases (Fig. 19).

When an appropriate flow rate is selected, the retention times differ only marginally between analytical and preparative columns. Preparative HPLC allows the isolation of enantiomerically pure compounds at up to the multi-kg range. However, if larger amounts are needed (10 kg up to 500 t), a continuous production method has to be implemented as described in the following paragraphs.

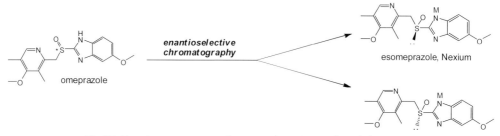

Fig. 17. Enantiomer separation of omeprazole, M: ½ Mg after salt formation.

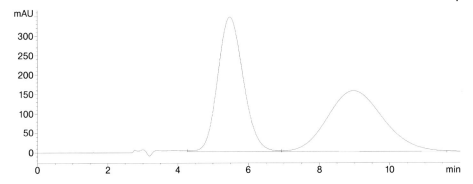

Fig. 18. Analytical enantiomer separation of omeprazole on Chiralpak AS
(column: 250 mm × 4.6 mm *i.d.*, d_p = 20 μm); eluent: isopropanol/ethanol
(70 : 30, *v/v*), injected amount: ~1.2 mg; flow: 1.0 mL min^{-1}; detection: 234 nm.

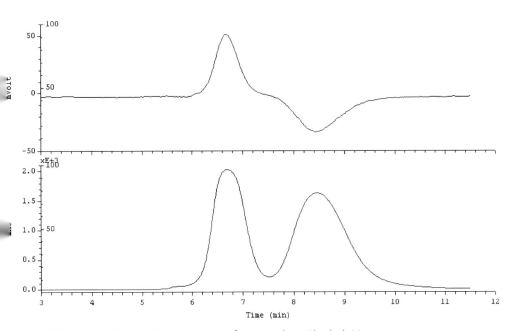

Fig. 19. Preparative enantiomer separation of omeprazole on Chiralpak AS
(column: 230 mm × 48 mm *i.d.*, d_p = 20 μm); eluent: isopropanol/ethanol
(70 : 30, *v/v*), injected amount: ~120 mg; flow: 100 mL min^{-1};
top trace: polarimetric detection; first eluting enantiomer: positive signal,
second eluting enantiomer: negative signal; bottom trace: UV detection: 234 nm.

2.6.6.3   Enantiomer Separation using Simulated Moving Bed (SMB) Chromatography

The technology of the Simulated Moving Bed (SMB) has been successfully used for more than 40 years in the petrochemical industry employing mainly zeolites as the stationary bed. Alongside the basic "Sorbex" process [30] of UOP (Universal Oil Products), a series of large industrial scale units are in use. The sugar industry also applies this technology to obtain fructose in large amounts through the use of ion-exchange resins.

During the last decade, this technique has been used for the scale-up of purification steps based on adsorption chromatography. Among the possible applications are pharmaceuticals, fine chemicals, biochemical products, flavors and fragrances, and especially enantiomers, which are separated by means of suitable chiral stationary phases (see Table 2).

Based on the analytical method developed in the laboratory, SMB separations can be implemented easily and straightforwardly. That stationary phase and mobile phase remain the same is one of the main reasons for the success of the SMB methodology and allows a speeding up of the development process for obtaining a fine chemical or a pharmaceutically active compound [31].

In addition, in comparison to conventional batch chromatography, the application of the SMB principle allows the continuous operation of processes, which makes it easier to obtain products of constant purity without analytical controls for each single batch. Even more importantly, a 90% reduction in the overall amount of solvents needed for a separation can be achieved [32]. The overloaded conditions make it possible to optimize productivity with regard to the inventory of stationary phases. This optimization requires knowledge of the adsorption isotherms (Eq. 8) and the use of special simulation programs [33].

2.6.6.3.1   Principles of Simulated Moving Bed Chromatography

The SMB technique is based on the technical simulation of a true counter-current separation, for which – in complete contrast to conventional HPLC – the "stationary phase" is moved in a counter current to the flow direction of the mobile phase and does not remain stationary (see Fig. 20, top).

An SMB unit consists of a number of chromatographic columns, separated by ports through which inlet and outlet streams can be fed or collected. The counter-current solid movement is simulated by periodically shifting the feed and withdrawal points of the unit in the same direction as the mobile phase flow (see Fig. 20, bottom). Four external streams are present: the racemic feed mixture; the desorbent, i.e. the eluent or the mixture of eluents constituting the mobile phase; the extract stream enriched in the enantiomer A; and the raffinate stream enriched in the enantiomer B. These streams divide the unit into four sections: section 1 between the desorbent inlet and the extract port, section 2 between the latter and the feed inlet, section 3 between this and the raffinate outlet, and section 4 between the raffinate port and the desorbent inlet.

Each of these sections plays a specific role in the process. The separation is performed in sections 2 and 3, where the less retained enantiomer B must be desorbed and carried by the mobile phase towards the raffinate, while A is

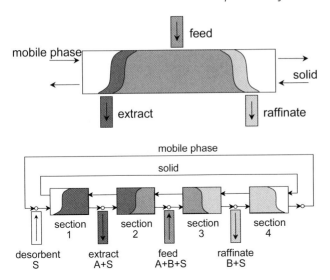

Fig. 20. Top: idealized true counter-current separation; components to be separated (feed), product streams (raffinate, extract); bottom: simulation of the counter current by segregation of the column and change of the positions of inlets and outlets after a pre-defined switch time.

retained by the stationary phase and carried towards the extract port through the simulated solid movement. In section 1, the stationary phase is regenerated by the fresh mobile phase stream through the recycling flow and A is conveyed towards the extract port. Finally, in section 4 the mobile phase is regenerated by adsorbing the amount of enantiomer B not collected in the raffinate. In this way, both the stationary and the mobile phase can be recycled to sections 4 and 1, respectively.

For an enantiomer separation, this means that the more strongly adsorbed enantiomer (the one eluted later under analytical conditions) is collected in the extract stream, while the less retained enantiomer (first eluting) is found in the raffinate steam. The two enantiomers are carried from the point at which they are added (feed), either upstream to the raffinate or downstream to the extract. A third component, the "desorbent", is used to desorb the more strongly adsorbed enantiomer in section 1 and to regenerate the stationary phase. Finally, the more weakly adsorbed enantiomer is adsorbed in section 4 in order to regenerate the desorbent itself.

2.6.6.3.2 Separation of Commercial Active Pharmaceutical Ingredients by SMB

A series of pharmaceutical active ingredients and intermediates are produced industrially by means of SMB separations (see Table 7). The quantities of enantiomerically pure compounds isolated by this approach currently amount to several hundred tonnes per annum. On a production scale, SMB systems consist of five or six dynamic axially compressed HPLC columns with internal diameters up to 1 m.

Table 7. Commercial pharmaceuticals produced by way of SMB chromatography.

Structure	Trade name/active pharmaceutical ingredient	Producer	CSP/solvent	Column i.d. [cm]	Cited according to
	Keppra Levetiracetam Anti-epileptic	UCB	Chiralpak AD n-heptane/ethanol	35 45 100	[34, 35]
	Zoloft Tetralon, Gestralin Anti-depressive	Pfizer	Various MeCN/MeOH	60	[36, 37]
	Citalopram Cipramil Cipralex Lexapro Anti-depressive	Lundbeck	Chiralpak AD Not known	80	[38]
	DOLE anti-cholesterol	Nissan (Daicel)	Chiralcel OF n-hexane/isopropanol	10	[39, 40]

2.6.7
**Enantioselective Chromatography by the Addition of Chiral Additives
to the Mobile Phase in HPLC and Capillary Electrophoresis**

In contrast to those direct methods for enantiomer separation using covalently bound or coated selectors described in the previous sections, it is also possible to dissolve some selectors in the mobile phase and to achieve a separation of analytes based on diastereomeric complexes that are formed in solution or at interfaces (silica particles or column walls). In addition to the chemical equilibria described in Section 2.6.2 and Fig. 3, more interactions between chiral additives and analytes have to be taken into account (see Fig. 21). Besides reversible formation of complexes between the analytes and chiral selector in solution, adsorption/desorption processes of the selector and adsorption/desorption processes of complexes between the analytes and chiral selector and the non-complexed analytes also play a role.

The complex interaction of thermodynamically and kinetically controlled parameters that contribute to a stereoselective recognition make it almost impossible to predict whether an enantiomer separation can be accomplished by the addition of a chiral selector to the eluent. Also, only some classes of selectors (e.g., cyclodextrins and macrocyclic antibiotics) have been successfully used as such chiral additives [41]. These restrictions stem from the facts that many aromatic substituents of other chiral selectors do not allow detection of the analytes against the background of the selectors and that the consumption of the expensive selectors does not give the impression of an economic advantage. Last but not least, some polymeric selectors show a very low solubility in aqueous eluents. Nevertheless, for use in capillary electrophoresis and capillary electrochromatography, a series of inexpensive water-soluble chiral selectors has been established, which allow for surprisingly effective enantiomer separations of charged and uncharged analytes [42].

To optimize such separation systems, on the one hand the nature and concentration of the chiral selector can be varied, while on the other hand the pH and organic additives (e.g., lower alcohols) make it possible to modify retention times and resolution. For electrophoretic applications, the electric field strength and the composition of the buffer ("background electrolyte", BGE) can also support enantiomer separations.

Dissolved chiral selectors have yet to find application in preparative separations.

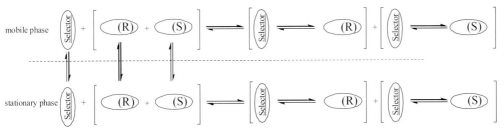

Fig. 21. The enantiomers form diastereomeric molecular associates with the dissolved chiral selector in solution and adsorbed on the stationary phase.

2.6.8
Determination of Enantiomeric Purity Through the Formation of Diastereomers

In the times before the development of modern CSPs, the chemical derivatization of analytes with chiral reagents and subsequent "achiral HPLC" was one of the most frequently employed methods for the determination of enantiomeric purity [43]. This method, often called "indirect" enantiomer separation, has lost importance during the last years. On the one hand, the derivatization requires an ideally complete conversion of the analyte with a reagent that is as enantiopure as possible, and the excess of this reagent must not disturb the subsequent chromatographic discrimination of the diastereomers. On the other hand, the structure of the analyte must be known in order to select a chiral derivatizing agent (CDA) that will react with one of the functional groups of the analyte in an understandable chemical reaction. Starting from chiral acid chlorides, such as O-methyl mandelic acid chloride or camphorsulfonic acid chloride, chiral amines or alcohols, for example, can be converted to diastereomeric amides or esters, which can be seperated by chromatography on achiral stationary phases. Besides chiral (activated) acids, amines, alcohols, and thiols can also be converted with isocyanates and thioisocyanates to give urea derivatives and urethanes, etc. [44]. Analytes without functional groups cannot be analyzed by indirect enantiomer separation. Whenever indirect enantiomer separation is used, one has to ensure through adequate experiments that during derivatization and chromatographic separation no racemization of the diastereomeric analytes will be observed. Finally, one should be aware that diastereomers, in contrast to enantiomers, do not show identical UV/Vis and fluorescence spectra and extinction coefficients at the same wavelength.

2.6.9
Indirect Enantiomer Separation on a Preparative Scale

The separation of derivatized enantiomers after a reaction with a CDA opens an interesting route to direct enantiomer separation for preparative separations. If a method for mild cleavage of the diastereomers can be found, which does not lead to racemization, the use of expensive CSPs can be avoided.

2.6.10
Enantiomer Separations Under Supercritical Fluid Chromatographic (SFC) Conditions

Above a critical temperature and a critical pressure, so-called supercritical fluids show physical properties which position them between liquids and gases. Like gases, they are easily compressible, and properties such as density and viscosity can be modified by pressure and temperature changes. Below the critical tempera-ture and at a pressure above the critical value, the fluid is transformed into a liquid that exhibits a low viscosity and high diffusion rates. Above the critical temperature and at a pressure below the critical value, the fluid turns into a gas [45].

For enantiomer separations, CO_2 is most often used as fluid, since it is relatively cheap and non-toxic. The polarity of super- and sub-critical CO_2 is equivalent to that of the lower alkanes and can be increased by adding modifiers (e.g., lower alcohols). In comparison to HPLC, SFC allows for higher plate counts and an improved resolution. Due to the lower viscosity, a smaller pressure drop is realized over the column, which allows the use of longer columns for analytical and preparative purposes.

Very often, analysis times are shortened by using SFC. CSPs compatible with SFC are cellulose and amylose derivatives, brush-type CSPs, as well as macrocyclic phases. The selectivity of the CSPs is in many cases similar to that found on these phases in the normal-phase mode.

Method development in SFC is significantly easier, since only one mobile phase (CO_2) with a few modifiers (methanol, ethanol, isopropanol, acetonitrile) instead of a multitude of solvent mixtures has to be tested. If the first experiments do not show at least a partial resolution, it is recommended that one should switch to another CSP. To optimize SFC separations, the amount of the modifier can be varied between 5 and 25%, as well as varying its nature. As in HPLC, acidic and basic modifiers can be used. Pressure changes have a greater influence on retention than on selectivity.

2.6.11
New Chiral Stationary Phases and Information Management Software

In the forthcoming years, the development of new chiral phases for HPLC will remain a dynamic topic for research in industry and academia.

More immobilized cellulose and amylose phases are currently under development at Daicel. Daicel is promoting the concept of a CSP library of more than 100 stationary phases that are not currently commercially available, which can be screened for enantiomer separations. Astec has recently presented an improved linking of antibiotic phases (Chirobiotic). As yet, no chiral monolithic stationary phases are commercially available, although one can assume that various companies are currently endeavoring to change the surface of monolithic material with chiral modifiers.

The group of Prof. Roussele in Marseille has been working for the past ten years on the assembly of a database for enantiomer separations in HPLC and GC called Chirbase [47]. So far, more than 100,000 enantiomer separations of more than 30,000 compounds have been described; every three months another 5000 entries are being added, some of which have not been published in the open literature.

2.6.12
Summary

Enantioselective HPLC has evolved in recent years to a routine method carried out on analytical, preparative, and industrial scale. The optimization of separations requires conforming to the conditions under which the chiral stationary phases are chemically stable. For each type of selector, a different and specific protocol has to be followed, since many different interactions and molecular forces can promote an enantiomer separation (see Table 1). In many cases, only a few types of stationary phases (see Table 4) allow the development of a robust method within a few hours or days, that will be reproducible for many years.

References

1 www.fda.gov/cder/guidance.stereo.htm, 5/1/92.

2 E. Francotte, *Chimia* 51 (1997), 717–725.

3 W. Lindner, in Houben-Weyl Methods of Organic Chemistry (Eds.: G. Helmchen, R. Hoffmann, J. Mulzer, E. Schaumann), Georg Thieme Verlag, Stuttgart, New York, Bd. E21a, 93–224.

4 V. Schurig, *Enantiomer* 1 (1996), 139–143.

5 F. Gasparrini, D. Misiti, C. Villani, *J. Chromatogr. A* 906 (2001), 35–50.

6 N. M. Maier, L. Nicoletti, M. Lämmerhofer, W. Lindner, *Chirality* 11 (1999), 522–528.

7 V. Meyer, Fallstricke und Fehlerquellen der HPLC in Bildern, Wiley-VCH, Weinheim, New York, Chichester, Brisbane, Singapore, Toronto, 1999, 8.

8 W. Lindner, M. Lämmerhofer, Grundlagen und Praxis der stereoselektiven Analytik mittels HPLC, CEC und CE, Incom 2003, Düsseldorf, 25.03.2003.

9 M. Schulte, J. Kinkel, R.-M. Nicoud, F. Charton, *Chem. Ing. Tech.* 68 (1996), 670–683.

10 E. Francotte in Chiral Separations (Ed.: S. Ahuja), American Chemical Society, New York, 1997, p. 272–308.

11 Y. Okamoto, E. Yashima, *Angew. Chem.* 110 (1998), 1072–1095; *Angew. Chem. Int. Ed. Engl.* 37 (1998), 1020–1043.

12 M. E. Andersson, D. Aslan, A. Clarke, J. Roeraade, G. Hagman, *J. Chromatogr. A* 1005 (2003), 83–101.

13 N. Mathijs, C. Perrin, M. Maftouk, D. L. Massart, J. Vander Heyden, *J. Chromatogr. A* 1041 (2004), 119–133.

14 S. Allenmark, S. Andersson, P. Möller, D. Sanchez, *Chirality* 7 (1995), 248–256.

15 W. Pirkle, D. House, J. Finn, *J. Chromatogr.* 192 (1980), 143–158.

16 N. M. Maier, G. Uray, O. P. Kleidernigg, W. Lindner, *Chirality* 6 (1994), 116–128.

17 J. Szeijtli, Cyclodextrins and Their Inclusion Complexes, Akademia Kiado, Budapest, 1982.

18 T. Ward, D. Armstrong, in Chromatographic Chiral Separations (Eds.: M. Zief, L. Crane), Marcel Dekker, New York, 1989, 131–163.

19 K. Ekborg-Ott, J. Kullmann, X. Wang, K. Gahm, L. He, D. Armstrong, *Chirality* 10 (1998), 627–660.

20 S. Allenmark, S. Andersson, *J. Chromatogr.* 666 (1994), 166–179.

21 EP 0297 901.

22 D. Arlt, B. Bömer, R. Grosser, W. Lange, *Angew. Chem.* 103 (1991), 1685–1687; *Angew. Chem.* 30 (1991), 1662–1664.

23 B. Sellergren, in Chiral Separation Techniques – A Practical Approach (Ed.: G. Subramanian), 2nd edition, Wiley-VCH, Weinheim, Chichester, New York, Toronto, Brisbane, Singapore, 2001, 153–186.

24 V. Davankov, *J. Chromatogr.* 666 (1994), 55–76.

25 M. Lämmerhofer, W. Lindner, *J. Chromatogr. A* 741 (1996), 33–48.

26 V. Meyer, Fallstricke und Fehlerquellen der HPLC in Bildern, Wiley-VCH, Weinheim, New York, Chichester, Brisbane, Singapore, Toronto, 1999, 39–128.

27 Y. Ye, B. Lord, R. Stringham, *J. Chromatogr. A* (2002), 139–146.

28 The chromatograms have been supplied by courtesy of PDR-Chiral, Inc., 1331 A South Killian Drive, Lake Park, FL 33403, USA.

29 K. Cabrera, M. Jung, M. Fluck, V. Schurig, *J. Chromatogr. A* 731 (1996), 315–321.

30 H. Colin, in Preparative and Production Scale Chromatography (Eds.: G. Ganetsos, P. Barker), Marcel Dekker, New York, Basel, Hong Kong, 1991, 2–45.

31 WO 2003051867.

32 M. J. Gattuso, B. McCulloch, J. W. Priegnitz, *Chem. Tech. Europe* 3 (1996), 27–30.

33 M. Mazzotti, M. Juza, M. Morbidelli, *Git Spez. Chromatogr.* 18 (1998), 70–74.

34 J. N. Kinkel, M. Schulte, R. M. Nicoud, F. Charton, Proceedings of the Chiral Europe '95 Symposium, Spring Innovations Limited, Stockport, UK, 1995.

35 M. Juza, M. Mazzotti, M. Morbidelli, *TIBTECH* 18 (2000), 108–118.

36 US Patent 6,107,492.

37 M. Hamende, E. Cavoy, *Chimie Nouvelle* 18 (2000), 3124–3126.

38 US Patent 6,444,854.

39 S. Houltou, *Manufacturing Chemist* 11 (2001), 23–25.

40 WO 2003006449.

41 WO 2002030903.

42 S. Nagamatsu, K. Murazumi, S. Makino, *J. Chromatogr. A* 832 (1999), 55–65.

43 C. Poole, in The Essence of Chromatography, Elsevier, Amsterdam, Boston, London, New York, Oxford, Paris, San Diego, San Francisco, Singapore, Sydney, Tokyo, 2003, 827.

44 B. Chankvetadze, Capillary Electrophoresis in Chiral Analysis, John Wiley & Sons, Chichester, 1997.

45 W. Lindner, in Houben-Weyl Methods of Organic Chemistry (Eds.: G. Helmchen, R. Hoffmann, J. Mulzer, E. Schaumann), Georg Thieme Verlag, Stuttgart, New York, Bd. E21a, 225–252.

46 M. Schulte, in Chiral Separation Techniques – A Practical Approach (Ed.: G. Subramanian), 2nd edition, Wiley-VCH, Weinheim, Chichester, New York, Toronto, Brisbane, Singapore, 2001, 187–204.

47 P. J. Schoenmakers, in Supercritical Fluid Chromatography (Ed.: R. M. Smith), Royal Society of Chemistry, London, 1988, 102–136.

48 E. Zarbl, M. Lämmerhofer, A. Woschreck, F. Hammerschmidt, C. Parenti, G. Cannazza, W. Lindner, *J. Sep. Sci.* 25 (2002), 1–6.

49 http://chirbase.u-3mrs.fr

2.7
Miniaturization

2.7.1
µLC/NanoLC – Optimization and Troubleshooting

Jürgen Maier-Rosenkranz

2.7.1.1 Introduction
There are two main reasons for entering the realm of micro- and nano-LC: to obtain necessary sensitivity or if only the smallest quantities of samples are available. The main fields of application are currently proteomics, pharmacokinetics, metabolism studies, microdialysis, and, increasingly, environmental analysis. Flow rates in micro- (2–50 µL min^{-1}) and nano-LC (200–2000 nL min^{-1}) place high demands on the HPLC system and the user. Continuous optimization with regard to robustness, sensitivity, detection limit, and resolution tends to be a feature of any application.

2.7.1.2 Sensitivity
Attainable sensitivity depends in particular on the column dimensions, the type of silica gel, and the detector used.

2.7.1.2.1 Influence of Column Length
The following relationship exists between sensitivity and variation of the column length:

$$F_{\Delta E} = \sqrt{\frac{L_1}{L_2}}$$

where
$F_{\Delta E}$ = factor for the change of sensitivity
L_1 = length of the first column
L_2 = length of the second column

Example: If a column of length 50 mm is used instead of one of length 250 mm, the sensitivity will increase as follows:

$$F_{\Delta E} = \sqrt{\frac{250}{50}} = \sqrt{5} = 2.24$$

2.7.1.2.2 Influence of Column Internal Diameter (*i.d.*)
The influence of column *i.d.* on the sensitivity is obviously greater than the influence of column length. Sensitivity will increase according to a square relationship with decreasing *i.d.*:

HPLC Made to Measure: A Practical Handbook for Optimization. Edited by Stavros Kromidas
Copyright © 2006 WILEY-VCH Verlag GmbH & Co. KGaA, Weinheim
ISBN: 3-527-31377-X

$$F_{\Delta E} = \left(\frac{d_1}{d_2}\right)^2$$

where

$F_{\Delta E}$ = factor for the change of sensitivity
d_1 = i.d. of the first column
d_2 = i.d. of the second column

Example: By replacing a 4 mm *i.d.* column with a 2 mm *i.d.* column, the sensitivity will increase by a factor of 4:

$$F_{\Delta E} = \left(\frac{4}{2}\right)^2 = 4$$

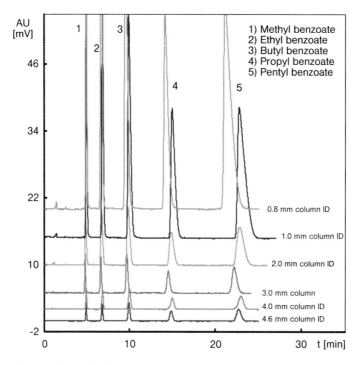

Fig. 1. Influence of column *i.d.* on sensitivity
(after Földi et al., University of Applied Sciences Krefeld, Germany).

Stationary phase:	GROM Sapphire 110 C18, 5 µm
Column length:	125 mm
Eluent:	45% H$_2$O, 55% MeCN (v/v)
Linear flow:	0.8 mm s^{-1} = 0.8 mL min^{-1} for 4.6 mm *i.d.*
Temperature:	ambient
Detection:	UV 254 nm; flow cell: 50 nL/0.2 mm
Injection:	300 nL benzoate test mix

2.7.1.2.3 Influence of Stationary Phase

The influence of the stationary phase on the sensitivity in terms of the selectivity, retention time, peak shape, and particle size, is very variable.

Assuming that only the particle size of the silica gel is changed and thus all other factors are constant, sensitivity changes according to the following formula:

$$F_{\Delta E} = \sqrt{\frac{dp_1}{dp_2}}$$

where

$F_{\Delta E}$ = factor for the change in sensitivity
dp_1 = particle size of the stationary phase of the first column
dp_2 = particle size of the stationary phase of the second column

Example: In place of a column with 5 μm packing, one is used with 3 μm packing; the sensitivity then changes as follows:

$$F_{\Delta E} = \sqrt{\frac{5}{3}} = \sqrt{1.67} = 1.29$$

All three factors cooperate. By changing from a classical column (250 × 4 mm, 5 μm) to a short, modern column (50 × 2 mm, 3 μm), the gain in sensitivity is given by:

$$F_{\Delta E \, complete} = F_{\Delta E} \, L \times F_{\Delta E} \, ID \times F_{\Delta E} \, dp = 2.24 \times 4 \times 1.29 = 11.55$$

2.7.1.3 Robustness

2.7.1.3.1 System Choice

With the acquisition of an HPLC system for micro- and nano-LC, significant influence on the robustness and stability of the subsequent analyses can be exercised through the selection of the system.

Three main systems are on the market today (January 2004):
- Syringe pumps (Eldex, MicroPro).
- Double syringe pumps (ProLab, Evolution 200; Gilson, nLC; Waters CapLC).
- Split systems: simple (unregulated), regulated (HP1100 CapLC), and splitting at a high pressure level (Dionex/LC packings).

The two great advantages of a syringe pump are the minimal pump abrasion and the absolutely pulse-free flow. Both advantages particularly affect the lifetime of the column. At the same time, the danger of a blockage is clearly smaller with these systems. A disadvantage of this principle is the additional time requirement for filling the syringes and the pre-compression of the eluent. The time required for the latter has been optimized in recent years by means of special software control, but it still takes a few minutes.

The consequence is that when they are used for short analysis times and high sample throughput (e.g., the determination of plasma levels in pharmacokinetic labs), these systems are not optimally applied. For separations with analysis times > 20 min (e.g., the determination of metabolites, proteomics), the time requirement for filling and pre-compression becomes negligible. In this case, the stability of the system for injecting valuable samples is more important.

Double syringe pumps do not have the disadvantage of the time requirement for filling and pre-compression. These have two syringes with an individual driving motor for each solvent channel. In this way, they differ from the piston pump. In piston pumps, the two pistons are usually driven by one motor via a shaft. Binary gradient systems thus have four syringes with four driving motors.

As the first syringe delivers the eluent, the second syringe is filled and pre-compressed. When this procedure is complete, a valve switches from the first to the second syringe. In this way, there is no delay in the delivery of the eluent. Depending upon the design of the syringe pumps, a pulse of variable magnitude results from the changeover. The strength of the pulse depends on the kind of switching, the frequency of the change from the first to the second syringe, and the delivery and/or syringe volume. Systems that operate with one pressure sensor on each syringe are subject to less strong pulsation because the changeover is carried out at identical pressure. Systems with syringes of larger volume have a lower frequency and therefore a less strong pulsation. The volumes of the syringes are smaller than that of a single piston pump. Thus, these have a higher piston frequency and therefore increase pump abrasion, which can lead to blockages. Double syringe pumps are particularly suitable for short analyses. If they are used for long gradients and separations, the disadvantage of shorter lifetime of the column compared to when a single syringe pump is used becomes apparent.

Split systems are favored because of their large working range. This does not apply to splitters working at a high pressure level, due to the fact these are optimized for certain ranges in each case.

Stable conditions with regard to the flow constant are only attained with regulated split systems (see Fig. 2).

When working with simple splitters (T-fitting, see Fig. 3), a small increase in the column pressure decreases the secondary flow rate because the split ratio is increased. The consequence is permanent control and adjustment of the flow rate.

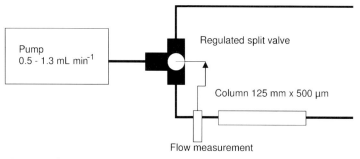

Fig. 2. Regulated split system.

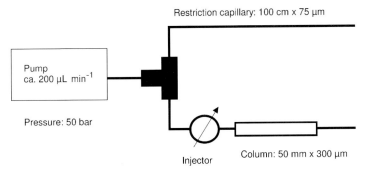

Fig. 3. Simple split.

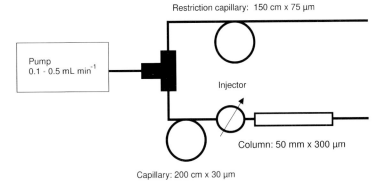

Fig. 4. Splitting at a high pressure level.

With splitting at a high pressure level, the influence of an increase of column pressure on the secondary flow rate is clearly smaller. This is achieved by the installation of a very thin, long capillary before the injector. Thus, the pressure and the split ratio are largely fixed (see Fig. 4). Control over and adjustment of the secondary flow rate remains essential.

Splitters work at the same time as a pulsation damper, which again has a positive effect on the lifetime of the columns and the stability of the system. The use of piston pumps is, however, unfavorable, as these have a very high abrasion. To protect the system, a suitable filter must be inserted (see, in addition, tip 89 in [1]).

The advantage of the split systems is that the gradient is generated under high flow-rate conditions and the volume influences regarding delay and accuracy are not serious. Nevertheless, the gradient delay volume (dwell volume) must always be controlled. Figure 5 shows the strong influence of the primary flow rate on the gradient.

In order to ensure stable gradient and flow conditions, it is necessary that each system is tested regularly. For the test gradients, the procedure to be used is described above. However, the acetonitrile in eluent A should be replaced by water. Only with water/acetonitrile or any other organic modifier it is possible to test a

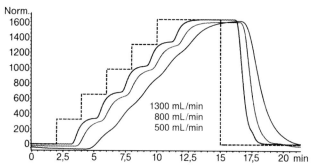

Fig. 5. Gradient profiles with different primary flow rates
(Source: Alexander Beck, UKT, Tübingen).
Conditions:
Gradient: 100% MeCN against 100% MeCN with 0.1% benzyl alcohol,
secondary flow rate 20 μL min^{-1}, UV 254 nm, no column –
rather a thin capillary, approx. pressure in each case 50 bar,
gradient delay volume approx. 1000 μL.
Gradient: 0–2 min: 0% B; 2–2.01 min: 0–20% B; 2.01–4 min: 20% B;
4–4.01 min: 20–40% B; 4.01–6 min: 40% B; 6–6.01 min: 40–60% B;
6.01–8 min: 60% B; 8–8.01 min: 60–80% B; 8.01–10 min: 80% B;
10–10.01 min: 80–100% B; 10.01–20 min: 100% B.

real gradient. The procedure described above can be applied if a regulated split system is used. Flow control measurements are made by means of thermal capacity, which is not only a function of the flow rate but also of the eluent composition. The software-regulated correction of the gradient is therefore not a component of the gradient profile. With these systems it is necessary to perform both runs to accomplish both the calibration and the gradient system control. The gradient response time should be < 2 min.

2.7.1.3.2 Capillary Connections

In micro- and nano-HPLC, the quality and the type of the connections is often crucial in determining the success or failure of a separation (see also Chapter 5.1). For micro- and nano-HPLC, connecting capillaries made from stainless steel, PEEK, fused silica, and PEEKSil are currently in use. Although there are a number of objective reasons for using each type of capillary or connection, it is usually individual preference or current availability that determine which is employed. In the following sections, the arguments in favor of the respective systems and the possible problems are presented.

Different units in use can lead to difficulties. A conversion table is therefore helpful (see Table 1).

Micro- and nano-HPLC require permanent control of all volumes in the system. For this, two further tables are helpful: the volumes of the capillaries and the columns (see Tables 2 and 3).

Ideally, a connection capillary should have a volume of no more than 1/50 of the column volume to avoid band broadening due to dead volumes in the system.

Table 1. Conversion table.

Inch		cm	mm	μm
	1.0000	2.5400	25.400	25,400
1/2	0.5000	1.2700	12.700	12,700
1/4	0.2500	0.6350	6.350	6,350
1/5	0.2000	0.5080	5.080	5,080
1/8	0.1250	0.3175	3.175	3,175
1/10	0.1000	0.2540	2.540	2,540
	0.0700	0.1778	1.778	1,778
1/16	0.0625	0.1588	1.588	1,588
	0.0550	0.1397	1.397	1,397
1/20	0.0500	0.1270	1.270	1,270
1/24	0.0420	0.1067	1.067	1,067
1/25	0.0400	0.1016	1.016	1,016
1/32	0.0313	0.0794	0.794	794
	0.0300	0.0762	0.762	762
1/40	0.0250	0.0635	0.635	635
1/50	0.0200	0.0508	0.508	508
	0.0150	0.0381	0.381	381
1/100	0.0100	0.0254	0.254	254
	0.0090	0.0229	0.229	229
	0.0080	0.0203	0.203	203
	0.0070	0.0178	0.178	178
	0.0060	0.0152	0.152	152
1/200	0.0050	0.0127	0.127	127
1/250	0.0040	0.0102	0.102	102
	0.0035	0.0089	0.089	89
1/400	0.0025	0.0064	0.064	64
1/1000	0.0010	0.0025	0.025	25

This concerns columns with *i.d.* ≥ 500 μm. For columns with *i.d.* ≤ 500 μm and capillaries with *i.d.* ≤ 75 μm, this rule no longer applies. Since band broadening is not a problem in capillaries ≤ 75 μm, such capillaries are optimized with regard to the time required to flush the capillary volume and the danger of blocking. It is possible to run a column of dimensions 100 mm × 75 μm with an output capillary of dimensions 250 mm × 50 μm; however, the dwell time of the sample in the capillary is larger than that on the column (for non-retained compounds).

Example

Column 100 mm × 75 μm, volumes: 442 nL empty, packed × 0.7 = 309 nL. Output capillary 250 mm × 50 μm, volume: 491 nL.

With a flow rate of 350 nL min^{-1}, the time for breakthrough is 53 s for the column and 84 s for the capillary. In nano-HPLC, time optimization and precautions

Table 2. Capillary volumes in µL.

Length [mm]	i.d.									
	10 µm	15 µm	20 µm	25 µm	30 µm	50 µm	64 µm	75 µm	100 µm	120 µm
50	0.004	0.009	0.016	0.025	0.035	0.098	0.16	0.22	0.39	0.57
75	0.006	0.013	0.024	0.037	0.053	0.147	0.24	0.33	0.59	0.85
100	0.008	0.018	0.031	0.049	0.071	0.196	0.32	0.44	0.79	1.13
150	0.012	0.026	0.047	0.074	0.106	0.294	0.48	0.66	1.18	1.70
200	0.016	0.035	0.063	0.098	0.141	0.393	0.64	0.88	1.57	2.26
250	0.020	0.044	0.079	0.123	0.177	0.491	0.80	1.10	1.96	2.83
300	0.024	0.053	0.094	0.147	0.212	0.589	0.96	1.32	2.36	3.39
400	0.031	0.071	0.126	0.196	0.283	0.785	1.29	1.77	3.14	4.52
500	0.039	0.088	0.157	0.245	0.353	0.981	1.61	2.21	3.93	5.65

Table 3. Column volumes in µL.

Column length [mm]	Column i.d.									
	50 µm	75 µm	100 µm	150 µm	200 µm	300 µm	500 µm	800 µm	1.0 µm	2.0 µm
10	0.020	0.044	0.079	0.18	0.31	0.71	1.96	5.0	7.9	31
20	0.039	0.088	0.157	0.35	0.63	1.41	3.93	10.0	15.7	63
30	0.059	0.132	0.236	0.53	0.94	2.12	5.89	15.1	23.6	94
50	0.098	0.221	0.393	0.88	1.57	3.53	9.81	25.1	39.3	157
75	0.147	0.331	0.589	1.32	2.36	5.30	14.72	37.7	58.9	236
100	0.196	0.442	0.785	1.77	3.14	7.07	19.63	50.2	78.5	314
125	0.245	0.552	0.981	2.21	3.93	8.83	24.53	62.8	98.1	393
150	0.294	0.662	1.178	2.65	4.71	10.60	29.44	75.4	117.8	471
250	0.491	1.104	1.963	4.42	7.85	17.66	49.06	125.6	196.3	785
300	0.589	1.325	2.355	5.30	9.42	21.20	58.88	150.7	235.5	942

against blocking of the capillaries are foremost. Capillaries with an *i.d.* ≥ 30 µm are not especially susceptible to blocking. With an *i.d.* ≤ 20 µm, the risk of blocking rises sharply.

A further key point is the quality of the plane of section of the capillaries. In order to achieve absolutely right-angled planes of section, cutting tools are used whereby the blade cuts around the capillary or otherwise the capillaries have to be polished.

Stainless Steel Capillaries

Stainless steel capillaries are interesting in two *o.d.*s, 1/32″ and 1/16″. The advantage of the 1/32″ capillaries is their great flexibility. The smallest *i.d.* is 0.1 or 0.12 mm. Thus, their use is limited to micro- and nano-HPLC. Some manufacturers use a 50 mm long 1/16″ capillary as an entrance for capillary columns with an *i.d.* between 300 and 800 µm. Advantages of this variant are that the column is reliably fixed in the injection valve and that the danger of blocking is low. When using PEEK "Fingertights", it must be noted that this connection is only stable up to approximately 200 bar.

PEEK Capillaries

In RP chromatography, PEEK capillaries have good chemical stability and are largely inert. The tailing due to interactions with silanol groups seen with fused silica capillaries does not arise here [2]. They are very flexible and are available at relatively low prices, in many dimensions. The different *i.d.*s can easily be determined from the color coding.

The 1/16″ capillaries have the advantage that they fit into each standard screw connection without the use of a sleeve. (Sleeves are 3–4 cm long, straight pieces of capillary that are used as adapters in order to attach thin capillaries in connections with larger *o.d.* The *o.d.* of the sleeves corresponds to the connection and the *i.d.* is such that the capillary that has to be connected can be pushed in.) The risk of these connections is that parts are used that do not have a small bore and are therefore not suitable for the micro and nano ranges. For the two most widely used PEEK capillaries, with *i.d.*s of 130 µm and 64 µm, the danger of a reduction in the cross-section due to the pressure of the ferrule or "Fingertight" is very small, because of the large wall thickness. The negative influence of a not right-angled and smooth surface of the cutting area is higher than with the 1/32″ or 360 µm capillaries due to the large *o.d.*

The 1/32″ capillaries are universally applicable. They can be connected directly with the 1/32″ screw connections or with a sleeve with a 1/16″ screw connection. The advantage of the 1/32″ connections is that they are more easily tightened than the 1/16″ screw connections and no tools are usually necessary. Due to small material thickness, this tubing requires that attention is paid to the grooved ring caused by the ferrule or "Fingertight" as this can lead to problems. The two main problems that can arise are a possible increase in pressure and incorrect positioning of the ferrule and/or "Fingertight", which again leads to dead volumes in the system. This danger is particularly large with 360 µm PEEK capillaries. An advantage of the 360 µm PEEK capillaries is that they can be used like fused silica capillaries in micro-screw connections.

Fused-silica Capillaries

Wide availability in almost all *i.d.*s and *o.d.*s as well as their low price made fused-silica capillaries the standard connection in micro- and nano-HPLC, particularly in the early days. In conjunction with different sleeves or special sleeve fittings,

Table 4. Available PEEK capillaries.

o.d.	i.d.				
		inch	cm	mm	μm
360 μm					51
					75
					102
					152
510 μm					65
					125
					255
1/32″	1/50	0.0200	0.051	0.508	508
		0.0150	0.038	0.381	381
	1/100	0.0100	0.025	0.254	254
		0.0090	0.023	0.229	229
		0.0080	0.020	0.203	203
		0.0070	0.018	0.178	178
	1/200	0.0050	0.013	0.127	127
		0.0035	0.009	0.089	89
	1/400	0.0025	0.006	0.064	64
1/16″	1/32	0.0313	0.079	0.794	794
		0.0300	0.076	0.762	762
	1/40	0.0250	0.064	0.635	635
	1/50	0.0200	0.051	0.508	508
		0.0150	0.038	0.381	381
	1/100	0.0100	0.025	0.250	250
		0.0090	0.023	0.229	229
		0.0080	0.020	0.203	203
		0.0070	0.017	0.170	170
		0.0060	0.015	0.152	152
	1/200	0.0050	0.013	0.130	130
	1/250	0.0040	0.010	0.102	102
		0.0035	0.009	0.089	89
	1/400	0.0025	0.006	0.064	64

these can be used in 1/16″, 1/32″, and 1/40″ micro connections. Non-deactivated fused-silica capillaries cannot be used for the chromatography of basic compounds because of the likelihood of strong peak-tailing [2].

The main problem when working with fused-silica capillaries is that upon wetting of the polyamide layer with mobile phase, this layer can be removed from the glass. The consequences are blockages and an extremely increased danger of fracture of the glass capillary. Broken glass or tears in the fused-silica capillary can destroy each separation. When working with fused silica, it should be ensured that all connections are made without flow, and that the ends of the capillaries are completely dry. When a leakage occurs, the connections have to be checked to see whether these are still in perfect condition.

The determination of the diameters is difficult. The o.d. can still be measured well with a calliper gauge or a micrometer screw. The i.d. can only be determined in a laboratory using a measuring magnifying glass with a certain variance. Capillary lengths cut with tungsten carbide or diamond blades with only one score and thereafter one break-off are to be completely avoided, since no smooth right-angled plane of section is likely to be obtained. When the end of a capillary is placed inside a sleeve, it has to be ensured that no small pieces of PEEK or Teflon are planed off by the sharp edge of the fused-silica capillary, since these can again cause dead volumes or blockages. For this reason, it is not advised to use short pieces of Teflon to connect fused-silica capillaries. A benefit of using fused-silica capillaries is that ferrules or Fingertights do not score the capillary.

PEEK-encased Glass Capillaries – PEEKSil

To a large extent, these capillaries represent the current optimum. Both 1/16″ and 1/32″ variants are available in all necessary i.d.s. These capillaries are supplied in fixed lengths and have an absolutely right-angled and smooth plane of section. The user in a lab cannot cut these capillaries.

They are also stable against imprinting of the sleeve fitting or Fingertight. They are pleasant to handle and the respective i.d.s are color-coded. However, with these capillaries there is a risk that the glass inside can fracture or break. Their high price has prevented ubiquitous deployment.

2.7.1.3.3 Precautions Against Blocking

Besides the causes of increased pressure and blockages described in the previous section, there are two other causes. Firstly, solvent impurities, pump abrasion, etc., and secondly the sample itself.

The primary side of the system is best protected by a cleaning guard column. Such a guard column is packed with rough (20–50 μm) type B silica and is placed before the injector. The volume of this column must be adapted to the flow rate used. Suitable columns have to be flushed in about 30 s, or at most 1 min. The i.d. should be as high as possible. The length is usually 5, 10 or 20 mm. A cartridge system is economical and can be rapidly replaced. Additionally, this column can serve the function of a mixing chamber. As regards the filter effect, a column is better than a frit made from stainless steel or polymer material. Additionally, it can take up dissolved impurities.

Classical sample preparation procedures such as filtration centrifugation, liquid-liquid extraction, and SPE are effective against blocking by the sample. The most important technology, however, is the guard column switching. The different variants are described in Section 2.7.1.3.5.

In the realm of micro- and nano-HPLC, it is usual to incorporate guard columns as protection for the main column. Due to the fact that the danger of blocking is increased with smaller i.d.s, it is a good idea to select the i.d. of the guard column to be about twofold larger than that of the main column. Moreover, it is better to use a larger particle size of the packing material in the guard column than in the main column, e.g. 3 µm material in the main column, 5 µm material into the guard column; in general, 5–7 µm is recommended for the guard column.

The use of in-line filters can be very helpful. Some stainless steel sieves and frits with a diameter of 1/16″ are available. These can be placed directly into a 1/16″ screw connection. Likewise, a set of micro couplings with filters inserted is available. Both the guard column and filter can worsen a separation after prolonged use, even if the pressure is only slightly raised.

2.7.1.3.4 Testing for Leakages

With ever smaller flow rates, recognizing leakages becomes increasingly difficult. Down to the lower µL range (2 µL min^{-1}), leakages can still usually be recognized visually, in particular if tissue paper is used to make the liquid visible.

The most important indication of a leakage is the injecting peak. With constant flow rate and identical capillaries, it must always come at an identical time. If the time increases without a visible leakage, an increase in the flow rate is recommended, until about 80–90% of the maximum pressure of the column is reached. In this way, the leakage should become larger and more visible.

Sometimes, it is helpful to individually test the sections of the system under high pressure. This procedure is particularly well-suited for injectors and switching valves. With detectors, attention must be paid to the ability of the flow cell to withstand pressure.

2.7.1.3.5 Guard Column Switching and Sample Loading Strategies

In micro- and nano-HPLC, sample loading is crucial and therefore a very important aspect. Usually, the sample is enriched at the column head and subsequently eluted by applying a gradient. It is reasonable to work with guard column switching. Thereby, the *i.d.* of the guard column is significantly larger (up to ca. 3×) than the *i.d.* of the main column and the length is between 5 and 10 mm. The particle size can also be larger (see Section 2.7.1.4.1). This principle offers a number of advantages over direct injection.

- Compounds with little or no retention (sugar, salts, acids, etc.) of the sample matrix are eluted only through the guard column and not through the main column.
- The larger *i.d.* makes it possible to load the sample at a higher flow rate (up to 20× higher than for the separation).
- The larger *i.d.* clearly lowers the risk of an increase in pressure due to impurities from the sample in the system.
- The higher particle size of the silica gel results in a lower pressure and therefore the system is not sensitive to blocking.
- In the case of a phase with weaker retention, e.g., a C$_8$ phase, the sample can be focused in the main column on a C$_{18}$ phase once more.

Within the analytical range, the sample is often eluted from the guard column to the main column by means of a back-flash procedure. In micro and nanoLC, this is not usually done, since impurities held back at the head of the guard column are otherwise rinsed onto the main column, greatly increasing the risk of a blockage there.

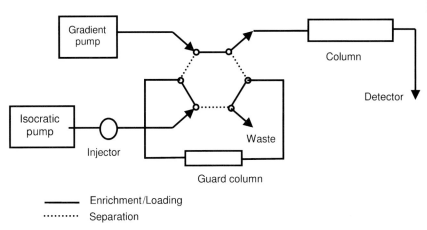

Fig. 6. Simple guard column switching.

Different variants of column switching are in use today.

This combination is particularly well-suited for columns with larger *i.d.*s (Fig. 6). With main columns of *i.d.* 300–800 μm, guard columns with *i.d.* up to 2 mm can be used. The flow rates in the guard column circuit are of the same order of magnitude as in analytical HPLC and therefore it is better to load the sample with an analytical HPLC pump.

A further operational area is working with two guard columns for enrichment purposes and one main column. As long as one guard column takes to load, the other one is in the separation cycle. For this, an eight-port valve is necessary. This variant allows an increase in sample throughput for rapid analyses.

Guard Column Switching For Micro Pumps

When using columns with smaller *i.d.*s and low flow rates, it is better to arrange the switching system in such a way that only one micro-HPLC system is necessary (Fig. 7). The high flow is switched to a lower one by means of a valve. This cannot be realized with all systems. If the decrease in flow rate is delayed, the pressure in

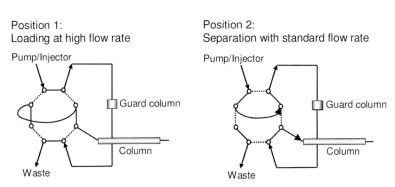

Fig. 7. Guard column switching for micro-LC pumps.

Position 1:
Loading at high flow rate

Position 2:
Separation at nano flow rate

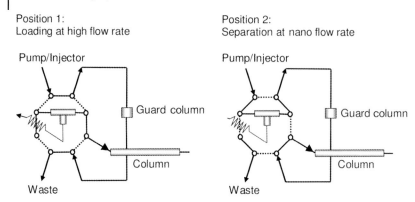

Fig. 8. Guard column switching for nano-HPLC.

the system can rise sharply during the changeover. In this case, the flow must first be lowered and then the valve switched.

As regards the dimensions of the guard and main columns, the same strategies as described above should be applied. As a result of the high starting flow rate, it is possible that gradient delays may be compensated for by large delay volumes. In this case, one starts the gradient with a high flow rate for the enrichment, and when the gradient arrives at the guard column the flow is decreased and switched to the main column.

In the nano-HPLC range with flow rates $\leq 1\ \mu\text{L min}^{-1}$, the following principle is preferable.

The enrichment takes place according to this principle on guard columns of up to 500 μm *i.d.* and at flow rates up to 50 μL min^{-1}. Afterwards, the valve is switched to the split (1 : 10 to 1 : 30) and the separation can be realized at low rates. Apart from the advantages already described, it is possible to perform highly sensitive analyses on nanoLC columns with a micro-HPLC system. The danger of a shift in the split ratio is relatively small for this arrangement, since the sample is pre-cleaned by passage through the guard column.

This circuit is optimally positioned directly at the detector, so that the post-column volume is minimal. The pre-column volume does not have a large influence on the resolution, since the sample is refocused on the main column.

Example (Fig. 9)

50 fmol absolute angiotensin I and angiotensin II/10 fmol Pepmix
1 μL injection (50 fmol μL^{-1})
(A) 0.1% FA in water (*v*/*v*)
(B) 0.1% FA, 80% MeCN in water (*v*/*v*/*v*)
gradient: 0–5 min 15% B; 10–60 min 60% B

Guard column: GROM-SIL 120 Octyl-5 CP, 5 μm, 5 mm × 170 μm,
5 min trapping and desalting, 6 μL min^{-1} (55 bar)

50 fmol Pepmix

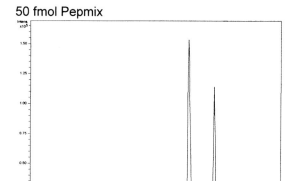

Extracted ion chromatogram for m/z = 523.8 and 530.8 [M + 2H]$^{2+}$ peptide ions

Mass spectra

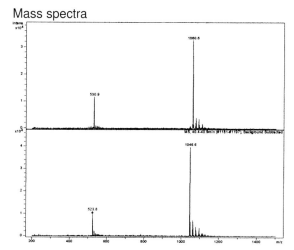

MS/MS Spectra

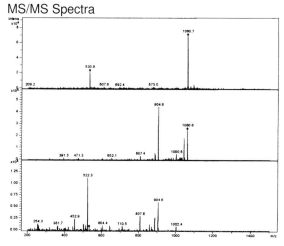

Fig. 9. Guard column switching in the nano-LC range.

10 fmol Pepmix

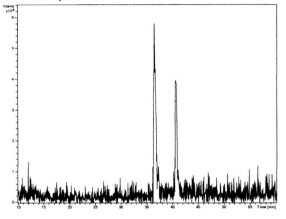

Ion chromatogram for m/z = 523.8 and 530.8 [M + 2H]$^{2+}$ peptide ions

Mass spectra

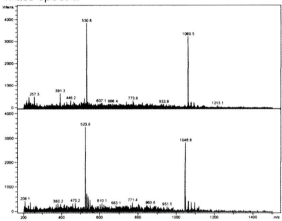

MS/MS Spectra

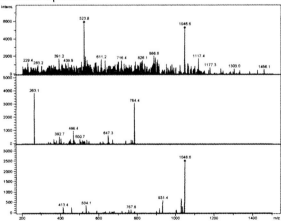

Fig. 9. (continued)

Column: GROM-SIL 120 ODS-3 CP, 3 μm, 100 mm × 75 μm,
flow rate: 250 nL min^{-1}
8 μm Spraytip (NewObjective)
Agilent 1100, CapLC, modified
Bruker Esquire 3000 plus ion trap MS (scan from m/z = 200 to 1500,
ICC 35000, 2 precursor, active exclusion after one spectrum, SPS (1.00 V))
(Source: Alexander Beck, UKT Tübingen)

2.7.1.4 Sensitivity/Resolution

2.7.1.4.1 Column Dimensions

Today, it is no problem to manufacture a column with 2 μm packing material, an *i.d.* of 50 μm, and 125 mm in length. Regarding efficiency, resolution, and sensitivity, this column represents the optimum. However, provided that the packing material has suitable selectivity for the selected application, for most applications it makes no sense to go into these ranges. The choice of the individual parameters depends on the requirements of the equipment, demands in terms of sensitivity and resolution, and on the stability and robustness of the separation.

The choice of the *i.d.* is mainly determined by the possibilities of the HPLC system to generate low flow rates and gradients with high accuracy and precision and the sensitivity needed. The use of 50 μm *i.d.* columns with flow rates in the range of 100–200 nL min^{-1} represents a considerable challenge for both the equipment and user. A further criterion is the choice of the detector, since the implementation clearly depends on the flow rate of the system (see Section 2.7.1.4.3).

The length of the column depends primarily on the type of chromatography.

RP chromatography: 10–250 mm
IEC: 10–100 mm
SEC: min. 150 mm
HIC: 30–250 mm
HILIC: 30–250 mm

As a rule of thumb, for RPC the following is considered to hold true: 1 cm of column length is required to determine one compound. When using ≤ 3 μm particles and gradients, the column length can usually be halved.

With suitable selectivity, ten compounds can usually be separated on a 50 mm long column; see also Chapter 2.7.3 on UPLC.

A further rule is that for compounds with ln K/%B curves with high slope, a shorter column can be selected. A high slope of the curve shows that even a small increase in the amount of eluent B will cause a strong reduction in the retention. This is particularly relevant for IEC, HIC, and HILIC.

Besides the efficiency and resolution, the particle size also determines the back-pressure in the system. Regarding the robustness of the system, it is quite useful to work with 4 and 5 µm materials when high resolution is required. Today, for proteomics analysis by RP chromatography, columns with 3–5 µm packing material, 75 µm *i.d.*, and lengths between 100 and 150 mm are used. For 2D-chromatography, an ion-exchange column 30–50 mm in length is used.

2.7.1.4.2 Packing Materials/Surface Covering

The selection of the packing material depends primarily on the selectivity that is required. Additionally, there are some further criteria by which a separation can be optimized. The use of modern, highly pure type B silica gels is usually the correct choice. In particular, when using a mass spectrometer or ELSD as detector, it is better to use a column that has no or very little bleeding, as otherwise this will clearly increase the noise and lower the sensitivity. Polymeric or dimer-bound alkyl chains are clearly more stable than monomer-bound phases. Under strongly acidic conditions, the bonding of shorter alkyl chains (C_1, C_4) is more easily broken than that of longer (C_8, C_{18}) ones. In former times, mainly 300 Å materials were used for the separation of peptides since with larger pores a better surface covering was attained. With the modern phases, 100–150 Å C_{18} phases can also be used. The disadvantage of the higher silanol activity caused by incomplete surface covering is no longer valid and the capacity is higher.

2.7.1.4.3 Detectors

MS Detector

The mass spectrometer is surely the most frequently used detector in the micro- and nanoLC range. This detector is universally applicable, highly selective, and sensitive.

To attain maximum sensitivity and selectivity, other detectors are also very efficient for certain applications.

UV/Vis Detector

Although this detector does not have a high specificity, it is very commonly used. This is due to the simple and favorable application and the large number of different small-volume flow cells. Unfortunately, the data for pathlength and volume do not show the same efficiency. With Z- and U-cells, the signal-to-noise ratio is clearly larger than with normal flow cells.

To assess which cell is practicable for a given column dimension, the following guidelines can be used:

- Maximum measuring cell volume: a peak must flush the flow cell 10 times.
- Optimal measuring cell volume: a peak must flush the flow cell 50 times.
- Example: With a peak width of 30 s and a flow rate of 5 µL min^{-1}, the peak volume amounts to 2.5 µL. Thus, the cell may have a maximum volume of 0.25 µL. The optimal volume is 0.05 µL = 50 nL.

Electrochemical Detector

This shows high specificity and sensitivity, and for some special analyses can be the detector of choice. One of the standard operational areas, for example, is in the analysis of catecholamines by means of microdialysis. Here, the flow cell volume must be known in order to select the appropriate column dimension.

Fluorescence Detector

Standard fluorescence detectors are not optimally applicable in micro-HPLC, since there are no suitable small-volume flow cells. To benefit from the high sensitivity and selectivity of this detector in micro- and nanoLC, laser-induced fluorescence (LIF) detectors need to be used. A disadvantage is that full wavelength coverage is not available, but depending upon the kind of laser in each case a certain wavelength is specified.

Available wavelengths:

Wavelength (nm)	Laser (type)
325	helium-cadmium
442	helium-cadmium
488	argon ion
514	argon ion
532	double diode
543	helium-neon
568	krypton
594	helium-neon
635	diode
650	diode
670	diode
785	diode

Typical compounds with primary fluorescence are:

pyrenes, aflatoxin:	325 nm
riboflavin:	442 nm
doxo/daunorubicine:	488 nm
rhodamine:	488 nm

By derivatizing the samples with a fluorescence marker, the area of application is considerably expanded.

Dansyl chloride	325 nm
OPA	325 nm
NDA	442 nm
CBQCA	442 nm
FITC	448 nm
DTAF	448 nm
NBD	448 nm

Evaporating Light-Scattering Detector (ELSD)

The development of these detectors has undergone considerable progress in recent years, so that today commercial devices are available that operate successfully at flow rates of down to 5 µL min^{-1}. Thus, columns with *i.d.* of up to 300 µm can be used.

The ELSD has several advantages. The detection principle is based on light scattering by compound particles and is thus independent of chromophores or electrochemically active groups. The response depends directly on the quantity of the eluted compound. Therefore, unknown compounds can also be quantified. The ELSD is also applicable in gradient methods.

References

1 J. Maier-Rosenkranz, in S. Kromidas (Ed.), "More Practical Problem Solving in HPLC", Wiley-VCH, Weinheim, 2004, ISBN 3-527-31113-0.

2 A. Prüß, C. Kempfer, J. Gysler, J. Maier-Rosenkranz, T. Jira, "Peak shape improvement of basic analytes in capillary liquid chromatography", *J. Sep. Sci.* 28 (2005), 291–294.

2.7.2
Microchip-Based Liquid Chromatography – Techniques and Possibilities

Jörg P. Kutter

2.7.2.1 **Introduction**
The last decade has seen a stormy development in the field of planar miniaturized analytical devices, also dubbed micro-total analysis systems (µ-TAS) or lab-on-a-chip systems [1, 2]. These devices have potentially a number of advantages over more conventional chemical equipment. The consumption of chemicals and sample volume as well as energy can be reduced, and consequently also the production of chemical waste and other unwanted by-products, such as heat. Many identical units can be fabricated and operated simultaneously and in parallel, thereby potentially reducing fabrication and running costs while at the same time increasing the throughput (e.g., analyses per time). From a separations point of view, miniaturization offers the potential to perform fast separations without compromising efficiency, while the planar format allows the easy combination and integration of other functional elements together with the separation, such as a sample preparation step and/or a post-column derivatization step [3, 4].

From the pioneering days of µ-TAS, separation methods, such as electrophoresis and chromatography, have been at the heart of many of these miniaturized systems. In fact, one of the first devices was a gas chromatograph realized on a silicon wafer, which was published as early as 1979 [5]. Later, it soon became evident that liquid separation systems would not only be less challenging to realize from an engineering point of view, but also more relevant for applications in the biochemical and biomedical fields. Also, because non-pulsating, reliable micropumps were, and still are, hardly available, many of the examples involving chromatographic separations on microchips make use of electroosmosis to drive the liquid phase through the system. Another reason for relying on electric fields rather than high pressures is the fact that the majority of planar miniaturized devices fabricated to date cannot withstand pressures of much more than a few tens of bars, even though much higher pressures are desirable for more efficient separations.

This chapter gives a very brief overview on the possibilities of how liquid-phase chromatographic separations can be realized on microchips, where the pros and cons of the various techniques lie, and how easy or tedious the implementation of optimization procedures might be. Because of the relatively young age of microchip-based chromatography, the focus in the research community has been on demonstrating different separation principles. Closer evaluation of the performance and, in particular, optimization is only just beginning to attract more attention.

2.7.2.2 Techniques

2.7.2.2.1 Pressure-Driven Liquid Chromatography (LC)

There are only very few examples of liquid chromatography on microchips using pressure to drive the mobile phase. This is mainly due to the limited pressure tolerance of most microchip designs to date, which, in turn, is mainly a result of the planar geometry and the materials used (glass, silicon, and polymers). One of the early examples featured a silicon-based chip with an integrated frit to retain a bed of packed particles, a split injector, and a long pathlength detection cell for absorbance measurements [6]. From a purely chromatographic point of view, the results were less than convincing, showing rather low plate numbers. This was most certainly due to the packing quality that could be achieved with this system, together with extra-column band-broadening mechanisms that had not been optimized at that stage. Moreover, because of the pressure-induced parabolic velocity profile, further band-broadening had to be accepted. Not surprisingly, therefore, to avoid these kinds of problems, researchers started to look at easier alternatives to realize chromatographic systems on chips, namely open-tubular or open-channel systems. To reduce the band broadening due to the parabolic flow profile, alternative pumping mechanisms, mainly involving electroosmosis, were also increasingly investigated.[1]

2.7.2.2.2 Open-Channel Electrochromatography (OCEC)

From a strictly engineering point of view, it is much easier to "just" derivatize the walls of the channels as opposed to creating a packed bed. In particular, if glass chips are to be used, all of the many already existing protocols for derivatizing glass capillaries or silica beads can potentially be used. Such systems have been realized on microchips and showed fairly decent separation performances, but also suffered from the same problems as their conventional counterparts, namely the limited loadability due to the small phase ratio. As expected from theory, smaller channel dimensions (typically, shallower channels) resulted in improved separation performance, with van Deemter minima at smaller plate heights and at higher linear velocities [7, 8].

2.7.2.2.3 Packed-Bed Electrochromatography

Despite the initial problems of packing a chip, the idea was never completely abandoned. Alternative packing procedures to pressure packing are available and have been used on microchips. Oleschuk et al. have used electrokinetic forces to pack silica particles into a specially designed chip [9]. The technique resembles the working of a snowplough ejecting snow towards the borders of the road. The packings resulting from this procedure proved to be fairly homogeneous and uniform. In order to keep the silica particles inside the channel, a special weir design was adopted. In this technique, the particles were packed from the side

---

[1] Just before type-setting Agilent Technologies presented a microchip LC/MS system featuring a polymer-based LC chip packed with particulate chromatographic material and operated in pressure-driven mode.

into a channel section that had weir structures upstream and downstream from the packing site. These weirs were only slightly etched rims in the channel leaving only a small gap between them and the cover plate. This gap was not wide enough to allow particles to squeeze through, but allowed liquid that was percolated through the packed bed to move on to the next compartment of the microsystem.

Another approach was pursued by Ceriotti et al. [10]. They used tapered channels in order to prevent particles from leaving the channel in the direction of the waste reservoir, both during packing and during the chromatographic separation. As packing material entered the tapered section during packing, the particles were jammed together, restricting their freedom to move and thereby creating a blockage. This prevented more packing material from moving further downstream, but still allowed the mobile phase to percolate through the interstices. With this approach, the researchers could avoid having to realize or integrate frits, which are more difficult to make and are known to be sources of band broadening and bubble generation.

2.7.2.2.4 Microfabricated Chromatographic Beds (Pillar Arrays)

Microfabrication offers the unique possibility of creating a packing that is both geometrically perfect in the arrangement of the individual "particles" and uniform with respect to the size distribution of the particles. In fact, it is potentially possible to design and determine all aspects of a packing, from the shape of the "particles" to their size and the distance between them. However, some restrictions are imposed by the microfabrication techniques, e.g., with respect to the minimal inter-particle distance. Of course, once these artificial packings have been fabricated, the surfaces need to be derivatized by standard chemical processes to yield chromatographically active phases. Electro-driven separations of peptides from a digested protein have been successfully demonstrated on such chips with C_{18}-modified surfaces [11]. Recently, theoretical studies have indicated that pressure-driven liquid chromatographic separation in systems with micro-fabricated chromatographic pillar arrays, where the "particles" are also porous, should result in very efficient separations [12]. This goes back to arguments originally expressed by Knox et al. [13]. Unfortunately, such phases with porous particles or pillars are not yet available, as they also pose a challenge to the microfabrication skills.

2.7.2.2.5 In Situ Polymerized Monolithic Stationary Phases

A powerful method to realize chromatographic beds on-chip without having to worry about packing procedures is to rely on different polymeric stationary phases, which can be created *in situ*, i.e., within the channel. To achieve this, a cocktail of monomers and other chemicals is introduced into the channel. This is usually accomplished quite easily since such a mixture has a relatively low viscosity and can be introduced by means of either pumps, syringes or even capillary forces. Once the channel is filled, polymerization is achieved, depending on the particular chemistry, by using initiator molecules and applying heat, or by utilizing photons (light of the appropriate wavelength) to start the process. An additional benefit of the latter method is that masks can be used to define those areas of the chip (and thereby the channel network) where light is allowed to start the polymerization. In this way, areas where no polymer is wanted can be kept free, for example, the injection region or the detection region. A range of different chemistries has been used to realize stationary phases in this way, mainly based on either acrylates, acrylamides, styrene–divinylbenzene co-polymers, or sol-gel chemistry (see, e.g., [14–18]).

2.7.2.3 Optimization and Possibilities

At the time of writing (winter 2004) chromatographic techniques on microchips are still relatively new and have not yet matured to the point where they can be routinely applied in a great number of applications. Therefore, the troubleshooting and optimization aspect has so far mainly been limited to tweaking the systems to the point that a separation could be demonstrated. Because a lot of emphasis has been placed on the technological side, not much focus has been directed towards the chromatographic aspects to date. Nevertheless, a number of advantages are available on chips, which, when properly used, can help boost the performance of miniaturized chips for chromatographic separations. Some of these issues will be mentioned below and discussed with respect to the optimization of overall performance.

2.7.2.3.1 Separation Performance

An important parameter determining the separation performance is the available separation length. Naturally, on a microfabricated device separation length is not available in arbitrarily large quantities. At least not in terms of straight channels. If more channel length is required than is available as a straight segment, the channel is often coiled up or arranged in a meander in order to fit a longer separation length into a small footprint. Closer investigations have shown, however, that the folding of the channels must fulfill certain requirements, otherwise it negatively affects the separation performance by introducing further band broadening. The geometry of the turns (radius of curvature, channel width) is related to the transition velocities of molecular bands around the turns and the diffusional transport across the width of the channel in the turn section. When following these design rules, virtually dispersion-free turns can be designed and very efficient separations on relatively long microchannels have been demonstrated (see, e.g., refs. [19, 20]). On the other hand, miniaturization can help to avoid very long separation lengths and still offer excellent separation performance. This is achieved on the one hand by improving mass transport through reduced diffusion pathlengths, but even more so by reducing extra-column band broadening. Typical sources of extra-column band broadening in traditional separation equipment are the injection step, the detection process, and (at least in electro-driven methods) Joule heating. Through the possibility of monolithic integration, injectors and detectors can be coupled directly to the separation column, with almost no dead volume, and the exquisite fluidic control allows the injection of very well-defined small plugs. Joule heating is reduced and affects separation performance much less in microchips, mainly because of the large surface-to-volume ratio of the microfluidic channels and the larger thermal mass of chips as compared to capillaries [21].

2.7.2.3.2 Isocratic and Gradient Elution

Optimization of liquid chromatographic separations is often achieved through changing the mobile phase composition to be used together with a given stationary phase. This variation in the mobile phase composition can be done in a discrete way (often by off-system mixing of two solvents in a certain ratio) or in a continuous way by the use of gradient formers. The excellent fluidic handling capabilities of microchips allows one to achieve almost any mixing ratio between two solvents directly on-chip, making the adjustment of the mobile phase composition very fast and readily available. This so-called solvent programming was successfully demonstrated in the optimization of micellar electrokinetic chromatography (MEKC) and open-channel electrochromatography (OCEC) separations on-chip [8, 22]. Figure 1 shows an example of a separation of four coumarin dyes in open-channel electrochromatography employing a very fast gradient elution. These procedures can allow fast optimizations without the need for further off-chip liquid handling, thereby potentially preparing such a system to be used in a computer-controlled environment for automated separation optimizations.

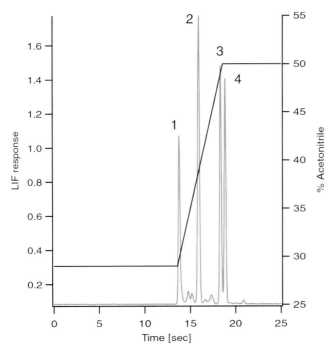

Fig. 1. OCEC separation of four coumarin dyes on C_{18}-modified channel walls using a 5 s gradient elution going from 29% to 50% acetonitrile in borate buffer. (Reprinted with permission from [8], copyright (1998) American Chemical Society).

2.7.2.3.3 Tailor-Made Stationary Phases

Through the possibilities of closely defining the properties of the stationary phase by choosing an appropriate cocktail of monomers and additives, the *in situ* generation of photoinduced polymeric monoliths offers a lot of design freedom and options in the optimization process. By carefully choosing a monomer with a certain functional group, the nature and the strength of the chromatographic interaction can be adjusted to the needs of the separation problem at hand. Additives can help to improve the electroosmotic flow that can be achieved on such phases, or they can help in defining the pore size, thereby tuning the kinetic properties and the back-pressure of the stationary phase [16]. By further taking advantage of the possibilities to locally determine the characteristics of a stationary phase through the use of chemical modifications combined with photomasking, columns with different functionalities seamlessly coupled together in the same channel can be realized. Finally, on microchips it is in principle as easy to realize a single channel as it is to make a row of parallel channels all equipped with an injector and connected to a detecting device. If different types of stationary phases are synthesized in these parallel channels, the optimization of a given separation problem could be automated and considerably accelerated. This is especially true if, at the same time, the above-mentioned liquid handling capabilities for generating various isocratic and gradient conditions are employed as well.

2.7.2.3.4 Sample Pretreatment and More-Dimensional Separations

One of the particular strengths of microchip devices is the possibility of coupling different functionalities together, without the need for complicated off-column couplers, which would only add extra dead volume and further contribute to band broadening. In this way, sample clean-up or pre-concentration steps can be performed ahead of the separation step, thereby helping to improve the quality of the separation by reducing interfering species and improving peak shapes and signal-to-noise ratios. Solid-phase extraction and isotachophoresis are but two techniques that have been demonstrated on microchips as pretreatment techniques coupled to on-chip separations [23, 24]. The ease of coupling on microchips by virtue of the monolithic integration possibilities combined with fast separation speeds also makes microchips predestined for two- or more-dimensional separation techniques. For comprehensive two-dimensional separations, where everything coming off the first dimension is "chopped" into segments, which get injected into the second dimension, there are stringent requirements for dead-volume-free coupling, efficient injection techniques, and very fast separations in the second dimension. All of these requirements can be fulfilled on chips, and excellent 2D separations of enzymatic digests of proteins have been demonstrated [25].

2.7.2.3.5 Issues and Challenges

One of the big issues in chip operation is the formation of bubbles in the channels. Once formed, bubbles can prove to be very hard to eliminate, unless special care is taken in the design of the channel geometry to avoid places where bubbles can get stuck in the first place, and to allow mobilization of bubbles with relatively moderate pressures. Bubbles can disrupt electrical conductance through a channel segment, thereby stopping liquid flow induced by electroosmosis. Unless current monitoring is a part of the experimental set-up, these failures might go unnoticed for a long time. Bubbles can also create voids or non-wetted areas, where no interaction between the analytes and the stationary phase takes place, thus reducing the separation performance. Finally, bubbles can also interfere with the detection process.

Another concern when utilizing electro-driven fluid manipulations on chips is electrolytic depletion of the background buffer solution. Apart from being a very likely source of the generation of gas bubbles, the electrolytic processes at the electrodes also cause a change in the pH of the solution. Because of the generally small volumes, the buffer capacity is soon exhausted so that the change in pH manifests itself in changes in, e.g., the electroosmotic mobility and the charge states of the analytes. Larger buffer reservoirs, lower operating voltages or specially designed electrode configurations are some of the strategies to address this issue.

In terms of dealing with failures caused by bubbles or other reasons leading to a malfunction of the chip, there is a potential luxury available when working with microchips that is typically not available with more conventional set-ups: multiple, identical systems and redundancy. Unless only very tiny amounts of sample are available, parallel systems can be used with identical separation columns so that identical analyses can be performed simultaneously. Thus, if there is a failure in one microchip, information can still be retrieved from the others. In a way, this approach can be compared to troubleshooting and maintenance of modern electronic equipment, where malfunctioning sub-units are identified and replaced instead of repairing the individual defective component. Troubleshooting on microfabricated separation systems in general is made more difficult by the fact that probably only a few generic chip solutions will be available, for which troubleshooting and optimization could follow a pre-determined set of rules and procedures or algorithms. Most of the chip solutions will most likely be tailor-made for a specific separation problem, making troubleshooting as individual as the chip design and operation itself.

2.7.2.4 Application Examples

Even though many liquid chromatographic realizations on chips merely prove that the concept is also feasible in this format, there are a few examples of work with a particular application in mind. In the following, some of these examples are briefly presented.

Fintschenko et al. have described the electrochromatographic separation of polycyclic aromatic hydrocarbons (PAHs) on microchips, using an *in situ* synthesized acrylate-based porous polymer monolith [17]. By employing masking tape and ultraviolet light, they initiated polymerization only in defined regions leaving the injection area and a detection window free of monolith. The degree of cross-linking and the average pore size was determined by carefully choosing the composition of the monomer solution. The authors demonstrated the separation of 13 PAHs, although not all peaks were baseline-resolved. Efficiencies of up to

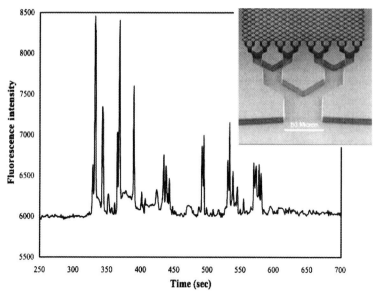

Fig. 2. Electrochromatographic separation of peptides from a tryptic digest of ovalbumin on C_{18}-modified pillars (main picture), and part of a microfabricated pillar array showing the inlet region of the array (upper left). (Reprinted with permission from [11] and [26], copyright (1998) American Chemical Society and (1999) Elsevier).

200,000 plates per meter using a 7 cm long channel were achieved, but relatively large standard deviations were still experienced. Some of the problems were traced back to pH changes due to electrolysis at the electrodes, and pressure build-ups before the monolith.

The group of Hjertén and co-workers has also reported on the use of continuous beds (porous monoliths) for electro-driven and pressure-driven chromatographic separations on chip [16]. Their stationary phases were based on methacrylamide chemistry and included functional groups for retention (C_3), ion-exchange interaction (ammonium moieties), and improved electroosmosis (sulfonate groups). Their investigations showed essentially flat van Deemter plots for higher linear velocities, indicating that *in situ* fabricated stationary phases on chips work very well and do not suffer from wall effects as might be expected from the other than circular cross-section of most microchannels. Alkylphenones were success-fully separated on a C_3-derivatized continuous bed using a 28.1 cm serpentine channel and a field strength of 500 V cm^{-1}. The plate numbers achieved were of the order of 200,000–300,000 per meter. Fast anion-exchange chromatography of four standard proteins was also demonstrated on a 4.5 cm channel using a pressure of 22 bar (flow rate 0.1 mL min^{-1}).

The applicability of microfabricated pillar arrays (see Section 2.7.2.2.4) for the electrochromatographic separation of peptides from a tryptic digest of ovalbumin was demonstrated by He et al. [11]. Their separation column was 4.5 cm long and

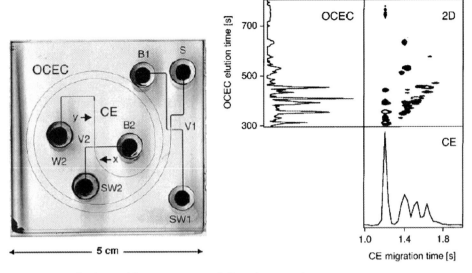

Fig. 3. Layout of a chip used for 2D separations (left), and an example of a 2D separation of peptides from a tryptic digest of β-casein (right). (Reprinted with permission from [25], copyright (2001) American Chemical Society).

consisted of a pillar array that created trenches between the pillars 1.5 μm wide and 10 μm deep. The surface of the pillars was modified with a C_{18}-silane. The separation resolved a number of peptide components within approximately 10 min under isocratic conditions, indicating the potential of microchip-based chromatographic techniques for more complex samples as well. In Fig. 2, part of a microfabricated pillar array is shown in the upper right-hand corner, while the main part shows the chromatogram from the aforementioned peptide separation.

A final example again concerns a separation of peptides. In this case, it is the two-dimensional separation of peptides from a tryptic digest of β-casein [25]. In the first dimension, separation was achieved by open-channel electrochromatography, i.e., a 25 cm microchannel coated with a C_{18} layer. The second dimension was then a very fast capillary electrophoretic separation in a channel just 1.2 cm long, allowing a rather comprehensive sampling of the first dimension (one injection every 3 s from the first into the second dimension). Even though the two separation techniques were not fully orthogonal, the 2D separation resulted in a large number of resolved spots and a 50% increase in information content compared to one-dimensional open-channel electrochromatography (see also Fig. 3). Such 2D chip separations could prove useful for fast fingerprinting of protein mixtures and protein digests.

Table 1. A qualitative comparison of the most common techniques used to realize liquid chromatography on a microchip.

Type of stationary phase	Ease of preparing stationary phase	Packing quality	Load-ability, phase ratio	Design flexibility, selectivity	Comments
Open channel, coated walls	easy	n.a.	low	limited	if detection sensitivity is not an issue
Particles, pressure-packed	cumber-some	low	high	moderate	standard HPLC particles can be used
Particles, electro-packed	moderate	moderate	high	moderate	standard HPLC particles can be used
Microfabricated pillars, coated	moderate	excellent	moderate	limited	porous pillars not yet available
Monoliths, polymerized *in situ*	moderate	n.a.	high	high	photomasking can be employed

2.7.2.5 Conclusions and Outlook

Liquid-phase separation techniques are essential components of lab-on-a-chip systems. Most of the traditionally used separation techniques have already been implemented on microchips. Additionally, this new format offers possibilities for performing separations and for increasing the separation performance (in terms of resolution, selectivity, speed, and throughput) that go beyond what is possible on conventional systems. Even so, while electrophoretic chip-based separation systems are used routinely and reliably in a number of applications (e.g., DNA fragment analysis), the liquid chromatographic systems are only about to enter that stage. As briefly outlined in this chapter, there is a choice of how to implement a chromatographic system on a chip, and all of these choices have advantages and disadvantages (see Table 1 for a qualitative comparison). It will therefore greatly depend on the particular application to determine which type of stationary phase and which variant of chromatography will be the most appropriate for the given problem. A range of possibilities for tuning and optimizing the performance is at hand, and the future will undoubtedly show ingenious implementations of many of the discussed optimization strategies plus a number of new ones.

References

1 D. R. Reyes, D. Iossifidis, P.-A. Auroux, A. Manz, *Anal. Chem.* **2002**, *74*, 2623–2636.

2 P.-A. Auroux, D. Iossifidis, D. R. Reyes, A. Manz, *Anal. Chem.* **2002**, *74*, 2637–2652.

3 G. J. M. Bruin, *Electrophoresis* **2000**, *21*, 3931–3951.

4 J. P. Kutter, *TrAC – Trends in Analytical Chemistry* **2000**, *19*, 352–363.

5 S. C. Terry, J. H. Jerman, J. B. Angel, *IEEE Trans. Electron. Devices* **1979**, *26*, 1880–1887.

6 G. Ocvirk, E. Verpoorte, A. Manz, M. Grasserbauer, H. M. Widmer, *Analytical Methods and Instrumentation* **1995**, *2*, 74–82.

7 S. C. Jacobson, R. Hergenröder, L. B. Koutny, J. M. Ramsey, *Anal. Chem.* **1994**, *66*, 2369–2373.

8 J. P. Kutter, S. C. Jacobson, N. Matsubara, J. M. Ramsey, *Anal. Chem.* **1998**, *70*, 3291–3297.

9 R. D. Oleschuk, L. L. Shultz-Lockyear, Y. Ning, D. J. Harrison, *Anal. Chem.* **2000**, *72*, 585–590.

10 L. Ceriotti, N. F. de Rooij, E. Verpoorte, *Anal. Chem.* **2002**, *74*, 639–647.

11 B. He, J. Ji, F. E. Regnier, *J. Chromatogr. A* **1999**, *853*, 257–262.

12 P. Gzil, N. Vervoort, G. V. Baron, G. Desmet, *Anal. Chem.* **2003**, *75*, 6244–6250.

13 J. H. Knox, *J. Chromatogr. A* **2002**, *960*, 7–18.

14 E. C. Peters, M. Petro, F. Svec, J. M. J. Fréchet, *Anal. Chem.* **1998**, *70*, 2288–2295.

15 K. Cabrera, D. Lubda, H. M. Eggenweiler, H. Minakuchi, K. Nakanishi, *Hrc-Journal of High Resolution Chromatography* **2000**, *23*, 93–99.

16 C. Ericson, J. Holm, T. Ericson, S. Hjertén, *Anal. Chem.* **2000**, *72*, 81–87.

17 Y. Fintschenko, W.-Y. Choi, S. M. Ngola, T. J. Shepodd, *Fresenius J. Anal. Chem.* **2001**, *371*, 174–181.

18 N. Tanaka, H. Kobayashi, N. Ishizuka, H. Minakuchi, K. Nakanishi, K. Hosoya, T. Ikegami, *J. Chromatogr. A* **2002**, *965*, 35–49.

19 C. T. Culbertson, S. C. Jacobson, J. M. Ramsey, *Anal. Chem.* **2000**, *72*, 5814–5819.

20 S. K. Griffiths, R. H. Nilson, *Anal. Chem.* **2002**, *74*, 2960–2967.

21 N. J. Petersen, R. P. Nikolajsen, K. B. Mogensen, J. P. Kutter, *Electrophoresis* **2004**, *25*, 253–269.

22 J. P. Kutter, S. C. Jacobson, J. M. Ramsey, *Anal. Chem.* **1997**, *69*, 5165–5171.

23 D. Kaniansky, M. Masar, J. Bielcikova, F. Ivanyi, F. Eisenbeiss, B. Stanislawski, B. Grass, A. Neyer, M. Johnck, *Anal. Chem.* **2000**, *72*, 3596–3604.

24 B. S. Broyles, S. C. Jacobson, J. M. Ramsey, *Anal. Chem.* **2003**, *75*, 2761–2767.

25 N. Gottschlich, S. C. Jacobson, C. T. Culbertson, J. M. Ramsey, *Anal. Chem.* **2001**, *73*, 2669–2674.

26 B. He, N. Tait, F. E. Regnier, *Anal. Chem.* **1998**, *70*, 3790–3797.

2.7.3
Ultra-Performance Liquid Chromatography

Uwe D. Neue, Eric S. Grumbach, Marianna Kele, Jeffrey R. Mazzeo, and Dirk Sievers

2.7.3.1 Introduction

There is a continued need to improve the performance of HPLC separations. On the one hand, people want faster analyses due to increased pressure to get rapid results. Such a situation is often encountered in the analysis of animal or human plasma or urine samples in the pharmaceutical industry. On the other hand, a higher separation performance is often needed. In some cases, the higher separation power simplifies method development. In other cases, such as complex peptide maps, it is desirable to obtain meaningful results in a reasonable time frame and analysis times of the order of 24 h, which could result in optimal separations, are not tolerable.

In Chapter 1.2 of this book, we discussed the issue of fast gradient separations and showed that one can accomplish high-quality rapid separations with short columns packed with small particles. How fast one can go at a still reasonable performance is limited by the available pressure. A similar situation arises if one wants to accomplish a truly high-powered separation within a reasonable time frame. Once again, the available pressure is the limiting factor. Thus, it is absolutely clear that better separations can be accomplished if more pressure can be applied to drive the LC separation.

However, this has not been possible until fairly recently [1]. As one is using higher pressures, i.e. pressures beyond 40 MPa (400 bar, 6000 psi), a significant amount of heat is generated in the column due to friction. This, in turn, results in a non-uniform flow over the cross-section of the column, which limits column performance. Nonetheless, as Jorgenson [1] has pointed out, this problem can be overcome by using a smaller column diameter, which enables dissipation of the heat generated inside the column. A smaller column diameter for such a high-performance column then imposes additional requirements on instrument design, such as extra-column bandspreading and sampling rate. The consequence of this is that the only way to take full advantage of the capabilities of columns packed with smaller particles is to create an instrument that has all the necessary features, such as pressure capability and flow-rate range, detector speed, low extra-column bandspreading, and related prerequisites.

To this end, a new branch of HPLC, termed Ultra-Performance Liquid Chromatography[TM] (UPLC[TM]) technology, was created [2, 3]. High-pressure fluidics and columns with a very small particle size are the key features of this new technology, but one should not neglect the high-speed, low-dispersion detectors that allow high-sensitivity detection in a small cell volume. The first-generation instrumentation, the ACQUITY UPLC[TM] instrument, is capable of working at pressures of 100 MPa (1000 bar, 15000 psi) and operates columns with 1 or 2 mm internal diameter packed with 1.7 μm particles. The combination of the simultaneous optimization of pump, injector, column, and detector technology results in increased speed, sensitivity, and resolution compared to classical LC. In the following, we outline the theoretical background and show the results that have been obtained with this UPLC technology.

2.7.3.2 Isocratic Separations

The principles of how to operate an HPLC column under optimal isocratic conditions were outlined a long time ago by Guiochon and coworkers [4]. The maximum column performance at the lowest pressure is always achieved around the minimum of the van Deemter curve. If we reduce the particle size, the back-pressure increases – at equal velocity – inversely proportionally to the square of the particle diameter. At the same time, the velocity at the minimum of the van Deemter curve increases with decreasing particle diameter. Thus, the pressure at the optimum Δp_{opt} rises with the third power of the reduction in particle diameter d_p.

$$\Delta p_{opt} \approx \frac{1}{d_p^3} \tag{1}$$

Note that the pressure at this optimum is, to a first approximation, independent of the viscosity of the mobile phase or of the temperature. The performance, on the other hand, increases by two separate factors. On the one hand, resolution increases inversely proportionally to the particle diameter, and on the other hand, the analysis time decreases with the reduction in particle size. Both of these features are highly desirable: higher performance in a shorter time. This is the fundamental driving force behind UPLC.

The thought process outlined in the last few sentences is the best and in practice most interesting way to gain from the availability of the smaller particles. In this approach, we reduce the column length and the particle size simultaneously. In this case, the maximum column performance remains constant [4–6], but we gain in the speed of the analysis. The pressure still increases with the reduction in particle size, but now only with the second power instead of the third power as discussed above.

This gain can be seen in plots of plate count versus analysis time [5, 6]. An example of such a plot for a high-performance separation is shown in Fig. 1. Note that the time axis is a logarithmic axis. The assumption in this plot is a retention factor of about 10, combined with a standard solvent composition with the same viscosity as water. The performances of three columns are shown: a 25 cm as well as a 30 cm 5 μm column as examples of the performance achievable by classical HPLC, and a 10 cm UPLC column showing the gain in speed at the same performance level as the 30 cm 5 μm column. The maximum plate count that is achievable with the UPLC column is about the same as that on a 30 cm column packed with 5 μm particles. However, this high performance (about 25000 plates) can now be achieved in about one-tenth of the time needed to do the job with the classical HPLC column. Conversely, if we force this column to be used for faster analyses, the column performance drops drastically. The 30 cm classical column

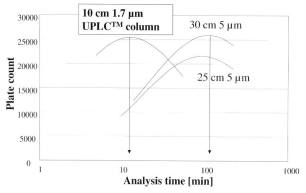

Fig. 1. Plot of plate count *vs.* analysis time for high-powered isocratic separations.

reaches the pressure limit of the HPLC instrumentation at a longer analysis time than the point at which the optimum performance of the UPLC column "kicks in". At this pressure limit, it achieves only roughly half the performance of the UPLC column. On the other hand, the UPLC column can achieve an even much shorter run time due to the higher pressure capabilities of the instrument.

An example of such a high-powered isocratic separation is shown in Fig. 2. The chromatogram shown is the degradation profile of terbenafine, degraded in 8.0 N HCl. The analysis was carried out isocratically at pH 10 using an ACQUITY UPLC™ BEH C$_{18}$ column. In Chapter 1.3 of this book, we pointed out the large selectivity differences that one can achieve with a pH change of the mobile phase. In this context, it should be mentioned that this column contains a packing of the inorganic-organic hybrid family. This type of packing is over an order of magnitude

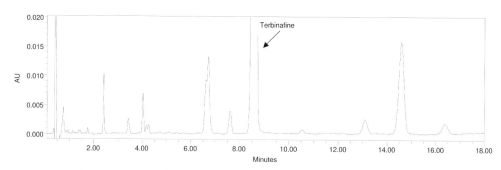

Fig. 2. High-powered isocratic UPLC separation.
Degradation profile of terbinafine (degradation in 8.0 N HCl).
Column: 100 mm × 2.1 mm 1.7 µm ACQUITY UPLC™ BEH C$_{18}$ column.
Conditions: flow rate: 0.5 mL min^{-1}; mobile phase: 35% 20 mM ammonium hydrogencarbonate (pH 10.0)/65% acetonitrile.
Temperature: 30 °C.
Instrument: Waters ACQUITY UPLC™ with TUV detector.
Detection: 210 nm.

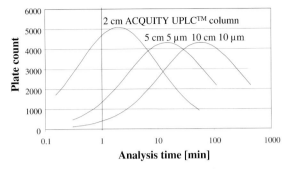

Fig. 3. Plot of plate count *vs.* analysis time for fast isocratic separations using short columns.

more stable than a silica-based C$_{18}$ under aggressive alkaline conditions [7]. The most important feature of the analysis shown in Fig. 2 is the partial resolution of the two peaks eluting just after 7 min. These two compounds would be unresolved on a column based on 5 μm particles. Note also that this high-resolution isocratic separation is complete in under 20 min.

If we are interested in very fast analyses, we can apply the same thought process as outlined above for the long column. However, as a starting point, we will now select a fast and short 5 cm 5 μm or a 10 cm 10 μm column as our reference point. This comparison is shown in Fig. 3. At the point of maximum performance, the 1.7 μm column is over 30 times faster than the 10 μm column, and the gain in speed is nearly an order of magnitude compared to the 5 μm column. With the UPLC column, we can achieve maximum performance in around 2 min, and

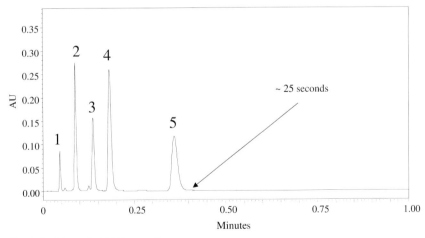

Fig. 4. Fast isocratic UPLC separation.
Column: 20 mm × 2.1 mm 1.7 μm ACQUITY UPLC^{TM} BEH C$_{18}$ column.
Conditions: flow rate: 1.4 mL min^{-1}; mobile phase: 70 : 30 acetonitrile/water.
Detection: 254 nm.
Samples: 1. thiourea, 2. toluene, 3. propylbenzene, 4. butylbenzene, 5. hexylbenzene.

even superior performance at a run time of under 1 min. If we examine the gain in separation power that UPLC offers for this 1 min analysis, we see that it is between three and four times what a 5 µm column can offer, and approximately ten times compared to a 10 µm particle column.

An example of such a high-powered fast isocratic separation is shown in Fig. 4. It was achieved on a 2.1 mm × 20 mm 1.7 µm ACQUITY UPLC^{TM} BEH C_{18} column. The separation was completed in less than 25 s. With a classical HPLC column on a classical instrument, a separation of the same quality will require about 5 min. This is a clear demonstration of how one can take advantage of the higher performance capabilities of the UPLC system to improve throughput. Fast separations of this type result in very narrow peak widths (in time units). The earlier peaks in this chromatogram are narrower than 1 s. This, in turn, requires a fast detector response as well as fast sampling rates. At the same time, the peak volume is very low, due to the small column volume and the high column performance. Thus, high-speed, low-volume detectors are an important feature of the overall ACQUITY UPLC^{TM} system desig.

2.7.3.3 Gradient Separations

These two examples – a high-performance separation and a very fast separation – demonstrate what UPLC can offer compared to HPLC in the case of isocratic separations. The same is true for gradient separations as well. However, gradient separations, especially fast gradient separations, work differently to isocratic separations. In Chapter 1.2 "Fast Gradient Separations", we have outlined these somewhat different principles for gradient separations. In general, a fast gradient separation requires a significantly higher linear velocity (= flow rate) than a fast isocratic separation in order to achieve the optimum separation power in a given analysis time. Theoretical estimates indicate that a good flow rate for a 1 min gradient separation on a 20 mm × 2.1 mm 1.7 µm ACQUITY UPLC^{TM} column is around 1 mL min^{-1} or higher. For a 3 cm column packed with the same particles, even higher flow rates are recommended for the same short 1 min gradient. Of course, method development may dictate slightly different values from these rough estimates, which are based on the assumption that a wide gradient is executed, and that the molecular weight of the analytes is in the range of 200 to 300. Note that the equivalent flow rate for a 4.6 mm i.d. column would be above 5 mL min^{-1}.

An example of a high-powered fast gradient separation is shown in Fig. 5. A group of pharmaceutical standards is separated with a rapid, 1 min gradient on a 20 mm × 2.1 mm 1.7 µm ACQUITY UPLC^{TM} column at a flow rate of 1.5 mL min^{-1}. The gradient is executed from 5% to 95% acetonitrile, i.e. it is a standard wide gradient as can be used for screening of the compositions of many different samples. The peak capacity is 89, which means that there is no compromise in the separation power of the gradient. The last peak elutes at 36 s, which means that a still faster gradient could be executed if one were to focus on just the separation of these standards. As a matter of fact, the same analytes can be baseline-separated in 9 s with a gradient modified specifically for this group of compounds (not shown).

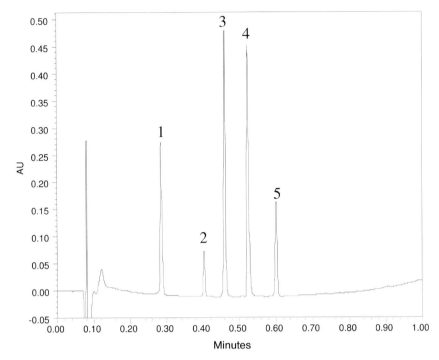

Fig. 5. Fast UPLC gradient separation.

Column:	20 mm × 2.1 mm 1.7 µm ACQUITY UPLC™ BEH C$_{18}$ column.
	Mobile phase A: 10 mM ammonium acetate, pH 5.
	Mobile phase B: 90 : 10 acetonitrile/water with 10 mM ammonium acetate, pH 5.
Gradient:	5% to 95% B over 1 min.
Flow rate:	1.5 mL min^{-1}.
Temperature:	30 °C.
Instrument:	Waters ACQUITY UPLC™ with TUV detector.
Detection:	210 nm.
Samples:	1. lidocaine, 2. prednisolone, 3. naproxen, 4. amitriptyline, 5. ibuprofen.

The workhorse column for UPLC separations is a 5 cm 2.1 mm column packed with 1.7 µm particles, as is a 15 cm × 4.6 mm 5 µm column for HPLC. As one can see, we have once again scaled the column length to the particle size. A very fast gradient separation for this column is a 3 min gradient, and it is normally used for 5 min or 10 min gradients. The recommended flow rates are between 0.5 mL min^{-1} for the slower gradient and about 0.75 mL min^{-1} for the faster gradient. Once again, experience with the actual sample may modify these suggestions somewhat. Also, if the organic modifier is acetonitrile, the flow rate may be slightly higher, or it may be slightly lower if methanol is used.

For gradient separations with a need to maximize the resolving power, the 10 cm × 2.1 mm ACQUITY UPLC™ column is used. A prime example of a separation that requires a high power is a peptide analysis, for example, a complex tryptic map. Such separations are typically executed over a time frame of 1–2 h,

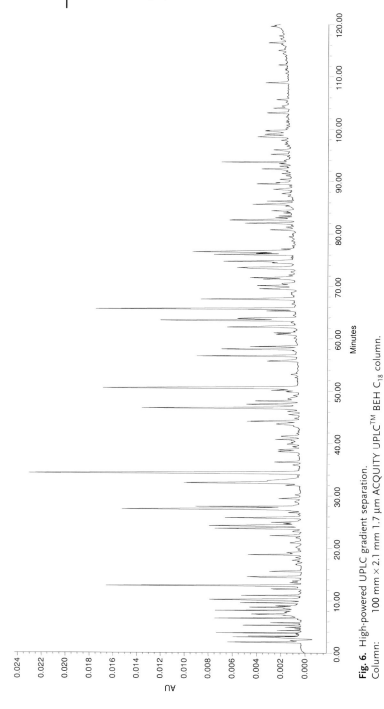

Fig. 6. High-powered UPLC gradient separation.

Column:	100 mm × 2.1 mm 1.7 μm ACQUITY UPLC™ BEH C_{18} column.
	Mobile phase A: 0.02% TFA in water.
	Mobile phase B: 0.016% TFA in acetonitrile.
Gradient:	5% to 35% B over 120 min.
Flow rate:	0.1 mL min^{-1} .
Temperature:	38 °C.
Instrument:	Waters ACQUITY UPLC™ with TUV detector.
Detection:	214 nm.
Sample:	MassPREP™ Phosphorylase b digestion standard.

but a more rapid separation can be carried out in about 0.5 h for less demanding samples. For a peptide analysis, lower flow rates are used, in the range of 0.1– 0.2 mL min^{-1} for a long gradient. An example of an analysis of this type is shown in Fig. 6. The peak capacity calculated for this separation was 589. Such a value cannot be reached with conventional HPLC columns in a reasonable time. An analysis of the principles of peak capacity in gradient chromatography [8] shows that such a separation power is only possible with a 5 µm 30 cm column with an analysis time of around 24 h at the same temperature (i.e., 30 °C). Even at elevated temperature (80 °C) an analysis of similar quality still takes about 8 h with a classical column. The important element of this comparison is that the use of elevated pressure in combination with very small particles enables us to achieve extraordinary separation power within reasonable time frames, and that other tools such as elevated temperature do not get us to the same spot. Of course, UPLC columns can also be used at elevated temperature to further speed up an analysis.

The need for increased separation performance is the driving force behind UPLC. This new technology will fulfill this need either for high-power separations in an acceptable time frame, or for very rapid separations with more conventional resolution, or for any intermediate scenario that can profit from both the increase in performance and the gain in speed. The ACQUITY UPLC™ system represents the first step into this new direction of chromatography.

References

1   J. E. MacNair, K. C. Lewis, J. W. Jorgenson, *Anal. Chem.* 69 (1997), 983.
2   Product Introduction at Pittsburgh Conference 2004.
3   J. R. Mazzeo, U. D. Neue, M. Kele, R. Plumb, Anal. Chem. A-pages 77 (2005), 460A-467A.
4   M. Martin, C. Eon, G. Guiochon, *J. Chromatogr.* 99 (1974), 357.
5   U. D. Neue, "HPLC Columns – Theory, Technology and Practice", Wiley-VCH, New York, Weinheim, 1997.
6   U. D. Neue, B. A. Alden, P. C. Iraneta, A. Méndez, E. S. Grumbach, K. Tran,

D. M. Diehl, "HPLC Columns for Pharmaceutical Analysis", in: Handbook of HPLC in Pharmaceutical Analysis (Eds.: M. Dong, S. Ahuja), Elsevier, Amsterdam, (2005), pp. 77–122.
7   K. D. Wyndham, J. E. O'Gara, T. H. Walter, K. H. Glose, N. L. Lawrence, B. A. Alden, G. S. Izzo, C. J. Hudalla, P. C. Iraneta, *Anal. Chem.* 75 (2003), 6781.
8   U. D. Neue, J. R. Mazzeo, "A Theoretical Study of the Optimization of Gradients at Elevated Temperature", *J. Sep. Sci.* 24 (2001), 921–929.

3
Coupling Techniques

HPLC Made to Measure: A Practical Handbook for Optimization. Edited by Stavros Kromidas
Copyright © 2006 WILEY-VCH Verlag GmbH & Co. KGaA, Weinheim
ISBN: 3-527-31377-X

3.1
Immunochromatographic Techniques

Michael G. Weller

Affinity-based methods and particularly immunochromatographic techniques are very promising for solving many demanding analytical tasks. If matrix problems or a lack of sensitivity are encountered, the use of immunochromatographic techniques is highly recommendable. It is likely that the improved availability of biochemical affinity ligands will lead to a strong increase in the use of this very selective approach. The optimization of such methods is very much dependent on the analyte and its complementary binding molecule. Hence, all optimization efforts should be preceded by the research of the physical and chemical properties of the molecules involved. In this chapter the topics affinity enrichment ("affinity SPE"), weak affinity chromatography and biochemical detection are described. Finally, three typical application examples will are shown.

3.1.1
Introduction

Today, affinity-based chromatographic techniques are not used very widely, in spite of their enormous analytical power [1]. This can be attributed to several causes, for example the limited availability of suitable affinity reagents or their high cost. In addition, some laboratories lack specific know-how on biochemical techniques, which is clearly different from more standard chromatographic knowledge. This leads to a relatively high "activation barrier" for the application of such techniques. In this chapter, the main points are discussed so as to facilitate the entry of prospective users into this promising field and to highlight potential optimization steps.

The fundamentals of affinity chromatography can be taken from the literature [2–5]. Since affinity-based analytical methods are extremely dependent on the analytes and their respective binding molecules, it is not possible to go into much detail with regard to practical advice. It has to be mentioned that "affinity chromatography" is rarely a true chromatographic technique, and in most cases has to be considered to be a selective extraction method. Hence, the term "affinity extraction" or "immunoaffinity extraction" should be preferred where appropriate.

3.1.2
Binding Molecules

In many affinity techniques, biomolecules such as DNA, oligonucleotides, proteins, peptides, cofactors, and other biochemical compounds are used as reagents. Mainly proteins and, in particular, antibodies are discussed herein, due to their broad applicability. However, many rules can be easily transferred to other binding molecules.

Antibodies belong to the immunological defence system in the bodies of vertebrates, and are present in human blood in concentrations of up to 10 g L^{-1}. Antibodies are glycoproteins with molecular weights of around 150,000 Da. They possess two binding sites, which are directed, for instance, against the surface structures of viruses, bacteria, and other pathogens. Antibodies of many different specificities can be obtained by methods similar to standard vaccination procedures. By these methods, it is not only possible to raise antibodies against proteins or other polymeric structures, but also against small molecules such as benzene, trinitrotoluene or polychlorinated biphenyls (PCB). Whether the analyte of interest is a natural compound or of synthetic origin is irrelevant. Compounds that are able to elicit an immune response directly are called antigens. Compounds of lower molecular mass (< 5000 Da), which are not directly immunogenic, are called haptens. These can be rendered immunogenic by covalent conjugation to a large carrier molecule (usually a protein). A human being is able to produce about 10 billion different antibodies. However, the concentrations of the respective antibodies are extremely low if the human body in question has never come into contact with the corresponding antigen. The antibody fraction (immunoglobulin fraction) of blood serum is always a mixture of different antibodies. In general, it is not possible to isolate well-defined antibody species from this crude mixture. However, this is not necessary for most applications.

Antibodies bind ("recognize") their corresponding antigen with very high selectivity. This is the most prominent feature of immunological techniques. However, it is possible to generate group-selective antibodies, which bind a whole substance group. The prerequisite for such antibodies is the existence of a common substructure, which is not too small or shielded, against which an immunization can be directed.

From the serum of an animal, only "polyclonal antibodies" can be obtained, i.e. a mixture of different antibody species in different concentrations. Drawbacks of polyclonal antibodies are their critical long-term supply and a lack of reproducibility from batch to batch. The composition of a serum changes all the time and the sera of two individuals are completely different in terms of antibody concentration and their selectivity. Hence, polyclonal antibodies cannot be considered as defined reagents. On the other hand, the production of polyclonal antibodies is much cheaper than any other method and hence is often the economically most sensible way to obtain an immunoreagent.

In contrast, "monoclonal antibodies" consist of only one protein species and are produced by cell culture. Hence, monoclonal antibodies can be considered to be pure chemical compounds (to a first approximation). The respective clones (cell lines) can be cultured "indefinitely" and can be stored for quite a long time by freezing in liquid nitrogen. Unfortunately, the generation and production of monoclonal antibodies (Mabs) is much more difficult and expensive than that of polyclonal antibodies (Pabs). There is an important misunderstanding in this respect, which should be mentioned here. It is definitely not true that polyclonal antibodies are less selective or show lower affinity than monoclonal ones. On the contrary, polyclonal antibodies often show a better performance than monoclonal antibodies, because it is more difficult to make good Mabs than good Pabs. Some problems of antibody production can be solved by recombinant techniques (genetic engineering). Unfortunately, these techniques are still in their infancy and are quite expensive. In the long term, some improvements can be expected in this field.

Antibodies form complexes with their respective antigens (here: analytes). These complexes can show a particular strength, which can be quantified by the affinity constant (or equilibrium constant). This affinity constant is about 10^7–10^{11} L mol^{-1} for most analytically useful antibodies. The higher this number, the more stable is the complex. The highest known affinity in the biochemical field is the interaction between avidin (an egg protein) and biotin (vitamin H), for which a value of around 10^{15} L mol^{-1} has been determined. The affinity constant plays an important role in immunoassays and other immunological techniques. The development of new methods is greatly facilitated if this constant is known.

3.1.3
Immunoassays

Often, the complex formation between antibody and antigen cannot be detected directly. In many cases, labeling of one of the reagents is necessary. Radioactive isotopes (e.g., ^{125}I, ^{14}C, ^{3}H), fluorescent dyes (e.g., fluorescein, rhodamine), gold or latex particles, or enzymes (such as horseradish peroxidase or alkaline phosphatase from calf intestine) can be used as labels. The differentiation between free and complexed reagent is often accomplished by a separation step (precipitation or a washing step at a surface).

There are innumerable formats for immunoassays, which, however, are not very different in their fundamental steps. The most important classes are the competitive (antibody limited) and the non-competitive (antibody excess) assays. The former is preferred for smaller molecules, whereas the latter is mainly used for larger ones (e.g., proteins).

3.1.4
Immunochromatographic Techniques

The nomenclature of immunochromatographic techniques is not consistent and may easily lead to misunderstandings. In addition, the respective hyphenations have their own designations. However, there are mainly three fundamental set-ups that have to be taken into consideration (Fig. 1).

(A) **Affinity Enrichment**
   (Affinity SPE)

(B) **Weak Affinity Chromatography (WAC)**
   (True affinity chromatography)

(C) **Biochemical Detector**
   (Biochemical post-column derivatization)

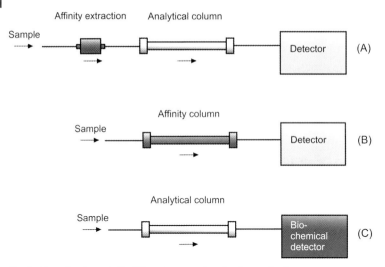

Fig. 1. Three fundamentally different approaches for the application of affinity-based methods for hyphenated analytical systems.

Table 1. Ready-to-use columns.

Company[a)]	Analytes
Abkem Iberia, Vigo, Spain	Microcystins, domoic acid
Affiland, Ans-Liege, Belgium	hCG-beta, hCH, GRF, IGF-1, EPO, clenbuterol, salbutamol
Cayman Chemical, Ann Harbor, USA	Prostaglandin E2, 8-isoprostane, cysteinyl-leukotriene, nitrotyrosine, CGRP, PPAR-gamma, sPLA2
Coring System Diagnostix, Gernsheim, Germany	Aflatoxins, ochratoxins, zearalenone
EY Laboratories, San Mateo, USA	Carbohydrates, glycoproteins (immobilized lectins)
Grace Davison, Deerfield, USA	Aflatoxins, lactoferrin, vitamin B12, testosterone, nortestosterone, ethinyl estradiol, estradiol, estrone, bisphenol A, chlorophenoxy acetic acid herbicides, phenylurea herbicides, organophosphate pesticides, vinclozolin
MIP Technologies, Lund, Sweden	Clenbuterol, beta agonists (class-selective), NNAL, riboflavine, triazines (class-selective)
R-Biopharm, Darmstadt, Germany	Aflatoxins, ochratoxin A, zearalenone, fumonisins, deoxynivalenol
Romer Labs Diagnostic, Herzogenburg, Austria	Aflatoxins, ochratoxin, zearalenone
Vicam, Watertown, USA	Aflatoxins, ochratoxin A, zearalenone, fumonisins, deoxynivalenol

[a)] This list is not exhaustive.

Variant A is by far the most frequently used method, and closely resembles the well-known solid-phase extraction (SPE). In addition, quite a few ready-to-use cartridges are available commercially (Table 1), which also support the popularity of variant A.

However, many of these applications belong to a (semi)-preparative approach, e.g., the isolation of specific proteins from raw extracts. If the term "affinity chromatography" is used in the literature, in nearly all cases it refers to variant A. Variant B is quite rare and more or less only of mechanistic interest at the moment. If a gradient elution is used, some mixed forms between A and B may occur. Non-specialists often suspect an approach similar to B when "affinity chromatography" is mentioned. However, technique B is generally known as "weak affinity chromatography (WAC)".

Variant C is also rarely used, although it shows considerable potential for the future. This set-up can be used for drug screening, toxicity tests, and other bioactivity-related purposes. Biochemical detectors are also useful for structural analysis, e.g., for the identification of structurally similar compounds.

Unfortunately, affinity techniques are referred to by many other names, such as immunoaffinity chromatography (IAC), bioaffinity chromatography, biospecific adsorption or high-performance affinity chromatography (HPAC). The latter denotes on-line techniques that do not use soft gels (for example, agarose) with large bead sizes, but hard materials, which can be used at higher pressures and flow rates. Biochemical detectors are also referred to as "post-column affinity detection" [6].

Chromatographic immunoassays are not mentioned here in detail because they do not lead to a chromatographic separation of analytes. They constitute a special method for performing an immunoassay, i.e., to separate bound and free reagent species.

3.1.4.1 Affinity Enrichment (Affinity SPE)

Besides classical immunoassays, affinity enrichment (immunoaffinity extraction) is one of the most important immunochemical techniques. Of major importance are off-line affinity enrichments, which are quite similar to the application of cartridges for solid-phase extraction (SPE). Hence, this approach is also called "affinity SPE". The main advantage of this technique is the extreme selectivity of the cartridges, which leads to very "clean" chromatograms. Therefore, this kind of sample preparation can be considered as one of the most powerful. Even samples in extremely difficult matrices can be processed directly. On the other hand, samples in a simple matrix, such as drinking water, rarely justify the added expenditure and complexity of an affinity technique. In these cases, a direct measurement (minimal sample pre-treatment) or the application of conventional SPE phases is faster and/or more economical. In a next step, one has to ascertain whether the necessary affinity cartridges are available commercially. In addition, one has to determine whether the respective antibodies are polyclonal or monoclonal. In the case of the former, sometimes the cartridges may no longer be available because the antibody ran out. Some producers/vendors of respective cartridges are listed in Table 1. This list is not intended to be comprehensive. It is often difficult to locate the typically small companies that offer such cartridges. Immobilized reagents or

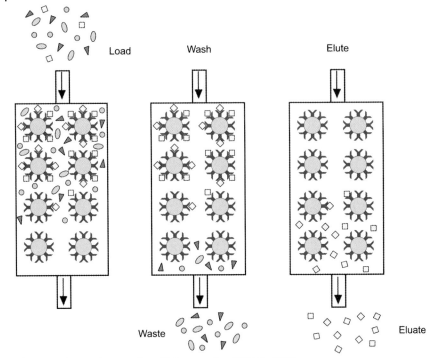

Fig. 2. Fundamental steps of an affinity extraction ("Affinity SPE").

cartridges that are relatively easy to obtain (e.g., from Sigma-Aldrich, Roche, GE Healthcare or Merck Biosciences), containing ligands such as protein A, protein G, avidin, streptavidin, lectins, biotin or heparin, have been omitted from Table 1.

If the desired column/cartridge is available commercially, its application is unlikely to be much more difficult than the performance of a conventional SPE. Figure 2 shows the general working steps.

First of all, the column should be conditioned. Stabilizers and preservatives that might have been used during storage are washed out and the cartridge is adjusted to the desired pH (and salt content, etc.). In addition, potential precipitation problems that might occur during the mixing of mobile phase and sample can be solved.

In step 2, the sample is applied to the cartridge. In general, the sample should be centrifuged and/or filtered to avoid clogging or fouling of the column. However, affinity beads with relatively large diameters are often used, which even allow the processing of raw samples with minimal pre-treatment. Due to the slow diffusion in coarse gel materials, the flow rates that can be applied are very limited. Some materials show improved kinetic properties (denoted, e.g., by "Fast Flow", "Ultra Flow", "Poros") to overcome this limitation. More detailed information can be found in the data sheets of the respective manufacturers. The samples should be buffered or mixed with particular "binding buffers" to enhance selective binding, which in most cases means suppression of non-specific binding. The addition of a low concentration of a solvent (often methanol or acetonitrile) or a surfactant (such as Tween 20) may suppress the non-specific interactions (hydrophobic) of matrix compounds with the stationary phase. The addition of salts, on the other hand, reduces non-specific binding based on ionic bonds, which is relevant for many biopolymers. For the selection of a carrier material, it should be kept in mind that all materials show some non-specific binding under certain conditions. Materials based on agarose, silica gel, and polystyrene/divinylbenzene, among others, are quite different in this respect. In comparison to standard materials for RP-SPE, the capacity of affinity columns or cartridges is quite low. To complicate matters further, this capacity is dependent on substance and on flow rate and other conditions.

The maximum capacity of an affinity cartridge can be estimated from the amount of immobilized ligand. Nevertheless, it is advisable to measure a breakthrough curve of the desired analyte. Competitive effects can be observed if the cartridge is overloaded or the flow rate is not adequate. A stronger binding analyte can "displace" a weaker binding one, for which the recovery can be strongly reduced.

The third step involves washing of the column to remove the rest of the sample volume and to elute weakly bound matrix compounds. During this step, a strong binding of the analyte to the ligand is favorable to avoid washing losses. Furthermore, if the capacity of the cartridge is lower than expected, increased washing losses can occur. In contrast, columns with excess capacity show a strong "rebinding", which reduces losses. Hence, the washing should be performed at relatively low flow rates to take full advantage of this "rebinding". As a rule of thumb, the washing flow rate should be not faster than the flow rate during the loading step. The washing should also be carried out with consideration of the following elution step. If salts have to be avoided (e.g., for mass spectrometry), the final washing solution should be salt-free. As for the loading step, one has to decide how rigorous the washing should be in order to achieve optimal selectivity without sacrificing recovery. This decision mainly depends on the strength of the analyte/ligand interaction.

In contrast to usual SPE cartridges, affinity columns are not blown dry after the loading step. Many gels would collapse and the regeneration of the biochemical ligand would be compromised. Elution is often performed with a small volume of aqueous solvent (e.g., 80% methanol/water). In principle, one may differentiate between specific and non-specific elution, the latter being far more popular. The former is based on the use of an analyte analogue, which "displaces" the analyte from the binding sites. However, this displacement does not have to be taken too literally. Mechanistically, the method is an inhibition of "rebinding" and hence it is strongly dependent on the kinetic dissociation rate. The specific elution has to be optimized for each application. The main advantage is its very mild effect on both the analyte and the ligand. The cartridges are easily regenerated.

Non-specific elution conditions (e.g., solvents, pH step) are used more frequently because of their broader applicability. Unfortunately, they often lead to some activity losses, even when optimized. The user has to decide whether to try a repeated use (regeneration) or to consider the cartridges as disposable. In routine applications,

Table 2. Activated columns for the immobilization of ligands.

Company[a]	Carrier material (Composition)	Conjugation chemistry[b]
3M	Empore Affinity Az Membrane	Azlactone
GE Healthcare	CNBr-activated Sepharose 4 Fast Flow (Derivatized, cross-linked agarose)	Cyanogen bromide
GE Healthcare	EAH Sepharose 4B (Amino-derivatized, cross-linked agarose)	Glutaraldehyde
GE Healthcare	ECH Sepharose 4B (Carboxy-derivatized, cross-linked agarose)	Carbodiimide activation
GE Healthcare	Epoxy-activated Sepharose 6B (Epoxy-derivatized, cross-linked agarose)	Epoxide
GE Healthcare	HiTrap NHS-activated HP (NHS- derivatized, cross-linked agarose)	N-Hydroxysuccinimide ester
GE Healthcare	NHS-activated Sepharose 4 Fast Flow (NHS- derivatized, cross-linked agarose)	N-Hydroxysuccinimide ester
GE Healthcare	Thiopropyl Sepharose 6B (Carboxy-derivatized, cross-linked agarose)	Addition or alkylation
Applied Biosystems	POROS 20 AL (Aldehyde-derivatized, cross-linked styrene/DVB copolymer)	Reductive amination
Applied Biosystems	POROS 20 EP (Epoxy-derivatized, cross-linked styrene/DVB copolymer)	Epoxide
Applied Biosystems	POROS 20 OH (Hydroxy-derivatized, cross-linked styrene/DVB copolymer)	Tresyl activation
BIA Separations	CIM CM (Carboxymethyl-derivatized polymer, monolith)	Carbodiimide activation
BIA Separations	CIM EDA (Ethylendiamine-derivatized polymer, monolith)	Glutaraldehyde
BIA Separations	CIM Epoxy (Epoxy-derivatized polymer, monolith)	Epoxide
Bio-Rad Laboratories	Affi-Gel 10 (NHS-derivatized, cross-linked agarose)	N-Hydroxysuccinimide ester
Bio-Rad Laboratories	Affi-Gel 102 (Amino-derivatized, cross-linked agarose)	Glutaraldehyde
Bio-Rad Laboratories	Affi-Gel Hz (Hydrazide-derivatized, cross-linked agarose)	Periodate
Fluka	Aminopropyl-CPG (Controlled Pore Glass; derivatized, porous glass)	Glutaraldehyde
Fluka	Carboxymethyl-CPG (Controlled Pore Glass; derivatized, porous glass)	Carbodiimide activation
Fluka	Glyceryl-CPG (Controlled Pore Glass; derivatized, porous glass)	Periodate
Millipore	AminoAryl-CPG (Controlled Pore Glass; derivatized, porous glass)	Diazonium method
Millipore	Aminopropyl-CPG (Controlled Pore Glass; derivatized, porous glass)	Glutaraldehyde
Millipore	Carboxyl-CPG (Controlled Pore Glass; derivatized, porous glass)	Carbodiimide activation
Millipore	Glyceryl-CPG (Controlled Pore Glass; derivatized, porous glass)	Periodate
Millipore	Hydrazide-CPG (Controlled Pore Glass; derivatized, porous glass)	Periodate
Millipore	Thiopropyl-CPG (Controlled Pore Glass; derivatized, porous glass)	Disulfide exchange
Millipore	Prosep-5CHO (Controlled Pore Glass; derivatized, porous glass)	Reductive amination

Table 2. (continued)

Company [a]	Carrier material (Composition)	Conjugation chemistry [b]
Chisso	Cellufine Amino affinity carrier (Amino-derivatized cellulose)	Glutaraldehyde
Chisso	Cellufine Formyl affinity carrier (Aldehyde-derivatized cellulose)	Reductive amination
Pierce/Perbio	AminoLink Plus Immobilization Kits (Aldehyde-derivatized, cross-linked agarose)	Reductive amination
Pierce/Perbio	CarboLink Coupling Gel (Hydrazide-derivatized, cross-linked agarose)	Periodate
Pierce/Perbio	ImmunoPure Epoxy-activated agarose (Epoxy-derivatized, cross-linked agarose)	Epoxide
Pierce/Perbio	PharmaLink Imm. Kit (Diaminodipropylamine-derivatized polyacrylamide)	Mannich reaction
Pierce/Perbio	SulfoLink Coupling Gel (Iodomethyl- derivatized, cross-linked agarose)	Alkylation
Pierce/Perbio	UltraLink Hydrazide Gel (Hydrazide-derivatized polyacrylamide)	Periodate
Pierce/Perbio	UltraLink Immobilization Kit (Azlactone-activated polyacrylamide)	Azlactone
Riedel-de Haën	Polymer Carrier VA-Epoxy (Epoxy-activated vinyl alcohol copolymer)	Epoxide
Riedel-de Haën	Polymer Carrier VA-Hydroxy (Vinyl alcohol copolymer)	Tresyl activation
Degussa/Röhm	Eupergit C, CM and C250L (Epoxy-activated methacrylamide copolymer)	Epoxide
Sterogene	Actigel ALD (Aldehyde-derivatized, cross-linked agarose)	Reductive amination
Sterogene	Actigel B Ultraflow (Epibromohydrin-derivatized, cross-linked agarose)	Alkylation
Sterogene	Actigel T (Activated thiol, cross-linked agarose)	Disulfide exchange
Sterogene	Activated ALD Cellthru BigBead (Aldehyde-derivatized, cross-linked agarose)	Reductive amination
Tosoh Bioscience	Toyopearl AF-Amino-650M (Amino-derivatized vinyl polymer)	Glutaraldehyde
Tosoh Bioscience	Toyopearl AF-Carboxy-650M (Carboxy-derivatized vinyl polymer)	Carbodiimide activation
Tosoh Bioscience	Toyopearl AF-Epoxy-650M (Epoxy-derivatized vinyl polymer)	Epoxide
Tosoh Bioscience	Toyopearl AF-Formyl-650M (Aldehyde-derivatized polymer)	Reductive amination
Tosoh Bioscience	Toyopearl AF-Tresyl-650M (Tresyl-derivatized vinyl polymer)	Alkylation
VitraBio, Steinach, Germany	Trisopor – aminopropyl (irregular porous glass, derivatized)	Glutaraldehyde
VitraBio, Steinach, Germany	Trisopor – diol (irregular porous glass, derivatized)	Periodate
VitraBio, Steinach, Germany	Trisoperl – aminopropyl (spherous porous glass, derivatized)	Glutaraldehyde
VitraBio, Steinach, Germany	Trisoperl – diol (spherous porous glass, derivatized)	Periodate
VitraBio, Steinach, Germany	Trisofil – aminopropyl (porous glass fibres, derivatized)	Glutaraldehyde
VitraBio, Steinach, Germany	Trisofil – diol (porous glass fibres, derivatized)	Periodate

a) This list is not exhaustive; b) typical approach.

the latter is often preferred, to guarantee specific column properties such as capacity and selectivity. In this context, it has to be kept in mind that weak elution conditions lead to even broader elution peak volumes. If a smaller elution volume is necessary, partial or total denaturation of the cartridge may be unavoidable. In the case of a mild elution in a larger volume, a subsequent concentration step might be advantageous, e.g., the evaporation of the solvent by a gentle nitrogen stream. This approach also has the advantage that a complete solvent exchange can be accomplished. A special variant of a non-specific elution is thermal denaturation by boiling in water, buffer or SDS (sodium dodecyl sulfate) solution. This method is often applied for immunoprecipitation (IP), which does not use an affinity matrix in a column but one that is added directly to the sample. The beads/particles are separated from the sample by centrifugation or a magnetic field (magnetic particles).

Those who do not have commercial affinity columns available, which will be the case with quite a few analytes, will have to deal with the question of the provision of a suitable ligand and its immobilization on carrier materials. The latter are available in many variants (see Table 2). The many different conjugation procedures cannot be discussed in detail here and the reader should refer to the appropriate literature [2–5, 7–9] or to the technical data sheets of the manufacturers. It should be stressed that the coupling of a ligand to a pre-activated material is quite simple and can be performed even by the inexperienced, if performed exactly according to the instructions of the manufacturer. The conjugation procedure should in general not need more than 24 h in total.

The choice of the ligand is so problem-specific that only some guidelines can be given. It should be noted that the antibody used needs to be available in a "pure" form (e.g., protein A or protein G purified) and should not contain any interfering additives (e.g., Tris buffer, which contains amino groups and these inhibit many conjugation reactions) or stabilizing proteins (competitive immobilization). Surfactants may also have negative effects. Suitable antibodies can either be sought through the free Internet [10] or in commercial databases [11, 12].

If a suitable ligand (e.g., antibody) is not available, one often has to decide against the use of an affinity method. In most cases, the in-house production of antibodies is out of the question. As a rule of thumb, a production time of 6–12 months is needed. A polyclonal serum costs roughly about 500 EUR, a monoclonal antibody at least 5000 EUR. Even more difficulties arise if the antigen (here: the analyte) is not available or if it is not suitable for (direct) immunization. The production of antibodies requires considerable experience and, even today, it is a risky venture without guarantee of obtaining a useful antibody at the end. One can conclude that access to a good antibody is a critical limitation of immunological techniques.

One solution may be so-called "molecular imprinting" [13]. By polymerization of monomers and cross-linkers in the presence of an analyte and subsequent leaching of the analyte, molecular cavities in the polymer are obtained, which lead to selective (re)binding of this compound. MIPs (molecularly imprinted polymers) can be considered to be fully synthetic equivalents of polyclonal antibodies. Hence, they are referred to as "artificial antibodies", although this name might be a little misleading. The relatively simple synthesis and the extraordinary robustness are important advantages; the latter enables multiple regenerations. Meanwhile, the first commercial suppliers of MIPs have appeared, such as MIP Technologies, Lund, Sweden and Oxonon, Emeryville, USA.

The separation and detection techniques that follow the affinity enrichment are not discussed here. Suitable techniques include HPLC-DAD, LC-MS(MS), CE, GC, immunoassays, and many others. The sample is introduced off-line as after a conventional enrichment or "clean-up" step.

The on-line variant, which is shown in Fig. 1a, is much more difficult to implement. Apart from the fact that suitable commercial cartridges are scarcely available, there are several factors disfavoring the on-line variant. Many compounds elute from the immuno-cartridges in very broad peaks because the affinity is too high, even under harsh elution conditions. In analogy to the off-line procedures, a compromise has to be found if the cartridges are to be regenerated. If too strongly eluting solvents are used, the antibody will be denatured and the cartridge can only be used once. For on-line techniques, a regeneration would be most desirable, as otherwise the cartridge would have to be changed (i.e., manually) after each run. This would hamper the automatic operation of such a system. However, an on-line set-up typically rules out the evaporation of the solvent, which might be used to concentrate and focus a broad elution peak. If the elution is performed with a high concentration of a strong solvent, enrichment on an RP column will be difficult. In some cases, dilution of the solvent with water is helpful. Usually, three columns are used: an affinity column, a trap column (RP) to concentrate the eluate, and an analytical column (RP) for the actual separation [14]. The relatively complex column switching system and the difficult optimization has limited the application of these on-line systems. The polarity of the analytes has to lie in a narrow window, otherwise the RP enrichment and the elution will not work properly. In some cases, the antibodies have been immobilized via protein G, which leads to a regular reloading of the cartridge with fresh antibody with each run. The elution not only removes the analyte from the column, but also the antibody. The use of a mass spectrometer strongly limits the possible use of solvents with salts, such as buffers, for the elution from the immunoaffinity column. However, a trap column would also separate salts from the analytes and would improve the compatibility with mass spectrometric techniques. To reduce the complexity of the system, the head of the analytical column itself may be used for the enrichment and trapping of the immuno-eluate.

3.1.4.2 "Weak Affinity Chromatography" (True Affinity Chromatography)

In contrast to affinity enrichment, where a high affinity for the ligand ($> 10^7$ L mol^{-1}) is desirable to avoid losses, the use of high affinity binders for chromatographic purposes leads to excessive band broadening. This is mainly due to the slow equilibrium (low dissociation rate constants). Hence, there are virtually no such applications reported in the literature; the separation power is simply too poor. Moreover, the limited availability of many biochemical reagents (price, amount) precludes the coating of a larger column. Two variants have been devised to overcome the limitation of the separation power. The first (more frequently used) is the use of an elution gradient after the binding of the analytes on the immunoaffinity column. Nevertheless, the peaks are often so broad that only the separation of two or three compounds can be performed. Usually, the application of an elution gradient is not motivated by the separation power, but more often to avoid a lengthy optimization of the elution conditions. If labile analytes (e.g., proteins) are to be separated, a gradient can help to approach the weakest elution conditions and hence to avoid denaturation by too harsh an elution solvent. It is possible to deduce from theory that usual chromatography is only possible with affinities below about 10^5 L mol^{-1}. Such antibodies are nearly always lost during the screening procedures, since the test systems are focused on high-affinity antibodies (e.g., for ELISA). To date, only model separations have been reported under "weak affinity" conditions. It is not yet clear whether routine applications might be possible in the future.

An interesting application is the affinity separation of a (polyclonal) antiserum. In general, there is little hope of separating polyclonal antibodies into monoclonal fractions by purely "physical" means. By the use of "weak affinity chromatography" on a column with immobilized hapten (small antigen), researchers have succeeded in separating a serum into twelve different fractions [15]. As expected, the affinity of the isolated antibodies increased with increasing elution volume. Crucial for the success of the separation was the choice of an immobilized ligand with an extremely low affinity (about 0.01% cross-reactivity in relation to the main analyte) for the antibodies to be separated. In addition, an eluent containing 30% dioxane at pH 3.4 was used to weaken the interactions even more. Also, pH, DMF, and 2-methoxyethanol gradients (also step gradients) were examined [16].

3.1.4.3 Biochemical Detectors

Similar to the situation with affinity enrichment (Fig. 1a), the off-line variant of biochemical "detectors" (Fig. 1c) is far more popular [17] than the on-line variants. In most cases, off-line coupling means fractionation of the eluate stream. This approach is frequently used, at least in the area of preparative HPLC. However, predominantly peak-controlled fractionation is applied based on UV (and MS) detection. In the field of drug screening, the chromatographic separation of, for example, extracts of natural products and their subsequent fractionation is routinely used. Today, the fractions are often no longer collected in small vials, but in microtitration plates (MTP) of various formats. Modern fraction collectors are often able to fractionate into one or even several MTP. The off-line approach is often considered to be less advanced; nevertheless, it shows some significant advantages:

1. Temporal decoupling of chromatography and detection.
2. Simple aliquoting for the performance of different assays.
3. Safeguarding of the fractions for subsequent, e.g. structural, analyses.

4. Simple implementation.
5. Low technological hurdles.
6. Well-suited for multidimensional assays [18].

The on-line approach is generally faster and more "elegant"; however, due to the high technological complexity, it is only used by one company (Kiadis, Leiden, The Netherlands). Some papers have been published that show the potential of on-line techniques [19].

For many users, the off-line variant seems to be the more suitable one. Setting up an appropriate system should be possible even for researchers with only limited experience of bioanalytical techniques. However, some critical points have to be taken into consideration. The selection of the mobile phase is critical. Since the mobile phase can only be removed with difficulty (e.g., by evaporation), it should be as compatible as possible with the planned biochemical assay(s). In addition, gradient methods cause more problems for detection than isocratic ones. If major problems should arise in this respect, evaporation of the solvent is often the best solution. To avoid the evaporation step, an isocratic separation with methanol/water (or neutral buffer) is a good choice. It has been repeatedly shown that methanol (compared to other solvents commonly used in chromatography) is the most "biocompatible", and hence can be used even in relatively high concentrations in biochemical assays (about 10% or even higher). If a dilution (e.g., $1:50 - 1:100$) is feasible, the problem of the mobile phase essentially disappears. The dilution buffer can be generally used to trap unwanted compounds in the sample and to adjust the pH to neutral (e.g., when acidic solvents have been used). In the case of high dilutions, even gradient separations are possible. On the other hand, if a strong dilution cannot be performed due to sensitivity limits, it is possible to tolerate a significant activity loss during gradient methods. As long as the base line is acceptable and the dynamic range remains sufficient, a strong solvent effect is no reason to preclude an experiment. Often, a so-called "sample buffer" is added to the wells of the MTP prior to the sample to achieve an immediate buffering. This is particularly important for unstable analytes (such as proteins) to avoid activity losses.

Chromatographic separations in this context are not limited to reversed phases. Normal phases, gel chromatography, and ion-exchange chromatography are possible options. Furthermore, (capillary) electrophoresis is a powerful separation method, although the small volumes make fractionation difficult.

Numerous biochemical tests are suitable as detection methods. The assay may be developed in-house. In this case, sufficient know-how should be available to perform and optimize the respective tests. In addition, enzyme inhibition tests are well-suited for HPLC coupling. Even bioassays (e.g., toxicity tests) should be suitable for this purpose. However, in this case, the toxicity of the solvent should be considered. Evaporation might be the best solution here as well. Users with less or even no experience of biochemical assays are not necessarily excluded from their application. For some analytical parameters, commercial test kits are available, which are not only useful for the analysis of the respective analytes in water or other matrices, but can also be used for the testing of eluates from HPLC after dilution or evaporation of the solvent. Admittedly, such test kits are relatively expensive and hence one would avoid testing complete HPLC runs. In these cases, an examination of specific peak fractions, as identified, for example, by UV absorbance, would be sensible. The solvent stability of test kits is often unknown. Therefore, evaporation of the solvent and redissolution of the analyte in a suitable buffer would be a good approach. These tests are often only semi-quantitative and hence they require only very basic calibration. However, it is advisable to use at least one negative and one positive control per plate for assays in MTP format. In general, a transfer of a calibration from one plate to another is not acceptable. If commercial test kits are used, the instructions issued by the manufacturer should be taken as a starting point. However, one is unlikely to get specific support from the manufacturer for the application to HPLC coupling. To be on the safe side, a high dilution of the eluate (> 1 : 50) or evaporation of the solvent should be performed. Only users with some immunoassay experience should dare to change the test conditions, since the connection between test parameters and results is quite complex. In principle, it is possible to perform immunoassays or other biochemical tests in water/solvent mixtures. It is possible to eliminate some solvent effects by using the same water/solvent mixture in the calibration solutions. The upper limit might be 20% with methanol; with other solvents, 10% might be more realistic. It has to be stressed that these limits are strongly dependent on the assay. Different antibodies (even if they are targeted against the same analyte) can display completely different stability behavior. Hence, it is very important to use the same batch of antibodies in the case of polyclonal ones or to switch to monoclonal or recombinant antibodies.

3.1.5
Examples

3.1.5.1 Example 1: Affinity Extraction (Affinity SPE)

Urine is generally considered to be a very difficult matrix. Different sample preparation schemes have hitherto been explored; however, many of them do not lead to satisfactory results. This is a typical application area for affinity extractions. Schedl et al. [20] reported on the successful application of so-called sol-gel glass for the immobilization of polyclonal pyrene antibodies for the analysis of hydroxyl metabolites of polycyclic aromatic hydrocarbons (PAHs). A sol-gel immobilization is not based on a coupling of the antibody to the solid matrix, but the buffered antibody is mixed with an acidically hydrolyzed solution of tetramethoxysilane. After mixing, a fast polycondensation takes place, which leads to a glassy solid in which the antibodies are sterically entrapped. The antibodies cannot be washed out, nor can proteases attack them. Small analyte molecules, however, such as PAHs, can enter the pores and will be bound selectively by the binding sites of the antibodies. Elution is performed with acetonitrile/water (1 : 1). The sol-gel glass is often smashed into small pieces for subsequent analysis. A suitable size fraction is obtained by sieving the glass particles.

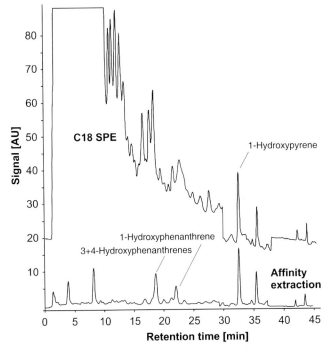

Fig. 3. Affinity extraction for the sample preparation of urine for the analysis of PAH metabolites. Comparison with conventional C18 SPE. (Figure courtesy of Prof. Dietmar Knopp).

In Fig. 3, two chromatograms are compared, which show the difference between sample preparation with classical C_{18} SPE and immunoaffinity extraction (affinity SPE). The main metabolite, 1-hydroxypyrene, can be examined with both methods; all other metabolites, which occur in lower concentrations, cannot be quantified without affinity purification. The detection limit with affinity extraction can be improved even further if a concentrating C_{18}-SPE is subsequently applied. Figure 3 clearly shows the improvement that can be achieved with affinity purification. Furthermore, it is much easier to identify new metabolites that have not been identified before because the chromatogram contains only very few peaks. Hyphenation with mass spectrometry would be particularly suitable to achieve a structural analysis.

3.1.5.2 Example 2: "Weak Affinity Chromatography" (WAC)

Due to the only moderate separation power, very few examples of WAC have been described in the literature, apart from enantioselective separations of racemates by "molecular imprints" (MIP), which typically require the separation of only two species.

In Fig. 4, the separation of some monosaccharides is illustrated [21]. For WAC this is a relatively complex sample. Wheatgerm agglutinin, a lectin (sugar-binding

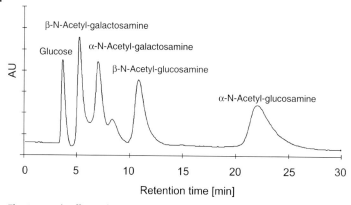

Fig. 4. "Weak Affinity Chromatography" (WAC) for the separation of sugars [21].

protein), was immobilized on aldehyde-activated silica gel (10 μm, 300 Å). Usually, wheatgerm agglutinin is only used for the binding of di- or polysaccharides. Hence, its affinity for monosaccharides is quite low. The authors estimated the affinity range used for this separation to be around 10^2–10^3 L mol^{-1}. These low affinities are a prerequisite for successful separation. Glucose and lactose (not shown) are not retained, since wheatgerm agglutinin only binds to acetylated sugars. The separation was performed on a 250 × 5 mm stainless steel column with a flow rate of 1 mL min^{-1} (phosphate buffer, pH 7.0, 20 mM sodium phosphate, 100 mM sodium sulfate). The eluate was mixed in a 1 : 1 ratio with 0.2 M NaOH for pulsed amperometric detection. The column had a capacity of 18 μmol of analyte. The addition of 5% serum did not lead to changed retention times. An analogous column without lectin did not show any retention of the analytes.

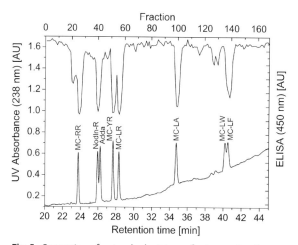

Fig. 5. Separation of a standard mixture of microcystins (hepatotoxic, cyclic peptides) by RP chromatography (acetonitrile/water gradient, 0.04 and 0.1% trifluoroacetic acid, respectively), detection at 238 nm (lower line) and detection by off-line enzyme immunoassay (upper line) [22]. (Figure courtesy of Dr. Anne Zeck).

3.1.5.3 **Example 3: Biochemical Detection**

These approaches can be designated as post-column derivatization methods and can be performed with the appropriate devices. Due to the relatively slow reaction rates of many biochemical reactions, off-line detection may be preferable. In this case, incubation and reaction times can be adjusted independently of the separation speed. In Fig. 5, the separation of eight microcystins (cyclic peptides) and similar compounds is illustrated. With a group-selective antibody against microcystins, the respective toxin peaks can be unambiguously detected. Moreover, the ELISA detection is about 100 times more sensitive and much more selective than UV detection [18].

References

1 M. G. Weller, Immunochromatographic techniques – a critical review, *Fresenius J. Anal. Chem.* 366 (2000) 635–645.

2 Affinity Chromatography – Principles and Methods, Amersham Pharmacia Biotech, Handbook, No. 18-1022-29, 2001.

3 P. Bailon, G. K. Ehrlich, W.-J. Fung, W. Berthold, Affinity Chromatography – Methods and Protocols, Humana Press, 2000.

4 P. Matejtschuk, Affinity Separations (Practical Approach Series), Oxford University Press, 1997.

5 D. S. Hage, Handbook of Affinity Chromatography, Taylor & Francis Group, 2005.

6 D. S. Hage, Affinity chromatography: A review of clinical applications, *Clin. Chem.* 45 (1999) 593–615.

7 G. T. Hermanson, P. K. Smith, A. Krishna Mallia, Immobilized Affinity Ligand Techniques, Pierce, 1992.

8 G. T. Hermanson, Bioconjugate Techniques, Academic Press, 1996.

9 M. Aslam, A. Dent, Bioconjugation, Macmillan Reference, 1998.

10 http://www.antibodyresource.com

11 Linscott's Directory of Immunological and Biological Reagents, http://www.linscottsdirectory.com

12 MSRS Catalog of Primary Antibodies, http://www.antibodies-probes.com

13 M. Komiyama, T. Takeuchi, T. Mukawa, H. Asanuma, Molecular Imprinting: From Fundamentals to Applications, Wiley, 2003.

14 J. Cai, J. Henion, On-line immuno-affinity extraction-coupled column capillary liquid chromatography/tandem mass spectrometry: Trace analysis of LSD analogs and metabolites in human urine, *Anal. Chem.* 68 (1996) 72–78.

15 G. Giraudi, C. Baggiani, Strategy for fractionating high-affinity antibodies to steroid hormones by affinity chromatography, *Analyst* 121 (1996) 939–944.

16 C. Parini, M. A. Baciagalupo, S. Colombi, N. Corocher, C. Baggiani, G. Giraudi, Fractionation of an antiserum to progesterone by affinity chromatography: Effect of pH, solvents and biospecific adsorbents, *Analyst* 120 (1995) 1153–1158.

17 P. M. Kraemer, Q. X. Li, B. D. Hammock, Integration of liquid chromatography with immunoassay: an approach combining the strengths of both methods. *J. AOAC Intern.* 77 (1994) 1275–1287.

18 A. Zeck, M. G. Weller, R. Niessner, Multidimensional biochemical detection of microcystins in liquid chromatography, *Anal. Chem.* 73 (2001) 5509–5517.

19 E. S. M. Lutz, A. J. Oosterkamp, H. Irth, Online coupling of liquid chromatography to biological assays, *Chimica Oggi* 15 (1997) 11–15.

20 M. Schedl, G. Wilharm, S. Achatz, A. Kettrup, R. Niessner, D. Knopp, Monitoring polycyclic aromatic hydrocarbon metabolites in human urine: Extraction and purification with a sol-gel glass immunosorbent, *Anal. Chem.* 73 (2001) 5669–5676.

21 L. Leickt, M. Bergström, D. Zopf, S. Ohlson: Bioaffinity chromatography in the 10 mM range of k_d, *Anal. Biochem.* 253 (1997) 135–136.

22 A. Zeck, Dissertation, Technische Universität München, 2002 (in German) (http://tumb1.biblio.tu-muenchen.de/publ/diss/ch/2001/zeck.pdf).

3.2
Enhanced Characterization and Comprehensive Analyses by Two-Dimensional Chromatography

Peter Kilz

3.2.1
Introduction

All separation methods suffer from limited chromatographic resolution (peak purity, limited peak capacity, determination of multiple property distributions, etc.). Many efforts have been made to overcome the intrinsic limitations in the chromatographic process (see Part 1: Fundamentals of Optimization). Even powerful detection methods such as the currently popular mass spectrometry techniques cannot resolve the separation dilemma. These detectors rely on the chromatography principally as a sample preparation step, which reduces the complexity of the sample but does not *separate* it. A more universal approach is the combination of different separation techniques into a single experiment (multidimensional chromatography; also called 2D chromatography, orthogonal chromatography or cross-fractionation).

Multidimensional chromatography tackles the root cause of limited chromatographic resolution by vastly improving the peak capacity, n_{total}, this being the product of the peak capacities, n_i, of the individual chromatographic separations (Eq. 1):

$$n_{total} = n_1 \cdot n_2 \cdot n_3 \tag{1}$$

Eq. (1) is a simplification for the sake of clarity based on isocratic elution and orthogonal separation techniques; a more general quantification is given in Ref. [1].

Chromatographic hyphenation has an important additional advantage: it allows the separation of samples that cannot be separated by any individual method alone. Figure 1 illustrates the precise mapping and identification of compounds for the case of a two-dimensional separation by a combination of HPLC with GPC. It shows three different samples, each of which have identical HPLC *and* GPC chromatograms. Any of the traditional separation optimization strategies would fail in each of these cases. However, the combination of HPLC with sequential GPC separation of HPLC fractions can easily show the differences in the sample. Figure 1 shows an example of a so-called contour plot (which can be read like a geographical map; the first dimensional separation is plotted on the *y*-axis, while the second separation axis is plotted as the abscissa).

This type of separation is extremely useful for any kind of complex samples that are difficult to separate, identify comprehensively, or fully deformulate. This is most often the case for synthetic, natural, and biopolymers (monodisperse proteins are one well-known exception), which possess coexisting multiple property

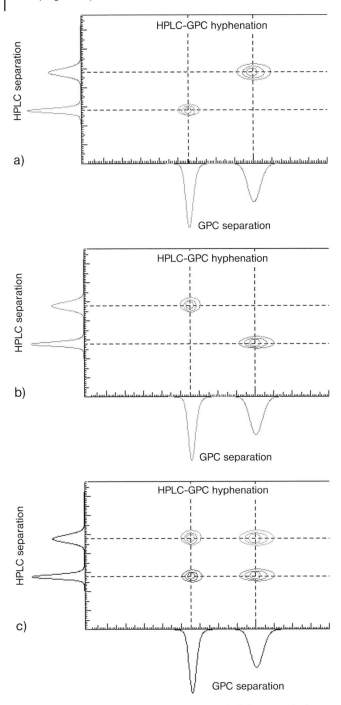

Fig. 1. 2D method combinations can show sample differences which cannot be detected by analysis in individual separation methods.

distributions (such as molar mass distribution, functionality type distribution, chemical composition distribution, molecular architecture (shape) distribution, to name just a few).

3.2.2
How Can I Take Advantage? – Experimental Aspects

Setting up a 2D chromatographic separation system is straightforward, if robust separation methods exist for each dimension. The experimental implementation can be handled fairly easily in most cases. Investment in new instrumentation and developing new methods will not be necessary in most laboratories (see Table 1).

The pivotal part in any multidimensional set-up is the fraction transfer step between the separation methods, which can be on-line or off-line, manual or automated. The individual separation techniques can be combined in many ways. Comprehensive 2D work employs complete transfer of the injected mass from the first to the final dimension. This ensures that the complete sample is analyzed and detected by the final separation method. Heart-cut approaches are not considered comprehensive for obvious reasons. The advantages and limitations of each approach are given in Table 2.

Off-line systems just require a fraction collection device with manual transfer and reinjection of the first dimension fractions into the second instrument. In on-line 2D systems, the fraction transfer is preferentially done by automatic transfer. Figure 2a shows the general layout and Fig. 2b shows a picture of an automated two-dimensional chromatography instrument as used by the author. The first dimension instrument is shown left to right, while the second dimension modules are depicted vertically. The first dimensional separation is selected to fractionate by a single property. This ensures that all transfer fractions only contain compounds which are heterogeneous in another property, which will be separated in the second chromatographic dimension.

Another important experimental aspect in multidimensional separations is the sensitivity of the detection in the final dimension. Consecutive sample dilution can lead to low final concentrations, which require sensitive detectors. In most

Table 1. Additional requirements for two-dimensional chromatography.

Requirements for single dimensional separations	Additional requirements for 2-dimensional separations	Specific for 2D work
Pump	Pump	–
Injector	2D transfer injector	+
Column(s)	Column(s)	–
(Detectors)	Detector	–
1st dimensional separation method	2nd dimensional separation method	–

Table 2. Summary of 2D transfer injection options.

Transfer	Mode	Advantages	Disadvantages	Example
Manual	Offline	Very simple fast setup	Time-consuming not for routine work not precise no correlation of fraction elution to transfer time not quantitative	Test tube
Automatic	Offline	Simple easy fast setup	Less precise no correlation of fraction elution to transfer time not quantitative	Fraction collector storage valve
Single-loop	Online	Correct concentrations correct transfer times automation	Transfer not quantitative setup time	Injection valve (with actuation)
Dual-loop	Online	Correct concentrations correct transfer times quantitative transfer automation	Setup time special valve	Actuated 8-port valve combination of 2 conventional 6-port injection valves

cases, evaporative light-scattering (ELS) detectors are used for this purpose, despite their limited linear response, which can render quantification difficult. Fluorescence and diode-array UV/Vis are also sensitive detection methods, which can detect samples at nanogram level. Mass spectrometers have a high potential in this respect too; however, they are currently not developed to such a level as to make them generally usable. As a general rule, the higher the inject band dilution of a given separation method, the more sensitive a subsequent detection method has to be.

Many chromatographic method combinations are possible and have indeed been used; e.g. SEC, LAC, LC-CAP, GC, SFC, CE, TREF [2]. Obviously, techniques such as GC and SFC, which destroy the mobile phase, can only be used as the final dimension. Most popular to date have been HPLCxGPC combinations, because comprehensive 2D chromatography was first established in synthetic polymer work. Details on the methodology and experimental implementation can be found in the literature [3].

In a simulation paper by Schure [4], different chromatographic method combinations are discussed on the basis of efficiency, sample dilution, and detectability. He investigated CE, GC, LC, SEC, and FFF in detail, while omitting some other potential techniques for method hyphenation, e.g. SFC and TREF.

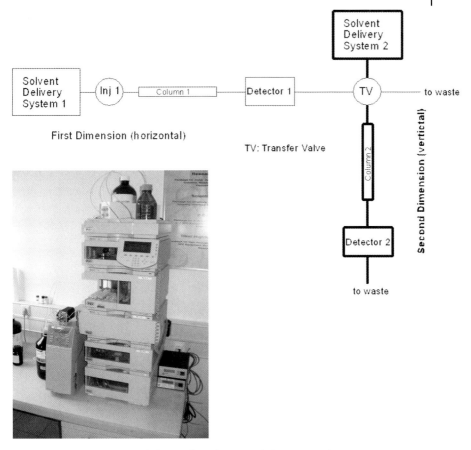

Fig. 2. (a) General experimental design of a 2-dimensional chromatograph.
(b) Completely automated 2D instrument comprising of gradient pump
(1st dimension, left), 2D transfer valve (on gradient pump) and GPC system
for 2nd dimension.

Schure pointed out several experimental factors (such as plate count, injection volume, injected mass, and injection band dilution) that should be considered when designing two-dimensional separations. He concluded that a low-resolution, low injected mass method with high dilution of the injection band would be a poor candidate for a two-dimensional experiment.

The proper sequence of separation methods is important for optimal resolution and accurate determination of property distributions. It has been shown that it is best to apply the method with the highest selectivity for one property as the first dimension. This ensures eluting fractions of the highest purity being transferred to the subsequent separation. In many cases, interaction chromatography is the best and most flexible choice as the first dimension separation method. From an experimental point of view, high flexibility is required for the first chromatographic dimension.

The compatibility of mobile phases that are transferred between chromatographic dimensions is an important issue in designing two-dimensional experiments. Complete miscibility of the mobile phases used in all dimensions is an obvious necessity. Otherwise, the separation in the second method is dramatically influenced and the fraction transfer is restricted or completely hindered. In gradient systems, this requirement has to be verified for the total composition range.

In SEC dimensions, the transfer of mixed mobile phases can affect molar mass calibration. In order to obtain reliable molar mass results, the calibration curves have to be measured using the extremes of mobile phase composition and tested for changes in elution behavior and pore-size influence in the SEC column packing. The better the thermodynamic property of the SEC eluent, the less influence is expected on the SEC calibration when the transfer of mobile phase from the previous dimension occurs. It has been shown to be advantageous to use the SEC eluent as one component of the mobile phase in the previous dimension to avoid potential interference and mobile phase incompatibility.

Obviously, two-dimensional separations take more time depending on the number of transfer injections from the first to the second dimension. Time requirements can be substantial if conventional columns are used in the second dimensional separation (this is the bottleneck in terms of time). However, the availability of high-throughput LC columns allows for much faster 2D results.

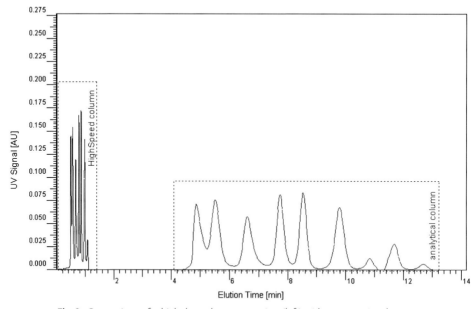

Fig. 3. Comparison of a high-throughput separation (left) with a conventional GPC separation (right). The efficiency of the GPC separation of PSS polystyrene standards is comparable in both cases; the specific resolution factor R_{sp} is 4.3 for the HighSpeed column as compared to 4.7 for the conventional analytical column.

For example, if a run on a traditional column takes 30 min and 20 2D-transfer injections are necessary, then the total 2D analysis time will be about 10 h. The same separation will require a total analysis time of only about 1 h if high-throughput columns are used in the second dimension.

The use of high-throughput columns has opened up new applications for two-dimensional work and has allowed it to be used not only for specialized R&D work, but also for more standard and QC tasks [5].

Figure 3 shows the time gains possible for a mixture of polymer standards when using HighSpeed GPC columns as compared to conventional GPC columns. The analysis time for the conventional system was about 15 min, but this was reduced to just 1.3 min when using a high-throughput column (PSS HighSpeed SDV 5 mm linear). Throughput increases by up to a factor of ten are possible, and allow GPC results to be obtained in less than 2 min.

3.2.3
2D Data Presentation and Analysis

Comprehensive 2D experiments yield a three-dimensional data array (similar to DAD data sets) of data sets in the form of ({dimension 1 property}, {dimension 2 property}, {concentration}), which can be expressed in many ways. The raw data view (in lattice or waterfall plot representation) consists of the chromatograms of the transferred fractions in the final dimension as shown in Fig. 4a. It can be used to investigate individual fractions that have been transferred, but is not quantifiable. The same is true for 3D surface plots, which are generated from the lattice plot by interpolation (see Fig. 4b). These can be regarded as virtual 3D landscapes, which can be viewed from any observation position in space and allow investigation of the peak shape or of trace amounts on the back side of major peaks. Figure 4c shows the corresponding contour plot (or contour map), which is the two-dimensional representation of the 3D surface plot. In such representations, different concentrations are shown as different colors. The chromatographic axis of the first dimension is plotted as the ordinate, while the properties determined in the second dimension are shown on the abscissa.

Comprehensive 2D data analysis is performed on the contour map by setting baselines (in fact, base planes) and selecting the regions for data integration in a similar way as in one-dimensional data processing. The retention axes of the different dimensions can be calibrated to transform chromatography times into physical properties of the samples. Figure 5 shows a base plane corrected contour plot in which the data analysis regions have been selected. Depending on the selected separation methods, a multitude of results for each region is obtained from the data analysis.

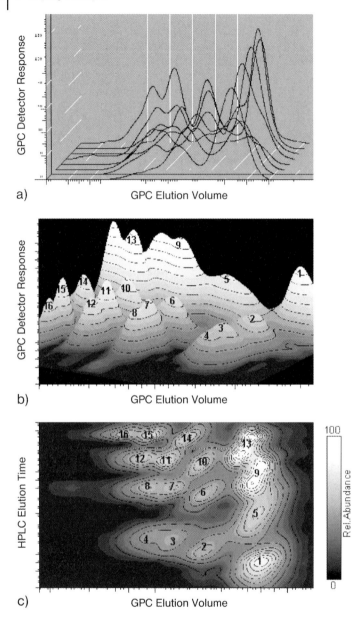

Fig. 4. Different representations of 2D results in an GPCxHPLC experiment.
(a) Pseudo 3D stacked transfer injections, (b) 3D surface plot with iso-lines,
(c) 2D contour plot with iso-lines

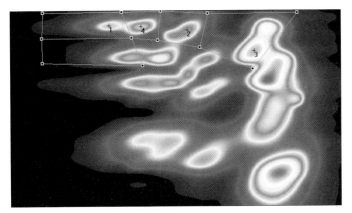

Fig. 5. Base plane corrected 2D contour plot with integration regions for 2D data processing.

3.2.4
The State-of-the-Art in 2D Chromatography

Comprehensive two-dimensional separations have become a classical method for analyzing and deformulating macromolecular samples. In the analysis of small molecules, two-dimensional chromatography is mostly used as a sample preparation step for advanced detection techniques (MS, NMR, etc.). Obviously, the analytical goal is very different and sample preparation techniques will not be discussed here.

The importance of in-depth analysis of star-block copolymers by HPLCxSEC is discussed as a representative example of a polymer application. Such macromolecules are widely used as models to investigate the improvement of the rheological properties of motor oils. In the cited example, the investigated sample consisted of 16 components with variation of molar mass and chemical composition. An in-depth overview of chromatographic methods for molar mass and chemical composition separations is given in Ref. [6].

As expected, SEC revealed four partially resolved peaks (Fig. 6a), which corresponded to the four molar masses present in the sample. Despite the adequate resolution, the SEC chromatogram did not indicate the co-elution of species of different chemical composition. Gradient HPLC analysis of the same sample gave poorly resolved peaks (Fig. 6b), which might have suggested the presence of four different compositions, but gave no indication of the molar mass and architecture variation in the sample.

On-line combination of these two methods dramatically increased the resolution of the separation system and gave a clear picture of the complex nature of the sample. A three-dimensional representation of the gradient HPLCxSEC separation is given in Fig. 7.

The complex nature of the sample is clearly visible in the contour map with coordinates of chemical composition and molar mass. Experimental evidence for the improved resolution in 2D separations is given in Fig. 8. This contour plot

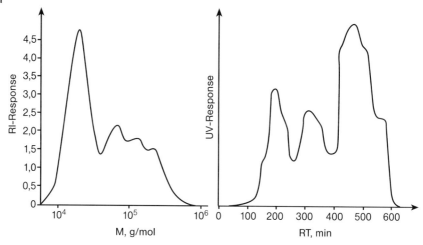

Fig. 6. SEC separation (a, left) and gradient HPLC analysis (b, right) of the star block copolymer with widely varying molar masses (separated predominantly by SEC) and butadiene content (mapped in the HPLC mode).

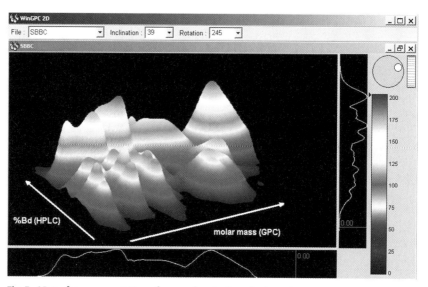

Fig. 7. 3D surface representation of a complex star-branched copolymer consisting of 16 species varying in butadiene content and molar mass.

was calculated from experimental data based on 28 transfer injections. The 16 expected peaks are easily correlated with the 16 well-resolved peaks in the 2D contour map.

The contour plot clearly reveals the broad heterogeneity of the sample (chemical composition plotted on the ordinate, molar mass distribution on the abscissa). The relative concentrations of the components are represented by colors.

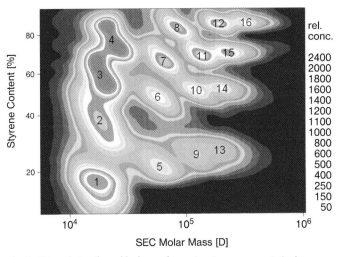

Fig. 8. 2D analysis of star block copolymer (contour representation) showing all expected 16 components as well separated peaks.

As expected, a small molar mass dependence of the HPLC separation is indicated by a drift of the peaks for components of similar chemical composition (e.g., with peaks nos. 1, 5, 9, and 13). This kind of behavior is normal for polymers, because pores in the HPLC stationary phase lead to size-exclusion effects, and these overlap with the enthalpic interactions at the surface of the stationary phase. The 2D experiment is sufficiently sensitive to reveal influences of this type. The amount of butadiene in each peak could be quantitatively determined through an appropriate calibration with samples of known composition. The molar masses could be calculated based on a conventional molar mass calibration of the second dimension.

Good agreement has been found between results determined by conventional analytical methods and the 2D separation with 2D data treatment as discussed in this chapter. Obviously, the 2D analysis yields much more detailed molecular information than the averaging of results from conventional 1D methods (see Fig. 8). In the cited example, results on all 16 components could be easily extracted from the 2D contour map.

Table 3. 2D results for the styrene/butadiene star block copolymer.

Sample	Reference		2D Results	
	Bd content [%]	Mw [kD]	Bd content [%]	Mw [kD]
A (peaks 4-8-12-16)	17	87	18	88
B (peaks 3-7-11-15)	39	88	37	82
C (peaks 2-6-10-14)	63	79	64	75
D (peaks 1-5-9-13)	78	77	81	72

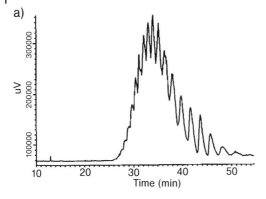

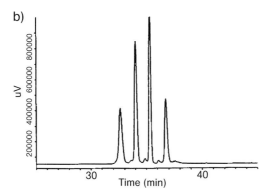

Fig. 9. Chromatograms from conventional (single dimensional) separations for the alcohol ethoxylate:
(a) isocratic NP separation measurement of EO units, and
(b) gradient separation of methylene units.

The 2D analysis of non-ionic surfactants by NPLCxRPLC shows the importance of investigations of this type in controlling product quality, predicting application properties, and elucidating structure–property relationships. The ethoxylated fatty alcohols in the sample possess various numbers of methylene groups (from the source of the fat) and a distribution of ethylene glycol (EO) units (from the oligomerization process). PEG separation was performed on a normal-phase system, while the alkyl chain length dependence was characterized using a C18-based phase system [7]. The results of the individual separations are shown in Fig. 9. In each case, neither the peak width nor the peak shape indicates that each narrow peak consists of several components varying in methylene group or EO content. However, the complexity of the sample is directly evident when the two separation techniques are combined in a two-dimensional experiment. Figure 10 shows the 2D contour map of the sample, which demonstrates the improved resolution in 2D experiments. The experimentally determined peak capacity is 78, which is in excellent agreement with the predicted 2D peak capacity based on

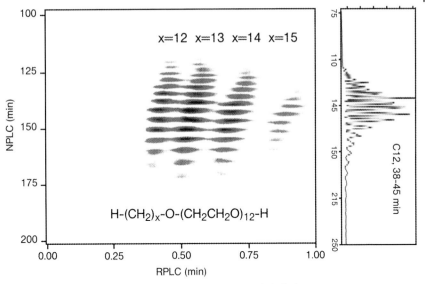

Fig. 10. 2D-Separation of non ionic surfactants (ethoxylated alcohols) with a combination of normal phase and reversed phase chromatography. The dramatic increase of resolution and peak capacity is visible in the contour plot. The conventional chromatogram of the normal phase (right) includes all co-eluting species with a varying number of methylene groups in the alcohol.

the individual separation techniques (4×19). From the 2D experiments, the surfactant activity and the quality of the product could be mapped onto the EO distribution and the appropriate composition in terms of fatty alcohol alkyl chain length.

3.2.5
Summary

Two-dimensional chromatography allows the in-depth analysis of complex samples by combining different separation techniques. It can be shown that the peak capacity in 2D separations is the product of the peak capacities of the individual chromatographic methods. Additionally, dual property distributions (e.g., molar mass, chemical composition or functionality) can be simultaneously determined in a single run. Comprehensive data analysis allows the fast and reliable measurement of multiple sample parameters.

References

1 P. Kilz, H. Pasch, Coupled LC
techniques in molecular
characterization, in *Encyclopedia of
Analytical Chemistry* (Ed.: R. A. Myers),
Wiley, Chichester, 2000.
2 P. Kilz, *Chromatographia* **59**, 3 (2004).
3 H. Pasch, B. Trathnigg, *HPLC of
Polymers*, Springer, Berlin, Heidelberg,
New York, 1997.

4 M. R. Schure, *Anal. Chem.* **71**, 1645 (1999).
5 H. Pasch, P. Kilz, *Macromol. Rapid
Commun.* **24**, 104 (2003).
6 G. Glöckner, *Gradient HPLC of
Copolymers and Chromatographic
Cross-Fractionation*, Springer, Berlin,
Heidelberg, New York, 1991.
7 R. E. Murphy, M. R. Schure, J. P. Foley,
Anal. Chem. **70**, 4353 (1998).

3.3
LC/MS – Hints and Recommendations on Optimization and Troubleshooting

Friedrich Mandel

LC/MS has developed from an "interfacing technique for experts" to a robust and routine detection method of a broad variety of compound classes. However, the bigger the "black box factor", the more easily the users tend to apply method parameters from literature or from colleagues without further thinking. Thus, missing the expected sensitivity, low robustness or linearity of the method often lead to big disappointment. The user easily gets blinded by all the high-tech features of the mass spectrometers – high selectivity, MS^n, accurate mass – and one forgets to consider the fundamental aspects of ionisation. The most common source of trouble in LC/MS is the ionisation process itself. Here and only here should the chemist start to investigate the reasons for poor or varying response, insufficient linearity or reproducibility. This is why in the following I would like to concentrate primarily on the aspect of ionisation of the analyte molecules.

3.3.1
Optimization of the Ionization Process

It might seem a trivial matter, but it is worth reiterating that only analytes in ionic form are monitored by a mass spectrometer. Recall that in electrospray ionization (ESI) the formation of cations or anions occurs in solution. This means that, on entering the ion source, the pH of the eluent has to be 2 pH units below (positive-ion ESI) or above (negative-ion ESI) the pK_a value of the analytes in order to ensure that 99.9% of the analyte molecules are protonated or deprotonated. Unfortunately, this is very often at variance with the aims of the chromatographer, who would like to develop methods with optimum analyte retention on RP stationary phase materials. Ideally, the analytes have to be in undissociated form for separations of this kind.

Fortunately, mass spectrometric detection is so selective that in most cases one can deviate from the ideal situation of baseline-separated LC peaks. Thus, one may perform the chromatographic separation at a slightly acidic pH while still being able to detect analyte anions in negative-ion electrospray mode, albeit at sub-optimal detection sensitivity. In the same way, one can detect analyte cations in positive-ion electrospray mode after having adjusted the pH to alkaline by adding ammonium hydroxide to the mobile phase, although again it must be emphasized that this is typically at the expense of a drastic loss in sensitivity.

In the case of an incompatibility between the optimal pH values for either chromatographic separation or ionization via acid/base equilibrium, there are several possible solutions. If an additional HPLC pump is available, acid or base may be added post-column in order to shift the pH value to a range that is optimal for electrospray ionization. This might even mean an inversion of the pH. It is import-

HPLC Made to Measure: A Practical Handbook for Optimization. Edited by Stavros Kromidas
Copyright © 2006 WILEY-VCH Verlag GmbH & Co. KGaA, Weinheim
ISBN: 3-527-31377-X

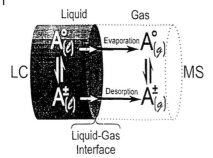

Fig. 1. Ionization principle at atmospheric pressure.

ant that the added buffer does not dilute the eluent too much (electrospray is a concentration-dependent detection technique) and that the analytes do not precipitate.

Besides the addition of acid or base after the separation column, many analytes are converted to cations or anions by the formation of adducts. For example, compounds bearing OH functional groups tend to form adducts with alkali metal cations. The recommended alkali concentration is below 1 mmol L^{-1} (i.e., sodium acetate) in order to avoid ion suppression effects. It is important to achieve the cation adduct formation by post-column addition and under no circumstances by addition to the HPLC buffer. This would cause a long-lasting contamination of the HPLC system. Carbohydrates can be detected as chloride adducts in negative-ion electrospray mode by adding 50 mm HCl (2 mmol L^{-1} to the ESI ion source).

Alternatively, one may switch to another ionization technique, whereby the cation or anion formation takes place in the gas phase at atmospheric pressure. These techniques are chemical ionization (Atmospheric Pressure Chemical Ionization, APCI) and photoionization (Atmospheric Pressure Photo Ionization, APPI).

With these ion sources, the eluent is entirely evaporated before the ionization of the analytes. This has the disadvantage of some thermal stress to the analytes because the mobile phase gets sprayed into a heated ceramic or quartz cartridge. Although these cartridges are heated up to 450 °C, the effective temperature inside the cartridge is in the range of only 120–150 °C. However, thermolabile compounds such as sugars and peptides do not survive this evaporation process. Another commonly encountered degradation route is the elimination of water (steroids) or CO_2 (carboxylic acids).

The advantage of these two ionization modes, in contrast to electrospray, is their lower tendency to interfere with the sample matrix.

3.3.2
Lost LC/MS Peaks

Let us assume that sample introduction, i.e. by the use of an automated injector, functions flawlessly. An amount of analyte that should give a reasonable signal intensity is then injected. The requisite amounts are in the range of picograms to nanograms, depending on the type of mass spectrometer and its operating mode. Assuming that an injection peak is seen in the UV or MS signal, the sample has supposedly been completely transferred onto the HPLC column. Sometimes, however, although the mass spectrometer monitors peaks, the expected analyte peak has "disappeared". Actually, there are only three reasons that can account for such an absence of an MS signal in electrospray ionization, which are outlined in the following sections.

3.3.2.1 Mobile Phase pH at the Edge of the Optimum Range

In order to obtain a good chromatographic separation, the pH of the mobile phase very often gets adjusted to a sub-optimal value for the subsequent ionization in electrospray. Moreover, it is possible that the pH of the mobile phase changes due to long-term storage in the HPLC system. This can readily be demonstrated by measuring the pH of a 0.1% aqueous formic acid or acetic acid solution directly after preparation and after several days of storage. In most HPLC systems, the solvent bottles are open to the laboratory atmosphere and this leads to a slow increase in pH in the example mentioned above. Even more unstable is the pH of an ammonium hydroxide solution for negative-ion electrospray. One can easily compensate for this kind of pH drift by post-column addition of freshly prepared acid or base at reasonable concentrations.

3.3.2.2 Ion-Pairing Agents in the HPLC System

Ion-pairing agents are poison to the electrospray ionization. The analyte ions that should be released during the electrospray process are locked in the ion pair and therefore do not give rise to an MS signal. Should ion-pair chromatography be unavoidable due to strong analyte polarity, the most weakly ion-pairing reagents (e.g., perfluorinated aliphatic carboxylic acids or aliphatic amines) should be selected and APCI should be the ionization method of choice. However, the unforeseen presence of ion-pairing agents in an HPLC system may also suppress the MS signal by 100%. Figure 2 shows a typical example of the negative influence

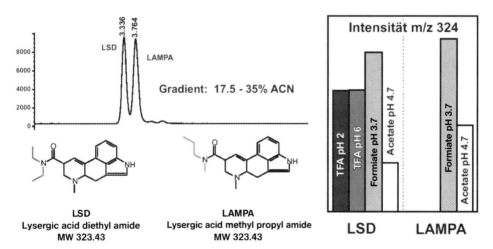

Fig. 2. ESI signal suppression by ion pairing with TFA.

of TFA on the LSD metabolite LAMPA. Irrespective of the pH value, TFA completely suppresses the LAMPA signal. As this example illustrates, even with minimal differences in chemical structure it is extremely difficult to predict the tendency of analytes to form ion pairs.

Even minimal residues of ion-pairing agents in solvent bottles, tubing, the vacuum degasser or the chromatographic column lead to a negative influence on the ionization in electrospray mode. In case of uncertainty, colleagues should be consulted to ascertain whether someone has unwittingly allowed ion-pairing agents to find their way into the jointly used HPLC system ("...I only used 0.001% aqueous TFA!"). If this has happened, all contaminated tubing, bottles, and columns should be taken out, the module with the worst memory effect – the vacuum degasser – should be bypassed, and the system should be flushed for several hours or days. If the chemical structure of the contaminant is known, then the decay of its signal (e.g., $m/z = 113$ for the TFA anion) should be monitored.

3.3.2.3 Ion Suppression by the Sample Matrix or Sample Contaminants

Interference with the ionization of an analyte by a disturbing component is widely called "ion suppression", even if the signal gets enhanced rather than suppressed. Most prone to this effect is of course electrospray, but APCI and APPI may occasionally also be affected. This interference of eluting compounds is a complex issue and is extensively discussed in Section 3.3.4.

3.3.3
How Clean Should an LC/MS Ion Source Be?

Those users of LC/MS with some experience in mass spectrometry tend to polish the ion source and ion optics before introducing an important series of samples. Of course, maximum signal intensities in API LC/MS ionization techniques will be achieved when all residues are completely eliminated from the ion source surfaces. The electrical fields that are necessary to form and/or transport the ionic analytes before they reach the mass analyzer are influenced by the degree of contamination with non-volatile sample constituents. On the other hand, experienced LC/MS users know that the long-term stability of an LC/MS system will be increased if the spray chamber is intentionally slightly contaminated. This is of particular importance when one wishes or has to quantify by means of external calibration. With the most commonly used ion sources it is sufficient to inject the sample matrix 10–20 times in order to achieve an almost constant analyte response after an initial exponential drop.

However, after long series of heavy sample matrix load, the response will drop significantly. This is strongly dependent on the instrument type and brand used. Those regions of the ion source that are exposed to atmospheric pressure are easily "regenerated" simply by flushing with solvents. If such treatment does not lead to recovery of the signal, then it is very likely that the ion optics (ion transfer capillary, skimmer, cone, ion guide) are seriously contaminated.

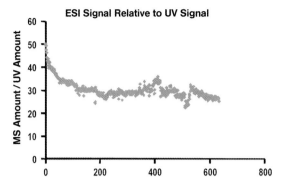

Fig. 3. Drop of signal intensity after extreme matrix stress of an ion source.

Before starting to vent an LC/MS system in order to clean the interior, a trick can be performed in order to recover the system for a few runs. Layers on the surfaces of the ion optics (skimmer, cone) cause electrostatic charging or an unwanted hysteresis when the operating voltage is applied. By changing the ion polarity for a few minutes, all DC-operated elements will be discharged. This may even be included in the acquisition method in order to increase the uptime of the instrument. The ion polarity can simply be changed at the end of a chromatographic run while the column is re-equilibrated. However, those elements in the ion optics that are operated at AC voltages (quadrupole/hexapole/octopole) will not be purged of contamination by this procedure. Then, unfortunately, the only solution will be venting of the system and cleaning. The best preventative measure against frequent instrument cleaning is diversion of the HPLC void volume to waste. Almost all involatile components in a sample will elute ahead of the chromatogram. One should thus use the diverter valves that are built into most of today's LC/MS systems.

3.3.4
Ion Suppression

The interference of co-eluting compounds with the ionization of the analytes of interest is the most frequent reason for insufficient response and reproducibility in LC/MS. The phrase "ion suppression" was introduced in the mid-1990s and describes the phenomenon only partially, as an enhancement of the signal may also occur. Methods with little sample preparation and/or little chromatographic separation are especially prone to this. Detection methods that monitor only single ions or MS/MS transitions (SIM, MRM) are quasi-blind to co-eluting compounds, while the detection of complete mass spectra allows the recognition of massive co-elution so that corrective action can be taken.

After the initial euphoria about the high selectivity of LC/MS-based methods had subsided, some disillusionment followed. It was immediately recognized that compounds that are invisible to MS detection are not always "inactive". The matrix effect is exerted solely in the ion source and is therefore independent of the selectivity of the subsequent MS(/MS) detection. Ion suppression occurs in all ionization modes – electrospray, APCI, and APPI. However, the magnitude of the effect may depend on the design of the ion source. The origin of the problem may not necessarily be limited to the sample; matrix effects may also be due to exogenous materials. In particular, in gradient HPLC methods, organic substances that are dissolved in the mobile phase or that bleed from parts of the HPLC system elute with the analytes of interest in the gradient and may massively influence the signal strength. Unfortunately, "HPLC grade" solvents are not necessarily suitable for use in LC/MS. It is highly recommended to test solvents from different suppliers as well as of various qualities and batches and then decide which to use based on the lowest background signal. LC/MS systems rapidly reveal if mobile phase bottles have been contaminated with phthalates from plastic caps or with tensides from the laboratory dish washer.

The matrix effect can be produced by various processes – hindering of the analyte ion release in electrospray, e.g., by alkali or phosphate, micelle formation by tensides, formation of ion pairs or even precipitation of the analytes. It is important that efforts are made during method development to recognize and exclude ion suppression or matrix effects.

Various experimental approaches are available for evaluating a matrix effect. The easiest way is to compare a "solvent standard" with a "matrix spike". The relevant effects are shown schematically in Fig. 4 (B. K. Choi et al., *J. Chromatogr. A* 907, 337–342 (2001)).

Ideally, the sample matrix should have no influence on the analyte signal. It is recommended that tests are carried out with matrix blanks of different origins to exclude random results. The approach described in Fig. 4 can be extended with little effort for the additional determination of the method recovery and extraction yield. The required steps are depicted in Fig. 5.

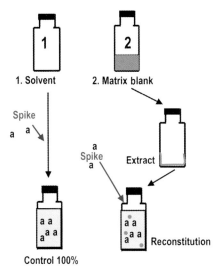

Fig. 4. Comparison of solvent standard with matrix spike.

Comparison of steps 1 and 3, i.e. the solvent standard without extraction and the extracted/reconstituted matrix spikes, provides insight into the method recovery. From steps 1 and 2, the matrix effect can be determined. To calculate the extraction yield, steps 2 and 3 are compared.

For the optimization of LC/MS parameters, the proven method is to continuously introduce the analyte into the ion source. This syringe infusion generates a constant analyte signal. In order to create ionization conditions identical to the intended LC/MS analysis, the analyte is usually infused into the eluent via a T-piece after the separation column (see Fig. 6).

With this experimental set-up, it is quite easy to evaluate the matrix effect. Instead of introducing the analyte into the neat mobile phase, we simultaneously run a chromatographic separation of a matrix blank. Without any matrix effect, we obtain a constant analyte signal. By infusion of the analyte into the eluent with its various

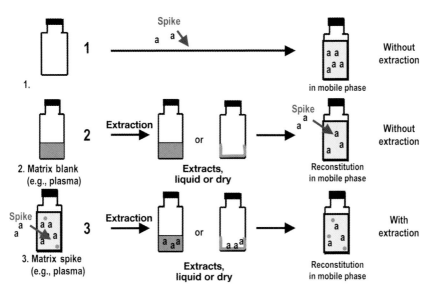

Fig. 5. Extended evaluation of the matrix effect.

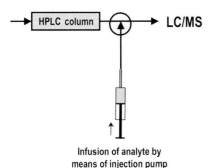

Infusion of analyte by
means of injection pump

Fig. 6. Method optimization by analyte infusion.

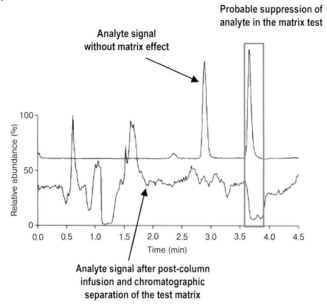

Fig. 7. Evaluation of a matrix effect by analyte infusion.

chromatographic peaks, we will observe an enhancement or a suppression of the analyte signal in retention time regions of high matrix content. In the final acquisition method, the analytes should elute at retention times that exhibit neither enhancement nor suppression. Figure 7 shows an overlay of the infusion signal of the analyte with the signal of a solvent standard.

Whatever approach is chosen to evaluate the ion suppression or matrix effect, the analytical method will have to be varied. The simplest step is probably a variation of the chromatographic system. The gradient slope and/or the mobile phase (acetonitrile vs. methanol, etc.) may be changed. If possible, the extraction procedure (SPE vs. liquid/liquid) should be changed. One may also try changing the ionization method. Matrix effects are less severe in APCI and APPI than in electrospray. If instruments with different ion source designs are available, then switching to a different LC/MS instrument should be investigated. Isotopically labeled internal standards should be used for method calibration. These have the advantage of having the same extraction yield, the same ionization yield, and the same chromatographic retention times (more so with ^{13}C than with ^{2}H) as the analyte. Even when matrix effects do occur, they will act on both the analytes and the internal standard in the same way.

References

1   Fenn, J. B., Mann, M., Meng, C. K., Wong, S. F., Whitehouse, C. M., "Electrospray ionization for mass spectrometry of large biomolecules", *Science* 246, 64–71 (1989).

2   Nohmi, T., Fenn, J. B., "Electrospray Mass Spectrometry of Poly(ethylene glycols) with Molecular Weights up to Five Million", *J. Am. Chem. Soc.* 114, 3241–3246 (1992).

3   Labowsky, M. J., Whitehouse, C. M., Fenn, J. B., "Three-dimensional Deconvolution of Multiply Charged Spectra", *Rapid Communications in Mass Spectrometry* 7, 71 (1993).

4   Fenn, J. B., "Ion Formation from Charged Droplets: Roles of Geometry, Energy and Time", *J. Am. Soc. Mass Spectrom.* 4, 524 (1993).

5   Bruins, A. P., "Mass Spectrometry with Ion Sources Operating at Atmospheric Pressure", *Mass Spectrom. Rev.* 10, 53–77 (1991).

6   Niessen, W. M. A., Tinke, A. P., "Liquid Chromatography-Mass Spectrometry. General Principles and Instrumentation", *J. Chromatogr.* A 703, 37–57 (1995).

7   Tomer, K. B., Moseley, M. A., Deterding, L. J., Parker, C. E., "Capillary Liquid Chromatography Mass Spectrometry", *Mass Spectrom. Rev.* 13, 431–457 (1994).

8   Wachs, T., Conboy, J. C., Garicia, F., Henion, J. D., "Liquid Chromatography-Mass Spectrometry and Related Techniques via Atmospheric Pressure Ionization", *J. Chromatogr. Sci.* 29, 357–366 (1991).

3.4
LC-NMR Coupling

Klaus Albert, Manfred Krucker, Karsten Putzbach, and Marc D. Grynbaum

3.4.1
NMR Basics

For structure elucidation of unknown compounds, nuclear magnetic resonance spectroscopy (NMR) is one of the most important spectroscopic techniques. Most commonly used is ^{1}H NMR spectroscopy; ^{13}C NMR measurements give additional information, but due to the low natural abundance of ^{13}C nuclei (1.1%; the isotope ratio of ^{1}H nuclei is 99.9%) these NMR measurements are relatively time-consuming. For structure elucidation, three major parameters can be derived from a ^{1}H NMR spectrum, namely chemical shifts (reflecting the structural neighborhood of each proton), integration values of signals, making their quantification possible, and coupling constants from the fine structure of the signals. From these coupling constants, structural information can be derived, e.g. to differentiate stereoisomers. In contrast, the only information from a ^{13}C spectrum is the chemical shift, although the values of these chemical shifts are within a 20-fold bigger range (increased spectral dispersion: 200 ppm for ^{13}C NMR compared to 10 ppm for ^{1}H NMR). Therefore, ^{13}C NMR spectroscopy is also an indispensable tool for structure elucidation purposes.

The major benefits of ^{1}H NMR spectroscopy are easy quantification and determination of steric arrangements. Disadvantages are the low sensitivity of the NMR detection and the fact that inorganic counterions and parts of the molecule without attached protons cannot be detected. Mass spectrometry, in contrast, is a very sensitive and selective detection technique; steric information, however, cannot readily be obtained. Therefore, these two detection methods are mutually complementary; for unambiguous structure elucidation purposes, a combined evaluation of both sets of data is highly desirable.

To acquire NMR spectra, the pulsed Fourier-transform (PFT) acquisition mode is utilized. For this purpose, a radiofrequency pulse of microsecond length excites all resonance transitions within a defined frequency domain. The interferogram obtained, with a time-dependent decline curve (free induction decay, FID), is then converted to an NMR spectrum by means of Fourier transformation. The big advantage of PFT-NMR spectroscopy is the possibility of accumulating any number of interferograms before processing the data. The signal-to-noise ratio (S/N) of the resulting NMR spectrum is dependent on the square root of the number of transients (scans) n collected: $S/N = \sqrt{n}$. Each single interferogram has to be optimized; the so-called pulse repetition time is dependent on characteristic time constants of the NMR experiment.

To measure NMR spectra, the molecules of interest are placed in a magnetic field generated by a cryo-magnet. In this way, the polarization of the nuclei is

HPLC Made to Measure: A Practical Handbook for Optimization. Edited by Stavros Kromidas
Copyright © 2006 WILEY-VCH Verlag GmbH & Co. KGaA, Weinheim
ISBN: 3-527-31377-X

divided into a parallel (ground state) and antiparallel (excited state) arrangement with respect to the magnetic field B_0. There are slightly more nuclei in the ground state (Boltzmann distribution), the equilibrium magnetization being dependent on the field strength of the cryo-magnet. The sensitivity of the NMR detection increases with increasing field strength. After pulse excitation, nuclei are transferred from the ground state to the excited one. The absorbed energy is released by relaxation (exchange processes); the spin-lattice relaxation time T_1 defines the time of interaction with the environment, while the spin-spin relaxation time T_2 is the time of interaction of atomic nuclei with each other. Repeated excitation of the spin system by another radiofrequency pulse is only possible after complete relaxation, that is, after the complete release of the spin system's absorbed energy. Therefore, T_1 defines the pulse repetition rate, T_2 the FID decay time and, according to $T_2 = 1 / \Pi \times W$, the signal half-width W of the NMR signals. In solution, T_1 and T_2 range from 0.1 to 1.0 s for ^{1}H nuclei, and from 1–10 s for ^{13}C nuclei. The pulse repetition interval for recording interferograms is between 0.7 and 1.0 s, hence about 60 interferograms can be accumulated per minute.

Valuable information can be derived from the recorded one-dimensional ^{1}H and ^{13}C NMR spectra with the parameters described above, but two-dimensional nuclear magnetic resonance experiments are additionally extremely important for the unambiguous structure elucidation of unknown compounds. By correlated spectroscopy, adjacent protons (^{1}H,^{1}H-COSY) and ^{1}H,^{13}C-pairs in immediate proximity (^{1}H,^{13}C-COSY) can be easily identified in the resulting contour plot. A time-saving and sensitive recording technique is the proton-sided detection of ^{1}H,^{13}C-correlations by the inverse HSQC (*Heteronuclear Single Quantum Coherence Spectroscopy*) method. Long-range ^{1}H,^{13}C correlations through several bonds are detected by the HMBC (*Heteronuclear Multiple Bond Coherence Spectroscopy*) program, and neighborhood through-space interactions by ROESY (*Rotating Frame Overhauser Effect Spectroscopy*), NOESY (*Nuclear Overhauser Effect Spectroscopy*), and transfer-NOESY techniques (see Chapter 2.1.7 by U. Skogsberg).

3.4.2
Sensitivity of the NMR Experiment

The sample requirements for the recording of NMR spectra depend on several factors. In general, two-dimensional experiments require a larger amount of sample. The molecular weight of the studied compound defines the correlation and relaxation time of the nuclei. High molecular weight compounds such as polymers, peptides, and proteins have long correlation times with unfavorable T_1 and T_2 times, whereas low molecular weight compounds show short correlation times with convenient T_1 and T_2 times. Higher external magnetic fields (B_0) lead to increased population differences between the ground and excited states and an increased spectral dispersion (number of NMR signals per chemical shift δ (ppm)). Magnetic field strengths between 4.7 Tesla (200 MHz) and 21.15 Tesla (900 MHz) are well-established. To record one-dimensional ^{1}H NMR spectra of molecules of

300 Daltons with a 400–600 MHz NMR spectrometer, the minimal sample amount is in the nanogram range; for ^{13}C and two-dimensional NMR spectra, microgram amounts are necessary.

3.4.3
NMR Spectroscopy in Flowing Systems

In contrast to conventional NMR measurements, in flowing systems the sample has a defined dwell time τ in the NMR coil. This dwell time is defined by the ratio of detection volume to flow rate and affects the relaxation times T_1 and T_2 as well as the signal intensity and signal half-width:

$$\frac{1}{T_{1\text{flow}}} = \frac{1}{T_{1\text{stat}}} + \frac{1}{\tau} \qquad \frac{1}{T_{2\text{flow}}} = \frac{1}{T_{2\text{stat}}} + \frac{1}{\tau}$$

For optimal detection, the dwell time should be at least 5 s, which will result in a tolerable line broadening of 0.2 Hz for the NMR signals.

3.4.4
NMR Probes for LC-NMR

Conventional NMR measurements are performed with samples in a 5 mm glass tube, which is positioned in the NMR probe by a rotor (Fig. 1a). The probe is located in the room temperature bore of the cryo-magnet in the position of the best homogeneity of the magnetic field. Glass tubes are interchanged by means of an air cushion, and this can be carried out automatically by robots.

For continuous-flow NMR measurements, two probe designs have become established. For LC-NMR with analytical HPLC columns (4.6 × 250 mm), probes with an active volume between 60 and 120 µL are used. In contrast to the conventional tube design, the double-saddle Helmholtz coil is arranged vertically, directly affixed to a U-shaped glass tube in a glass Dewar vessel. The inlet and outlet capillaries are connected by shrinkable Teflon tubings (Fig. 1b).

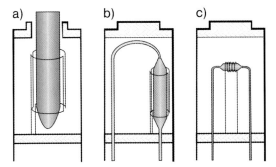

Fig. 1. NMR probes: (a) 5 mm tube probe, (b) flow probe in double-saddle Helmholtz design, c) solenoidal micro flow probe.

For capillary-HPLC-NMR, the use of solenoidal probes with an active detection volume of 1.5 µL has proven satisfactory. The detection capillary, with a microcoil wound around it, is placed horizontally (transverse to the magnetic field) and – in order to eliminate inhomogeneities in the magnetic field and susceptibility – stored in a container filled with a perfluorotributylamine (FC-43) solution (susceptibility matching fluid) (Fig. 1c). This solenoidal design shows high sensitivity down to the low nanogram range, matching the small amounts of sample eluting from capillary separations.

3.4.5
Practical Realization of Analytical HPLC-NMR and Capillary-HPLC-NMR

For LC-NMR, the HPLC system – preferentially installed on a movable table – is positioned at a defined distance from the cryo-magnet. Conventional cryo-magnets have a strong magnetic stray field, which decreases sharply with increasing distance from the cryo-magnet. Depending on the field strength of the cryo-magnet, the 5 Gauss line, up to which the HPLC system can be brought without problems, lies at a distance of 1.5 to 2 m (Fig. 2). Actively shielded cryo-magnets have an additional coil that compensates for the stray magnetic field, so that the 5 Gauss line is in the region of the outer Dewar vessel. With these, the HPLC system can be installed directly next to the cryo-magnet (this is the preferred set-up due to minimized peak broadening arising during transport in the transfer capillaries).

The analytical HPLC system is connected to the flow probe by stainless steel capillaries; the capillary HPLC system is connected by a fused-silica transfer capillary.

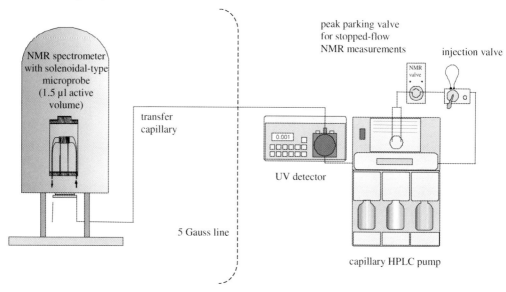

Fig. 2. Capillary-HPLC-NMR system with solenoidal micro flow probe (1.5 µL active detection volume).

In the continuous-flow mode it is possible to analyze a chromatographic separation; additional 2D-NMR measurements can be performed in the stopped-flow mode.

3.4.6
Continuous-Flow Measurements

In the continuous-flow mode, a chromatographic separation can be monitored in characteristic time intervals that are defined by the pulse repetition rate. For chromatographic peaks of 30 to 120 s peak width and pulse repetition times of 1.2 s, up to 36 individual interferograms can be accumulated. For analytical separations with acceptable signal-to-noise ratios the detection limit is 100–200 μg of each component; for capillary separations it is 1–2 μg. On-line HPLC-NMR separations are depicted in contour plots (pseudo 2D plots). The abscissa shows the ^{1}H chemical shift and the ordinate the chromatographic retention time. Characteristic structural elements can be easily recognized in the contour plot, as exemplified for the separation of a mixture of tocopherol homologues in Fig. 3. This experiment was performed in continuous-flow mode and the recorded plot contains 128 accumulated interferograms, each one recorded with a pulse repetition time of 39.6 s. Recording and processing procedures were in accordance with those of 2D NMR measurements. The contour plot shows information about the stereochemistry of the analytes. The isomers β- and γ-tocopherol can be unambiguously distinguished by the chemical shifts of their aromatic protons H-5 and H-7 at $\delta = 6.2$ and 6.4 ppm. This information can only be gained from NMR measurements, not by mass spectrometry. The contour plot allows the detection of specific substances in complex mixtures through their characteristic chemical shifts. The intensities of the proton signals in a time frame are plotted against the retention time and are summed to give the ^{1}H NMR chromatogram on the left ordinate with a resolution of 40 s per data point.

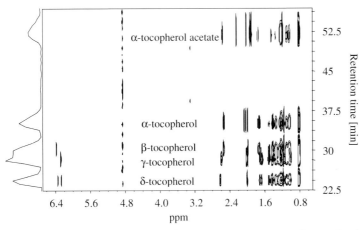

Fig. 3. Contour plot of the continuous-flow HPLC-NMR separation of a tocopherol mixture.

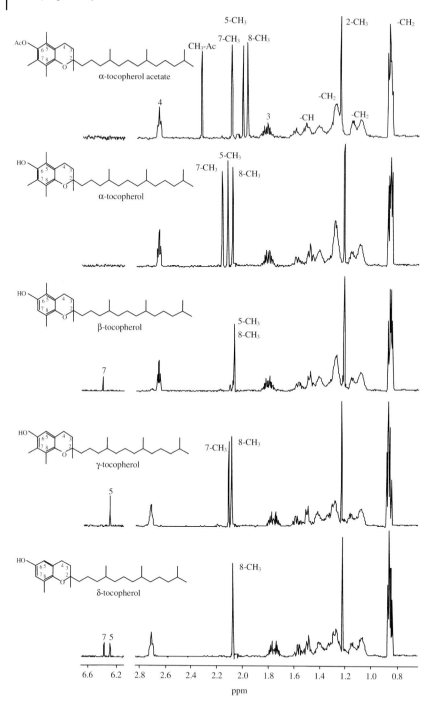

Fig. 4. ^{1}H NMR spectra of tocopherols, extracted from the contour plot of the continuous-flow measurement (Fig. 4) at each peak maximum.

Figure 4 shows the extracted ^{1}H NMR spectra, taken from the contour plot at the eluting peak maxima. These NMR spectra are of the same quality as conventionally recorded ones; chemical shifts, coupling constants, and integration values can be unambiguously determined.

There are special prerequisites for the solvents (mobile phase) used in continuous-flow NMR measurements. HPLC-grade solvents contain traces of impurities and often stabilizers. For continuous-flow NMR, solvents of higher purity are needed as these contaminants would give rise to additional signals in the ^{1}H NMR spectrum that may be coincident with the resonances of the analyte. Another important point is the number and position of the solvent signals in the ^{1}H NMR spectrum. Many solvent signals lead to a risk of overlap with those of the analytes, which would complicate the interpretation (integration, signal pattern). Solvents that generate only one NMR signal, such as water, acetonitrile, acetone, and methanol, are commercially available in high purity. THF, in contrast, cannot/should not be used as it has signals throughout the whole spectrum. If acids, bases or organic buffers are added to the mobile phase in order to improve the separation, their deuterated or halogenated analogues should be utilized. The addition of inorganic buffers or salts causes no problems for continuous-flow NMR, as long as the salt concentration is reasonably low (see Table 1). To record spectra in the continuous-flow mode, the intensity of solvent signals has to be reduced, because otherwise these signals can lead to dynamic problems in the receiver (A/D converter) of the spectrometer, in addition to the challenges mentioned above. Selective solvent signal suppression (elimination) is achieved by so-called pre-saturation techniques with shaped-pulses that selectively override the Boltzmann distribution for the mobile phase signals so that no NMR signals appear. This kind of solvent suppression technique is easy to handle, implementable in different pulse programs, and extremely efficient even for the suppression of several NMR signals. A disadvantage of this method is that analyte signals in the region of the solvent signals cannot be used for interpretation of the ^{1}H NMR spectra as they will also be suppressed.

3.4.7
Stopped-Flow Measurements

Besides the continuous-flow mode, there also exists the possibility of stopped-flow NMR experiments. For this, the chromatographic run is stopped as soon as the peak maximum reaches the active detection volume of the NMR probe. By increasing the accumulation time for a spectrum, the sensitivity can be increased ($S/N = \sqrt{n}$), which is a major advantage of this method. It allows the investigation of secondary components in mixtures with concentrations down to 0.1% of that of the actual analyte. Another benefit is that 2D experiments can be performed. Often, only 2D spectra enable an unambiguous and full structure elucidation. Figure 5 depicts such a 2D NMR spectrum of estradiol. There are several practical ways of performing stopped-flow measurements. The simplest is to stop the solvent flow through the system by means of a peak parking valve. In order to prevent

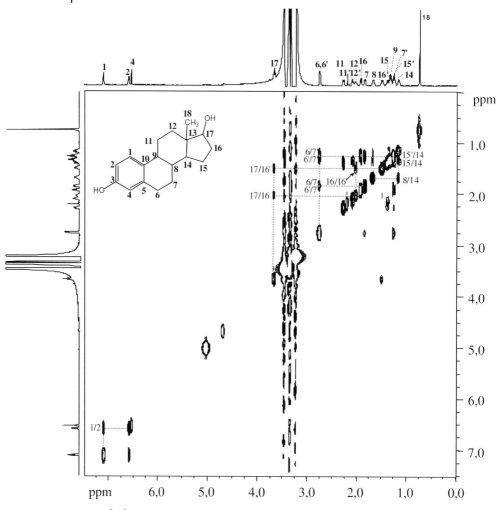

Fig. 5. ^{1}H,^{1}H-COSY spectrum of estradiol.

peak diffusion during the acquisition, the outlet should also be closed by a valve. As soon as the measurements have been performed, the separation can be continued and the next peak analyzed. However, after having stopped the separation for several hours, the analyte mixture should be newly injected, because diffusion processes will inevitably take place. To circumvent the problem of diffusion processes, the peaks can be stored in loops. Loops are capillaries having the same volume as the NMR probe. The analytes are separated before the actual NMR experiments. After storing all of the peaks in loops, they are transferred to the probe and investigated. An alternative is the use of a solid-phase-extraction NMR unit (SPE-NMR). The analytes are pre-concentrated by multiple trapping on an SPE column and, after drying, are eluted from the SPE cartridge with deuterated solvents. The drawback of this system is the complex and expensive equipment.

3.4.8
Capillary Separations

A further development in the field of HPLC-NMR is miniaturization to capillary-HPLC-NMR. The resolution of NMR spectra obtained in continuous- and stopped-flow mode using capillary-HPLC-NMR is comparable to that of conventionally recorded spectra. The great advantage of capillary-HPLC-NMR is its high sensitivity in the nanogram range and the minimal consumption of deuterated solvents, which typically amounts to no more than about 50 mL per week. In comparison, classical HPLC-NMR consumes more than 1 L of solvent per day and even for a single NMR measurement in a conventional 5 mm NMR tube about 750 μL of deuterated solvent is necessary. These figures highlight the fact that from the economic viewpoint only the hyphenation of a miniaturized capillary-HPLC system to microprobe NMR and conventional tube measurements allows the use of deuterated solvents. The major advantage of the capillary separation technique is that the interpretation of NMR spectra is tremendously simplified due to the use of deuterated solvents. Solvent suppression may thus no longer be necessary. As the solvent signals are much smaller (resulting from partially non-deuterated solvent), the overlay with the analyte signals is often diminished.

Figure 6 depicts the contour plot of a capillary-HPLC-NMR separation of unsaturated fatty acid methyl esters. The injected sample amount was about 1–2 μg. As a representative example, Fig. 7 shows the capillary ^{1}H NMR spectrum of eicosenic acid methyl ester.

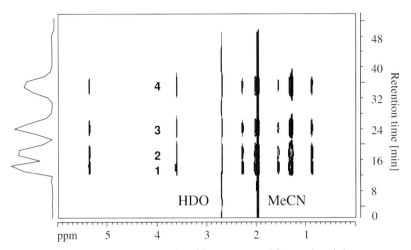

Fig. 6. Capillary-HPLC-NMR contour plot of four unsaturated fatty acid methyl esters: (1) palmitoleic acid-, (2) oleic acid-, (3) eicosenoic acid-, and (4) erucic acid methyl ester.

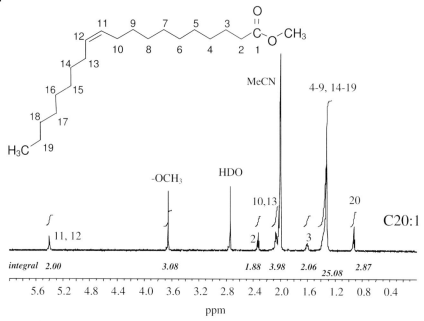

Fig. 7. ^{1}H NMR spectrum of eicosenoic acid methyl ester with integration of the proton signals.

The advantage of the extremely high sensitivity in the nanogram range only applies for compounds that are well-soluble in the solvent mixture used. For less soluble compounds, the use of conventional HPLC-NMR with detection volumes in the range 40 to 120 μL can be recommended.

In the structure elucidation of unknown compounds, ^{13}C NMR is routinely used in conjunction with ^{1}H NMR spectroscopy. An important two-dimensional technique is inverse ^{1}H,^{13}C-correlation to assign direct correlations between adjacent protons and carbon atoms. Due to the low natural abundance of the ^{13}C isotope of 1.1%, several hours of accumulation and injected sample amounts in the low μg range in the 1.5 μL detection volume are required for successful 2D NMR experiments.

3.4.9
Outlook

On-line HPLC-NMR using analytical columns has become established as a routine method. Several problems in pharmaceutical, biomedical, and environmental analysis have been solved by simply hyphenating LC with NMR. For sparingly soluble compounds, the existing probes with detection volumes in the range 40 to 120 μL are the best solution.

Hyphenation of capillary-HPLC to microcoil-NMR with detection volumes of 1.5 μL is the method of choice for samples of well-soluble compounds of limited availability. Meanwhile, through advances in chromatographic separation and NMR detection, sample amounts in the low nanogram range can be measured.

Microcoils, as used for capillary-HPLC-NMR, are 10 mm long and the diameter of the NMR probe is 50 mm. With these, a capillary cross-coil arrangement with four microcoils in the *xy*-plane of the probe is geometrically possible.

The problem of cross-talk between several microcoils has been solved by Professor Andrew Webb and his coworkers, but application in parallel HPLC-NMR detection has still yet to be achieved. For this, adequate NMR instrumentation will be necessary. Parallel HPLC-NMR detection with four microcoils in a hyphenated system will surely be realized within the next few years, and in the longer term a configuration of 16 microcoils in four parallel horizontal detection planes, as introduced by Professor Raftery, will be realized. Besides parallel detection, the capillary technique in principle opens the door for the hyphenation of NMR spectroscopy with other capillary separation techniques such as gas chromatography and supercritical fluid chromatography (SFC). The long spin-lattice relaxation times in the supercritical and gaseous states can be reduced by relaxation agents present in the separation capillary prior to the actual NMR detection.

This technique has been successfully applied to record continuous-flow ^{13}C NMR spectra (Fig. 8) and has drastically increased the sensitivity per time unit. The potential of this method still needs to be further explored.

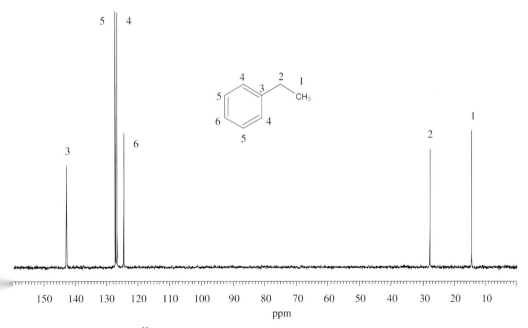

Fig. 8. Continuous-flow ^{13}C NMR spectrum of ethylbenzene using paramagnetic relaxation agents.

To point out future developments in LC-NMR: ^{13}C NMR spectroscopy will be increasingly applied in on-line coupling. Besides HPLC-MS and the ion sources ESI and APCI, ^{1}H and ^{13}C NMR detection will be more commonly used in the future for structure elucidation of unknown compounds. Sensitivity in the picogram range will never be reached for HPLC-MS, but NMR is indispensable as a complementary technique to mass spectrometry. Highly selective enrichment and separation methods in hyphenation with mass spectrometry and NMR spectroscopy will be efficient tools for the analysis of complex samples and the identification of their components (Table 1).

Table 1. Important features of LC-NMR-coupling.

Parameter	*What is important*	*Possible solution*
Sample preparation	• High enrichment of the analytes during sample preparation	• SPE (liquid samples) • MSPD (viscous and solid samples)
Solvents and additives	• Purity of solvents • Number, position, and width of solvent signals • Additives such as acids, bases or buffers should not give rise to additional signals in the NMR spectrum	• Use of NMR-pure solvents • Use of solvents that show only one signal, such as acetonitrile, acetone, and water • Use deuterated solvents (capillary techniques) or at least replace water with D_2O for classical HPLC-NMR • Use deuterated or perhalogenated analogues of additives
Continuous-flow measurements	• Sufficient residence time of the analytes in the NMR probe • Sufficient selectivity (α value, resolution) for two closely eluting analytes	• Reduced flow-rates in the HPLC system • Adjust the solvent composition
Stopped-flow measurements	• Minimization of diffusion-processes • The analyte peak must be perfectly trapped in the active detection volume of the NMR probe	• Closed system (valve) • Peak storage of the separated analytes in loops • Accurate determination of the transfer time from UV detector to NMR probe
Solvent suppression	• Reduction of the intensity of dominating solvent signals • Analyte signals with similar excitation frequencies must not be suppressed with the solvent signals • The solvent suppression method should be easily implementable in the pulse program	• Pre-saturation methods applying shaped pulses are easy to handle and implementable • The energy for suppression of solvent signals should be chosen as low as possible

References

1   K. Albert (Ed.), *On-line LC-NMR and Related Techniques*, John Wiley and Sons Ltd., Chichester, England, 2002.
2   D. A. Jayawickrama, J. V. Sweedler, *J. Chromatogr. A 1000* (**2003**), 819.
3   B. Behnke, G. Schlotterbeck, U. Tallarek, S. Strohschein, L.-H. Tseng, T. Keller, K. Albert, E. Bayer, *Anal. Chem. 68* (**1996**), 1110.
4   D. L. Olson, T. L. Peck, A. G. Webb, R. L. Magin, J. V. Sweedler, *Science 270* (**1995**), 1967.
5   G. Fisher, C. Petucci, E. MacNamara, D. Raftery, *J. Magn. Reson. 138* (**1999**), 160.
6   H. Fischer, M. Seiler, T. Ertl, U. Eberhardinger, H. Bertagnolli, H. Schmitt-Willich, K. Albert, *J. Phys. Chem. B 107* (**2003**), 4879.
7   M. Krucker, A. Lienau, K. Putzbach, M. D. Grynbaum, P. Schuler, K. Albert, *Anal. Chem. 76* (**2004**), 2623.
8   S. A. Barker, *J. Chromatogr. A 885* (**2000**), 115–127.

4
Computer-Aided Optimization

4.1
Computer-Facilitated HPLC Method Development Using DryLab® Software

Lloyd R. Snyder and Loren Wrisley

"Computer simulation" refers to the use of a computer for predictions of separation as a function of experimental conditions. In most cases, computer simulation requires two or more experimental runs with a given sample, in order to "calibrate" the computer prior to making predictions. Computer simulation can be used in HPLC method development to select optimum final conditions for the separation of a given sample, so that fewer actual experiments are required and better HPLC methods result. Ideally, computer simulation should (a) allow for the simultaneous variation of as many as two experimental conditions that affect separation selectivity, (b) be applicable for both isocratic and gradient elution, and (c) be able to predict the further effect on separation of changes in flow rate, column dimensions and particle size.

Software for computer-facilitated HPLC method development has been commercially available for the past two decades. The present chapter describes one such computer program: DryLab® (Rheodyne LLC, Rohnert Park, CA, USA). The detailed operation of this software is presented here, with a half dozen examples of its real-world application. The present version of DryLab is the result of two decades of continuing development and use by practicing chromatographers.

4.1.1
Introduction

HPLC method development can be carried out in a variety of ways [1]. Systematic, trial-and-error changes in conditions are generally used to achieve adequate resolution R_s of the sample, with a common goal of baseline resolution ($R_s \geq 1.5$) for all peaks. Changes in conditions for improved resolution may involve (a) adjusting the retention range of the sample for $0.5 < k < 20$, (b) varying column efficiency (values of the plate number N), or (c) varying separation selectivity (values of the separation factor α); i.e.:

$$R_s = \frac{1}{4}(\alpha - 1) N^{1/2} \frac{k}{k+1} \tag{1}$$

Table 1 summarizes various conditions that are used to control k, N, and α. Because so many factors can influence separation, and because challenging samples often require the simultaneous adjustment of several variables, today many workers use computer-facilitated method development (CMD). Software for CMD uses a small number of experimental runs to simulate what will happen when any of several conditions are changed. That is, two or more runs are used to "calibrate" the software for a given sample, after which simulated runs can be

HPLC Made to Measure: A Practical Handbook for Optimization. Edited by Stavros Kromidas
Copyright © 2006 WILEY-VCH Verlag GmbH & Co. KGaA, Weinheim
ISBN: 3-527-31377-X

(a)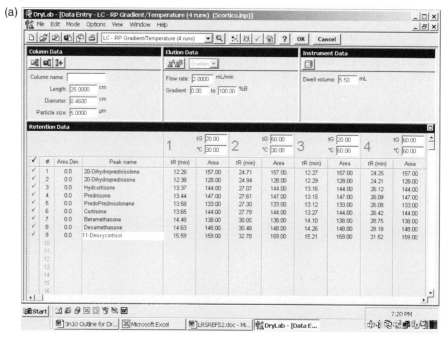

(b)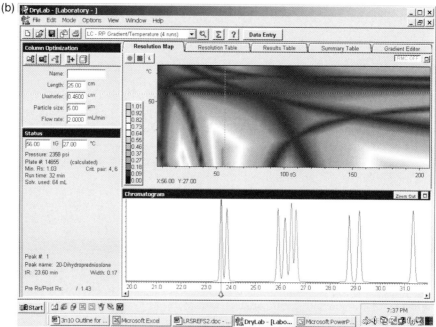

Fig. 1. DryLab data input (a) and calculation (Laboratory).
(b) Screen of the separation of 9 corticosteroids with optimization of
temperature T and gradient time t_g.

Table 1. Conditions for controlling resolution in reversed-phase HPLC. DryLab also allows the user to simulate separations by normal-phase or ion-exchange chromatography.

Factor to be controlled	Approach
Retention (k)	Usually %B
Column efficiency (N)	Column length, particle size, flow rate
Selectivity (α)	%B, temperature, pH, organic solvent (MeCN, MeOH, THF), buffer, ion-pair or other additive concentration, column

carried out by entering new conditions into the computer. Alternatively, the computer may summarize the results of a large number of such simulations in the form of convenient resolution maps (Fig. 1), without further effort on the part of the user.

4.1.1.1 History

CMD was first introduced for gas chromatography (GC) by Laub and Purnell [2], and later extended by them to isocratic HPLC [3]. Experiments for two different separation temperatures were used to predict values of α for any temperature, based on the well-known relationship:

$$\log k = \frac{a + b}{T_K} \tag{2}$$

where T_K is the absolute temperature, and a and b are constants for a particular compound and experimental conditions other than temperature; values of a and b are measurable from experimental values of k at two different temperatures. A temperature could then be predicted for a maximum value of α (and approximately maximum R_s) for the critical band pair. A few years later, similar CMD approaches were presented for optimizing reversed-phase HPLC separations by varying the mobile phase: (a) simultaneously changing the concentrations of methanol, acetonitrile, and tetrahydrofuran [4] or (b) varying pH and ion-pair reagent concentration [5]. Predictions of retention were based either on empirical or fundamental equations for retention as a function of a given variable. In the early 1980s, several instrument companies introduced CMD software that allowed the user to make use of the Glajch–Kirkland approach [4], and by the late 1980s there was sufficient literature to justify several books on CMD [6–9]. DryLab® software for CMD ("computer simulation"), which became available in late 1985 [10], is discussed in the present chapter. During the past 20 years, DryLab® (Rheodyne LLC, Rohnert Park, CA) has evolved into a multifunctional product that provides a wide range of features and capabilities.

4.1.1.2 **Theory**

DryLab® initially made use of the following semi-empirical concepts:

(a) In reversed-phase HPLC (RP LC), retention can be approximated by Eq. (3):

$$\log k = \log k_{\mathrm{w}} - S\Phi \tag{3}$$

where k_{w} and S are empirical constants for a given solute and experimental conditions, and Φ is the volume fraction of organic solvent B in the mobile phase [11].

(b) Based on Eq. (2), with only two experimental runs it is possible to predict isocratic retention as a function of Φ and gradient retention as a function of gradient conditions, column size, and flow rate [12]; together, (a) and (b) comprise what has come to be called the linear-solvent-strength model of HPLC [13].

(c) For either isocratic or gradient elution it is possible to approximate bandwidth W by means of the Knox equation, when the sample molecular weight, temperature, mobile phase composition, and "column conditions" (column size, particle size, flow rate) are known [12–14].

The accuracy of DryLab, based on the above relationships, has been analyzed in several publications [15–21], and comparisons of separation *vs.* DryLab predictions have been reported in dozens of other papers, many of which are summarized in Ref. [13]. In most cases, the accuracy of resolution predictions is of the order of $\pm 10\%$, which is usually more than adequate. Expressions such as Eqs. (2) and (3), which are suitable for predicting retention as a function of certain conditions, require only two experimental runs for DryLab implementation. Other variables, such as pH or buffer concentration, require three or more runs, in which cases a cubic spline fit to the data can be used for predictive purposes.

4.1.2
DryLab Capabilities

4.1.2.1 **DryLab Operation**

A separation mode is first selected (Section 4.1.2.2), which also defines the number of experimental calibration runs that are required; e.g., four runs at two different gradient times t_G and temperatures T (°C) for the simultaneous optimization of t_G and T. The following experimental data are then entered into the computer: column conditions, gradient range, the equipment delay ("dwell") volume, the mobile phase composition, and retention times and peak areas for each compound in each of the four chromatograms (see Fig. 1a, "Data Entry" screen). Data can be entered either manually or by electronic transfer from a data system. Finally, it is necessary to carry out peak matching, where data for each peak are matched with a given compound in the sample. Peak matching can be carried out either manually or with the help of the DryLab automatic peak-tracking module.

With the completion of data entry, a *Laboratory* screen can be displayed (Fig. 1b). Several display options are shown at the top of the screen: R_s map, R_s table, results table, etc. Fig. 1b displays a resolution map, where values of R_s are shown in color for various combinations of T (*y*-axis) and t_G (*x*-axis). In this case, maximum R_s occurs for $T = 27\ °C$ and $t_G = 56$ min (note the cross-hairs in Fig. 1b). A chromatogram for the selected conditions is displayed at the bottom of the screen. The user can carry out further simulations, which might involve changes in (a) the gradient program (including the use of segmented gradients), (b) column conditions, or (c) values of T or t_G.

4.1.2.2 Mode Choices

With the exception of the column, any one of the selectivity conditions in Table 1 can be modeled, or two of them can be modeled simultaneously, by DryLab (if the column is changed, new calibration runs must be carried out). The number of experimental runs required for DryLab simulation varies from two to nine [21], depending on which selectivity variables are chosen. When a wide range in some condition (e.g., pH) is to be covered, additional runs can be carried out for improved predictive accuracy [20]. Additional modes are available for normal-phase and ion-exchange HPLC, as well as GC. The user can also create virtually any mode desired.

Resolution Map Options

In some cases, not all peaks in the chromatogram will be of interest. DryLab allows the user to flag compounds of interest, in which case only these peaks are considered in constructing the R_s map. Another R_s map option is the *robust resolution map* (illustrated later), which allows the user to select a minimum acceptable value of R_s and a maximum allowable run time. An R_s map is then displayed that highlights conditions that meet the latter requirements.

Gradient Elution Options

When the *gradient editor* of the *Laboratory* screen (Fig. 1b) is selected, the user can enter any desired gradient program, including segmented gradients. This module also features the *optimum linear gradient* (OLG) option, in which the computer automatically determines gradient conditions for maximum R_s in minimum run time. The OLG option can also be extended to the selection of optimized multi-segment gradients [22]. Finally, given gradient input data, it is also possible to simulate isocratic separation. Thus, in the example of Fig. 1, based on experimental gradient runs, it is possible to predict isocratic separation as a function of %B and T, as well as to display a resolution map for the latter variables.

Plate Number Options

Predictions of bandwidth by DryLab require an estimate of the plate number N for each band, which can be provided by any of three options. First, DryLab can calculate a value of N based on mobile phase composition, temperature, column conditions, and an estimate of sample molecular weight that is based on sample retention. Resulting calculations of bandwidth based on this approach are usually reliable to about ±25%, which can be adequate for many separations. Second, the user can enter an average value of N into the computer, based on varying N while observing changes in one of the simulated input runs compared to the corresponding experimental run. The resulting accuracy of simulated bandwidths is usually of the order of ±10% or better. The third option is entry of bandwidths for each of the experimental runs, which is the most accurate procedure. However, this approach may require editing bandwidth values in the event of overlapped bands. It is also possible to enter information on band asymmetry (as either asymmetry factor or tailing factor values for each band).

4.1.3
Practical Applications of DryLab® in the Laboratory

The effective use of DryLab requires some reasonable first choices of experimental conditions (column packing, mobile phase solvent, buffer, pH, organic modifier, etc.) [1]. Also, the goals of the separation should be considered in advance. Is the separation for a "strength" (potency) assay, an impurity assay, or a preparative separation? In the first case, it is necessary to separate the component of interest from related impurities, degradation products, excipients, etc., but the separation of peaks other than that of the compound of interest is unnecessary. In the second case, it may be necessary to separate several peaks with adequate resolution for quantification ($R_s > 1.5$). The third case would suggest maximum resolution ($R_s \gg 1.5$, usually of a single peak) in order to achieve the highest column loading and sample throughput. Some of these considerations will be guided by the final use of the separation; e.g., an LC-MS method will limit the mobile phase to volatile components and may require a mobile phase pH that promotes sample ionization.

Laboratory Example 1

In this example, the separation goal was the resolution of a single sulfated impurity from a complex mixture for quantification by UV detection. Ideally, an isocratic separation would be desired for simplicity. Previous trial-and-error method development had failed to yield a suitable separation, but did suggest the initial use of a C_{18} column with a mobile phase based on pH 3, 20 mM phosphate buffer plus acetonitrile (MeCN). The results of prior experiments also pointed to the use of a narrower gradient range (25–60% MeCN) for the initial software calibration experiments.

Next, two DryLab calibration runs were carried out, as shown in Fig. 2; a spiked sample was used with gradient times of 15 and 90 min. Peak tracking by area matching and photodiode array showed that the separation of the peak of interest was not achieved in either run (Fig. 3). Also, significant changes in peak elution order were observed. Retention data from the two runs were modeled by DryLab to yield the resolution map of Fig. 4, for *isocratic* separation of the sample as a function of %B (i.e., % MeCN). In Fig. 4, only the resolution of peak #3 is

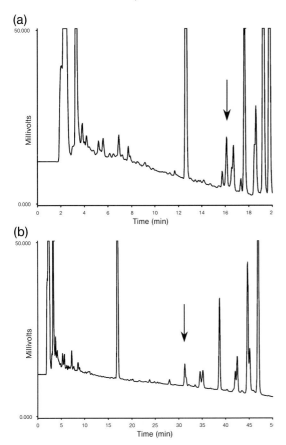

Fig. 2. Laboratory Example 1:
Initial gradient chromatograms for the development of a quantitative assay
for impurities. Sample spiked with compound of interest (see arrows).
Conditions: 25 × 0.46 cm C_{18} column, 1.0 mL min^{-1}, 35 °C.
(a) 25–60% acetonitrile in 15 min; (b) 25–60% acetonitrile in 90 min.

considered. Further modeling with the software showed that the best separation
would be achieved with a mobile phase of (about) 36% MeCN. A mobile phase
with < 28% B would also provide good separation, but retention times would
then be unreasonably long (> 30 min). The predicted separation of Fig. 5 shows
that peak 3 is well separated from the other peaks, with a resolution $R_s \approx 2$. Peaks
1 and 2 co-elute, but this is of no consequence. The experimental (verification)
chromatogram shown in Fig. 6 agrees well with that predicted by DryLab and
resulted in an acceptable routine method. The experimental resolution seen in
Fig. 6 is somewhat lower than that predicted by Fig. 5, due to minor peak tailing
(a tailing factor T of about 1.3) and a lower plate number N than predicted by the
software. Peak tailing and values of N (see Laboratory Example 5 below) can be
adjusted so as to refine the model during DryLab simulations, but in practice this
is seldom necessary.

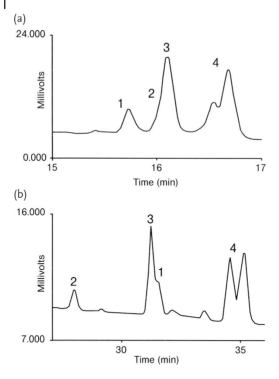

Fig. 3. Laboratory Example 1:
Detailed view of separations (A) and (B) shown in Fig. 2.
Peak 3 is the peak of interest.

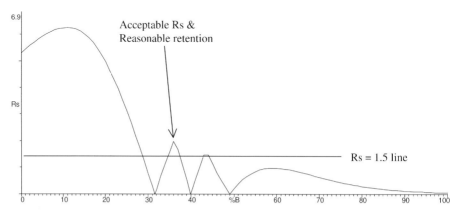

Fig. 4. Laboratory Example 1 (see Fig. 2):
Resolution map for the separation of peak 3 from other peaks as a function
of % acetonitrile (B) in the isocratic mobile phase.

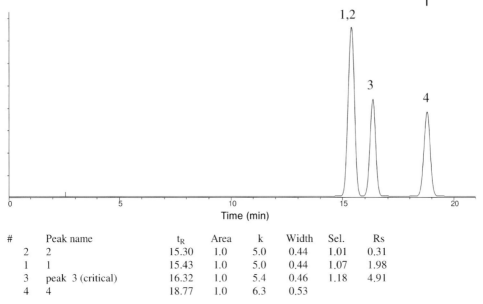

#	Peak name	t_R	Area	k	Width	Sel.	Rs
2	2	15.30	1.0	5.0	0.44	1.01	0.31
1	1	15.43	1.0	5.0	0.44	1.07	1.98
3	peak 3 (critical)	16.32	1.0	5.4	0.46	1.18	4.91
4	4	18.77	1.0	6.3	0.53		

Fig. 5. Laboratory Example 1 (see Fig. 2):
DryLab-predicted separation at 36% acetonitrile.

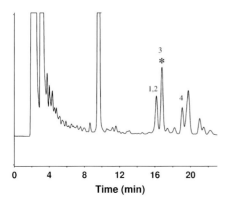

Fig. 6. Laboratory Example 1:
Experimentally verified optimized isocratic separation.

Laboratory Example 2

In a second example, the separation of several impurity peaks and a parent drug substance was required. Because the method was likely to be used for several years, robustness and a reasonable run time were important considerations. Nothing was known about the potential separation, other than the structure of the parent compound. Therefore, more or less "standard" conditions were used for the initial DryLab runs: 20 and 80 min gradients from 10–90% MeCN, with a buffer of pH 2.5, 20 mM phosphate buffer (Fig. 7).

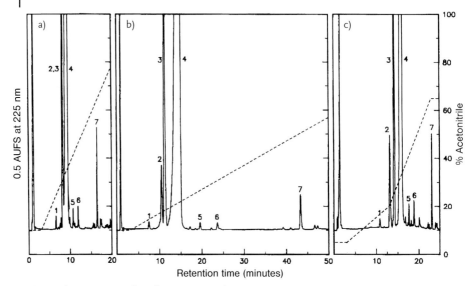

Fig. 7. Chromatograms for Laboratory Example 2:
Conditions: 15 × 0.46 cm C₈ column; 1.5 mL min⁻¹.
(a) 10–90% acetonitrile in 20 min; (b) 10–90% acetonitrile in 80 min;
(c) 5/20/65/65%-acetonitrile in 0/10/20/25 min.

In the 20 min gradient (Fig. 7a) peaks 2 and 3 co-elute. In the 80 min gradient (Fig. 7b), peaks 2 and 3 are marginally resolved, but the run time is excessive. Using the automatic linear optimization feature of DryLab, a faster separation can be achieved, albeit still with only marginal resolution of peaks 2 and 3. At this point, interactive gradient modeling using segmented gradients was begun. Fourteen segmented gradients were rapidly modeled on the computer to arrive at the optimized separation shown in Fig. 7c. The software allows promising simulations to be saved for reference, which enables the user to keep track of progress toward a final separation. In Fig. 7c, peaks 2 and 3 are better resolved, overall band spacing is optimized, and the run time is acceptable for routine use. These computer simulations also correctly predicted that an improved resolution of peaks 2 and 3 would require a reduction in %B at the start of the gradient.

Laboratory Example 3

DryLab software can be effectively used for preparative or semi-preparative HPLC method development and optimization. In this example, the laboratory was asked to provide at least 1 g of a chromatographically purified steroid, with a three day turnaround time. The desired compound only differs from the major impurity in the spatial orientation of a specific hydroxyl group, and synthetic chemists had been unable to synthesize the pure steroid in a purity of greater than 88%.

Six different separation requirements were defined. First, the method should be developed for an analytical scale column, then scaled-up for a larger, semi-preparative column that contained the same stationary phase. Second, the separation would be optimized for maximum resolution, so as to permit maximum column overload. Experience with similar preparative separations suggested that a resolution of $R_s > 4$ would be necessary. Third, the separation should not exceed 20–30 min in cycle time, because a number of injections would be required in order to isolate sufficient material. Fourth, an isocratic separation was required, if possible. Fifth, the separation should be performed at room temperature, since temperature control of the large semi-preparative column at high mobile phase flow rates was not feasible. Finally, the mobile phase should contain no involatile buffers or modifiers, so as to facilitate the removal of solvent from resulting HPLC fractions.

Fortunately, the laboratory separation chemist had some experience with the analytical separation of these isomers. It was known that good separation could be achieved on a C_{18} stationary phase with methanol as the organic modifier, whereas separation was inadequate using acetonitrile. Optimization of the separation using methanol/water was pursued on an analytical scale column (C_{18}) containing an 8 μm particle packing material for preparative separation.

Gradient retention data for the steroid and the isomers were used to calibrate the software for isocratic method optimization (Fig. 8). Prior experience guided the choice of the initial gradient range (50–80% B). Fig. 9 shows the resulting isocratic *robust resolution map* (RRM) for this separation. An RRM is the usual resolution map (as in Fig. 4), but with the option to enter the maximum desired run time (20 min) and a minimum acceptable resolution ($R_s = 4$). Conditions which are unacceptable then appear as the cross-hatched region of Fig. 9.

From the map, it appears that a mobile phase of 64–65% methanol will yield the desired separation. A mobile phase of 64% methanol was predicted to give the separation shown in Fig. 10, which can be compared with the experimental run of Fig. 11. The resolution of $R_s = 4.4$ in Fig. 11 agrees well with that predicted in Fig. 10 ($R_s = 4.3$).

Following typical loading experiments, a maximum sample size for the preparative column was calculated: 400 mg of crude product. Taking into account that (a) the crude material contained approximately 72% of the desired product

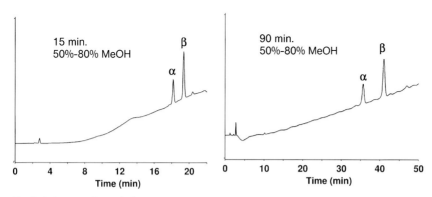

Fig. 8. Laboratory Example 3: Initial separations.
Conditions: column: 25 × 0.46 cm IB-Sil C18-BD, 8 μm particles, flow = 1.0 mL min⁻¹.

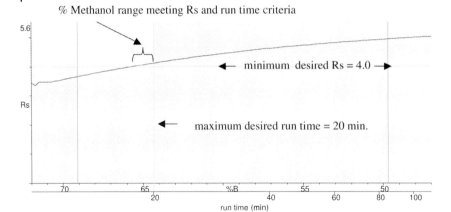

% Methanol range meeting Rs and run time criteria

← minimum desired Rs = 4.0 →

← maximum desired run time = 20 min.

Fig. 9. Laboratory Example 3:
Robust resolution map.

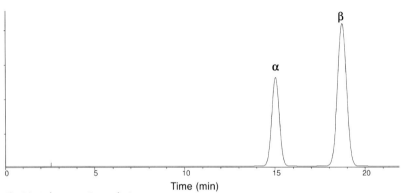

Time (min)

Fig. 10. Laboratory Example 3:
DryLab-predicted isocratic separation for 64% methanol ($R_s = 4.3$).

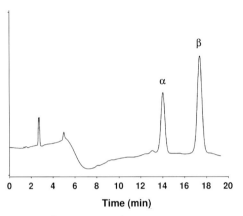

Time (min)

Fig. 11. Laboratory Example 3:
Experimental confirmation of the DryLab prediction from Fig. 10 ($R_s = 4.4$).

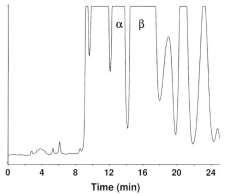

Fig. 12. Laboratory Example 3:
Semipreparative separation under optimized conditions.
Condition: column: 25 × 5 cm C_{18}, 8 μm particles; 118 mL min^{-1}.; 64% methanol
(isocratic); 400 mg crude product injected (72% β in mixture, β : α ratio = 12 : 1).

(β-isomer), (b) the cycle time was about 30 min, and (c) isolation losses were
expected, the predicted throughput was a (satisfactory) 500 mg h^{-1}. The chromatogram shown in Fig. 12 illustrates the optimized full-scale semi-preparative
separation of the crude mixture. This example shows the value of systematic
method development based on computer-assisted modeling. From the first trial
experiment to the recovery of more than 2 g of pure product (ready for roto-vap
isolation), less than a single 8 h shift had elapsed. The use of computer simulation
with no guesswork or wasted laboratory experiments led to the fastest possible
result.

Laboratory Example 4

The following example illustrates the value of an initial negative result. The sample
was a pure drug substance spiked with five possible impurities; the goal was
separation of all peaks. The effects of both gradient time and temperature on
separation were first determined by means of just four experimental runs: 10–
90% MeCN/water in 15 min and 90 min at 25 °C and 45 °C. The DryLab resolution
map and best resulting separation are shown in Fig. 13, indicating no possibility
of a satisfactory separation in this way.

A need for major changes in the starting conditions is suggested by Fig. 13,
since minor changes are not likely to provide a much-improved separation (i.e., a
significant change in selectivity is required). Very often, a simple change of the
organic modifier (to methanol or tetrahydrofuran, for example) and repeat of the
initial experimental runs will prove successful. In this case, the organic modifier
was switched to methanol for the gradient calibration runs. This led directly to
the successfully optimized isocratic separation shown in Fig. 14. In the end, the
final separation required only eight experimental runs (in two automated screening
sets) over the course of two days.

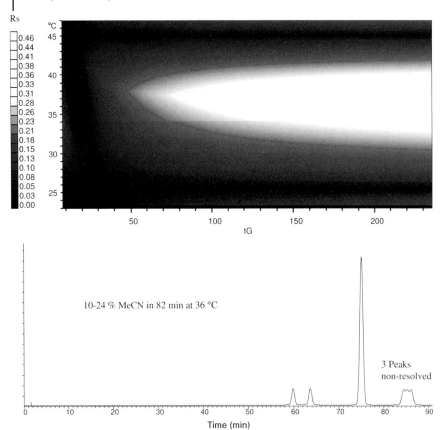

10-24 % MeCN in 82 min at 36 °C

3 Peaks
non-resolved

Fig. 13. Laboratory Example 4:
Resolution map and optimal separation.

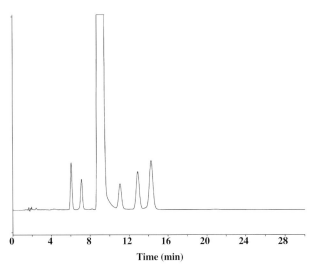

Fig. 14. Laboratory Example 4:
Optimized separation.
Conditions: 47% methanol
(isocratic); 75 × 4.6 mm C_{18}
column; 3 μm-particles, 30 °C;
1.0 mL min^{-1}.

Laboratory Example 5

The optimization of a separation using DryLab can extend beyond simple changes in mobile phase composition or gradient profile. In the present example, the separation of two active components, two potential impurities, and a formulation stabilizer (BHA) was required, using isocratic conditions if possible. Four trial runs were carried out, varying gradient time and temperature (150×3 mm C_{18} column, 15 and 40 min gradients, 30–90% MeCN/water, 20 °C and 40 °C).

The DryLab resolution map (Fig. 15) shows that a promising result can be achieved within these experimental boundaries, with maximum resolution of the two impurity peaks (the first peaks in the chromatogram) at a temperature of 35 °C. A best (interim) DryLab-predicted isocratic separation is shown in Fig. 16. Because of excess resolution in this separation, it is clear that a reduction in run time is possible with further changes in the conditions. The effects of shorter columns (10 cm instead of the original 15 cm) and higher flow rates (0.75 mL min^{-1} as opposed to the original 0.45 mL min^{-1}) were modeled using DryLab, leading to the result shown in Fig. 17. The total run time was reduced from about 9 min to less than 4 min, while maintaining adequate resolution between the peaks of interest.

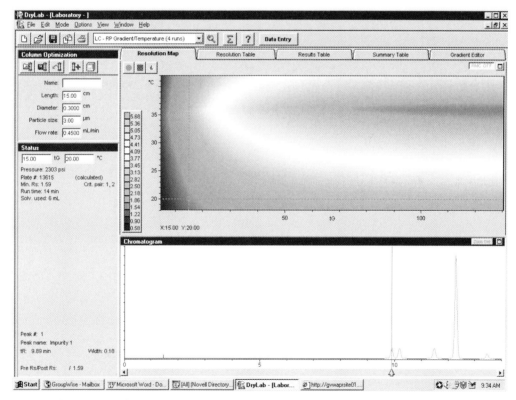

Fig. 15. Laboratory Example 5:
Initial DryLab resolution map and predicted separation.

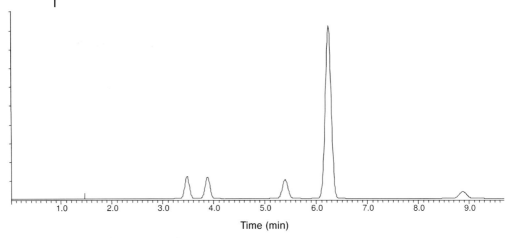

Fig. 16. Laboratory Example 5:
Interim optimized separation from DryLab.
Conditions: 0.45 mL min^{-1}, 35 °C, isocratic elution at 50% acetonitrile.

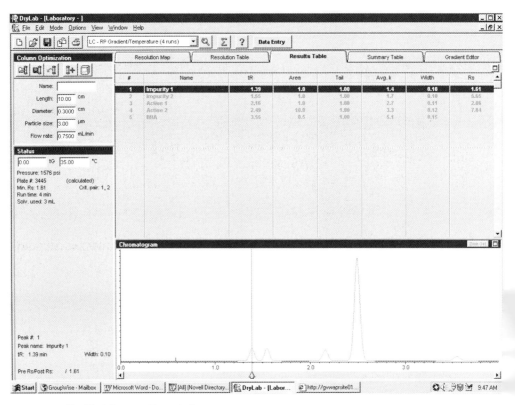

Fig. 17. Laboratory Example 5:
Final predicted optimal isocratic separation after column and flow optimization.

Laboratory Example 6

In this example, the goal was to develop a method for the quantification of three components in a mixture of as many as ten closely related compounds. Thus, the method had to provide resolution of the three components from all the other compounds, but it was not necessary to resolve the other compounds from each other. An isocratic method with a run time of no more than 30 min was desired.

The starting conditions included use of an aqueous buffer and acetonitrile, a 250×3 mm, 3 µm particle C_{18} column, a flow rate of 0.6 mL min^{-1}, and UV detection at 200 nm. Four DryLab calibration runs with varying gradient time and temperature were selected (20–40% MeCN in 15 min or 60 min, with a temperature of 25 °C or 45 °C). Previous experience with similar compounds helped to narrow the gradient range.

For this set of experimental runs, peak tracking was achieved by evaluating peak areas, UV spectra, and fluorescence detection. Fortunately, three of the components could be tracked by means of their fluorescence detection, while two other peaks had distinctive UV spectra. The resulting DryLab resolution map (Fig. 18) shows that all ten bands can be separated with a resolution of $R_s = 1.43$, but an isocratic separation is preferred. Again, the original goal is baseline separation of only three components (peaks 3, 7, and 9). At this point, manual interactive modeling was carried out in an isocratic mode. An acceptable separation resulted for 22% MeCN at 40 °C (Fig. 19). The run time slightly exceeds the 30 min goal, but a small increase in flow rate would remedy this while not compromising separation.

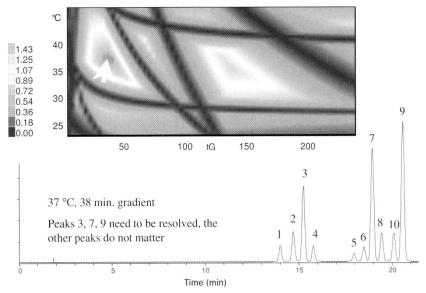

Fig. 18. Laboratory Example 6:
Initial DryLab resolution map and predicted gradient separation.

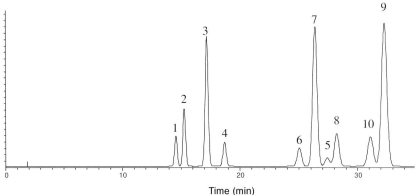

Fig. 19. Laboratory Example 6:
DryLab-optimized isocratic separation using 22% acetonitrile at 40 °C.

The DryLab modeling of this separation also indicated that it was especially sensitive to small changes in % MeCN and temperature. These results were incorporated into the method development procedure as recommendations for allowed variations in conditions.

4.1.4
Conclusions

Method development can be made easier, quicker, and more effective by the use of computer simulation. DryLab, which has been in use in hundreds of laboratories around the world during the past 20 years, has been developed to the point where the user can simulate any combination of any two experimental conditions that affect separation selectivity. At the same time, this software also allows the user to study the effects of changing column dimensions, particle size or flow rate, in either an isocratic or gradient mode. More recently, DryLab has been incorporated into HPLC systems which allow unattended method development (Waters Corp., Milford, MA). In all of these applications, predictions by DryLab have been shown to be essentially equivalent to actual experiments – while greatly reducing the cost and time required in method development.

The future development of computer simulation will be shaped in part by the limitations of presently available software. The major shortcoming of present computer programs is their limited ability to carry out peak tracking, which is a prerequisite for the use of computer simulation. Peak tracking becomes more difficult when (a) the number of compounds in the sample exceeds about ten and/or (b) trace components are present. This tends to complicate the use of computer simulation for many samples; e.g., impurity assays, peptide digests, etc. The increasing use of LC-MS should in time both minimize difficulties in peak tracking and further simplify method development.

A continuing trend is the use of computer simulation for "automated method development" (AMD). Initial method development experiments are carried out, the results are fed to a computer simulation module, conditions for optimized separation are predicted, and the optimized separation is then carried out by the HPLC system. The result is unattended method development. Recent versions of AMD allow for only a limited choice of separation conditions, but it can be anticipated that more complex and powerful method development strategies will be provided by future software.

The "Holy Grail" of computer simulation is the ability to predict separation from just a description of experimental conditions and the molecular structures of sample components. If successful, this approach could eliminate the need for actual experimental runs. Preliminary attempts in this direction have been reported, but so far these appear insufficiently reliable to be of much help in method development [23]. In the opinion of one of the present authors (LRS), accurate predictions of separation in this way are unlikely to be possible in the foreseeable future.

References

1   L. R. Snyder, J. J. Kirkland, J. L. Glajch, *Practical HPLC Method Development*, 2nd ed., Wiley-Interscience, New York, 1997.

2   R. J. Laub, J. H. Purnell, P. S. Williams, *J. Chromatogr.* 134 (1977), 249.

3   R. J. Laub, J. H. Purnell, *J. Chromatogr.* 161 (1978), 49.

4   J. L. Glajch, J. J. Kirkland, K. M. Squire, J. M. Minor, *J. Chromatogr.* 199 (1980), 57.

5   B. Sachok, R. C. Kong, S. N. Deming, *J. Chromatogr.* 199 (1980), 317.

6   J. C. Berridge, *Techniques for the Automated Optimization of HPLC Separation*, Wiley, New York, 1985.

7   P. J. Schoenmakers, *Optimization of Chromatographic Selectivity*, Elsevier, Amsterdam, 1986.

8   Sz. Nyiredy (Ed.), *J. Liq. Chromatogr.* 12, Nos. 1 and 2 (1989).

9   J. L. Glajch, L. R. Snyder (Eds.), *Computer-Assisted Chromatographic Method Development*, Elsevier, Amsterdam, 1990 (also appeared as volume 485 of the Journal of Chromatography).

10 L. R. Snyder, J. W. Dolan, *LC/GC Mag.* 4 (1986), 921.

11 K. Valko, L. R. Snyder, J. L. Glajch, *J. Chromatogr.* 656 (1993), 501.

12 L. R. Snyder, M. A. Stadalius, in *High-Performance Liquid Chromatography. Advances and Perspectives*, Vol. 4 (Ed.: Cs. Horvath), Academic Press, New York, 1986, p. 195.

13 L. R. Snyder, J. W. Dolan, *Adv. Chromatogr.* 38 (1998), 115.

14 R. W. Stout, J. J. DeStefano, L. R. Snyder, *J. Chromatogr.* 282 (1983), 263.

15 M. A. Quarry, R. L. Grob, L. R. Snyder, *Anal. Chem.* 58 (1986), 907.

16 L. R. Snyder, M. A. Quarry, *J. Liq. Chromatogr.* 10 (1987), 1789.

17 D. D. Lisi, J. D. Stuart, L. R. Snyder, *J. Chromatogr.* 555 (1991), 1.

18 J. A. Lewis, D. C. Lommen, W. D. Raddatz, J. W. Dolan, L. R. Snyder, I. Molnar, *J. Chromatogr.* 592 (1992), 183.

19 P. L. Zhu, L. R. Snyder, J. W. Dolan, N. M. Djordjevic, D. W. Hill, L. C. Sander, T. J. Waeghe, *J. Chromatogr. A* 756 (1996), 21.

20 J. W. Dolan, L. R. Snyder, L. C. Sander, P. Haber, T. Baczek, R. Kaliszan, *J. Chromatogr. A* 857 (1999), 41.

21 T. H. Jupille, J. W. Dolan, L. R. Snyder, W. D. Raddatz, I. Molnar, *J. Chromatogr. A* 948 (2002), 35.

22 T. Jupille, L. Snyder, I. Molnar, *LC.GC Europe* 15 (2002), 596.

23 T. Baczek, R. Kaliszan, H. A. Claessens, *LC.GC Europe* 14 (2001), 304.

4.2
ChromSword® Software for Automated and Computer-Assisted Development of HPLC Methods

Sergey Galushko, Vsevolod Tanchuk, Irina Shishkina, Oleg Pylypchenko, and Wolf-Dieter Beinert

4.2.1
Introduction

Method development processes in chromatography can be considered in terms of studying the empirical relationships between the quality of a chromatogram and the chromatographic conditions. A chromatographer changes conditions and evaluates the quality of chromatograms in order to try to find acceptable conditions to achieve a separation in a reasonable time. The time required to find acceptable conditions or to make any conclusion can be substantially reduced by using computer programs for method development. HPLC method development programs can be used interactively (off-line) and for automated and unattended optimization (on-line).

4.2.1.1 Off-Line Mode
Computer programs enable the user to use well-established relationships between separation, retention, and chromatographic conditions to allow the prediction of results of different experiments and optimization of separation conditions, working mainly with a PC rather than with an HPLC system. In this case, a user enters some information such as run data and conditions, and works with the software interactively.

4.2.1.2 On-Line Mode
The next step in the computer science is when an intelligent method development program is connected to an HPLC system. In this case, it plans and performs experiments autonomously.

ChromSword® supports both off-line interactive procedures for method optimization and on-line unattended method development.

4.2.2
ChromSword® Versions

ChromSword® for computer-assisted HPLC method development was developed between 1990 and 1995 as an extension of ChromDream® HPLC method development software [1]. In 1999, the first version for automatic HPLC optimization was developed and launched by S. Galushko in collaboration with Merck KGaA (Darmstadt, Germany). As a result of cooperation with VWR International Scientific Instruments, Darmstadt, Germany, Hitachi High Technologies

HPLC Made to Measure: A Practical Handbook for Optimization. Edited by Stavros Kromidas
Copyright © 2006 WILEY-VCH Verlag GmbH & Co. KGaA, Weinheim
ISBN: 3-527-31377-X

America, San Jose, CA, USA, and Scientific Software Inc., Pleasanton, CA, USA, versions for Hitachi LaChrom and LaChromElite hardware and EZChromElite software were launched. Partnership with Agilent Technologies resulted in the creation of a powerful version that supports Agilent 1100 LC and LC/MS systems with six columns and twelve solvent selectors. Further systems supported by ChromSword® Auto are Waters Alliance LC systems with Millenium and Empower softwares. ChromSword® Auto also supports 2–8 column and 2–16 solvent switching valves for all HPLC systems described.

The software versions are now available from Dr. S. Galushko Software Entwicklung, Muehltal, Germany, VWR Scientific Instruments GmbH, Darmstadt, Germany, Agilent Technologies GmbH, Waldbronn, Germany, Hitachi High Technology America, San Jose, CA, USA, and Iris Technologies, Lawrence, KA, USA.

4.2.3
Experimental Set-Up for On-Line Mode

ChromSword® 3.X contains special modules that are necessary for automatic HPLC method development with different hardware and HPLC software. In developing methods, the program does not control instrument modules directly, but interacts with the corresponding HPLC software, modifying methods and evaluating run data. The result is that all instrument methods and data are stored by the corresponding HPLC software and can be re-evaluated by a user. The automation modules of the software control:

- Agilent 1100 LC and LC/MS systems with Agilent ChemStation HPLC software (Agilent Technologies GmbH, Waldbronn, Germany).
- Agilent 1100 LC systems with EZChromElite HPLC software (Scientific Software Inc., USA).
- Agillent 1100 LC with Empower HPLC software (Waters Inc., Milford, USA).
- Hitachi LaChrom systems with HPLC System Manager software and Hitachi LaChromElite LC systems with EZChromElite HPLC software (VWR Scientific Instruments GmbH, Darmstadt, Germany).
- Waters Alliance LC system with Millennium and Empower software (Waters Inc., Milford, USA).

4.2.4
Method Development with ChromSword®

4.2.4.1 Off-Line Mode (Computer-Assisted Method Development)
ChromSword® works with different retention models. The retention model is a type of experiment-based mathematical expression that describes the relationship between the retention of a compound and its properties, as well as the conditions appertaining to the chromatographic experiments. The determination of retention models that adequately describe the effect of chromatographic conditions on the retention of compounds in a sample is very much the focal point in method development software. In this case, on the basis of only a few experiments, the

method development software can predict the results of many other experiments under different conditions, thus allowing a chromatographer to simulate these experiments with a computer and to find the conditions for acceptable or optimal separation.

ChromSword® supports three approaches for the determination of retention models in reversed-phase HPLC:

(1) A traditional formal approach, which applies linear, quadratic, cubic or other polynomial models to describe the relationship between the retention of solutes and the concentration of an organic solvent in a mobile phase:

$$\ln k = a + b(C)$$
$$\ln k = a + b(C) + d(C)^2$$
$$\ln k = a + b(C) + d(C)^2 + e(C)^3$$

Here, k is the retention factor value of a compound, C is the concentration of an organic solvent in the mobile phase, and a, b, d, e are parameters of equations that must be determined by the software for each compound from the retention data obtained with different concentrations of the organic solvent in the mobile phase. The first linear model is the simplest. It requires two initial experiments to start optimizing, but sometimes it does not predict completely correctly the effect of the concentration of an organic solvent in a mobile phase. Additional experiments do not as a rule lead to improvement in the accuracy of the linear model. The quadratic model describes retention more adequately. Additional experiments improve the accuracy, and three initial experiments are required to start computer optimization. The higher the power of a model, the more complex retention behavior can be described and the more initial experiments must be performed to start optimization of the separation.

ChromSword® supports the optimization of separation for polynomial models up to a power of six. Thus, the most complex retention–concentration effects can be described and the separation optimized. All polynomial models predict the retention of solutes rather precisely in the interpolation region of those concentrations studied. These models are less reliable in the extrapolation region. For example, if experiments were performed with 40% and 50% of organic solvent in a mobile phase, one can expect rather good prediction of retention and separation in the region between these concentrations and less accuracy in the regions of 30–35% and 50–55%. Extrapolation to wider limits very often leads to substantial deviations between predicted and experimental data.

(2) An approach that takes into account both the features of the solutes being separated and the characteristics of the type of reversed-phase HPLC column being used [1–4]. In this approach, retention of a solute is described by:

$$\ln k = a(V)^{2/3} + b(\Delta G) + c$$

Here, V is the molecular volume of a solute, ΔG is the energy of interaction of the solute with water, and a, b, and c are parameters which are determined by the characteristics of the reversed-phase column in the eluent being used. This approach is more precise and rapid than that based on formal linear and quadratic polynomial models, but it requires that both the parameters of the solutes (volume; energy of interaction with water) and the characteristics of the reversed-phase column under the experimental conditions be known.

The characteristics of various commercially available reversed-phase HPLC columns were initially determined with a wide range of concentrations of methanol and acetonitrile in water. ChromSword® contains a database of column parameters in these eluents; they are loaded automatically when a user chooses a column and an eluent from the menu. ChromSword® calculates the parameters of compounds from their structural formulae. If a chromatographer does not know the structural formulae of the compounds being studied or decides not to draw them, these parameters can be determined by ChromSword® from two chromatographic experiments with different concentrations of an organic solvent in a mobile phase. This approach enables ChromSword® to predict regular or irregular retention behavior of the solutes to be separated. It also enables a user to start optimizing retention without any preliminary experiments if the structural formulae of the compounds are known, and to start optimization of the separation after entering the retention data for only one run. For solutes with unknown or undefined structures, this approach can also be used very successfully after entering the retention data and chromatographic conditions for two runs. The main advantage of the structure and column properties related approach is that it addresses both column and compound features. It works rather precisely in the interpolation region and reliably in the extrapolation regions. This opportunity of building a retention model with practically acceptable accuracy using as input only structural formulae and retention data from one run allows the optimization of retention and separation in minimal time.

(3) An approach that applies special functions such as:

$$\ln k = f(\text{pH}, \text{p}K, T, C)$$

These functions are used for optimization of pH and temperature, and also allow the determination of the nature of unknown compounds (neutral, acid, base) and an experimental determination of $\text{p}K$ values and other physico-chemical characteristics of solutes under HPLC conditions.

The software contains several modules for drawing or importing chemical structures, and for storing and processing data.

- The **first guess** module is used for the prediction of initial conditions in isocratic and gradient reversed-phase HPLC (RP HPLC) from structural formulae of compounds and the properties of different reversed-phase columns [1–4]. The module also contains a database of properties of more than 75 types of commercial reversed-phase columns.

- The **simulation** and **optimization** modules are used for the simulation of various experiments and for the optimization of isocratic and gradient HPLC methods. Parameters that can be optimized are concentration of organic solvent in the mobile phase, linear and multi-segment gradient profile, pH value, temperature, type of reversed-phase column, and organic solvent.

Table 1. Input/output of ChromSword® 2.0

Minimal input	Expected output
Structural formulae are considered	
Structural formulae (up to 100 in a file)	Starting conditions for RP HPLC (column type, eluent)
Structural formulae and data of one run	Optimal eluent for separation of a mixture in isocratic RP HPLC on the column being used (first approximation). Starting conditions for RP HPLC on another column type with another eluent. Optimal gradient profile.
Structural formulae are not considered	
Data for two runs with different concentrations of an organic solvent in the MP (RP HPLC)	Optimal eluent for separation of a mixture by isocratic RP HPLC on the column being used. Starting conditions for RP HPLC on another column type with another eluent. Evaluation of analyte parameters (molecular volume, polarity).
Data for two runs with different concentrations of an organic solvent or a buffer in the MP	Optimal eluent for separation of a mixture by isocratic RP, NP or IE HPLC. Optimal gradient profile.
Data for two runs with different gradient profiles	Optimal gradient profile for separation of a mixture by gradient HPLC. Optimal eluent for separation of a mixture by isocratic HPLC.
Data for two runs with different column temperatures	Optimal temperature for separation of a mixture by isocratic HPLC. Enthalpy sorption of analytes.
Data for two runs with different pH of the MP	Optimal pH for separation of a mixture by isocratic RP HPLC.
Data for three runs with different pH of the MP	Optimal pH for separation of a mixture by isocratic RP HPLC. Nature of analytes (base, acid, neutral). pK value of analytes.
Two-variable optimizations: Data for three or four runs with different concentrations, pH, temperatures, columns, solvents, gradient profiles	Optimal gradient profile and temperature; concentration and pH; concentration and temperature; pH and temperature; concentration of two different organic solvents; optimal connection of two columns with different selectivity and concentration, gradient profile, pH or temperature.

The modules to be used and the generated results depend on information which a user enters into the software (Table 1).

ChromSword® can work with substantial masses of data. One sample file can contain up to 100 compounds, including structural formulae and data for up to 20 runs in the each module.

4.2.4.2 On-Line Mode – Fully Automated Optimization of Isocratic and Gradient Separations

Two approaches are most frequently applied for method development and optimization:

1. intelligent optimization, whereby a chromatographer uses his or her experience and reasoning to evaluate the results of an experiment and to plan the next one, and

2. screening conditions, whereby a chromatographer runs one or more generic methods with different columns and solvents, pH value, temperature, etc.
   In this case, the results of a completed chromatographic run are not taken into account when planning the next experiment. Normally, a set of runs is evaluated by a chromatographer in planning the next screening sequence.

Both approaches are applied rather successfully in practice and can be automated, although automation of the intelligent approach is substantially more difficult. ChromSword® supports both approaches and several types of problems can be solved automatically. Typically, these are:

- Development or optimization of final HPLC methods when all components of a mixture to be separated are available as standards.
- Development of HPLC gradient methods when only some components of a mixture are known and are available as standards.
- Development of HPLC methods when all components of a mixture are unknown and no standards are available.
- Rapid method development and optimization.
- Rapid method screening with different columns, solvents, and buffers.

Typical samples in the pharmaceutical and chemical industries consist of a main ingredient and impurities or degradation products, natural extracts, and other complex mixtures. The system is applied for automated optimization of a solvent concentration, gradient profiles, pH, and temperature in reversed-phase, normal phase, ion and ion-pair chromatographies, including optimization of chiral separations. Four examples of automatic method development with ChromSword® and three different hardware configurations (Merck, Agilent, Waters) are given in Figs. 1 to 4.

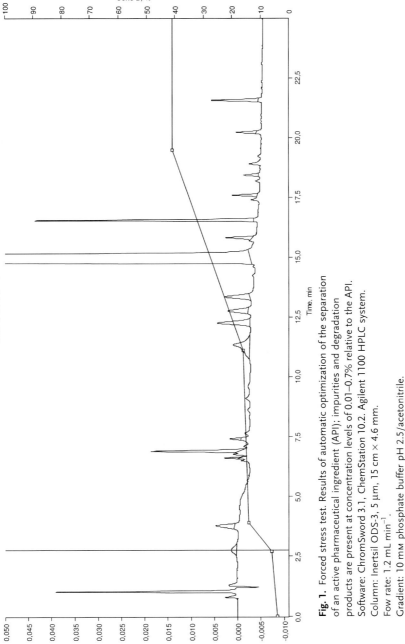

Fig. 1. Forced stress test. Results of automatic optimization of the separation of an active pharmaceutical ingredient (API); impurities and degradation products are present at concentration levels of 0.01–0.7% relative to the API.
Software: ChromSword 3.1, ChemStation 10.2. Agilent 1100 HPLC system.
Column: Inertsil ODS-3, 5 μm, 15 cm × 4.6 mm.
Fow rate: 1.2 mL min⁻¹.
Gradient: 10 mM phosphate buffer pH 2.5/acetonitrile.

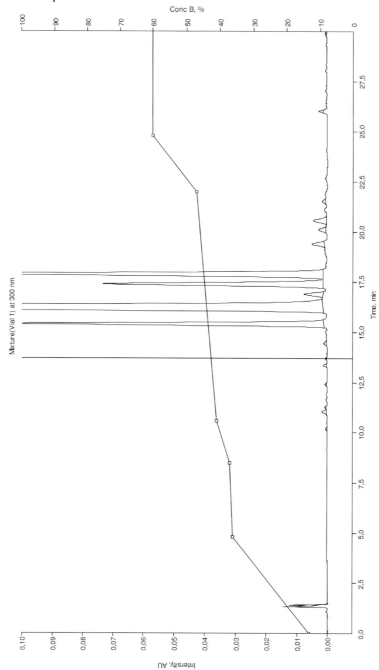

Fig. 2. Forced stress test. Results of automatic optimization of the separation of an active pharmaceutical ingredient (API); impurities and degradation products are present at concentration levels of 0.005–9% relative to the API. Software: ChromSword 3.2, EZChromElite. LaChromElite HPLC system. Column: XTerra RP18, 3.5 μm, 10 cm × 4.6 mm. Flow rate: 1.0 mL min⁻¹. Gradient: 20 mM phosphoric acid/acetonitrile.

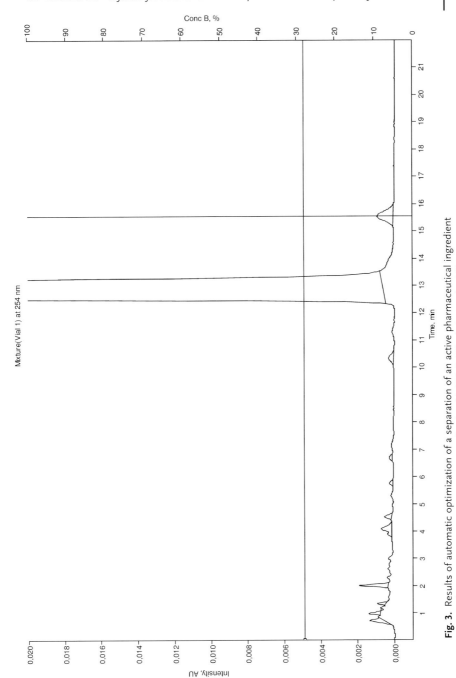

Fig. 3. Results of automatic optimization of a separation of an active pharmaceutical ingredient (API) and impurities present at concentration levels of 0.008–0.2% relative to the API. Software: ChromSword 3.1, Waters Empower. Waters Alliance HPLC system.
Column: XTerra RP 18, 3.5 μm, 5 μm, 10 cm × 4.6 mm.
Fow rate: 1.5 mL min^{-1}.
Gadient: 10 mM phosphate buffer pH 3.5/acetonitrile.

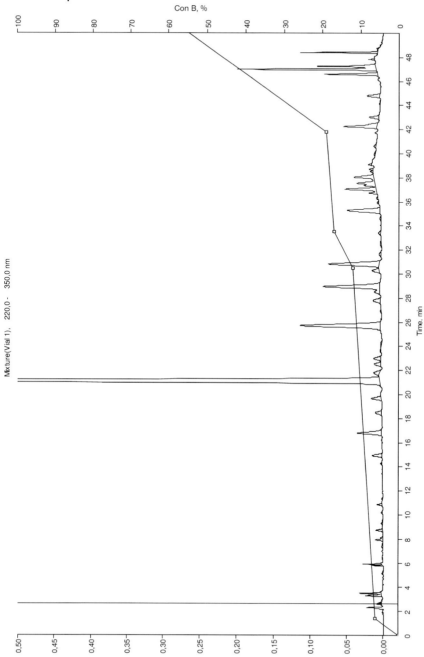

Fig. 4. Fast screening result of automatic optimization of the separation
of a natural sample (apple juice); 81 components have been separated.
Software: ChromSword 3.2, EZChromElite. LaChromElite HPLC system.
Column: Purospher RP 18,e 5 μm, 25 cm × 4.0 mm.
Flow rate: 1.0 mL min^{-1}.
Gradient: 0.1% TFA/acetonitrile.

4.2.4.2.1 Software Functions for Automation

Computative numerical methods, automation technology, and artificial intelligence are combined in the software to carry out HPLC method development completely automatically.

The software for automatic optimization consists of different modules:

- A computative and optimization module.
  This module models chromatographic behavior of analytes from compound structures, column and solvent properties or, alternatively, from the retention data of two initial runs under different conditions. Moreover, it searches for and predicts optimal conditions.

- A module for automation and data exchange.
  This serves to control the HPLC software and hardware systems and provides the necessary data exchange facility.

- A module for artificial intelligence.
  This module is a real-time expert system or an intelligent agent, which serves to simulate the actions of a human chromatographer in a method development procedure, such as in evaluating results, taking decisions, and solving problems.

4.2.4.2.2 How Does the System Optimize Separations?

Initial Set-up and Start of Optimization

For the initial set-up, a chromatographer has to specify only method development goals to develop isocratic gradient or isocratic and/or gradient methods. It is not necessary for the user to input any additional information about a sample, such as the presence of basic or acidic compounds, small or large molecules, isomers, etc. The software evaluates solute properties automatically under the conditions being used and plans the next experiments taking into account the results of previous ones.

Prediction of Starting Conditions from Analyte Structures and Reversed-phase Column Properties

The selection of starting conditions is the first step in HPLC method development. Instead of starting with arbitrary conditions or conditions derived from the chromatographer's experience or intuition, the software can predict suitable initial conditions for reversed-phase methods from analyte structures and properties of the sorbent/solvent system [1–3]. If structures are known, the theoretical approach has the potential to save time and effort, since in this way the experimental method development process will start under the theoretically predicted optimum conditions.

After input of the compound structures and a column type, the software immediately calculates the theoretical retention model ($\ln k$ in dependence on the concentration of the organic modifier) for the compounds to be separated.

This theoretical retention model is used to predict the optimum starting conditions for the experimental method optimization process. Experimental results show that in most cases the predicted optimum organic modifier concentration differs from the real optimum by at most only ±5–6% [4].

After these calculations, which are performed immediately after entering the structures, the system automatically commences wthe experimental runs. The results of the experimental runs are used to fine-tune the first theoretically set up retention model. Thus, the retention model becomes accurate enough to finally predict the optimum isocratic separation conditions.

The Empirical Approach if no Structures are Available

If the structures of the compounds to be separated are unknown or if the desired sorbent/eluent system is not available from the software's database, reversed-phase HPLC methods are optimized using the empirical approach.

The system starts with an arbitrary user-selected set of starting conditions, collects data, and makes a decision on the second run conditions. After the second run, an empirical linear retention model is set-up from the observed retention data, which is used to calculate the next conditions to be applied. Once the retention data for three different sets of chromatographic conditions are available, the quadratic retention model can be set-up (curve-fitting using a polynome of the power two).

Further such optimization process cycles are performed until the retention model is accurate enough to predict the optimum separation conditions.

Optimization of Linear and Multi-step Gradients [5]

Once the retention models have been established by the procedures described above, no further experiments are conducted. The software automatically generates tens of thousands of linear and multi-step gradient profiles in a very short time. It predicts the retention of compounds for each gradient profile generated, evaluates the predicted separation using an optimization function, and searches for the best gradient profile using a super-fast Monte Carlo optimization method [5].

In this way, the often intuitive and very time-consuming process of gradient optimization is replaced by an effective, fast, and completely automatic procedure using the power of computing methods without the need to perform chromatographic experiments.

Peak Tracking for Substance Mixtures Containing Impurities

The easiest way to unequivocally solve the task of peak assignment and peak tracking is to perform experimental method development using single pure substances, i.e. standards of the compounds to be separated. This procedure is utilized by the software when optimizing the separation of known target substances. However, in most cases, suitable standards are not available for all compounds. For example, in the analysis of the impurities in a product, usually only a few (if any) of the impurities will be available as pure substances. ChromSword® is equipped with special "intelligent peak tracking" functions, which ensure peak assignment in compound mixtures, with or without using spectral data. If UV or mass spectra are available and consistent, the software uses these data; however, if no reliable spectra are available the system applies a combination of artificial intelligence reasoning and numerical procedures to build retention models using area, peak shape, efficiency, and other available information. By recording and tracking the peaks that emerge under different chromatographic conditions, the system ensures that all peaks are clearly isolated and optimized.

Column and Solvent Switching

The software has been equipped with a function such that it is not only able to conduct experiments with one column/organic modifier/buffer combination, but to automatically optimize a method by trying different column/organic modifier/buffer combinations. The system provides unattended HPLC method development and performs autonomous development and optimization of isocratic and gradient methods for selection of the best variant: column, pH value, solvent. Two typical hardware configurations and other mixed combinations are supported by ChromSword®: standard and powerful.

1. Standard configuration
For the standard configuration, an automatic column selection valve for accommodating two different columns is required, and solvent switching is performed using the channels of a quaternary low-pressure gradient pump. This configuration can be realized for all HPLC systems supported by ChromSword®.

If reversed-phase columns with different properties (e.g., RP-C_8 and RP-C_{18}) and three different organic modifiers (e.g., methanol, acetonitrile, and a mixture of acetonitrile with tetrahydrofuran) are applied, the system carries out an automatic method optimization procedure for each column/eluent combination. In relatively short time and in an unattended manner, the most suitable column/eluent combination can be found completely automatically.

After the computational and experimental optimization process has been completed, the software delivers as a result three to ten optimum sets of separation conditions. In addition, all of the chromatograms related to method development are archived and can be browsed. The method development history is documented automatically.

2. Powerful configuration
The powerful configuration can be realized with ChromSword® and HPLC systems described, and external column selection/solvent selection valves. The six-position selection valve is designed for use with up to eight columns. The 10–16 position port selection valve is used as a solvent selector. Valve control is integrated in ChemStation software and in the ChromSword® switching columns/solvents algorithm for unattended HPLC method development and rapid screen-

ing. Therefore, ChromSword® can try up to three different organic solvents, 2–16 different aqueous eluents, and 2–8 columns in searching for the optimal conditions fully automatically – corresponding in total up to 216 different optimization experiments. Control area network (CAN) – the standard communication interface for 1100 Series modules – provides high-speed data transfer for real-time valve control. This technology offers the highest flexibility for developing new methods.

4.2.5
Conclusion

ChromSword® is a software tool that is able to find the optimum conditions for HPLC automatically. Most optimizations may be achieved overnight or in weekend runs, resulting in a substantial reduction in working time and effort compared to that normally involved in method development. Since the system offers several different isocratic and gradient optimum solutions, the user can select the solution which fits best for his or her particular application. The system optimizes for optimum peak resolution in minimum analysis time. Minimization of the run time of routine methods offers the potential to substantially increase throughput and productivity in the analytical laboratory. The program also makes complex HPLC method development accessible to those with little HPLC experience.

References

1 S. V. Galushko, The calculation of retention and selectivity in RP LC, *J. Chromatogr.* 552 (1991), 91–102 (1991).

2 S. V. Galushko, A. A. Kamenchuk, G. L. Pit, The calculation of retention and selectivity in RP LC. IV. Software for selection of initial conditions and for simulating chromatographic behavior, *J. Chromatogr.* 660 (1994), 47–59.

3 S. V. Galushko, The calculation of retention and selectivity in RP LC.

II. Methanol-water eluents, *Chromatographia* 36 (1993), 39–41.

4 S. V. Galushko, A. A. Kamenchuk G. L. Pit, Software for method development in reversed-phase liquid chromatography, American Laboratory, 27, N3 (1995), 421–432.

5 S. V. Galushko, A. A. Kamenchuk, Fast optimizing of the multi-segment gradient profile in reversed-phase high performance liquid chromatography using the Monte Carlo method, *LC-GC Int. Mag.* 8, N 10 (1995), 581–587.

4.3
Multifactorial Systematic Method Development and Optimization in Reversed-Phase HPLC

Michael Pfeffer

During the process of developing new HPLC methods, the chromatographic conditions, i.e. the factors influencing chromatographic separation, are varied and optimized until the analytes have been satisfactorily separated from one another. Such influencing factors are found in the column (type of stationary phase, particle size, temperature and dimensions of the column), mobile phase (percentage of organic solvents, pH, gradient run time, as well as the slope of gradient, etc.) and the type of detection (UV, fluorescence, etc.). If accessories are not available, such as a column selection system and/or other tools, the individual influencing factor is usually developed and/or optimized gradually. This is typically very time-consuming and does not always result in actually determining optimal conditions.

The idea is to examine the influence of chromatographic factors systematically in real time and to a large extent automatically and to seek out and verify their "global optimum", i.e. the optimum constellation.

Commercially available software packages have been compared according to their suitability for this purpose. At last, HEUREKA, a new software tool, which is based on the principle of multifactorial development and optimization, was introduced. It is meant to spare the programming of numerous method files and to help with finding and verifying the optimum constellation of chromatographic conditions. As opposed to other software aids, HEUREKA places a very strong emphasis on the influence of stationary phases on chromatographic separation. Up to 12 stationary phases can be tested. That sets HEUREKA apart from other systems. The advantage is that overlooking a peak is not likely in purity analysis. The mode of operation is highlighted by mixing a steroid with contaminants.

4.3.1
Introduction and Factorial Viewpoint

Chromatographic separation of an analyte mixture depends on numerous conditions. In reversed-phase chromatography, the retention and thus the resolution of peaks is invariably influenced by varying the percentage of organic solvents or the column temperature. Both are regarded as "factors" that considerably affect the elution of the peaks and with it the chromatographic separation. A multiplicity of chromatographic conditions can therefore be recognized as factors and their influence purposefully used during method development.

This approach is based on publications by Preu et al. [1, 2] on strategy during the optimization of chemical reactions. They pointed out that usually one parameter after another is gradually optimized. A disadvantage of this OVAT (one

HPLC Made to Measure: A Practical Handbook for Optimization. Edited by Stavros Kromidas
Copyright © 2006 WILEY-VCH Verlag GmbH & Co. KGaA, Weinheim
ISBN: 3-527-31377-X

variable at a time) method is that interactions between the various influencing parameters cannot be recognized and taken into consideration. Usually, only a very small segment of all possible combinations is examined. Due to this limitation, the determined optimum is in many cases not the "global optimum". An important conclusion is that the actual optimum can only be found if all (relevant) parameters can be varied at the same time. Therefore, chemometric methods are preferably used in chemistry.

When transposed onto the problem of developing chromatographic methods, this means that during sequential optimization of chromatographic parameters there is a risk of overlooking the interactions of numerous influencing factors, thus failing to take them into account. As a result, optimization may overshoot the "global optimum", i.e., the constellation of factors under which all (relevant) peaks are separated.

Figure 1 shows how one can imagine the optimization effects, together with the positioning of "local optima" (good separation in a narrow range of all the relevant factors) and a "global optimum" for chromatographic resolution. Actually, it is necessary to imagine a multi-dimensional array, in which one ineluctably finds the precise constellation of factors that results in the best separation of all the peaks.

Under normal conditions, one systematically changes one factor after another by first trying to separate the sample on a standard stationary phase eluting with an acetonitrile/water mixture, then with a second solvent, such as methanol.

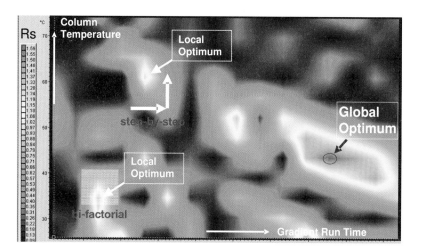

Fig. 1. Simulated representation of the effects that the factors gradient run time and column temperature have on the quality of a chromatographic separation. Separation is insufficient in the blue zones, which is an indication of co-elution of peaks; all peaks are sufficiently well separated in the red zones. On the left-hand-side of the figure, there is a color scale for correlating the worst separated peak in the chromatogram. Possible effects (overshooting the target) during gradual optimization of chromatographic factors (stepwise; gradient run time and column temperature changed successively) or when simultaneously changing two factors (bifactorial change), a "local optimum" is achieved. As is represented here, the "global optimum" (best separation) was not found.

Table 1. Factors that potentially influence the separation of substance mixtures in reversed-phase HPLC.

Column	*Type of stationary phase, particle size, column dimensions, column temperature*		
Mobile phase	Type of solvents, percentage of organic solvents, flow (with gradient elution), gradient rise and gradient time, pH, modifier, buffer type and ionic strength		
Detection	Type of detection:	UV/Vis:	wavelength of the absorption
		Fluorescence:	excitation and emission wavelengths
		MS:	molecular weight

Afterwards, one lowers the temperature, perhaps from ambient to 15 °C, in order to improve separation. If one does not hit the mark, a further stationary phase is tested in the same manner. In this way, the influence of the factors "type of stationary phase", "solvents", and "column temperature" are tested, but without being able to recognize the influences of the three factors on each other. We do not learn whether the individual factors mutually negate themselves, strengthen one another, or jointly have some completely different unimaginable effects on the chromatographic separation.

Therefore, it is important not to test experimentally the individual factors in isolation, but rather as they interact with one another. With this sequential approach, there is a risk of not finding the constellation of chromatographic conditions for optimal separation of the peaks. For example, the expected additional selectivity may not be manifested on lowering the column temperature, since the factors "column temperature" and "solvent" do not positively correlate, but rather mutually interfere.

Table 1 summarizes a set of factors that potentially influence chromatographic separation.

4.3.2
Strategy for Partially Automated Method Development

The exploratory practical work in the laboratory begins during method development after conclusion of the preliminary theoretical work on a separation problem. First, one specifies the possible influencing factors that are to be tested, e.g. six columns with different types of stationary phases, ideally with identical dimensions, the solvents acetonitrile and methanol, the column temperatures (15 °C and 45 °C), and if necessary, the influence of the pH (e.g., 2, 7, and 9). Handbooks that have already been published on the strategic changes of chromatographic parameters can afford valuable assistance [3, 4] and should always be consulted to save time.

This so-called scouting can be simply automated in the case of switching valves for columns [5] and solvents [6]. Software products (see Chapters 4.1 and 4.2 in this book) such as AMDS (Waters), ChromSword Auto (VWR), and HEUREKA (AnaConDa) additionally facilitate the programming work by automatically providing sequences with the necessary control methods.

A general flow chart for partly automated method development is presented in Fig. 2. From this, it is clear that the analyst repeatedly intervenes in order to evaluate intermediate results to steer the direction of development. At the outset, according to the available information as well as the experience of the analyst, the chromatographic factors that are to be examined are defined and their limits are delineated. Thereafter, the HPLC system takes over the execution of the necessary experiments and/or the chromatographic runs. The analyst evaluates the runs once they have finished. The number of peaks, and their form, resolution, and sequence are important. As a result, conditions can be hypothesized that should have the highest potential for optimal separation of all of the peaks, or at least of all the relevant ones.

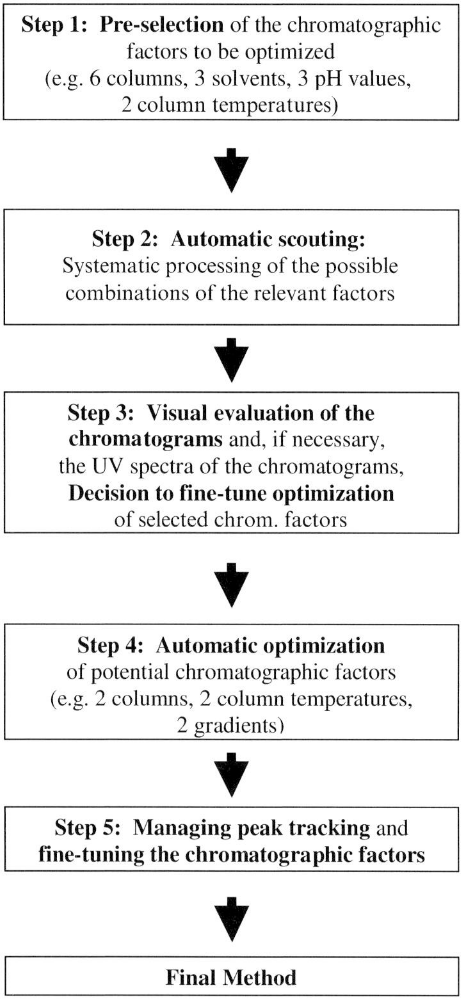

Step 1: Pre-selection of the chromatographic
factors to be optimized
(e.g. 6 columns, 3 solvents, 3 pH values,
2 column temperatures)

Step 2: Automatic scouting:
Systematic processing of the possible
combinations of the relevant factors

**Step 3: Visual evaluation of the
chromatograms** and, if necessary,
the UV spectra of the chromatograms,
Decision to fine-tune optimization
of selected chrom. factors

Step 4: Automatic optimization
of potential chromatographic factors
(e.g. 2 columns, 2 column temperatures,
2 gradients)

Step 5: Managing peak tracking and
fine-tuning the chromatographic factors

Final Method

Fig. 2. Sequence of partly automated method development in reversed-phase HPLC.

Table 2. Comparison of commercially available automatic method development systems for HPLC.

Characteristic	AMDS	ChromSword Auto	HEUREKA
Manufacturer	Waters/LC Resources	VWR	AnaConDa
Chromatography Software Optimization Utility	Empower (Waters) DryLab 2000 Plus (LC Resources)	Chemstation (Agilent), Empower (Waters) or EZChrom (VWR) Expert system (Galushko, VWR)	Chemstation (Agilent) Development (Prof. Otto, Univ. Freiberg; AnaConDa, Schering)
Configuration of a method development system	Autosampler Quaternary pump Photodiode array detector Column thermostat Switching valve/solvent (6 positions) Switching valve/columns (6 positions)	Autosampler Quaternary pump Photodiode array detector Column thermostat Switching valve/solvent (12 positions) Switching valve/columns (6 positions)	Autosampler at present, binary pump Photodiode array detector Column thermostat Switching valve/solvent (plan: 12 positions) Switching valve/columns (12 positions)
User-friendliness	Very good for users of Millennium and/or DryLab (training necessary, among other reasons, because of numerous hidden adjustments)	Good (training is absolutely necessary, due to the high number of varying possibilities of the software)	Very good for users of Chemstation software (however, training necessary, due to the innovative processing approach)
Capacity expenditure for a simple method development	**Scouting:** 3 h (preparation of solutions, programming, evaluation of chromatograms) optimization (1 h programming, evaluation lasts quite a long while with numerous peaks)	**Scouting:** 3 h (preparation of solutions, programming, evaluation of chromatograms) 1 h (conducted with ChromSword)	**Scouting:** 3 h (preparation of solutions, programming, evaluation of chromatograms) optimization (0.5 h programming, evaluation lasts quite a long while with numerous peaks)
Duration of a method development (run time) (depending upon the complexity of the problem)	2–5 days	2–5 days	2–5 days

Table 2. (continued)

Characteristic	AMDS	ChromSword Auto	HEUREKA
Processing Approach	**Scouting:** Control methods are provided automatically according to the data entered and systematically processed. **Fine-tuned optimization:** Optimization runs are conducted within the entered limits (e.g., two column temperatures and gradient run times each) and imported into DryLab (with more than 10 peaks and peaks of < 0.5% (area) problems arise); optimal method with DryLab automatically simulated and then realized.	**Scouting:** Suggestion for first method by evaluation of the chemical structure or according to the data entered by the user. **Fine-tuned optimization:** Automatic optimization starting with the stronger and working toward the weaker elution of the mobile phase, intermittent with separation simulation.	**Scouting and optimization:** Control methods are automatically compiled according to the data entered by the operator and systematically processed. The chromatograms are imported into a database. At present, the peak identification is still essentially manual. From the data, the optimal separation is determined and then realized. Scouting and optimization are conducted in one step.
Automatically compiled method of development	Not available at present	Available	Planned
Laboratory-specific aspects	• Waters/LC Resources offers security and continuity in availability and advancement of the software on a long-term basis • Software and hardware supplied from one source • Good service • At present, control of devices from other manufacturers only possible in certain cases or not at all	• Due to the increased distribution, there is security and continuity of availability and advancement of the software • High degree of flexibility (solutions for special problems possible) • Concerned, helpful team • Control of devices made by Agilent 1100, VWR, and Waters possible	• System not yet completely developed • High degree of flexibility and open-mindedness in relation to customer's requests; however, uncertainty due to the small size of the manufacturer • Concerned, helpful team • At present, control only of Agilent 1100 apparatus possible; control of other manufacturers' apparatus planned.

Table 2. (continued)

Characteristic	AMDS	ChromSword Auto	HEUREKA
Possibilities of the Software/Advantages	• Goal: optimization of the resolution or • Goal: optimization of the run time • Range of all possibilities of DryLab, although not automatically • Data can be applied to other problems • Optimization of the column parameters • Can be used in normal- and reversed-phase HPLC • Representation of the peaks by "max plot" (all peaks are presented as their absorption maximum), which has the advantage that no relevant peak is overlooked	• Structural formula as starting point for optimization • Search for impurities beneath the main peak • Limits optimization to defined analytes • Optimization with the goals of best resolution, short run time, maximum number of peaks • existing HPLC methods can be optimized automatically • Optimization with LC/MS • Very efficient developing tool • Can be used with normal- and reversed-phase HPLC	• Multifactorial approach • Optimizing the resolution • Optimizing the run time • Data can be used with other problems • Extensive graphics to represent global optimum • Can be used with normal- and reversed-phase HPLC as well as for chiral separation problems
Limits/Disadvantages	• max. approx. 10 peaks, which should exhibit > 0.5% (area) • maximal bifactorial function • at present, restriction on HPLC by waters	• Stepwise approach	• Presently limited to Agilent 1100 • Presently only small distribution
Potential ranges of application in the context of producing medicinal products	• Quality control of highly pure substances; drug substance analysis • Analysis of mixtures with few components • Optimization of existing methods • Determinable ruggedness of the method	• Analysis of samples with numerous contaminants • Search for contaminants beneath the main peak • Optimization of existing methods	• Analysis of samples with numerous (up to 20–30) components (crude materials in intermediate and final stages of synthesis) • Search for contaminants beneath the main peak (reference substances or suspected)

In the next step, experiments aimed at optimizing the separation results are planned. The work can again be facilitated with software aids (Table 2). In the simplest case, one uses DryLab (see Chapter 4.1), which, for example, requires four runs (one short and one long gradient at two column temperatures for each). The sample is doubly-injected in order to improve the retention time precision and that of the peak. The AMDS system also uses DryLab for optimization and generates the basic runs automatically. ChromSword Auto automatically conducts runs by starting with the stronger and working toward the weaker elution. Intermediate conditions are simulated. Only runs with promising chromatographic conditions are carried out.

Ultimately, the method development systems suggest one or several constellations of chromatographic conditions that are identified as being optimal according to their particular algorithms. The analyst now needs to rigorously test the correctness of the peak tracking in order to assure the stability of the discovered HPLC methods. Otherwise, less robust or useless separations (co-elution of peaks) may be indicated because the software has "erred". In the final analysis, the suggested HPLC methods are verified for the actual sample, examined with respect to suitability as well as ruggedness, and, should the need arise, manually readjusted slightly.

None of the systems presently available on the market is capable of conducting a fully automatic method development without supervision. In order to save analysis time, it is advisable, as represented in Fig. 2, to critically and repeatedly analyze the intermediate results obtained during the development process and to judge the further course of development on the basis of one's own experiences and to realign it carefully.

4.3.3
Comparison of Commercially Available Software Packages with Regard to Their Contribution to Factorial Method Development

If one plans to purchase an automatic method development system, one will first have to get an idea of the systems that are presently available on the market. Essentially, only ChromSword Auto and, since recently, AMDS (based upon DryLab) are available, along with HEUREKA, which is presently being established on the market. Characteristics and experiences with the three systems are compared in Table 2. This table can also be used as a check list in order to facilitate the decision in favor of one system or the other.

ChromSword Auto is presently being marketed as the most effective and mature system. As mentioned above, none of the systems, however, can accomplish a development with optimization in a completely unsupervised manner. Nevertheless, the goals can be achieved with all of them. Crucial for the purchase, however, is what software is already used for controlling HPLC devices in the laboratory, whether DryLab has already been installed, whether apparatus of a certain manufacturer is already in operation, or whether management prescribes a certain manufacturer. One must also decide whether one trusts only the security and continuity of a large manufacturer or whether one wants to speculate on the innovative strength, flexibility, and speed of a smaller company.

In the context of scouting, AMDS and ChromSword Auto can be used for testing: e.g., six columns, two column temperatures, three solvents, and, with a six- or twelve-position switching valve at the entrance port of the pump: water, different buffers, concentrations, and pH values (e.g., 2, 7, and 9). The great advantage is that the majority of the necessary control methods do not have to be programmed manually, but are taken over by the method development system. Thereby, errors are avoided and time is saved. All factors are systematically screened. During evaluation, however, the factors are normally not correlated with one another. Interactions between individual factors cannot be recognized.

The fine-tuned optimization is based on a maximum of two factors (e.g., DryLab with gradient and column temperature, AMDS). Further factors (such as pH) are optimized and/or adapted, when necessary, in a stepwise manner downstream. ChromSword Auto works likewise in a linear framework, in which it is meant to optimize. It collects data during the optimization runs that represent empirical values, and these are included as empirical data in the next experimental step, until the system finds the best method.

Both systems are essentially capable of successively optimizing the influencing factors in a sequential manner, with the risk of overshooting the goal, as represented in Fig. 1.

4.3.4
Development of a New System for Multifactorial Method Development

None of the commercially available systems is capable of testing more than two simultaneously varied chromatographic factors. DryLab, which uses AMDS for optimization, needs four basic runs in order to find the optimal constellation in the dimensions "column temperature" and "gradient run time". All other potential influencing variables have to be determined beforehand. The separation capacity of the chromatographic system refers to all optimizable factors, although these are simulated only within a relatively small range. AMDS searches the "local optimum" therein until the process is complete.

When attempting to find the optimum in multi-dimensional space, one quickly arrives at the limits of the conventional method development systems. The sequential approach, however, harbors the danger that the factors that are subsequently optimized enter into unexpected interaction with those previously optimized and that the optimization conducted thereafter becomes worthless. One may have unknowingly optimized beyond the actual optimum and possibly even landed in a cul-de-sac.

Therefore, it was considered how more than two chromatographic factors, e.g. stationary phase, column temperature, gradient run time, solvent, gradient rise, etc., could be tested systematically in one experiment. With column thermostats, which were specifically developed for this purpose (Model HELIOS, AnaConDa), a maximum of twelve columns and an arbitrary number of temperatures can be tried out within the range 10–80 °C. Besides water, a quaternary pump makes available acetonitrile, methanol, and a third organic solvent in order to test the factor "solvent". Using a solvent switching valve during method development, up to twelve further solutions can be made available to automatically examine the effect of buffers and pH values on the separation problem.

The HPLC system that is to be used for such a universal development of a chromatographic separation method consists in the optimal case of a quaternary pump with a solvent switching valve (6–12 positions), an autosampler, a programmable column thermostat (for 6–12 columns), a photodiode array detector and other detectors, depending upon the need. A personal computer with suitable software is needed for the automatic control system. The method development software controls the HPLC system via this software and produces the chromatograms necessary for development and optimization.

Such an HPLC instrument is capable of systematically varying chromatographic conditions in one sequence, of converting to control methods, and of generating manifold chromatograms. An example of a standard sequence is given in Table 3.

Finally, chromatograms and UV spectra are automatically imported and integrated. Peak area ratios and UV data are evaluated by the system for subsequent peak tracking. The correct assignment, however, frequently succeeds only with relatively large peaks (> 0.5% area), peaks with clearly differing UV spectra and chromatograms. These should exhibit no more than ten peaks. Since the peak assignment can be very difficult, it is recommended that the result is checked in any case. It is also advisable to look at the form of the peaks, their number in the chromatogram, and their sequence.

Table 3. Example sequence for the systematic investigation of the influences of the solvents acetonitrile and methanol, of six stationary phases, and of two gradients on chromatographic separation with 24 chromatographic runs.

Gradient with acetonitrile/water	Column 1	Short gradient	Run no. 1
		Long gradient	Run no. 2
	Column 2	Short gradient	Run no. 3
		Long gradient	Run no. 4
	...	...	...
	Column 6	Short gradient	Run no. 11
		Long gradient	Run no. 12
Gradient with methanol/water	Column 1	Short gradient	Run no. 13
		Long gradient	Run no. 14
	Column 2	Short gradient	Run no. 15
		Long gradient	Run no. 16
	...	...	...
	Column 6	Short gradient	Run no. 23
		Long gradient	Run no. 24

The purpose of the optimization software is to find the "global optimum" from the large number of chromatograms. In order to accomplish this, qualitative parameters, such as the type of stationary phase or the solvent used, have to be brought into a form that is usable by the computer. This was achieved by Otto et al. [7], for example, using a set of equations to weight the share of the individual types of stationary phases to the retention time of a peak. Thus, for instance, a coefficient is computed for column A, while the coefficients of the other columns are defined as zero; this means that column 1 supplies a share of the retention time of a peak, but columns 2 to 12 do not. This is how the influence of qualitative factors is handled, such as the influences that the type of stationary phase and solvent have on the retention times and peak resolution in multidimensional space. The interpolation between quantifiable values of factors, such as between the column temperature and the gradient run time, is linear. These algorithms are implemented in HEUREKA (Greek: "I have found it!") [8].

4.3.4.1 Selection of Stationary Phases

It is not easy to select a representative range of columns for the test, considering the inscrutably large number of stationary phases available for reversed-phase HPLC. The grouping in Table 4 was set up according to pragmatic criteria and a study of the literature [9] (see Chapters 2.1.1 to 2.1.6 of this book). The resulting

Table 4. Grouping reversed phases for testing stationary phases with column switching.

Main group	Subgroup	Representative examples
Reversed phases with increased polarity	C_{18} phases with high silanol activity	Hypersil ODS, Platinum C_{18} EPS
	C_{18} phases with polar groups ($\pi \rightarrow \pi$-interactions for special selectivity towards molecules with C=C or aromatic structures; steric selectivity)	Prontosil 120-C_{18} ace EPS, XTerra RP18, YMC Hydrosphere
	Reversed phases with short side chains (polar columns with steric selectivity recognize highly hydrophobic compounds (only in stationary phases with low carbon content))	YMC Pro C_4, YMC Pro C_8
C_{18} phases with lower hydrophobicity	(separation of substances with different polarities, separation of acids)	YMC Pro C18, Ultrasep ES RP18e Pharm, Luna C_{18}, YMC J'sphere ODS-H80, MP Gel C_{18}
Special phases	for fast separations	Chromolith Performance RP18e (fast separations with high flow rates)
	for special selectivity	Fluofix (separation of substances containing fluorine)
	for the separation of large molecules	Vydac C18 Protein & Peptide

set of columns has served satisfactorily in dozens of method development processes for the purity analysis of steroidal and heterocyclic compounds.

If, in order to save time, one wishes to examine six or less stationary phases, it may be advantageous to make a representative selection from each group on offer for installation in the column thermostats. If, however, a column switching valve with twelve positions is available, one can install several representative selections from the groups, as desired, particularly from the group of "C_{18} phases with lower hydrophobicity" and still afford oneself the "luxury" of testing one or two new entrants onto the market or other promising columns. The more columns that are tested, the less likely it is that a contaminant will be overlooked during purity analysis (see also Chapter 2.1.1). At the same time, an acceptable separation can usually be found on one of the twelve columns. The mobile phase may then be optimized more rapidly and with far less effort.

The dimensions of the columns are selected in relation to the goal of the method development and the number of the relevant peaks that are to be expected. When the goal is a method purely for assay purposes, with a maximum sample throughput, one installs columns of length 50 mm into the thermostats. For purity testing and the goal of optimum resolution of all peaks, then columns of length 250 mm should be used. If, however, one expects only a small number of peaks ($n \approx 7$), it is recommended that the shorter columns are used.

A set of columns that has been systematically set up according to Table 4 has the advantage that one can easily determine correlations between the peak patterns and the peculiarities of the stationary phases and can use them beneficially in optimizing the method.

4.3.4.2 Optimizing Methods with HEUREKA

First, one selects those factors that can be varied (see Step 2 in Fig. 2). Table 5 displays the chromatographic conditions that can be regarded as factors of HEUREKA, and that can be changed and tested simultaneously.

Subsequently, one defines the limits within which the factors are to be varied. For example, the system might receive the instruction (Fig. 3) to test columns 1

Table 5. Chromatographic factors that can be varied simultaneously with HEUREKA.

Chromatographic factor	Possible number of factor conditions
Stationary phases	12
Gradient run time	2
Temperature of the column	2
Concentration of the organic solvent at the start of the gradient (% B initial)	2
Concentration of the organic solvent at the end of the gradient (% B final)	2
Flow rate	2
Solvent	1 (3)

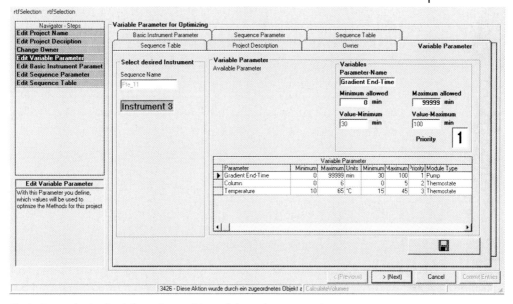

Fig. 3. Screen display for defining the conditions of those chromatographic factors that are to be examined.

to 6 at temperatures of 15 and 45 °C with gradient run times of 30 and 100 min in each case, and the sequence in which this is to proceed. With six to twelve columns, however, it is advisable to change at most only two or three other factors since the optimization run time is doubled for each additional factor and can quickly extend to a period of several days.

Accordingly, the constant fundamental values are entered, such as wavelengths, flow rate, initial and final condition of the gradients, etc. At the push of a button, HEUREKA produces a sequence of 24 control methods, which is then started manually. The methods are processed successively. As soon as one chromatographic run is completed, chromatograms and UV spectra are entered into a database for further evaluation. Figures 4 to 6 show a series of chromatograms recorded from a mixture of steroid contaminants.

As early as the visual comparison of the chromatograms produced with six different stationary phases, it can be seen that columns nos. 1 and 5 (Synergi POLAR RP and XTerra RP18, respectively) can be considered capable of separating the steroid mixture more effectively than the rest. The primary component is very well separated from the secondary component. The other columns shall no longer be considered for further use in method development due to their inability to separate the mixture.

If one takes a look at the chromatograms recorded with both of the "front runners" (Figs. 5 and 6), then one can see that as expected with SynergiPOLAR RP short gradients and high temperatures have an adverse effect on the separation of the mixture. Those small peaks resulting from late elution give an impression

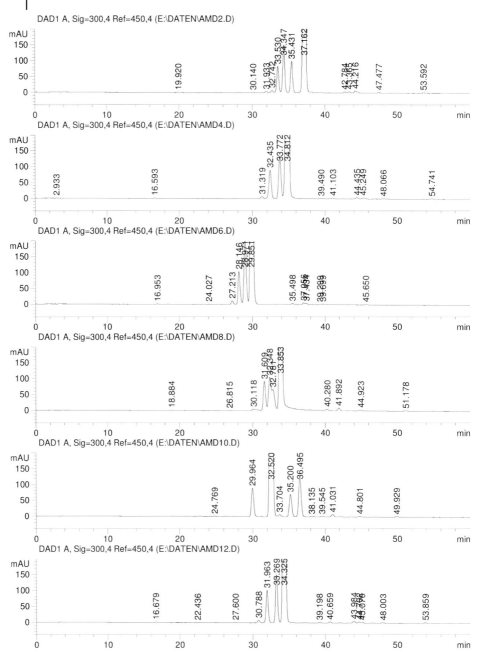

Fig. 4. Chromatograms of a mixture of steroid contaminants obtained with various stationary phases. From top to bottom: SynergiPOLAR RP 4 µm (250 × 4.6 mm), Prontosil ODS AQ 5 µm (250 × 4.0 mm), YMC J'sphere L80 4 µm (250 × 4.0 mm), Luna Phenyl Hexyl 5 µm (250 × 4.6 mm), XTerra RP18 5 µm (250 × 4.6 mm), and Symmetry 5 µm RP18 (250 × 4.6 mm).

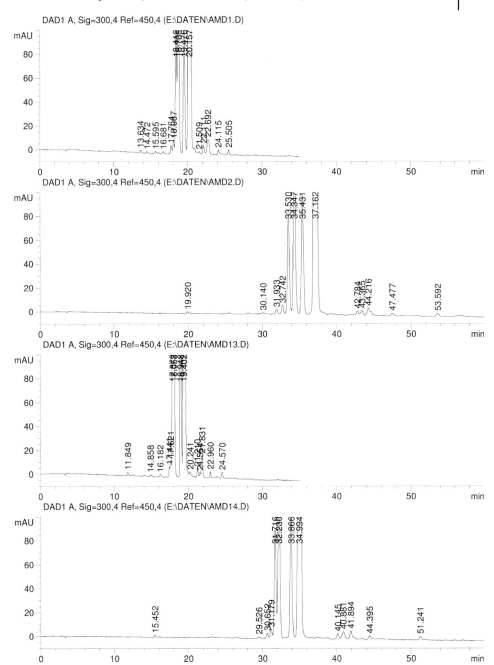

Fig. 5. Chromatograms of a mixture of steroid contaminants obtained with the reversed phase SynergiPOLAR RP. From top to bottom: column temperature: 15 °C, gradient run time: 30 min and 100 min; column temperature: 45 °C, gradient run time: 30 min and 100 min.

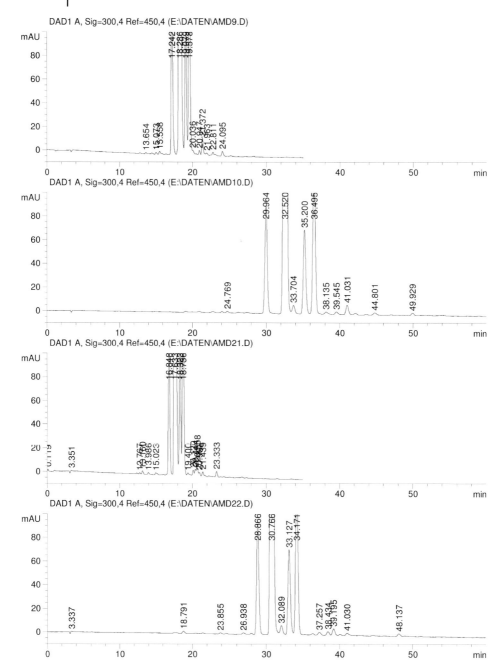

Fig. 6. Chromatograms of a mixture of steroid contaminants obtained with reversed-phase XTerra RP18. From top to bottom: column temperature: 15 °C, gradient run time: 30 min and 100 min; column temperature: 45 °C, gradient run time: 30 min and 100 min.

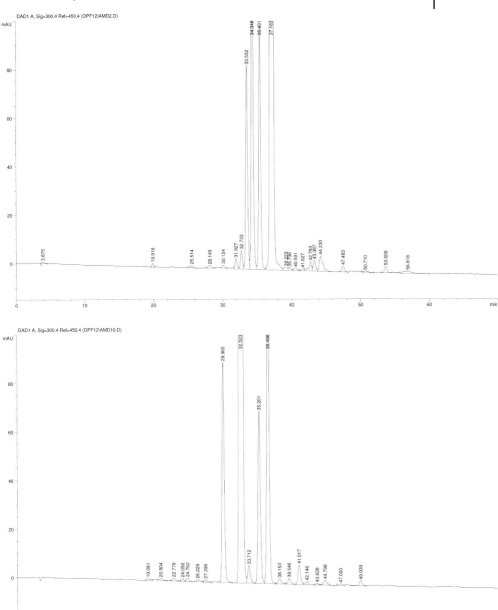

Fig. 7. Chromatograms of a mixture of steroid contaminants obtained with reversed-phase SynergiPOLAR RP (upper chromatogram) in direct comparison to XTerra RP18 (lower chromatogram).

of better separation at higher temperature. With the short gradient, they partially co-elute. Using XTerra RP18, two of the largest secondary components are only eluted after the main component. The four peaks eluted with SynergiPOLAR RP with long gradients in the range of 40 to 45 min are baseline-separated with XTerra RP18 with long gradients and at high temperature. The peak at 32.089 min (Fig. 6, lower chromatogram) is presumably separated with SynergiPOLAR RP into two peaks (t_R = 31.933 and 32.742 min; Fig. 5, second chromatogram from top). Only LC/MS analyses can offer an explanation here.

In the long runs (Fig. 7), all peaks are eluted symmetrically with both columns at low temperature. The peak pattern seen with XTerra gives a more harmonious impression, since all peaks are symmetrical and there are no signs of shoulders, etc. With SynergiPOLAR RP, however, 26 peaks are eluted as opposed to the 21 peaks seen with XTerra. With the latter, some small peaks have evidently further merged or are only slightly separated. If one wants to resolve these more precisely, because the aim of the method is to provide the complete impurity profile or full information regarding degradation products of the substance, then as the next step one would try to increase the separation capacity of the system by chromatographing with two serially arranged columns and adapting the gradient.

Testing of six stationary phases with two gradients at two temperatures showed that the problem with SynergiPOLAR RP could be solved. As expected, the peaks of the primary and secondary components could be separated satisfactorily at low column temperature and with flat gradients.

This example confirmed that with the large number of chromatograms generated with the suggested approach, some of them lead us to believe that the goal of method development, i.e. finding the "global optimum" for the relevant peaks has, at least nearly, been achieved.

4.3.4.3 Evaluation of Data with HEUREKA

HEUREKA provides some aids for visualizing the separation sequence. They can be used after the relevant peaks in all chromatograms have been integrated, assigned, and designated. The minimum values are then calculated for the resolution of each chromatogram. A maximum of two dimensions and/or factors can be selected and their effect on the quality of separation visualized.

Column no. 1 (SynergiPOLAR RP) and column no. 5 (XTerra RP18) come to the fore when compared with the others (Fig. 8). The dependence of the separation quality on the column temperature and gradient run time for column no. 1 is shown in Fig. 9. Figure 10 presents an example of a stationary phase on which the separation does not improve by varying the two parameters of column temperature and gradient run time.

Finally, HEUREKA can provide a list (Fig. 11) of the calculated optimal separation results. By clicking on a line, HEUREKA produces a method file for the HPLC system against which the optimal separation that was found can be checked for correctness in physical reality.

In this case, one forwent the realization of the method since the conditions found were almost identical to those used during the systematic scouting.

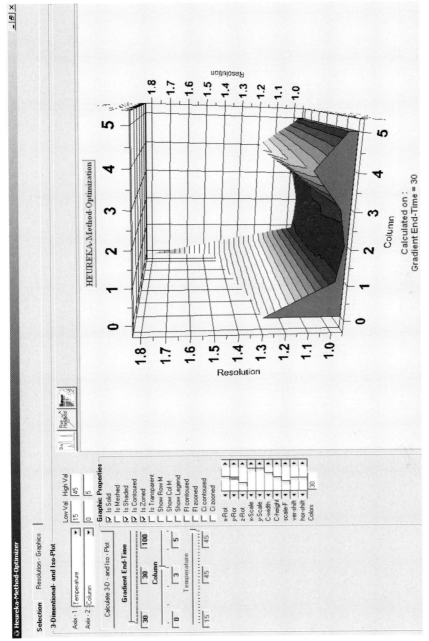

Fig. 8. Representation of the resolution in dependence on the stationary phases.
x-axis: columns nos. 1 to 6 (for mathematical reasons, from no. 0 to 5);
y-axis: column temperature;
z-axis: resolution as a measure of the quality of the separation.

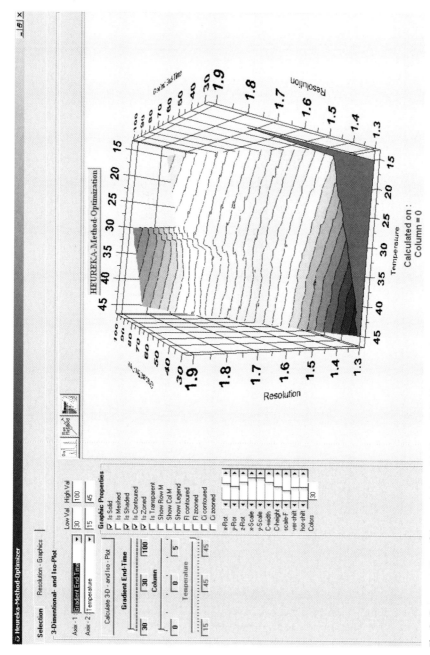

Fig. 9. Representation of the resolution with SynergiPOLAR RP as a function of column temperature and gradient run time.

x-axis: column temperature;

y-axis: gradient run time;

z-axis: resolution as a measure for the quality of separation.

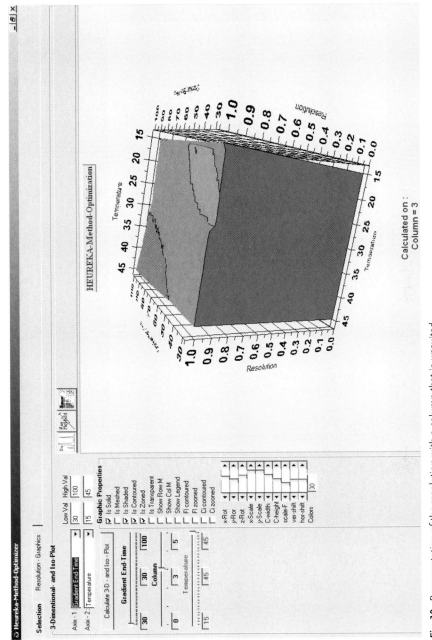

Fig. 10. Representation of the resolution with a column that is unsuited to solving the separation problem.

x-axis: column temperature;

y-axis: gradient run time;

z-axis: resolution as a measure of the quality of separation.

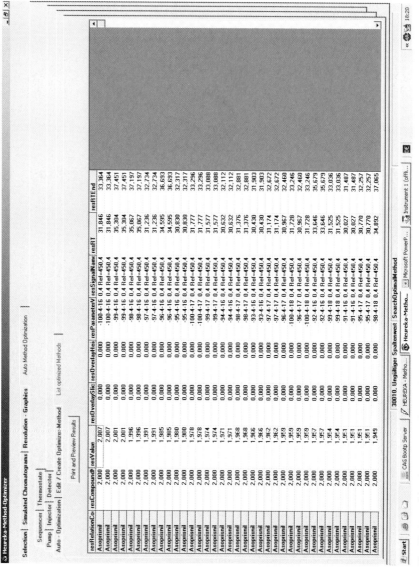

Fig. 11. List of the chromatographic conditions that suggest optimal separation success (best list).

4.3.5
Conclusion and Outlook

It has been pointed out with the aid of an example how the factors that influence chromatographic separation can be systematically examined. The inclusion of the stationary phase as a factor offers an insight into the complexity of the interaction of factors in chromatography. With HEUREKA, a tool has been created with which method development in HPLC can be accomplished multifactorially. The "global" optimum for a separation problem can be better found and with fewer manual tasks than previously. Method development should less frequently end in an inextricable cul-de-sac. With HEUREKA, the earlier ideas of automating method development in reversed-phase HPLC have been consistently developed further and put into practice [5].

In a future development phase, it could become reality that the data on retention times stored in the HEUREKA database are linked with the chemical structural formulae of the analytes that occur. It will then be possible to use search programs to look for actual structures or partial structures and to find the related chromatographic conditions. Using the fast-growing and complex database, a learning expert system might be attached that recognizes correlations between chemical structures and their chromatographic characteristics and that could provide predictions for solving new separation problems. Finally, thought can be given to further developing the systematic examination of chromatographic factors, as described above, and to apply the principle of statistical experimental design [1, 2] to method development. The number of chromatographic experiments would thereby be drastically reduced and the actual time-in-use of the HPLC instruments would be shortened.

References

1   M. Preu, M. Petz, Experimental Design zur Optimierung komplexer chemischer Derivatisierungsreaktionen – Prinzip und Anwendungsbeispiele, *Lebensmittelchemie* 53 (1999), 31–32.

2   M. Preu, M. Petz, D. Guyot, Experimental Design zur Optimierung einer Derivatisierungsreaktion, *GIT Labor-Fachzeitschrift* 1998, 854–857.

3   M. A. Briese, *Effiziente Methodenentwicklung in der HPLC Laborpraxis* 21 (1997), 26–32.

4   B. A. Bidlingmeyer, A. D. Broske, W. D. Snyder, Establishing a selectivity survey as part of the method development activity for an isocratic reversed-phase liquid chromatographic method, *J. Chromatogr. Sci.* 42 (2004), 184–190.

5   M. Pfeffer, H. Windt, Automatization for development of HPLC methods, *Fresenius J. Anal. Chem.* 369 (2001), 36–41.

6   U. Huber, Method development – solvent and column selection with Agilent 1100 Series valve solutions, Agilent Publication No. 5900-8347EN (2002).

7   M. Otto, A. Schirmer, U. Clausnitzer, M. Pfeffer, Systematic optimisation of high-performance liquid chromatographic separation by varying the temperature, gradient, and stationary phase, *Anal. Bioanal. Chem.* 372 (2002), 341–346.

8   K. Bürkle, J. Paschlau, Methodenentwicklung bequem und leistungsstark, *LaborPraxis* 26 (2002), 56–58.

9   M. R. Euerby, P. Petersson, Chromatographic classification and comparison of commercially available reversed-phase liquid chromatographic columns using principal component analysis, *J. Chromatogr. A* 994 (2003), 13–36.

5
User Reports

HPLC Made to Measure: A Practical Handbook for Optimization. Edited by Stavros Kromidas
Copyright © 2006 WILEY-VCH Verlag GmbH & Co. KGaA, Weinheim
ISBN: 3-527-31377-X

5.1
Nano-LC-MS/MS in Proteomics

Heike Schäfer, Christiane Lohaus, Helmut E. Meyer, and Katrin Marcus

5.1.1
Proteomics – An Introduction

The term "proteome" was first introduced in 1996 by Wilkins and Williams [1]. It was designed to denote the protein complement of a genome and indicates the quantitative expression profile of a cell, an organism, or a tissue under exactly defined conditions at a given point in time. In contrast to the temporally constant genome, the proteome is dynamic and variable depending on intracellular and extracellular parameters. Consequently, the analysis of a proteome represents an important supplementation to the genome analysis as states can be detected that are not ascertainable on the basis of the genome. It is now well established that one gene may produce a variety of protein products as a consequence of alternative splicings of the pre-mRNA and post-translational modifications. An additional introduction of post-translational modifications, e.g. sugars and phosphates, or controlled proteolysis contributes to a complex protein pattern. Thus, quantification has to be carried out at the protein level as there is only a poor correlation between mRNA and the corresponding gene product levels. Consequently, proteome analysis must incorporate analysis of protein expression patterns, protein quantification, and post-translational modifications. In a differential proteome analysis, a huge variety of information about differentially expressed proteins is obtained by comparing the proteomes of different cell, tissue or organism states, e.g. disease-specific proteins can be detected by comparison of the protein patterns of diseased and healthy tissue.

Presently, various techniques are available for an effective separation of proteins: the classical 1D- and 2D-PAGE [2, 3], other two-dimensional separation techniques such as BAC or CTAB electrophoresis [4, 5], protein-chip technology [6], and one-, two- or multi-dimensional chromatographic methods [6]. In 1D-PAGE, the proteins are separated according to their molecular weights. When analyzing complex protein mixtures, usually more than one protein is detected in one gel band and supplementary analyses should be performed, including a chromatographic peptide separation step (e.g., by reversed-phase HPLC) prior to mass spectrometry measurements. Classical 2D-PAGE is a combination of isoelectric focusing in the first dimension and SDS-PAGE in the second dimension, and provides the capability for the simultaneous resolution of 10,000 protein species. The quality of the gels depends on the characteristics of the proteins; high molecular weight and hydrophobic proteins (e.g., membrane proteins) may precipitate in the gel of the first dimension and then the transfer to the second dimension will be hampered. These "problematic" proteins are generally separated using chromatographic protein or peptide separation methods, e.g., high-performance liquid chromatography (HPLC). A differential analysis has hitherto not been possible as imaging systems for these separation methods are lacking. One of the cutting-edge developments in proteomics today is the application of protein chips. The surface of the chips varies depending on the scientific question to be addressed, e.g. antibodies are immobilized on the chips for the binding of specific proteins in the sample, or proteins can be immobilized on the chip surface to study protein–protein interactions [7, 8]. The identification of the proteins in all of the above described techniques is generally achieved by mass spectrometry. For a sensitive detection and identification of the proteins or peptides, different types of highly sensitive mass spectrometers (e.g., time-of-flight (TOF), triple quadrupole, or ion trap) with various ionization modes (electrospray ionization (ESI) [9], matrix-assisted laser desorption/ionization (MALDI) [10]) are used today. ESI-MS is well-suited for on-line coupling with liquid chromatography [11] since the analyte molecules are available in solution. MALDI systems can also be combined with HPLC separations; the coupling in this case has to be done off-line. By introducing additional pre-concentration columns prior to the actual peptide separation, the sensitivity of the whole system, and hence the detection and identification rate, can be increased substantially. The combination of nano-LC and MS is described in this chapter, including on-line coupling with an ESI ion-trap mass spectrometer as well as off-line coupling with a MALDI-TOF MS. The crucial steps to be aware of executing protein identification using LC-MS are summarized in Table 3 at the end of the section.

5.1.2
Sample Preparation for Nano-LC

Today, most differential proteomic studies are based on separation by 2D-PAGE. This technique comprises two different separation steps: first, the proteins are separated according to charge by isoelectric focusing (IEF) and in the second step – as in standard SDS-PAGE – according to their molecular weight (Fig. 1) [2, 3]. The differential gel images are analyzed with the help of suitable image analysis software tools, such as Proteomweaver (Definiens, Munich, Germany), Decider (GE Healthcare, Munich, Germany) or Progenesis (Nonlinear, Newcastle, Great Britain). After separation, the protein spots of interest are proteolytically digested. In most studies, the protease trypsin is used to cleave the amino acid chain of a protein at the C-terminal site of lysine and arginine residues. The ratio of trypsin to protein should be at most 1 : 20. Trypsin (e.g., porcine, sequencing grade modified, Promega, Madison, USA) is solubilized in digestion buffer (10 mM NH_4HCO_3) and added to the sample, which is preferably highly concentrated and in a minimal volume. Generally, 2 µL with a concentration of 0.03 µg µL^{-1} is sufficient for in-gel protein digestion. The sample is digested at 37 °C for at least 6 h. Tryptic peptides are extracted from the gel using HPLC-compatible solvents such as 5% formic acid (FA) or 5% trifluoroacetic acid (TFA) in doubly-distilled H_2O and used for further analyses [12]. Often, acetonitrile (MeCN) is added to the solvents to improve the peptide extraction. In this procedure, the MeCN has to be

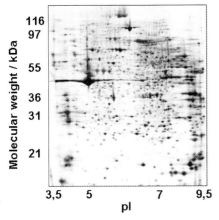

Fig. 1. 2D-PAGE. 180 µg of platelet proteins is separated. For isoelectric focusing, immobilized pH gradient strips (pH 3–10) have been used. The separation in the second dimension has been achieved using a Tris-glycine gel with an acrylamide concentration of 11%. The separation of proteins with molecular weights between 10 and 150 kDa is possible.

removed after extraction to ensure efficient peptide binding on the column and to avoid direct elution of the peptides.

5.1.3
Nano-LC

The attractiveness of HPLC in proteomics increased with the development of micro- and nano-LC methods (columns with inner diameters between 180 and 75 µm). By using miniaturized HPLC systems, the sensitivity of the analysis was drastically improved. Thus, femtomole amounts of proteins/peptides are sufficient for a successful identification of electrophoretically separated and proteolytically digested proteins. Additionally, chemically inert HPLC systems avoid non-specific, irreversible peptide binding and sample loss. Moreover, the development of adequate pumps and systems has enabled the handling of low flow rates for nanoHPLC (200 nL min^{-1}). Here, two switching valves can be combined: the peptide sample can be first pre-concentrated and desalted on a short column with a wider inner diameter (5 mm × 0.3 mm i.d.) at a higher flow rate (30 µL min^{-1}). After pre-concentration, the pre-column is directly switched in line to the separation column and the peptides are separated by RP-HPLC using an RP C$_{18}$ column with an inner diameter of 75 µm (flow rate 200 nL min^{-1}). With this set-up, volumes of several microliters can be injected into the system within minutes and peptide samples can be detected and analyzed without a loss of sensitivity or resolution. A direct injection of the samples onto the nano-LC separation column is not possible as several hours would be necessary for sample transfer to the column head due to the low flow rates. The system described above can be completed by a second pre-column. Thus during peptide separation from the first precolumn with a low flow rate, the other precolumn is rinsed with 50% MeCN/0.1% TFA for 20 min and 84% ACN/0.1% TFA for 10 min. Subsequently columns are equilibrated with 0.1% TFA for 20 min after each run (Fig. 2). Using this analytical strategy a considerable reduction of memory effects is obtained allowing the reliable identification of proteins in complex mixtures, especially when separating high hydrophobic peptides [13]. The HPLC system consists of the following components:

- an autosampler with integrated sample loop (e.g., Famos™, LC Packings Dionex, Idstein, Germany), to allow round-the-clock analyses;
- a component including switching valves and a loading pump for concentration of the samples prior to their separation (e.g., Switchos™, LC Packings Dionex);
- an HPLC pump with integrated splitting device for minimization of the flow for nanoHPLC and a UV detector with a Z-cell (with an optical path length of 1 cm and volume 3 nL) to control the HPLC system (e.g., Ultimate™, LC Packings Dionex).

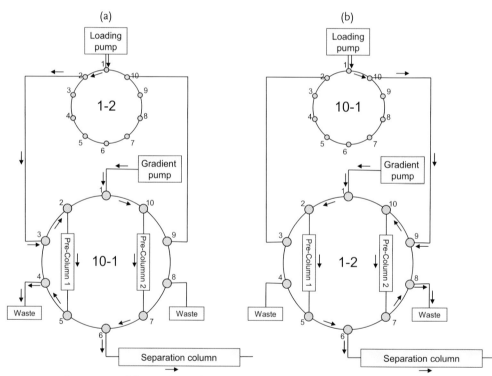

Fig. 2. Set-up of the switching valves including two pre-columns, two ten-port switching valves (two-position valves: in position 1-2 port 1/2, 3/4, 5/6, 7/8 and 9/10 are connected, in position 10/1 port 2/3, 4/5, 6/7, 8/9 and 10/1 are connected) and one separation column. Peptides are pre-concentrated on the right pre-column for 6 min. (a) The right pre-column (connection 10-7) is placed into the separation system, while the left pre-column (connection 2-5) is washed with organic solvents and equilibrated. (b) After one separation cycle the whole system is switched.

In RP-HPLC, peptide separation is accomplished by gradient elution with increasing amounts of an organic solvent – preferably acetonitrile (MeCN). An additional important component of the solvent system is the ion-pairing reagent (typically TFA or FA), which ensures more effective peptide separation. In the majority of cases, a binary system is used consisting of, e.g., solvent A 0.1% FA in doubly distilled H_2O and solvent B 84% MeCN and 0.1% FA in doubly distilled H_2O [14]. The sample is transferred to the pre-column and pre-concentrated using 0.1% TFA in doubly distilled H_2O as the binding efficiency is much higher than with FA. In contrast, during peptide separation FA is used instead of TFA in the gradient system, because the suppression of ionization in mass spectrometry is much lower [15]. Depending on the separation problem, various elution programs are used differing, e.g., in duration or gradient course: the more complex the sample the more flat the gradient needed to achieve optimal separation. Besides the intrinsic gradient, steps for washing (high amount of organic solvent, MeCN) and equilibration (usually corresponding to the starting conditions) of the column are included in the elution program. Table 1 and Fig. 3 show a typical 1 h elution

Table 1. Summary of a 1 hour gradient.

Time [min]	%A	%B
0	95	5
35	50	50
40	5	95
45	5	95
46	95	5
64	95	5

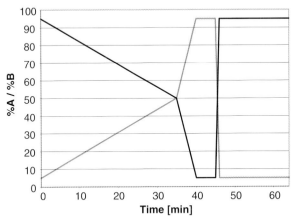

Fig. 3. The gradient program. The gradient starts with 5% B. Most of the proteins are eluted between 5% and 50% of solvent B; therefore, the flow rate of B is slowly increased in this range (1,3% B/min). After this step, the proportion of solvent B is increased to 95% within 5 min and is maintained for 5 min to ensure elution of highly hydrophobic peptides and efficient removal of contaminants from the system. Then the column is equilibrated with 5% B for 18 min.

profile for the separation of up to ten protein components, whereas the elution is performed in 30 min. When performing nano-HPLC, one important point is the prevention of dead volumes in the systems. Even small dead volumes of the order of 1 μL at a flow rate of 200 nL min^{-1} act as "mixing chambers", causing a decrease in sensitivity and separation efficiency: important information about the composition of the analyzed sample will be lost. With the simple procedure described below, the system can be tested for void volumes:

- The separation column is equilibrated using the starting conditions of the elution program until a base line is reached.

- The pre-column, previously flushed with 0.1% TFA, is switched to the separation system run with FA. Due to the different UV absorptions of TFA and FA, a distinct decrease in absorption is observed on changing the pre-column system from TFA to FA. In the case of a system free from void volumes, the UV absorp-

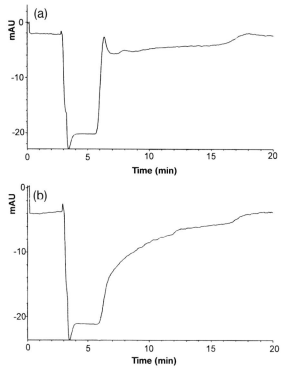

Fig. 4. Detection of void volumes. The chromatograms show the first 20 min after starting the separation using the described gradient.
(a) No void volume inside the system is detected. The UV absorption is reduced within a short time frame to a constant value and remains there until the TFA plug has passed the Z-cell. Subsequently, the UV absorption increases rapidly to the original value.
(b) Void volumes in the pre-column or the connections between the individual HPLC components are observed. The UV absorption slowly increases to the original value before the TFA plug has passed the Z-cell.

tion is reduced to a constant value within a short timeframe and remains there until the TFA plug has passed the pre-column, the separation column, and the Z-cell. Subsequently, the UV absorption increases rapidly to the original value (Fig. 4a) [15]. Void volumes present in the pre-column or the connections between the individual HPLC components can be identified from the appearance of a slowly ascending base line after the TFA plug has passed the Z-cell. Connections containing void volumes cause a mixing of the two solvent systems, so that the original absorption of the FA solvent is reached much more slowly (Fig. 4b).

- The pressure of the separation column is an additional parameter for checking the system: it decreases after connection of the pre-column to the whole separation system. It should remain reproducible over all analyses and a pressure difference of 5 bar should not be exceeded. If this difference is exceeded, it is evidence for the presence of void volumes in the system. The columns and connections have to be thoroughly tested and, if necessary, changed to remove the dead volumes and to ensure the highest sensitivity and resolution.

5.1.4
On-Line LC-ESI-MS/MS Coupling

When the HPLC is coupled on-line to an ESI mass spectrometer [11], the separated, solubilized analyte is directly transferred into the ion source of the mass spectrometer, ionized, and the resulting singly-, doubly-, and triply-charged ions are detected in the mass analyzer. A nanospray ion source is used to avoid sample loss as it is compatible with the small flow rates and sample amounts handled by the nanoHPLC. The ion source (e.g., PicoView™ 100, New Objective Inc., Woburn, USA) consists of:

- a voltage connecting device,
- a movable XYZ stage for exact positioning of the nanoESI tip,
- a nanoESI tip (e.g., PicoTips™, New Objective Inc.): a metal-coated glass capillary for transfer of the applied voltage to the solubilized analyte.

In the presence of a strong electric field, a fine spray of small, highly charged droplets is created at the tip and these are transferred to the ESI mass spectrometer (e.g., LCQdeca XP, ThermoElectron, San Jose, USA). Peptides are evaporated completely in a heated capillary and the resulting ions are focused into the entrance to the end-cap electrode of the ion trap. The peptide ions are kept in the trap either until a certain ion density is reached (2×10^7 ions), or for a defined period of time (200 ms), if the ion density was not reached before. Subsequently, the peptide ions are ejected from the trap to the detector, a secondary electron multiplier, according to their mass-to-charge ratio (m/z). For the analyses, initially one spectrum is produced over the whole mass range (full mass spectrum) followed by, e.g., three tandem mass spectra (MS/MS) of the three most intensive ions (Fig. 5). These ions have to exceed a specific intensity

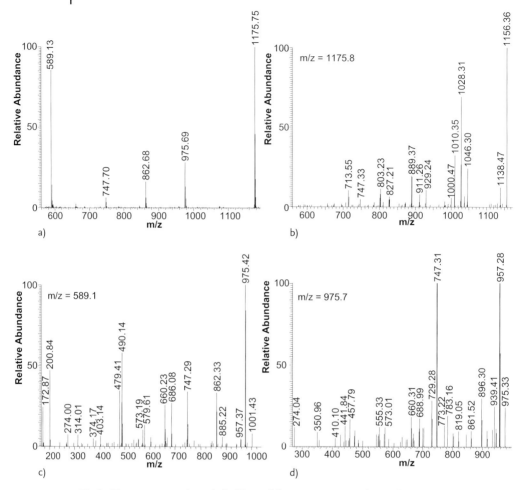

Fig. 5. Mass spectrometric analysis. First, a full mass spectrum is detected (a). The three ions with the most intense signals (m/z = 1175.8 Da, m/z = 589.1 Da, and m/z = 975.7 Da) are selected for fragmentation analyses and are successively fragmented in the following process (b–d).

value (e.g. 3×10^5) to be chosen for fragmentation. With the integrated feature "dynamic exclusion", already fragmented peptides are excluded from MS/MS selection during the next minutes. In this way, many even less concentrated peptides are preferably detected by MS/MS analyses, providing very specific sequence data of the analyzed protein.

It is of the utmost importance to clean the mass analyzer at regular intervals using a mixture of organic solvent, acid, and water in order to remove residues from the heated capillary. Additionally, the ion focusing system has to be cleaned with an alcohol, because deposits in this system result in a decrease in sensitivity and resolution.

5.1.5
Off-Line LC-MALDI-MS/MS Coupling

5.1.5.1 Sample Fractionation

Although on-line HPLC-MS using an electrospray interface is a straightforward technique nowadays, it can be of advantage to analyze specific samples off-line. In the case of MALDI-MS, the nanoHPLC has to be coupled to the mass spectrometer off-line. After separation, the sample is transferred to a sample plate (target), e.g. AnchorChip™ target (Bruker Daltonics, Bremen, Germany), equipped with a special coating that provides improved crystallization of MALDI samples and a 10- to 100-fold sensitivity increase. Small hydrophilic positions/anchors (600 µm diameter) are applied equidistantly on the hydrophobic surface of the AnchorChip™ target thereby reducing sample spreading to only a small area. This results in co-crystallization of matrix and sample and therefore higher sample concentration focused at the distinct hydrophilic anchor points [16]. One of the most commonly used matrices is a saturated solution of α-cyano-4-hydroxy-cinnamic acid in 97% acetone/3% water containing a small amount of TFA to support the ionization of the sample molecules in the MALDI source. The matrix is directed to the AnchorChip™ target and this is then dried. The sample fractionation after HPLC can be performed automatically using special sample fractionation systems (e.g., Probot, LC Packings Dionex). The target is introduced to a special desk and can be adjusted in the three XYZ planes so that the small nano-LC volume is placed directly onto the matrix. Using this set-up, fractions of volume less than 40 nL can be collected automatically. After collection, the individual fractions are recrystallized from 0.01% TFA, 30% acetone, and 60% ethanol in water and analyzed.

5.1.5.2 MALDI-TOF-MS/MS Analyses

During the MALDI process, charged peptide ions are desorbed from the target, together with matrix molecules, by a nanosecond laser pulse, resulting in predominantly singly-charged peptide ions. Most of the laser energy is absorbed by the matrix, preventing unwanted fragmentation of the biomolecule. The ionized biomolecules are accelerated in an electric field and enter the flight tube. During the flight in this tube, molecules are separated according to their mass-to-charge ratio and reach the detector at different times. In this way, each molecule yields a distinct signal. By the incorporation of delayed extraction (DE) into MALDI-MS, a dramatic improvement of performance in terms of sensitivity, mass resolution, and mass accuracy of precursor ions up to approximately 10 kDa has been achieved [17]. The ions are detected by means of a secondary electron multiplier. Due to metastable decay of the peptide ions (PSD, post-source decay) in the field-free drift zone [18, 19], fragmentation analyses of single peptides can be performed (Fig. 6). For these PSD studies, one peptide (parent mass) is selected for fragmentation and all other peptide ions are deflected by application of an electric field. On their way in the drift tube, peptides are separated according to their mass-to-charge ratio and are transferred to the detector.

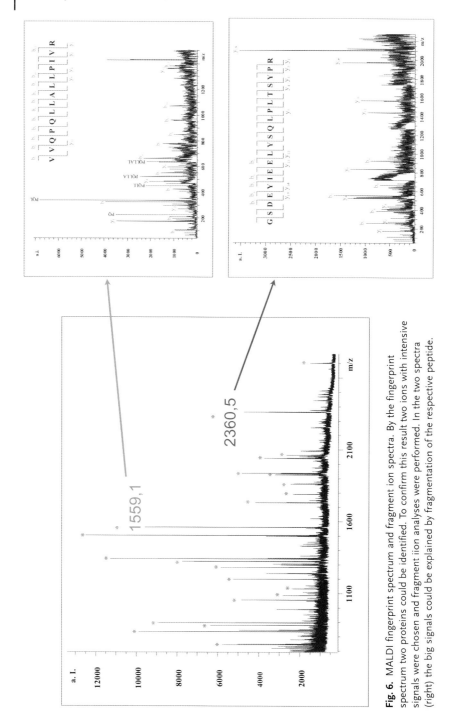

Fig. 6. MALDI fingerprint spectrum and fragment ion spectra. By the fingerprint spectrum two proteins could be identified. To confirm this result two ions with intensive signals were chosen and fragment iion analyses were performed. In the two spectra (right) the big signals could be explained by fragmentation of the respective peptide.

Because of the energy diversity the fragments show, first single spectra in the different mass ranges are detected. These fragment spectra are summarized to one full range spectrum. The new developed LIFT-technology (Bruker, Daltonic, Bremen, Germany) potentiates the acquisition of only one fragment ion spectrum over the whole mass range. For that purpose the potential energy of all fragment ions is raised in a LIFT cell directly behind the source. The difference of the kinetic energy of the precursor ion and the smallest fragment ion does not exceed 30% [20].

5.1.6
Data Analysis

The amino acid sequence differs for each individual protein. Therefore, protein digestion using a specific protease results in a specific, distinct peptide pattern – the peptide mass fingerprint (PMF) – which is individual for every single protein. The PMF data thus obtained are evaluated automatically using search algorithms that compare measured data with theoretically estimated data from protein and DNA databases. The data evaluation is performed using different search programs, which are freely accessible on the Internet, such as

ProFound (prowl.rockefeller.edu/profound_bin/WebProFound.exe) [21], Mascot (http://www.matrixscience.com/search_form_select.html) [22], or MSFit (http://prospector.ucsf.edu/) [23].

Often, protein analysis based on the PMF has to be extended to fragmentation analyses, for example, in cases where the sample contains more than one protein an assignment and unequivocal identification of all protein components in parallel is problematic using MALDI-MS. Therefore, it is essential to analyze individual peptides by fragmentation experiments based on MALDI-PSD, as the fragmentation of peptides generally provides more detailed sequence information. These fragmentation data are also analyzed by means of special database search programs such as SEQUEST™ [24, 25] or Mascot [22], which compare theoretically produced MS/MS fragmentation patterns with the experimentally obtained data.

5.1.7
Application in Practice: Analysis of α-Crystallin A in Mice Lenses

In this final section, we highlight an application of the above-described methods. The eye lens contains many different proteins, 90% of which are crystallins. The amino acid sequence of the crystallins is largely constant; in the case of α-crystallin A only 3% of amino acids changes during evolution. Due to the dense protein formation in the eye lens, the crystallins exhibit a specific structure with high stability and a capacity for protein–protein interactions. The aim of the described analysis was to obtain preferably complete sequence coverage of α-crystallin A [12]. For this purpose, the whole protein extract from mice eye lenses was separated by means of 2D-PAGE [26]. Detected spots considered to be α-crystallin A were cut from the gel and a tryptic digest was performed. The resulting

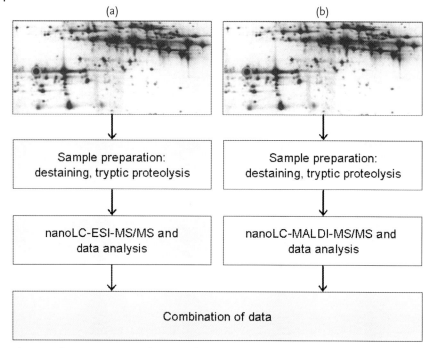

<table>
<tr><td>(a)</td><td>(b)</td></tr>
</table>

| Sample preparation: destaining, tryptic proteolysis | Sample preparation: destaining, tryptic proteolysis |

| nanoLC-ESI-MS/MS and data analysis | nanoLC-MALDI-MS/MS and data analysis |

| Combination of data |

Fig. 7. Application in practice. The spots of both gels are cut out after 2D-PAGE and Coomassie G-250 staining. The protein spots of one gel are analyzed by on-line nano-LC-ESI-MS/MS after tryptic digestion (a), whereas protein spots of the other gel are analyzed using off-line nano-LC and MALDI-TOF-MS/MS analysis (b).

Table 2. Comparison of the identified α-crystallin A peptides using nano-LC-ESI-MS/MS or nano-LC-MALDI-MS/MS, respectively. The peptides exclusively identified by nano-LC-ESI-MS/MS are marked red, those detected by nano-LC-MALDI-MS/MS are labeled blue. The number of identified peptides is explicitly enlarged by a combination of both methods (marked purple).

nano-LC-ESI-MS/MS	nano-LC-MALDI-MS/MS
MDVTIQHPWF KRALGPFYPS RLFDQFFGEG LFEYDLLPFL SSTISPYYRQ SLFRTVLDSG ISEVRSDRDK FVIFLDVKHF SPEDLTVKVL EDFVEIHGKH NERQDDHGYI SREFHRRYRL PSNVDQSALS CSLSADGMLT FSGPKVQSGL DAGHSERAIP VSREEKPSSA PSS 58%	MDVTIQHPWF KRALGPFYPS RLFDQFFGEG LFEYDLLPFL SSTISPYYRQ SLFRTVLDSG ISEVRSDRDK FVIFLDVKHF SPEDLTVKVL EDFVEIHGKH NERQDDHGYI SREFHRRYRL PSNVDQSALS CSLSADGMLT FSGPKVQSGL DAGHSERAIP VSREEKPSSA PSS 54%

MDVTIQHPWF KRALGPFYPS RLFDQFFGEG LFEYDLLPFL SSTISPYYRQ SLFRTVLDSG
ISEVRSDRDK FVIFLDVKHF SPEDLTVKVL EDFVEIHGKH NERQDDHGYI SREFHRRYRL
PSNVDQSALS CSLSADGMLT FSGPKVQSGL DAGHSERAIP VSREEKPSSA PSS
70%

peptides were separated by nano-LC and subsequently either transferred to an
ESI ion-trap MS on-line or automatically spotted onto a MALDI AnchorChip™
target off-line. Fragmentation analyses were performed as described above.
Database searches were performed with the SEQUEST™ algorithm. Two α-
crystallin A spots with the same position in two independent separations were
analyzed by both techniques in parallel (Fig. 7). Table 2 shows the respective
sequence coverages detected by nano-LC-ESI-MS/MS and nano-LC-MALDI-MS
analyses. The results clearly demonstrate the complementary nature of the two

Table 3. Crucial steps in proteomics. The most important points and methods
to be aware of are summarized in this table.

Analysis Steps	Critical Point	Solution
General	Contaminations	Usage of preferably **pure solvents**; do not handle the samples without gloves, to prevent sample contamination (e.g., by keratin).
Sample preparation	Consistent sample quality	Preparation and application of **standardized protocols**, to avoid variances within different samples and to ensure reproducible results.
2D-PAGE	Reproducibility	Preparation of **standardized protocols**, to avoid variances within the different separations.
HPLC preparation	Handling of protein spots	**MS-compatible staining; destaining of spots and pH adjustment** prior to digestion.
	Sample dilution	**Addition of minor volumes** of protease and extraction solution to avoid needless sample dilution.
	Unspecific adsorption	Usage of **borosilicate glass** vials for sample preparation; utilization of **bio-inert systems** and spare parts to prevent unspecific adsorption.
HPLC	Sensitivity	**Prevention of void volumes** and adjustment of column parameters and gradient systems to ensure optimal sensitivity.
ESI-MS	Sensitivity, Resolution	**Control of device parameters at regular intervals** by analysis of peptide standards to guarantee optimal detection conditions.
MALDI-MS	Preparation	**Optimization** of the sample concentration and the target preparation to obtain optimal results.
	Sensitivity, Resolution	**Control of device parameters at regular intervals** by analysis of peptide standards to ensure optimal detection conditions.
Data analysis	Software	**Database choice** allowing for attention to sample- and device-specific parameters.
	Duration of data-base searches	**Choice of a proper computer system** appropriate to the generated data sets.

methods. In the MALDI-TOF-MS/MS analyses, an improved detection of singly-charged ions and therefore of mainly short peptides was observed. Furthermore, MALDI-TOF-MS provides additional sequence information through the generation of immonium ions that are characteristic of a particular amino acid. These immonium ions yield supplementary information for the database searches that facilitate the peptide/protein identification. In the described example, more highly charged ions (doubly or triply), and thus peptides with a molecular weight > 3000 Da, were not detected by MALDI-MS. These larger peptides, in turn, could be identified using ESI-MS/MS as a result of the generation of multiply-charged ions. The demonstrated results were confirmed by analyses of further α-crystallin A spots using both techniques (results not shown) [12].

One highly important advantage of "on-line" LC-ESI-MS/MS is the automated production of a huge amount of fragmentation data during one run, providing very specific sequence data for the identification of the analyzed protein. In contrast, off-line LC-MALDI-MS analyses are very time-consuming as single fractions first have to be spotted onto the target. This can, in fact, be a big advantage, as the samples can be stored for up to several days prior to further analyses and complementation of the results. This is, of course, not possible in the case of on-line measurements.

To sum up, nano-LC-ESI-MS/MS and -MALDI-MS prove to be complementary, depending on the actual problem. Comprehensive mass spectrometric analysis is ensured by the application of a combination of both methods, resulting in satisfactory information about the proteins in a given sample.

References

1 M. Wilkins, J. C. Sanchez, A. A. Gooley, R. D. Appel, I. Humphrey-Smith, D. F. Hochstrasser, K. L. Williams, Progress with proteome projects: why all proteins expressed by a genome should be identified and how to do it, *Biotechnol. Genet. Eng. Rev.* 13 (1996), 19–50.

2 A. Görg, W. Postel, S. Günther, J. Weser, Improved horizontal two-dimensional electrophoresis with hybrid isoelectric focusing in immobilized pH gradients in the first dimension and laying-on transfer to the second dimension, *Electrophoresis* 9 (1988), 531–546.

3 J. Klose, Protein mapping by combined isoelectric focusing and electrophoresis of mouse tissues. A novel approach to testing for induced point mutation in mammals, *Humangenetik* 26 (1975), 231–243.

4 J. Hartinger, K. Stenius, D. Hogemann, R. Jahn, 16-BAC/SDS-PAGE: a two-dimensional gel electrophoresis system suitable for the separation of integral membrane proteins, *Anal. Biochem.* 240 (1996), 126–133.

5 C. Navarre, H. Degand, K. L. Bennett, J. S. Crawford, E. Mortz, M. Boutry, Subproteomics: identification of plasma membrane proteins from the yeast Saccharomyces cerevisiae, *Proteomics* 12 (2002), 1706–1714.

6 P. G. Righetti, A. Castagna, F. Antonucci, C. Piubelli, D. Cecconi, N. Campostrini, G. Zanusso, S. Monaco, The Proteome: *Anno Domini* 2002, *Clin. Chem. Lab. Med.* 41(4) (2003), 425–438.

7 A. M. Lueking, H. Horn, H. Eickhoff, H. Lehrach, G. Walter, Protein microarrays for gene expression and antibody screening, *Anal. Biochem.* 270 (1999), 103–111.

8 C. A. K. Borrebaeck, S. Ekstrom, A. C. Molmborg Hager, J. Nilson,

T. Laurell, G. Marko-Varga, Protein chips based on recombinant antibody fragments: a highly sensitive approach as detected by mass spectrometry, *BioTechniques* 30 (2001), 1126–1132.

9 J. B. Fenn, Electrospray ionization for mass spectrometry of large biomolecules, *Science* 246 (1989), 64–71.

10 M. Karas, M. Gluckmann, J. Schafer, Ionization in matrix-assisted laser desorption/ionization: singly-charged molecular ions are the lucky survivors, *J. Mass Spectrom.* 35 (2000), 1–12.

11 M. Linscheid, Die Kopplung von Flüssigkeitschromatographie mit Massenspektrometrie, *CLB Chemie in Labor und Biotechnik* 41(3) (1990), 125–133.

12 H. Schaefer, K. Marcus, A. Sickmann, M. Herrmann, J. Klose, H. E. Meyer, Identification of phosphorylation and acetylation sites in αA-crystallin of the eye lens (*mus musculus*) after two-dimensional gel electrophoresis, *Anal. Bioanal. Chem.* 376 (2003), 966–972.

13 H. Schaefer, J.-P. Chervet, C. Bunse, C. Joppich, H. E. Meyer, K. Marcus, A peptide preconcentration approach for nano-high-performance liquid chromatography to diminish memory effects, *Proteomics* 4 (2004), 2541–2544.

14 M. Serwe, M. Blüggel, H. E. Meyer, High Performance Liquid Chromatography, in Microcharacterization of Proteins (Eds.: R. Kellner, F. Lottspeich, H. E. Meyer), 2nd Edition, Wiley-VCH, Weinheim, 1999.

15 G. Mitulovic, M. Smoluch, J.-P. Chervet, I. Steinmacher, A. Kungl, K. Mechtler, An improved method for tracking and reducing the void volume in nanoHPLC-MS with micro trapping columns, *Anal. Bioanal. Chem.* 376 (2003), 946–951.

16 M. Schuerenberg, C. Luebbert, H. Eickhoff, M. Kalkum, H. Lehrach, E. Nordhoff, Prestructured MALDI-MS sample supports, *Anal. Chem.* 72 (2000), 3436–3442.

17 R. Kaufmann, P. Chaurand, D. Kirsch, B. Spengler, Post-source decay and delayed extraction in matrix-assisted laser desorption/ionization-reflectron time-of-flight mass spectrometry. Are there trade-offs?, *Rapid Commun. Mass Spectrom.* 10 (1996), 1199–1208.

18 B. Spengler, D. Kirsch, Peptide sequencing by matrix-assisted laser desorption mass spectrometry, *Rapid Commun. Mass Spectrom.* 6 (1992), 105–108.

19 R. Kaufmann, D. Kirsch, B. Spengler, Sequencing of peptides in a time-of-flight mass spectrometer: evaluation of post-source decay following matrix-assisted laser desorption/ionization (MALDI), *Int. J. Mass Spectrom. Ion. Proc.* 131 (1994), 355–385.

20 D. Suckau, A. Resemann, M. Schuerenberg, P. Hufnagel, J. Franzen, A. Holle, A novel MALDI LIFT-TOF/TOF mass spectrometer for proteomics, *Anal. Bioanl. Chem.* 376 (2003), 952–965.

21 W. Zhang, B. T. Chait, ProFound: an expert system for protein identification using mass spectrometric peptide mapping information, *Anal. Chem.* 72(11) (2000), 2482–2489.

22 D. N. Perkins, D. J. C. Pappin, D. M. Creasy, J. S. Cottrell, Probability-based protein identification by searching sequence databases using mass spectrometry data, *Electrophoresis* 20 (1999), 3551–3567.

23 K. R. Clauser, P. R. Baker, A. L. Burlingame, Role of accurate mass measurement (±10 ppm) in protein identification strategies employing MS or MS/MS and database searching, *Anal. Chem.* 71(14) (1999), 2871–2882.

24 J. K. Eng, A. L. McCormack, J. R. Yates III, An approach to correlate tandem mass spectral data of peptides with amino acid sequences in a protein database, *J. Am. Soc. Mass Spectrom.* 5 (1994), 976–989.

25 J. R. Yates III, J. K. Eng, A. L. McCormack, D. Schieltz, Method to correlate tandem mass spectra of modified peptides to amino acid sequences in the protein database, *Anal. Chem.* 67 (1995), 1426–1436.

26 P. R. Jungblut, A. Otto, J. Favor, M. Löwe, E.-C. Müller, M. Kastner, K. Sperling, J. Klose, Identification of mouse crystallins in 2D protein patterns by sequencing and mass spectrometry. Application to cataract mutants, *FEBS Letters* 435 (1998), 131–137.

5.2
Verification Methods for Robustness in RP-HPLC

Hans Bilke

5.2.1
Introduction

During the daily routine in an HPLC laboratory, problems often arise that may result from a lack of robustness of the RP-HPLC methods used.

The robustness of an RP-HPLC method can generally be established in two different ways, either during a systematic method development [1, 6, 10, 11] or by the application of a statistical experimental design [2–5, 13, 15–36].

A good approach is the systematic change of the important chromatographic parameters and the measurement of the resulting effects on the separation. This procedure applies to both ways of establishing robustness.

5.2.2
Testing Robustness in Analytical RP-HPLC by Means of
Systematic Method Development

Computer-facilitated HPLC method development using a simulation program (e.g., DryLab, ACD LC & GC Simulator) can be very useful for the investigation of the influence of the respective chromatographic parameters on the separation and consequently on the robustness of a given RP-HPLC method.

Part of a systematic method development is an investigation of the influence of the important chromatographic parameters (percent of the organic mobile phase B, temperature, pH of the aqueous mobile phase A, buffer concentration of the aqueous mobile phase A for the isocratic mode, and additionally gradient run time/gradient slope for the gradient mode) on the critical peak resolution in the chromatogram.

For a statement of robustness it is important to know which changes of the experimental conditions are acceptable. This information can be obtained during a systematic method development, without additional experimental effort, from a "map of critical resolution", which represents the critical resolution as a function of one or two chromatographic parameters.

Various simulation programs for computer-facilitated HPLC method development make such resolution maps for one or two simultaneously varied parameters [6].

Figures 1, 5, 9, 10 and 11 show examples of such resolution maps.

Figure 1 shows critical resolution as a function of gradient run time. This map of the critical resolution makes it possible to estimate the tolerance of a chromatographic working parameter (here, the gradient slope, indicated in minutes of gradient run time). The most robust working range is obtained with a gradient

HPLC Made to Measure: A Practical Handbook for Optimization. Edited by Stavros Kromidas
Copyright © 2006 WILEY-VCH Verlag GmbH & Co. KGaA, Weinheim
ISBN: 3-527-31377-X

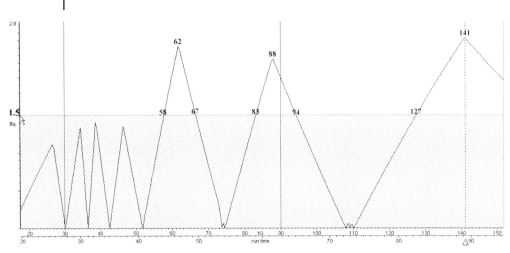

Fig. 1. One-dimensional map of critical resolution; illustration of critical resolution with respect to the gradient run time.

run time of about 141 min. However, robust ranges are also obtained with shorter gradient run times (88 min and 62 min). This means that, for example, the gradient run time can vary between 58 min and 67 min or between 83 min and 94 min, without influence on the baseline separation (Rs ≥ 1.5) or the robustness of the method.

The following chromatograms, derived from the simulation, confirm this statement (Figs. 2 to 4). The robustness of the method at various gradient run times is exclusively determined by a selective separation of the peaks 13 to 16.

Figure 5 shows the critical resolution as a function of the temperature.

A robust working range is obtained in the temperature interval between 57 °C and 69 °C. These variations of temperature can be accepted without influence on the baseline separation (Rs ≥ 1.5) or robustness of the method (Figs. 6 to 8).

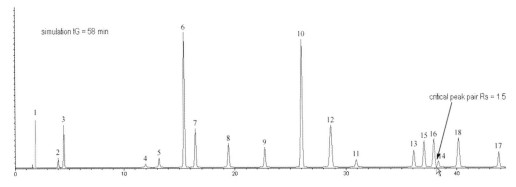

Fig. 2. Chromatographic separation with a gradient run time t_G of 58 min.

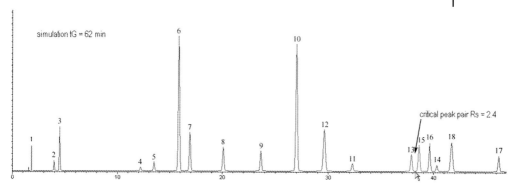

Fig. 3. Chromatographic separation with a gradient run time t_G of 62 min.

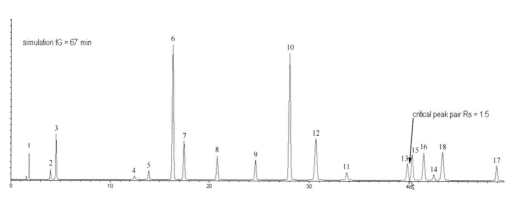

Fig. 4. Chromatographic separation with a gradient run time t_G of 67 min.

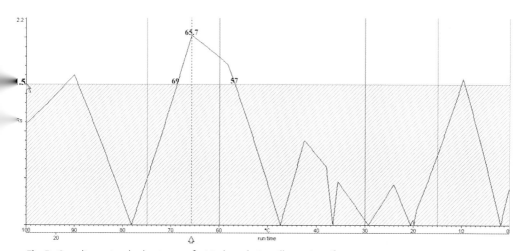

Fig. 5. One-dimensional robust map of critical resolution; illustration of critical resolution with respect to the temperature.

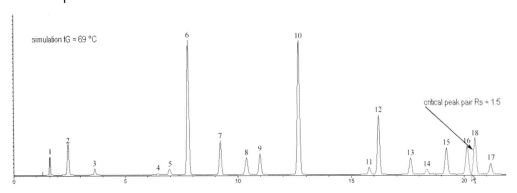

Fig. 6. Chromatographic separation with a column temperature of 69 °C.

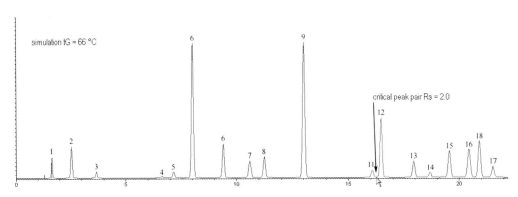

Fig. 7. Chromatographic separation with a column temperature of 66 °C.

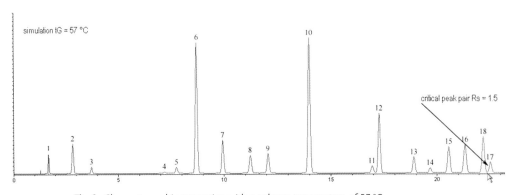

Fig. 8. Chromatographic separation with a column temperature of 57 °C.

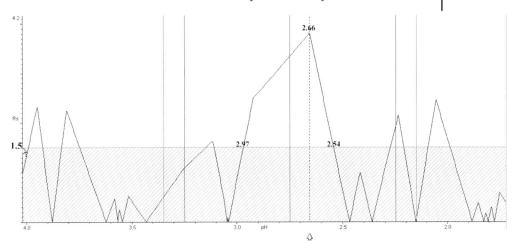

Fig. 9. One-dimensional robust map of critical resolution; illustration of critical resolution with respect to the pH of the aqueous mobile phase A.

When changing the parameter "temperature", the robustness of the method is again determined by the selective separation of the pairs of peaks 11/12, 16/18, and 18/17 at the end of the chromatogram.

With charged compounds (weak acids or bases), the adjustment of the pH of the aqueous mobile phase has the strongest influence on the change in selectivity of a chromatographic separation. This is due to the fact that organic acids, basic compounds, and hybrid ionic compounds (zwitterions) show very different behavior in response to pH variations of the eluent [7–9].

A statement about the robustness of the HPLC method for the parameter "pH value" is given by the map of critical resolution in Fig. 9.

Figure 9 shows critical resolution as a function of the pH value. The most robust working range for the chromatographic parameter "pH of the aqueous mobile phase" is obtained at pH 2.5–3.0. The variation range of the pH value thus includes ±0.5 pH units. Within this range, a baseline separation (Rs ≥ 1.5) for the most critical peak pair is obtained even in the most unfavorable case.

Other parameters, such as ionic strength, concentration of the ion pairing reagent, etc., can be described easily and transparently, and can take any value within the allowed variation range. A precondition for this is that none of the checked parameters influence one another.

The determination of robustness criteria is facilitated by maps of critical resolution for two simultaneously varied parameters (Figs. 10 and 11). Such resolution maps can be derived from a systematic method development and can be represented without additional experiments using appropriate simulation software packages.

Figure 10 shows a two-dimensional resolution map of critical resolution with respect to the gradient slope t_G and pH of the aqueous mobile phase A.

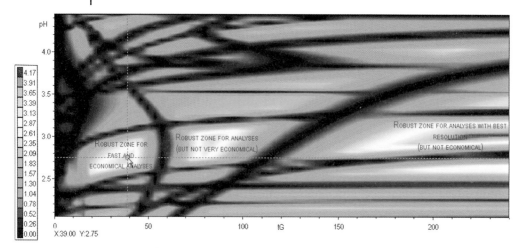

Fig. 10. Two-dimensional map of critical resolution; illustration of critical resolution with respect to the gradient slope t_G and the pH of the aqueous mobile phase A.

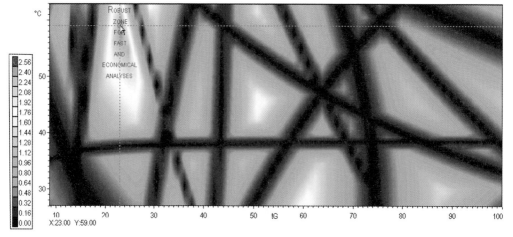

Fig. 11. Two-dimensional resolution map of critical resolution; illustration of critical resolution with respect to the gradient slope t_G and temperature.

The values of the simultaneously varied chromatographic parameters gradient run time t_G and pH are plotted on the respective axes. The color values of the resolution map describe the critical resolution. The best resolution (Rs = 4.1) is obtained with the uneconomical gradient run time of 230 min and at a pH of 3.0.

Significantly more economical analyses with similarly robust working ranges (Rs ≥ 3.0) are to be expected with gradient run times of 83 min and 39 min at lower pH values (2.95 and 2.75).

Figure 11 depicts a two-dimensional resolution map of critical resolution with respect to the gradient slope t_G and temperature.

The values of the simultaneously varied chromatographic parameters gradient run time t_G and temperature are plotted on the respective axes. The color values of the resolution scale describe the critical resolution. The best robustness is obtained with a gradient run time of 23 min and a temperature of 59 °C.

The illustrated resolution maps permit immediate statements to be made with regard to both maximum resolution and robust working range and offer the HPLC user the possibility of recognizing the critical chromatographic parameters. Exact knowledge of the peak movement also facilitates the adaptation to the RP-HPLC method with determined and/or casual changes of the experimental parameters and points out the limits of the permitted changes.

Jupille [10] for isocratic methods and Molnar and Rieger [11] for gradient methods have set out simple calculations by means of Excel macros enabling quantitative statements to be made about the robustness. They focused on effects arising from changes in the chemistry of the separation.

Jupille defined R_b as a robustness parameter for an isocratic function:

$$R_b = \frac{R_{s(test)}}{R_{s(ref)}}$$

$R_{s(ref)}$ is the critical resolution of the separation under "reference" conditions (e.g., chromatographic conditions of the available RP-HPLC method)

$R_{s(test)}$ is the lower value of the two critical resolutions, obtained when one of the chromatographic variables is changed by "one unit"

Only seven experimental runs are necessary for an isocratic procedure in order to obtain all of the required data to determine simulation models (DryLab models) for the four most important chromatographic parameters:

- reference run,
- % B (1 additional run),
- temperature (1 additional run),
- pH value (2 additional runs),
- buffer concentration (2 additional runs).

Figure 12 illustrates an Excel spreadsheet for the calculation of the robustness under consideration of the following one-dimensional simulation models for the four most important chromatographic parameters:

- LC RP isocratic % B (2 runs),
- LC RP isocratic temperature (2 runs),
- LC RP isocratic pH value (3 runs),
- LC RP isocratic ionic strength (3 runs).

With the condition that all parameters (physical parameters such as particle size, column dimension, flow rate) which are not changed in the model must be identical, the method robustness is evaluated for each chromatographic parameter, both individually or overall.

	A	B	C	D	E	F	G	H	I	J
1	→ Robustness Evaluation Spreadsheet									
2										
3		Project Name:		DryLab Training Course						
4		Date:	28.11.2003							
5		Chemist:		Tom Jupille						
6		Comments:								
7										
8										
9		1.Enter the method parameters				2. Enter the "unit changes" in variables				
10										
11			Variable	Value	Units		Unit Change	Units		
12			%B =	55%			1%	(absolute)		
13			T =	35	deg C		3	deg C		
14			pH =	2.7	units		0.2	units		
15			[buffer] =	25	mM		5%	(relative)		
16										
17										
18		Start	3. Start DryLab							
19										
20										
21		4. Load the Files								
22										
23			Variable	Rb	File Name					
24		Load	%B:	94%	C:\Documents and Settings\Tom Jupille\Desktop\advanced drylab course\Automation Toolkit examples\robustness\robustness example %B.DLB					
25		Load	T:	95%	C:\Documents and Settings\Tom Jupille\Desktop\advanced drylab course\Automation Toolkit examples\robustness\robustness example T.DLB					
26		Load	pH:	98%	C:\Documents and Settings\Tom Jupille\Desktop\advanced drylab course\Automation Toolkit examples\robustness\robustness example pH.DLB					
27		Load	[buffer]:	96%	C:\Documents and Settings\Tom Jupille\Desktop\advanced drylab course\Automation Toolkit examples\robustness\robustness example buffer.DLB					
28										
29			Overall:	84%						
30										
31										
32		Close	5. Close DryLab							
33										

Fig. 12. Excel spreadsheet for the quantification of the method robustness according to Jupille.

The percentage values which show the maximum decrease of the robustness within the specified changes of parameters are listed for the robustness parameter R_b as a result of the fully automatic simulation and calculation of the resolution values. The changes in the resolution are determined with respect to one parameter while the value of the other parameter is kept constant, in order to obtain a quantitative statement from two-dimensional resolution maps.

According to Molnar and Rieger, the necessary simulation and calculation can be carried out with DryLab software by means of Excel spreadsheets as shown in Figs. 13 and 14. an evaluation enlarged fundamentally is given with now available Peak Match software. The results of the calculation of robustness of the critical resolution with respect to gradient run time t_G and pH of the aqueous mobile phase A are illustrated in Fig. 13.

On varying the pH by ±0.1, the resolution decreases to 76.5% of its maximum value. Varying the gradient time by ±1 min results in a substantially smaller reduction of the resolution. For this, a value of 96.9% is calculated. The small changes of the resolution Rs with the gradient time t_G show that the method is quite robust with regard to the gradient time; variations of the gradient time t_G

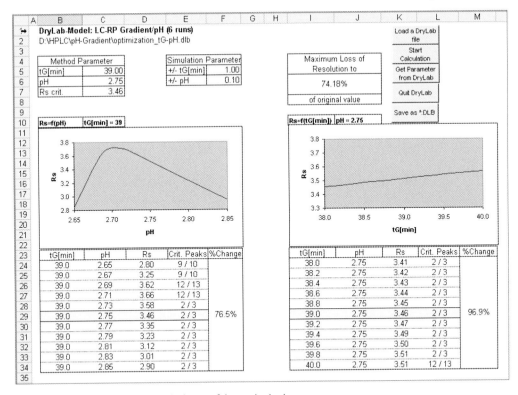

Fig. 13. Excel spreadsheet for the calculation of the method robustness carried out from a two-dimensional map of resolution.

within the indicated range do not show significant effects on the resolution Rs. The pH value of the aqueous mobile phase has a bigger influence and has to be carefully adjusted for robust analysis, a well-known issue for numerous applications with buffers.

Quantitative statements on the calculation of robustness for the critical resolution with respect to the gradient time t_G and the temperature and the corresponding graphical results are illustrated in Fig. 14.

When varying the temperature by ±3 °C, the resolution decreases to 85.9% of its maximum value. Varying the gradient time by ±1 min results in a slightly larger reduction of the resolution. A value of 80.3% is calculated. The small changes in the resolution Rs with the temperature T and gradient time t_G show that the method is quite robust with regard to the temperature and the gradient. Variations of the temperature and gradient time t_G within the indicated range do not show significant effects on the resolution Rs.

With systematic, computer-aided method development it is possible – *without additional experiments* – to find not only optimal chromatographic conditions for the best and/or an economical resolution, but one can also assess the method

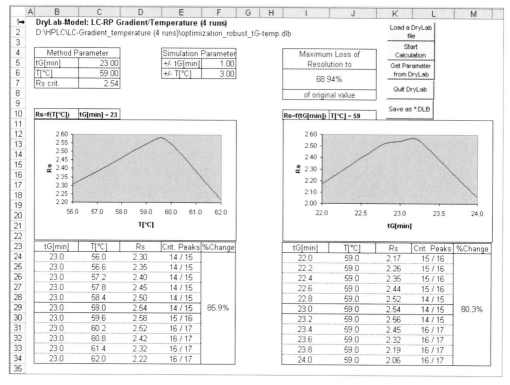

Fig. 14. Excel spreadsheet for the calculation of the method robustness from a two-dimensional map of resolution.

robustness. Jupille and Mólnar/Rieger indicated approaches for simple calculations of quantitative robustness values, both from several one-dimensional maps and from two-dimensional resolution maps.

5.2.3
Robustness Test in Analytical RP-HPLC by Means of Statistical Experimental Design (DoE)

Statistical experimental design is a strategic instrument for the optimization of processes and offers an empirical processing concept [12] for the connection between the influencing and disturbing variables in the process and the resulting target values. Here, there is pronounced distinction between target values (responses) and factors. The target value y is a function $f()$ of the factor(s) x. Their character is determined by the "functioning" of the process.

One assumes that changed factors can also change the target values during the process. Unfortunately, the connections between factors and target values can be overlaid or interfered with by measurement errors and by unknown or unidentified disturbance factors.

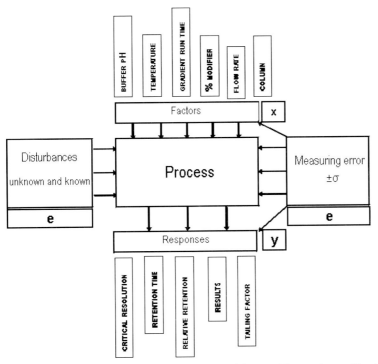

Fig. 15. Empirical process model illustrated as a black-box-model (according to Wember [12]).

Statistical experimental design without appropriate software is laborious, time-consuming, and not state-of-the-art. In the last years there has been enormous progress in the algorithms, and in the functionality and the practice/user-friendliness of the software for experimental design. Some solutions are combined with general statistics packages, while others are exclusively used for experimental design. There is software for all kinds of users, some for those with basic knowledge and some for those with advanced statistical knowledge.

It is an unquestioned objective of the robustness test using statistical experimental design (DoE = Design of Experiments) to perform a systematic investigation of the connections between target values y and factors $x_1, x_2, \ldots x_p$ with very few experiments. In most cases, these can be adequately described by a linear model (plan of first order):

$$y = c \qquad\qquad \text{constant or y-part of axes}$$
$$+\, a_1 \cdot x_1 + a_2 \cdot x_2 + a_3 \cdot x_3 + \ldots + a_p \cdot x_p \qquad \text{main effect or linear term}$$
$$+\, b_{12} \cdot x_1 \cdot x_2 + b_{13} \cdot x_1.x_3 + \ldots \qquad \text{double-alternating effect}$$
$$+\, b_{23} \cdot x_2 \cdot x_3 + b_{24} \cdot x_2 \cdot x_4 + \ldots + b_{p-1p} \cdot x_{p-1} \cdot x_p \quad \text{(total } p \cdot (p-1) \text{ term)}$$
$$+\, e \qquad\qquad \text{error}$$

The robustness of an analytical RP HPLC method can be determined through the following six steps [5]:

- selection of the factors (chemical and physical),
- fixing the level of the factors and their steps,
- selection of the model for experimental design (DoE),
- implementation of the experiments and measurement of the analytical target values,
- identification of effects of factors,
- statistical analysis and interpretation of the results.

The factors can be quantitative (continuously adjustable) or qualitative (not continuously adjustable).

According to Hund et al., the factors for the robustness test can be divided into two groups: method-related factors and non-method-related factors [13]. Method-related factors cover generally quantitative factors such as pH of the eluent, temperature of the column, buffer concentration or ionic strength of the eluent, and quantity of organic solvent in the eluent. Non-method-related factors are frequently qualitative, such as the column factors: batch, age, manufacturer. The factors to be selected for the robustness test should be those that can be changed during daily routine work, or that change if an HPLC method is transferred, e.g., to another laboratory or to other equipment.

It is not always simple to decide which factors have an effect on the target values and which do not. The most relevant factors are:

- the pH of the buffer-containing mobile phase (pH unit),
- the quantity of organic solvent in the mobile phase (% B),
- the temperature of the column (°C),
- the buffer concentration or ionic strength of the aqueous mobile phase (mM),
- for an isocratic mode of operation, and additionally,
- gradient run time t_G (min),
- gradient slope,
- quantity of the organic solvent in the mobile phase at the beginning of the gradient (% B),
- quantity of the organic solvent in the mobile phase at the end of the gradient (% B),
- for a gradient mode of operation.

According to Jimidar [4], the following factors should be determined for an HPLC method:

- pH of the buffer solution,
- column temperature,
- column (two different batches, used and unused columns),
- % organic solvent at the beginning of the gradient or under isocratic conditions,
- % organic solvent at the end of the gradient,
- gradient slope, expressed in terms of gradient run time t_G,
- three dummy factors.

The "dummy" factors are simulated variables and include the nominal parameter values of the HPLC method to be tested, i.e. they do not cause any change within the method. The estimated effects of "dummies" can either be used as a measurement of the experimental error of an effect, or in the statistical evaluation of the factors tested [31].

Additional factors are:
- concentration of additives in the mobile phase, e.g. ion-pairing reagents (mM),
- flow rate,
- detector wavelength.

As a physical factor, the flow rate could also be of interest. However, with isocratic operation the flow rate of the mobile phase can be modified without a change in the relative peak distance. The retention factor k remains unaffected by the flow rate, since the retention and the dead time change proportionally with a change therein. A slight, usually insignificant, change in the resolution arises if the flow rate is changed by less than 20%. The choice of the flow rate as an factor in the robustness test is not reasonable if the chromatographic resolution Rs has been selected as a target value and not the peak area.

With gradient procedures, the situation is not so well-defined. The elution power changes continuously and therefore the retention factor k for the mobile phase is changed during the separation, i.e. the retention factor k is an unsuitable parameter for the description of the retention with gradient elution. According to Dolan [14], an average retention factor k^* should be used. If all peaks are well separated, i.e. the resolution is > 2, slight changes in the flow rate of the order of about 10–15% have only an insignificant influence on the separation (resolution) [14]. Otherwise, if the resolution is marginal, the use of the flow rate can be considered as an factor in the robustness test.

Only method-related factors are examined in most robustness tests [15–18]. The choice of the "correct" factors is a key step in the implementation of the robustness test. The level of each factor should be selected in such a way that the maximum differences in the values are included, as might occur in daily routine work or during a method transfer [19]. The test level of the factors always depends on the purpose of the method and should always be practicable. For quantitative factors, a "low" and a "high" step should be set, so that the nominal values of the given HPLC method are included. The maximum displacement of the factor in robustness studies is relatively small and a linear correlation of the quantity of the target value y of most factors within the intervals to be checked can be assumed; thus, the choice of two adjusting stages (two-level design) for each individual factor (low/high level) is absolutely sufficient. Three adjusting stages (three-level design, low/high/nominal level) should always be selected if the possibility of nonlinear behavior between target value(s) and factor(s) cannot be ruled out [20–22].

Literature examples of some factors are listed in Table 1.

Due to the fact that the factors (as shown in Table 1) have a characteristic adjustment range, the range of the experimental design can easily be defined.

Table 1. Settings of the influencing parameters for the test of robustness of an HPLC method.

pH value of the buffer solution	nominal ± 0.1 Units [23, 24] nominal ± 0.2 Units [4] nominal ± 0.2 Units [5] nominal ± 0.2 Units [13] nominal ± 0.2 Units [27] nominal ± 0.3 Units [28] nominal ± 0.5 Units [25, 26]
Column temperature	nominal ± 10% rel. [23, 24] nominal ± 14% rel. [4] nominal ± 5% rel. [5] nominal ± 10% rel. [13] nominal ± 17% rel. [28] nominal ± 17% rel. [25] nominal ± 7% rel. [29]
Buffer concentration	nominal ± 10% rel. [23, 24] nominal ± 10% rel. [4] nominal ± 10% rel. [5] nominal ± 7% rel. [13] nominal ± 10% rel. [28] nominal ± 20% rel. [25]
Concentration of additives	nominal ± 10% rel. [4] nominal ± 8% rel. [5]
% Organic solvent at the beginning of the gradient or isocratic conditions	nominal ± 10% rel. [23, 24] nominal ± 10% rel. [4] nominal ± 7% rel. [13] nominal ± 4% rel. [28] nominal ± 3% rel. [29]
% Organic solvent at the end of the gradient	nominal ± 10% rel. [23] nominal ± 4% rel. [28]
Gradient slope, expressed in gradient time t_G	nominal ± 10% rel. [20, 21] nominal ± 5% rel. [4]
Flow rate	nominal ± 0.1 mL min^{-1} [23, 24] nominal ± 0.1 mL min^{-1} [4] nominal ± 0.1 mL min^{-1} [28] nominal ± 0.1 and 0.2 mL min^{-1} [13] nominal ± 0.05 mL min^{-1} [29]
Detector wavelength	nominal ± 3 nm [4] nominal ± 2 and 3 nm [13] nominal ± 5 nm [5] nominal ± 5 nm [28] nominal ± 1 nm [29]

The number of corners within the range of the experimental design in relation to the number of factors is illustrated in Fig. 16 [12].

Assuming that for all factors the setting can be determined independently of the others, the following experimental design ranges are determined: in the case of two factors a rectangle with four edges, with three factors a cuboid with eight corners, and with four factors a hyper-cuboid with 16 corners. The greater the number of factors, the more complex the system becomes (see the illustration for six factors). If ten factors are chosen, the number of corners is $2^{10} = 1024$. This means that the number of experimental runs exceeds any reasonably realizable limit, even if one wishes to examine the experimental design range only at the corners. The experimental effort with 3 to 11 test factors [30], which is usual for the robustness test, can only be realized through the selection of suitable experimental design strategies.

According to Wember [12] and Jimidar [4], the following conditions need to be considered during experiment design, since they strongly influence the actual planning:

- number of influencing parameters determined,
- independence of the changes in the influencing parameters,
- experience relating to the planned determination,
- knowledge about the reproducibility of the results,
- costs for a single determination,
- simultaneous studies concerning the most important method parameters (factors),
- systematic operation with a minimum number of experiments,
- preferably minor expenditure of time for the experimental work and analysis,
- application of a user-friendly statistical software tool (e.g., Statgraphics Plus, Cornerstone, Design-Expert, etc.).

Various types of statistical experimental design (DoE) can, in principle, be used for the robustness test of an HPLC method. However, only a few of them are practicable, e.g. fractional factorial test schedules [13, 27, 31–33] and Plackett–Burman test schedules [15, 16, 18, 31].

The advantage of factorial experimental plans (factorial designs) in the form of full factorial experimental plans (full factorial designs) is that regression models with main effects and reciprocal effects can be treated with them. The full factorial experimental plans, which consider all corners, show a statistical resolution of V. This means that the risk of misinterpretations is very small. This type of experimental plan, in which all possible combinations of the factors are varied, usually in two stages (in each case, an upper and lower stage), is typically used only with a few factors (e.g., two or three); this is due to the fact that the number of corners in a (high-dimensional) cuboid grows exponentially with the number of factors. For example, seven factors produce 2^7 corners, i.e. 128 experiments.

A reduction in the number of necessary test runs, also with the possibility of a larger number of factors (e.g., 11 [30]), can be achieved with fractional factorial experimental plans (fractional factorial designs), which are derived from full

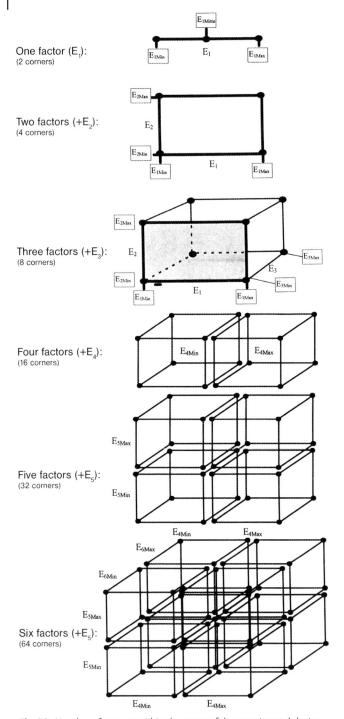

One factor (E_1):
(2 corners)

Two factors (+E_2):
(4 corners)

Three factors (+E_3):
(8 corners)

Four factors (+E_4):
(16 corners)

Five factors (+E_5):
(32 corners)

Six factors (+E_5):
(64 corners)

Fig. 16. Number of corners within the range of the experimental design.

factorial plans. Here, not all corners are occupied. Only involution models with main effects can be examined with such experimental plans of the first order, and interactions must be avoided. A disadvantage of all these experimental plans is the fact that they can only be defined for power-of-two numbers (2, 4, 8, 16, 32, 64, ...) with the given number of test runs. This means that with a further factor the number of experiments is doubled.

The most important alternative to the fractional factorial experimental plans are Plackett–Burman plans [31]. With these, plans are now available that show similar characteristics to full and fractional factorial experimental plans; in addition, they are definable for all multiples of 4 (12, 20, 24, 28, ...), which are not power-of-two numbers. A disadvantage, as with the fractional factorial experimental plans, is that only a statistical resolution of III is given for these experimental plans; this means that only main effects can be estimated. A mixing of main effects and interactions cannot easily be determined, since very few show an overlapping of a two-fold reciprocal effect with a main effect; thus, misinterpretations can occur in practice [36]. Table 2 shows a Plackett–Burman experimental plan for $n = 12$.

For the design of a Plackett–Burman experimental plan for less than 11 factors, the appropriate number of columns are omitted.

Due to the fact that it is more reasonable for the robustness test of an HPLC method to determine whether a method is robust in relation to small specific changes of the method parameters, instead of making statements about main effects and intreractions as well as their mixing, the Plackett–Burman plans are the most frequently applied experimental plans for the robustness test of an HPLC method.

Table 2. Plackett–Burman experimental plan for $n = 12$.

Experiments	Factors									
	A	B	C	D	E	F	G	H	I	J
1	−	−	−	+	+	+	−	+	+	−
2	−	−	+	+	+	−	+	+	−	+
3	+	−	+	+	−	+	−	−	−	+
4	+	−	−	−	+	+	+	−	+	+
5	−	+	+	+	−	+	+	−	+	−
6	+	−	+	−	−	−	+	+	+	−
7	+	+	+	−	+	+	−	+	−	−
8	+	+	−	+	+	−	+	−	−	−
9	−	+	−	−	−	+	+	+	−	+
10	+	+	−	+	−	−	−	+	+	+
11	−	−	−	−	−	−	−	−	−	−
12	−	+	+	−	+	−	−	−	+	+

In the following example, concerning the determination of the content of active components in pharmaceutical formulations, robustness tests of HPLC methods by means of statistical experimental design (DoE) of the Plackett–Burman type are explained in more detail.

For this, factors that could have a potential influence on the chromatographic separations were examined (Table 3). Figure 17 shows a chromatogram of the computer-optimized separation of the critical pair of peaks 6/7 (nominal chromatographic conditions).

For the determination of the influence of these six factors, a Plackett–Burman test plan with 12 experiments was chosen (Table 4).

In order to make statements about the correlation between factors and target values, considering the very limited number of experiments, all individual tests must be carried out in a randomized order.

With any systematic approach, there is the risk that systematic errors can be wrongly interpreted as main effects. With a randomized order of the tests, such systematic errors would only distort the connections between factors and target value(s) [35].

The tests were carried out in a randomized order (as proposed by the software) and the target values (Table 5) were determined for additional statistical evaluation.

Analysis of variance and effects were calculated for the analysis of the experimental design (Tables 6 and 7), along with a graphical analysis of the results with standardized Pareto diagrams, probability network, and main effect illustration.

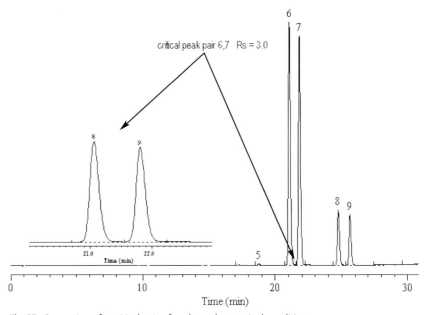

Fig. 17. Separation of a critical pair of peaks under nominal conditions.

Table 3. Important factors of an HPLC method with regard to the robustness test.

Factor	Unit	Limits	Level (–1)	Level (+1)	Nominal
pH value	–	± 0.1 Units	2.4	2.6	2.5
Temperature	°C	± 5% rel.	47.5	52.5	50.0
[Modifier]	%	± 10% rel.	4.5	5.5	5.0
Slope t_G	min	± 5% rel.	31.35	34.65	33.00
Flow	mL min^{-1}	± 10% rel.	0.9	1.1	1.0
Dummy	–	± 1	–1	+1	0

Table 4. Plackett–Burman test plan for $n = 12$.

Exp.	Factor							
	A Dum1	B pH	C Temp.	D Dum2	E Slope	F [Modifier]	G Dum3	H Flow
1	+1	+1	–1	+1	–1	–1	–1	+1
2	–1	–1	+1	+1	+1	–1	+1	+1
3	+1	–1	+1	+1	–1	+1	–1	–1
4	+1	+1	–1	+1	+1	–1	+1	–1
5	+1	–1	–1	–1	+1	+1	+1	–1
6	–1	+1	–1	–1	–1	+1	+1	+1
7	–1	–1	–1	+1	+1	+1	–1	+1
8	–1	–1	–1	–1	–1	–1	–1	–1
9	+1	–1	+1	–1	–1	–1	+1	+1
10	+1	+1	+1	–1	+1	+1	–1	+1
11	–1	+1	+1	+1	–1	+1	+1	–1
12	–1	+1	+1	–1	+1	–1	–1	–1

Table 5. Determined target values "resolution Rs".

Experiment	Response (resolution Rs)
1	2.97
2	2.96
3	2.73
4	2.85
5	2.99
6	2.76
7	3.18
8	2.70
9	2.83
10	3.02
11	2.75
12	2.97

Table 6. Calculated effects for each individual factor and their 95% confidence interval.

Factor	Resolution Rs		
	Effect	Confidence level (α = 5%)	P-value
B: pH	−0.012	± 0.111	0.8178
C: Temperature	−0.032	± 0.111	0.5358
E: Slope t_G	0.205	± 0.111	0.0024
F: [Modifier]	0.025	± 0.111	0.6235
H: Flow	0.122	± 0.111	0.0350

Table 7. Calculated effects for each factor and their 99% confidence interval.

Factor	Resolution Rs		
	Effect	Confidence level (α = 1%)	P-value
B: pH	−0.012	± 0.160	0.8178
C: Temperature	−0.032	± 0.160	0.5358
E: Slope t_G	0.205	± 0.160	0.0024
F: [Modifier]	0.025	± 0.160	0.6235
H: Flow	0.122	± 0.160	0.0350

The changed target value is one effect resulting from the change of the factor from the lower to the upper step. The calculated effects indicate the intensity of the influence of the respective factor on the change in the target value.

The factor gradient slope has the strongest effect on the resolution. The analysis of variance shows that two effects have P values smaller than 0.05; i.e., they show a significant difference of zero with a selected confidence interval of 95%. The factor slopes for t_G and flow rate show a significant effect on the resolution.

The strength and significance of the effects are clearly shown in the Pareto diagrams, and the direction of action of the effects is shown in the presentation of the main effects (Fig. 21).

A Pareto diagram clearly represents the standardized effects. The indicated line in a standardized Pareto diagram is the significance line for the test effect = 0. While the effects to the left of the line are not significantly different from zero (C, F, B), the effects to the right of the line are significantly different from zero (E, H). It is quite reasonable to delete non-significant contributions from the model, i.e. the bars that end to the left of the significance line.

The decision as to the significance of the effects should not be left to a mere algorithm. According to Jimidar [18], a 1% probability of error should be selected in order to reduce the possibility of false positive decisions during the determination of the significance of effects.

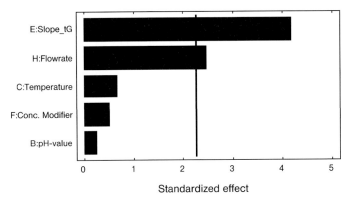

Fig. 18. Illustration of the effects in a standardized Pareto diagram (resolution) for the error possibility α = 5%.

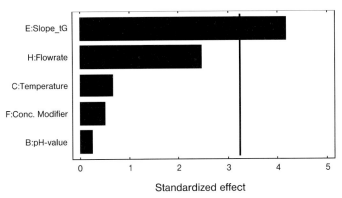

Fig. 19. Illustration of the effects in a standardized Pareto diagram (resolution) for the error possibility α = 1%.

As is evident from the analysis of variance with a probability of error α = 1%, only one effect has a P value smaller than 0.01, i.e., it shows a significant difference from zero with a selected confidence interval of 99%, because the factor "slope" has a significant effect on the resolution. This is also very clearly shown in the Pareto diagram (Fig. 19).

An additional possibility for the assessment of the significance of effects is the presentation of the standardized effects within the probability network (Fig. 20).

It must be taken into consideration that significant members in the model do not behave as random variables, i.e., they lie close to the straight lines. However, significant members lie far from the straight line (factor slope). No clear correlation is apparent for the factor "flow".

The presentation of the main effects is suitable for the evaluation of the influence of the individual factors. The impact of the main effects on the target value is very clearly demonstrated in this illustration. In contrast to the Pareto diagram, the directions of action (signs) can also be recognized here.

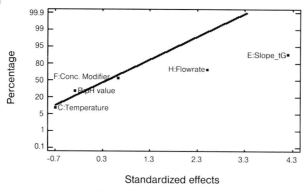

Fig. 20. Illustration of the effects in the probability network.

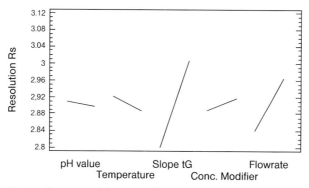

Fig. 21. Illustration of the main effects (for the resolution).

An increase of slope and flow rate result in a better separation of the critical pair of peaks 6 and 7. In addition, an increase in the modifier concentration produces an improvement of the resolution, but with clearly smaller strength. Increases of the pH value and temperature cause the opposite; a degradation of the resolution is to be expected. Depending on the viewpoint (probability of error $\alpha = 1\%$ or $\alpha = 5\%$), the slope has a statistically significant influence on the target value resolution.

The "worst case" combination of the significant factors are shown in Table 8.

The resolution of Rs = 2.8 for the critical pair of peaks 6 and 7 (Fig. 22), obtained under these chromatographic conditions, clearly exceeds the resolution Rs = 2.0 that is often required for routine analysis. As a result, this method is a robust one.

Table 8. "Worst case" combination of the factors.

Factor	Unit	"Worst case" value	Nominal value	Resolution Rs (x = 5)
pH	–	2.5	2.5	
Temperature	°C	50.0	50.0	
[Modifier]	%	5.0	5.0	2.8
Slope t_G	**min**	**31.35**	**33.0**	
Flow	mL min^{-1}	1.0	1.0	

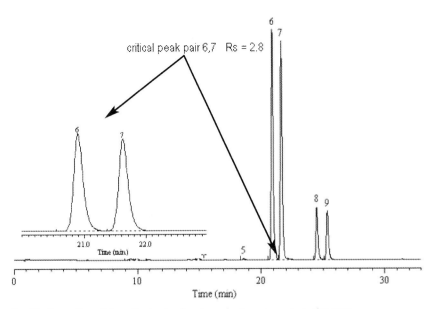

Fig. 22. Separation of the critical pair of peaks under "worst case" combination of the influencing variables.

5.2.4
Conclusion

Increasing requirements for the quality of analysis methods are accompanied by a need for new, efficient procedures for the examination of robustness. The working conditions of HPLC should be expressly adapted to the practice, and the observed changes in selectivity should consistently be used for a sensible evaluation in order to obtain both economical and robust HPLC methods.

Relevant robustness statements about an RP-HPLC method can generally be made in two different ways: either by a computer-facilitated systematic method development or by application of a statistical experimental design (DoE).

References

1 L. R. Snyder, J. J. Kirkland, J. L. Glajch, Practical HPLC Method Development, 2nd ed., Wiley-Interscience, New York, 1997.

2 Y. Vander Heyden, D. L. Massart, M. Mulholland, in Robustness of Analytical Chemical Methods and Pharmaceutical Technological Products, Data Handling in Science and Technology (Eds.: M. M. W. B. Hendrics, J. H. de Boer, A. K. Smilde), Elsevier, Amsterdam, 1996, Ch. 3 and Ch. 5.

3 E. J. Klein, S. L. Rivera, A Review of Criteria Functions and Response Surface Methodology for the Optimization of Analytical-Scale HPLC Separations, Marcel Dekker Inc., 2000, p. 2097–2121.

4 I. Jimidar, Lecture presented at the workshop: Computer-Assisted RPLC Method Development, Darmstadt, Nov. 2000.

5 R. Romero, D. Gazquez, M. Sanchez-Vinas, L. Cuadros Rodriguez, M. G. Bagur, *LCGC North America* 20 (2002), 72–80.

6 I. Molnar, *J. Chromatogr. A* 965 (2002), 175.

7 C. Horvath, W. Melander, I. Molnar, *Anal. Chem.* 49 (1976), 2295.

8 J. A. Lewis, J. W. Dolan, L. R. Snyder, I. Molnar, *J. Chromatogr.* 592 (1992), 197.

9 H. W. Bilke, C. Gernet, I. Molnar, *J. Chromatogr.* 729 (1996), 189.

10 T. Jupille, in DryLab 2000 Plus Automation Toolkit, 2002.

11 H.-J. Rieger, I. Molnar, *LaborPraxis* 3 (2003), 40.

12 Th. Wember, Technische Statistik und statistische Versuchsplanung, Eigenverlag, 2000, Version 6.

13 E. Hund, Y. Vander Heyden, M. Haustein, D. L. Massart, J. Smeyers-Verbeke, *J. Chromatogr. A* 874 (2000), 167–185.

14 J. W. Dolan, *LC·GC Europe* 5 (2003), 252.

15 Y. Vander Heyden, K. Luypaert, C. Hartmann, D. L. Massart, J. Hoogmartens, J. de Beer, *Anal. Chim. Acta* 312 (1995), 245–262.

16 Y. Vander Heyden, A. Bourgeois, D. L. Massart, *Anal. Chim. Acta* 347 (1997), 369–384.

17 L. M. B. C. Alvares-Ribeiro, A. A. S. C. Machado, *Anal. Chim. Acta* 355 (1997), 195–201.

18 M. Jimidar, N. Niemeijer, R. Peetres, J. Hoogmartens, *J. Pharm. Biomed. Anal.* 18 (1998), 479–485.

19 J. A. Van Leeuwen, L. M. C. Buydens, B. G. M. Vandeginste, G. Kateman, P. J. Schoenmakers, M. Mulholland, *Chemometrics and Intelligent Laboratory Systems* 10 (1991), 337–347.

20 M. Mulholland, J. Waterhouse, *J. Chromatogr.* 395 (1987), 539–551.

21 M. Mulholland, J. Waterhouse, *Chromatographia* 25 (1988), 769–774.

22 J. A. Van Leeuwen, L. M. C. Buydens, B. G. M. Vandeginste, G. Kateman, P. J. Schoenmakers, M. Mulholland, *Chemometrics and Intelligent Laboratory Systems* 11 (1991), 37–55.

23 H. W. Bilke, Lecture presented at the workshop: Computer-Assisted RPLC Method Development, Darmstadt, Nov. 2000.

24 H. W. Bilke, Lecture presented at the Novia HPLC 2003, Mannheim, Nov. 2003.

25 Z. Yongxin, J. Augustinjs, E. Roets, J. Hoogmartens, *Pharmeuropa* 9 (1997), 323–327.

26 D. Song, E. Roets, J. Hoogmartens, *Pharmeuropa* 11 (1999), 432–436.

27 R. Ragonese, M. Mulholland, J. Kalman, *J. Chromatogr. A* 870 (2000), 45–51.

28 Y. Vander Heyden, M. Jimidar, E. Hund, N. Niemeijer, R. Peetres, J. Smeyers-Verbeke, D. L. Massart, J. Hoogmartens, *J. Chromatogr. A* 845 (1999), 145–154.

29 R. B. Waters, A. Dovletoglou, *J. Liq. Chromatogr.* 18 (2003), 2975–2985.

30 J. A. van Leeuwen, B. G. M. Vandeginste, G. Kateman, M. Mulholland, A. Cleland, *Anal. Chim. Acta* 228 (1990), 145–153.

31 Y. Vander Heyden, D. L. Massart, in Robustness of Analytical Chemical Methods and Pharmaceutical Technological Products, Data Handling in Science and Technology (Eds.: M. M. W. B. Hendrics, J. H. de Boer, A. K. Smilde), Elsevier, Amsterdam, 1996, pp. 79–147.

32 M. Jimidar, M. S. Khots, T. P. Hamoir, D. L. Massart, *Quim. Anal.* 12 (1993), 63–68.

33 E. Morgan, Chemometrics – Experimental Design, Analytical Chemistry by Open Learning, Wiley, Chichester, 1991, pp. 118–188.

34 R. L. Plackett, J. P. Burman, *Biometrika* 33 (1946), 305–325.

35 C. Krüll, *LaborPraxis* 11 (2003), 58–60.

36 S. Johne, Praxishandbuch STATGRAPHICS, Teil 3/1: Statistische Versuchsplanung Teil 1 (2002), 27–28.

5.3
Separation of Complex Sample Mixtures

Knut Wagner

5.3.1
Introduction

Special solutions are required for the separation of very complex sample mixtures, which are the daily challenge in contemporary bioanalytics. The main focus of this chapter is an introduction to the theory and technical solutions in multi-dimensional chromatography. Furthermore, a selected real example is used to demonstrate the significant increase in separation power that may be achieved by a clever coupling of different separation mechanisms. In order to encourage the reader to realize unconventional approaches, the key issues are discussed and it is shown that reproducible results can be achieved.

The need for complex analyte separations has tremendously increased in recent years with the progress in life sciences such as genomics and proteomics. The decoding of the human genome has attracted most attention since 2001, with the finalization of the first draft sequence of its 3 billion base pairs. The genome encodes hundred thousands of proteins and peptides, which have a molecular weight range from 100 Da to 300 kDa. However, the genome does not provide any information about the actual expression of the proteins and post-translatorial modifications. Therefore, the scientific interest is increasingly moving towards proteomics. Proteomics is defined as the identification, characterization, and quantification of all the proteins expressed by a genome. The proteome, unlike the genome, is not a fixed feature of an organism. Instead, it changes with the state of development, the tissue, or even the environmental conditions under which an organism finds itself. It is evident that proteins are central for the understanding of cellular function and disease processes. Cell regulations by proteins can be monitored by measuring the protein expressions and changes thereof using differential approaches. The search for new drug compounds, such as inhibitors for proteases, kinases, and phosphatases, as well as agonists/ antagonists of growth factors, cytokines, and receptors, is the driving force for the pharmaceutical industry to make huge investments and to allocate resources to the field of proteomics.

It is standard that more than a thousand proteins are present in a biological sample such as a cell lysate. Besides the state-of-the-art separation technique, 2D-gel electrophoresis, liquid-phase separation techniques are beginning to play a major role in proteomics. Their advantages are reproducibility, automation, speed, and their suitability for direct coupling with mass spectrometry.

HPLC Made to Measure: A Practical Handbook for Optimization. Edited by Stavros Kromidas
Copyright © 2006 WILEY-VCH Verlag GmbH & Co. KGaA, Weinheim
ISBN: 3-527-31377-X

5.3.2
Multidimensional HPLC

One-dimensional separations have limited resolving power when applied to complex sample mixtures. This is evident from the fact that a seemingly single peak does not necessarily represent a single component but a significant number of substances. Increases in chromatographic resolution can be achieved by variations in the plate number, N, the selectivity, α, or the retention factor, k. However, adjustment of the retention factor has only a limited influence on resolution and this is only true at low values. An increase in the number of plates by extending the column length often results in marginal increases in resolution, but probably also leads to an increase in the analysis time. Since selectivity has the greatest influence on resolution, it is the variable that attracts the most attention.

The limitations of single-stage separation systems have been recognized for many years. In order to describe separations of a multicomponent mixture, the concept of peak capacity was introduced, which is defined as the maximum number of components, n, that can be placed side by side in the available separation space.

$$n = \left(1 + \frac{\sqrt{N}}{r}\right) \ln(1 + k_i) \tag{1}$$

In this formula, r is the number of standard deviations taken as equaling the peak width (typically 4), and k is the retention factor of the last peak in a series. Modern, high-resolution chromatographic systems yield peak capacities which are calculated to be in the range of 100–300. The peak capacity is a theoretical value that will never be attained since components in a complex mixture are usually not uniformly distributed, appear randomly, and are overlapping each other. Assuming that the number of components in a mixture can be estimated, when the number of components equals the peak capacity of the system used, the maximum number of visible peaks will be equivalent to 37% of the system's peak capacity.

A practical means of effecting changes in resolution is the introduction of a different fundamental mechanism of interaction by the use of two (or more) separation stages (multidimensional chromatography, or coupled-column chromatography). Several powerful combinations of linear liquid chromatographic techniques based on miscellaneous displacement mechanisms, e.g. ion-exchange, reversed-phase, hydrophobic interaction, and affinity chromatographies, are available. The judicious choice of separation combination is not only restricted by their orthogonality, but chromatographic parameters such as selectivity, resolution, peak capacity, load capacity, bio recovery, and speed of separation also have to be taken into consideration in order to attain a comprehensive multidimensional separation. To obtain the maximum benefit from multidimensional chromatographic systems, the basic mechanisms controlling separation in each dimension should be different (orthogonal). Furthermore, the maximum peak capacity can only be achieved when the entire analyte is subjected to the various separation dimensions. A separation of two components which has already been achieved in the first separation mode must be maintained throughout the whole separation. Therefore, the fractionation intervals for the analyte transfer into the second dimension need to be suitably short. All the fractions taken from the first dimension have to independently undergo a second separation that relies on a different mechanism. The maximum peak capacity attainable in a system of this type is given by the product of the peak capacities of each dimension (neglecting the additional band broadening of the migrating components in the second dimension). In coupled-column chromatography, utilization of the total available separation space requires a large number of secondary column separations, so that all sample cuts taken in the first dimension can be transferred for subsequent separations.

$$n_{tot} = n_a \times n_b \qquad (2)$$

A representation of the peak capacity of a planar two-dimensional system is presented in Fig. 1. A multidimensional HPLC system in which the entire first dimension column effluent (not merely an interesting region) is reanalyzed as discrete fractions at regular intervals in the second dimension is referred to as using the "comprehensive" concept. If only the components of interest from the first column effluent are subjected to separation on the second column, this is referred to as the "heart-cutting" technique. Heart-cutting techniques require that the retention properties of the analytes in the first dimension are known in advance. This technique is appropriate when only one or a few components need to be isolated.

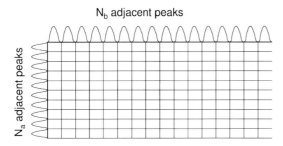

Fig. 1. Statistical model of overlap. The rectangular boxes of the gridwork correspond to the peak capacity of a 2D system with Gaussian distribution, in this case *n* equals approximately 160 ($n_a \times n_b$).

5.3.3

Techniques for Multidimensional Separations

The transfer of a discrete fraction to the second dimension can be accomplished in two different ways:

5.3.3.1 Off-Line Technique

The first technique is the off-line mode, whereby fractions are collected and later reinjected into the second dimension. This can either be done manually or in an automated fashion. This approach puts low demands on the instrumental set-up and there are no limitations on separation speeds in each dimension. Off-line techniques are prone to sample losses by vial contamination, low reproducibility, and long analysis times. Furthermore, sample dilution and eluent incompatibilities may have a negative impact. However, many off-line experimental approaches have been described throughout the literature and fully automated equipment is now commercially available. Figure 2 displays the principle of fractionation and reinjection in 2D-HPLC, as used in the off-line approach.

5.3.3.2 On-Line Technique

The more elegant technique is the on-line coupling of different separation modes, of which two variations are in common use. One technique uses a continuous flow in the first dimension and different separation speeds in either dimension. To maintain a high sampling rate after the first dimension, the separation in the second dimension should be very fast compared to that in the first dimension in order to analyze on-line enough samples across each peak without any fraction storage. To satisfy this demand, two or more parallel columns can be applied in the second dimension, even though this places stringent demands on the high-speed separation. The advantages are no sample losses, excellent reproducibility, highest resolution power, and fast 2D-separations. High separation speeds in the second chromatographic dimension can be achieved by using either monolithic silica columns or columns filled with small, non-porous or porous silica particles (e.g., 1.5 µm particles).

The second technique uses an interrupted flow and gradient elution in the first dimension. Discrete fractions are transferred to one single secondary column at

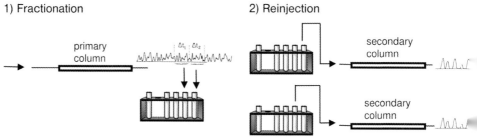

1) Fractionation   2) Reinjection

Fig. 2. Off-line technique in 2D-HPLC.

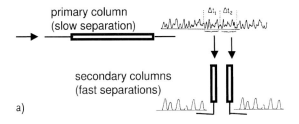

a)

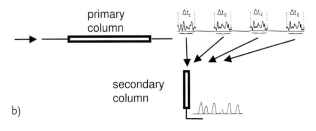

b)

Fig. 3. On-line technique in 2D-HPLC.
a) Continuous flow – different separation speeds.
b) Discontinuous flow – step gradient elution.

intervals that match the separation cycle in the second dimension. A step gradient is used for the transfer of discrete sample fractions to the secondary column. After the transfer of the fractions to the subsequent column, the flow in the first column is completely stopped until the secondary column is eluted and regenerated by an independent HPLC system. This approach is used, for example, in the separation of biomolecules by capillary chromatography with MS detection. Advantages are low demands on the instrumental set-up and no restrictions in terms of separation speeds, although resolution is sacrificed. The two different on-line techniques are illustrated in Fig. 3.

The interfacing between the first and second dimensions in a multidimensional HPLC system is a crucial feature. It can be achieved by filling suitably sized storage loops connected to automated switching valves for transferring the fraction onto the next column, which places stringent demands on adjusting the loop size, flow rate, and column dimensions in both dimensions. Furthermore, the eluents in the two dimensions have to be compatible. Alternatively, the fractions eluting from the first dimension can be enriched on the column head of a secondary column (on-column focusing). During the loading time of one column, a second parallel column can be eluted. A set of fully automated switching valves can be used for alternate linking of the first dimension effluent to the parallel columns. The latter interface construction is especially preferred for biomolecules since it is more flexible to use, is more rugged, and sample desalting and enrichment is performed automatically.

5.3.4
On-Line Sample Preparation as a Previous Stage of Multidimensional HPLC

The simplest form of column coupling is HPLC-integrated size-selective sample fractionation using "restricted access materials" (RAM). However, using RAM columns is not necessarily a multidimensional HPLC separation in a classical sense. In the first step, the target analytes are separated from the unwanted sample compounds (matrix). In the second step, the RAM column with the enriched analyte is eluted in-line with an analytical HPLC column. RAMs combine two chromatographic separation modes in one column, namely size-exclusion and adsorption chromatography. Besides a defined pore size, the specific feature of RAM is the topochemically bifunctional surface of the particles. The basic principle of RAM is the size-exclusion process due to the physical diffusion barrier determined by the pore diameter (see also Chapter 2.4). The RAM is based upon LiChrospher 60, which is a spherical (25 μm diameter) based silica with an average pore size of 6 nm. This provides a molecular weight cut-off of approximately 15 kDa for globular proteins. Simultaneously, a selective retention of low molecular mass analytes occurs in the interior of the pore surface by hydrophobic, ion-exchange, or affinity chromatography mechanisms, depending on the bonded ligand functionality. The internal pore surface is covered with, for example, reversed-phase C_4, C_8, or C_{18} moieties, sulfonic acid cation-exchange groups ($-SO_3H$), or diethylaminoethyl anion-exchange (DEAE) groups. Diol groups are introduced as a hydrophilic non-adsorptive outer surface coating on the particles. This prevents irreversible protein binding and accumulation, thereby enabling the frequent re-use of the RAM columns. The synthesis of RAMs can involve several steps, including specific enzyme reactions. High molecular weight sample components, e.g. matrix proteins, cannot penetrate into the pores and thus are eluted quantitatively within the interparticle volume of such columns. After sample injection, the RAM column is washed with eluent in order to completely flush the matrix direct to waste. Only analytes of low molecular mass have access to the binding centers on the inner pore surface and thus can be retained and selectively extracted. In the second step, a valve is switched and the RAM column with an inverse flow direction is connected in series with the corresponding analytical column. The desorption and separation of the target analytes is performed by gradient elution of the coupled RAM and analytical columns. RAM columns with reversed-phase properties are used as pre-columns in coupled-column systems for HPLC-integrated, extractive sample clean-up of low molecular weight target analytes in crude biofluids. RAM columns readily implement the inherent advantages of on-line column switching in HPLC schemes, e.g. automation and the use of a closed system, for sample clean-up procedures. RAMs allow the discrimination of higher molecular weight proteins, such as albumin, and enable the enrichment of lower molecular weight components. They allow for direct and repetitive injection as well as size-selective fractionation of untreated biofluids such as plasma, serum, urine, fermentation broth or the supernatant of tissue homogenates. A schematic illustration of a restricted access material is depicted in Fig. 4.

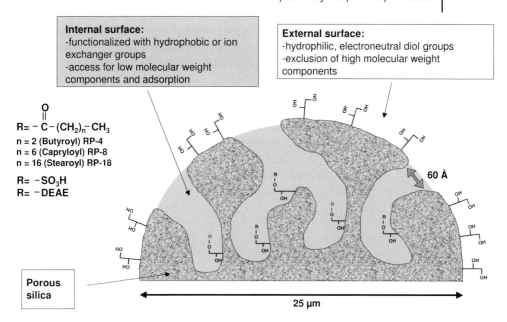

Internal surface:
-functionalized with hydrophobic or ion exchanger groups
-access for low molecular weight components and adsorption

External surface:
-hydrophilic, electroneutral diol groups
-exclusion of high molecular weight components

$R= -\overset{\overset{\displaystyle O}{\|}}{C} - (CH_2)_n - CH_3$
n = 2 (Butyroyl) RP-4
n = 6 (Capryloyl) RP-8
n = 16 (Stearoyl) RP-18

$R= -SO_3H$
$R= -DEAE$

60 Å

Porous silica

25 µm

Fig. 4. Principle of restricted access material (RAM) based on the silica LiChrospher 60.

5.3.5
Fields of Application of Multidimensional HPLC

Besides some applications in which different reversed-phase mechanisms are combined, such as tetrachlorophthalimidopropyl interactions and classical octadecyl interactions, the coupling of fundamentally different interactions is very common in bioanalytics.

Besides affinity chromatography, HPLC separations of four different modes are frequently used for proteins and peptides. These are size-exclusion, ion-exchange, reversed-phase, and hydrophobic interaction chromatographies. Each mode has a different separation mechanism with its own inherent advantages and disadvantages.

Size-exclusion chromatography (SEC) separates molecules by size and shape, differentiating them according to their ability to enter the pores of the column packing material. SEC is limited primarily to separating mixtures of large molecules ($M_w > 5000$ Da) in which there is a significant size difference between the components of interest. The advantages of SEC for proteins and peptides are that aqueous, non-denaturing eluents can be used, elution times and orders are predictable, and molecular size information can be obtained. A primary disadvantage of SEC is its inability to separate highly complex mixtures due to low efficiency and low peak capacity. A typical SEC column can resolve only a few peaks per run.

Ion-exchange chromatography (IEC) separates proteins and peptides according to their ionic charge. IEC provides good resolution and high capacity under conditions that are generally non-denaturing. Ion exchange is one of the most frequently used chromatographic techniques for the separation and purification of proteins, polypeptides, nucleic acids, polynucleotides, and other charged biomolecules. The main disadvantages of IEC are that it requires gradient elution with salt-containing eluents and that preparative separations require desalination of the samples.

Hydrophobic interaction chromatography (HIC) separates proteins or peptides according to their surface hydrophobicity. In HIC, an eluent with a high salt concentration causes the hydrophobic portions of proteins or peptides in their native conformations (non-denatured) to interact with a weakly hydrophobic stationary phase. The proteins and peptides can then be resolved by eluting with a mobile phase gradient of decreasing ionic strength. HIC provides good resolution under non-denaturing conditions. Its main disadvantages are that it requires gradient elution with salt-containing eluents and some samples may have low solubility in the initial highly ionic mobile phase.

Reversed-phase chromatography (RPC), like HIC, separates peptides and proteins according to their hydrophobicity. However, unlike HIC, RPC interactions are not necessarily limited to just hydrophobic groups on the surface of proteins. RPC eluents are usually a mixture of aqueous and organic solvents. The main disadvantage of RPC is that denaturation can occur, depending on the protein and the percentage of organic solvent in the eluent. However, the ability to unfold proteins and peptides also gives RPC its greatest advantage, namely unparalleled resolution. RPC can resolve peptides that differ only by a single amino acid. For the analysis of proteins and peptides, RPC is the most commonly used mode of HPLC. Although the denaturing effects of RPC can make it unfavorable for preparative separations, for peptides and proteins that are not denatured by it, preparative RPC has the advantage of generally providing samples in solvents that are relatively easy to remove.

5.3.5.1 What can be Realized? – A Practical Example

A real example, which might seem slightly exotic, is used to demonstrate how a highly efficient multidimensional HPLC system can be set-up by using commercially available components. The key feature for success is the clever combination of column switching valves and their interfacing. The emphasis in the cited case was on the construction of a multidimensional HPLC system for the analysis of proteins and peptides of a molecular weight smaller than 15 kDa by implementing novel approaches and system components to significantly improve resolution, speed, repeatability, and robustness.

The first step was HPLC-integrated sample preparation using "restricted access materials" (RAM) with cation-exchange functionalities. This was followed by cation-exchange chromatography, whereby the analytes were pre-separated in the first dimension. Subsequently, ultra-high-speed reversed-phase separations using non-porous *n*-octadecyl-modified 1.5 μm silica beads were applied in the second dimension as a highly resolving orthogonal chromatographic mode. A continuously working fully automated two-dimensional HPLC system was set-up in order to avoid sample losses by multiple re-injections, to benefit from speed, and to achieve reproducible results. The number of reversed-phase columns in the second dimension is limited by the application of different separation speeds. The ion-exchange separation proceeds slowly, while high speeds are made possible by using short, non-porous 1.5 μm particle diameter reversed-phase columns. A novel column switching technique including four parallel reversed-phase columns was developed and employed in the second dimension for on-line fractionation and separation. A schematic of the comprehensive on-line 2D-HPLC set-up, including size-selective sample fractionation, is shown in Fig. 5.

Besides the RAM column for sample preparation, the multidimensional separation platform comprised one analytical ion-exchange column and four parallel reversed-phase columns. At any given time, two reversed-phase columns were being gradient eluted simultaneously, one column was being loaded with an analyte fraction from the ion-exchange dimension, and the fourth column

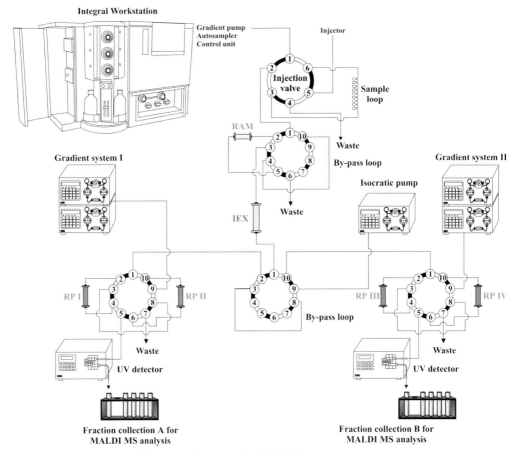

Fig. 5. Schematic overview of the comprehensive on-line 2D-HPLC set-up, including size-selective sample fractionation and four parallel reversed-phase columns.

was being regenerated. The column arrangement was fully automated and controlled by an HPLC workstation. In the HPLC-integrated sample preparation step, the lower molecular weight target analytes were enriched on the internal surface of the RAM packing while the matrix and larger proteins were eluted directly to waste. After washing the matrix from the column, the ten-port valve was switched. This allowed elution of the RAM column in an inverse flow direction in series with the analytical ion-exchange column. The effluent from the ion-exchange mode with the partially separated analytes was enriched on the column head of a reversed-phase column and therefore automatically desalted. The aqueous ion-exchange eluent allowed complete adsorption of the biomolecules on the hydrophobic phase and was directed to waste. Two other reversed-phase columns were eluted in parallel by two separate gradient systems. A second, external UV detector was used to transmit the analogous detector signal as a second channel to the software of the workstation. At the same time, the fourth column was regenerated with eluent A, provided by an isocratic HPLC pump. Each of the four identical reversed-phase columns was subjected to cyclic sample enrichment (sample loading), elution, and regeneration. The cyclic column iteration was performed by a total of five two-position ten-port valves. The first ten-port valve served as an injector, while the second was used for column switching in the sample fractionation step. During the 96 min gradient time of the ion-exchange column, a total of 24 fractions were further analyzed. Each fraction was collected over 4 min (corresponding to 2 mL of effluent) and was subsequently transferred to the second dimension. The lower three ten-port valves in Fig. 5 were used for column iteration. The central valve was switched after each fraction left the ion-exchanger column, leading to intervals of 4 min. The two peripheral valves were switched alternately after every second time interval leading to 8 min intervals. In a total analysis time of 96 min, 24 chromatograms in the reversed-phase dimension corresponding to 24 fractions were generated. The complete procedure for controlling the 2D-HPLC system was programmed by means of the method editor software of the workstation.

The coupling was either made up of a RAM pre-column with cation-exchange functionality in series with an analytical cation-exchange column or a RAM pre-column with anion-exchange functionality in series with an analytical anion-exchange column, each interfaced with the four parallel reversed-phase columns. The columns applied and their operating conditions are shown in Fig. 6.

The separation power of the multidimensional HPLC system is demonstrated by its application to human hemofiltrate, which consists of more than 1000 proteins and peptides of different molecular weights. Figure 7 displays the partial separation of human hemofiltrate on the analytical cation-exchange column in the first dimension after being subjected to selective enrichment on a cationic RAM. The fractions (24 in total) were continuously transferred to the second dimension at 4 min intervals for subsequent analysis by reversed-phase chromatography. The numbers of the fractions are given in the chromatogram, which was generated in a total analysis time of 96 min. As examples, 2 of the 24 reversed-phase chromatograms are displayed in Fig. 8.

a)

Sample fractionation:

column type:	LiChrospher 60 Å XDS-SO$_3$ 25 x 4.0 mm (MERCK)
injection volume:	100 µL (50 µg µL^{-1})
eluent:	0.01 m KH$_2$PO$_4$, pH 3.0
flow rate:	0.2 mL^{-1} min

1st dimension:

column type:	TosoHaas TSK-gel SP-NPR
column size:	35 x 4.6 mm
eluent A:	0.01 m KH$_2$PO$_4$, pH 3.0
eluent B:	1.0 m KH$_2$PO$_4$, pH 3.0
gradient:	0% to 100% B in 96 min
flow rate:	0.5 mL^{-1} min

b)

2nd dimension:

Sampling rate from IXC: every 4 min (central 10-port valve)

column type:	MICRA ODS I (non-porous 1.5 µm)
column size:	14 x 4.6 mm
eluent A:	water (0.1% TFA)
eluent B:	acetonitrile (0.1% TFA)
gradient:	4% to 40% B in 6 min
	40% to 100% B in 0.66 min
	100% B for 0.15 min
regeneration:	a) 4% B for 1.17 min
	b) 100% water for 4 min (0.5 mL^{-1} min)
flow rate :	2.0 mL^{-1} min
detection :	UV, 215 nm

Fig. 6. Columns and their operating conditions.

More than 60 peaks were resolved in the reversed-phase chromatograms, clearly demonstrating the high peak capacity for small proteins and peptides at an analysis time of 7 min. Comparison of two consecutive reversed-phase chromatograms from a single first-dimension run confirmed the orthogonal separation power of the system. The chromatograms are different, with only a few peaks common to both. However, despite a total number of 1000 resolved peaks in the 2D system, it does not necessarily signify that 1000 different components are resolved, as would be expected in a single-mode run. This is an intrinsic feature of the column-coupling approach. One peak of a single component can appear in two or more consecutive chromatograms due to partial separation in the ion-exchanger mode or as a result of splitting of a first dimension peak into two adjacent fractions. The 3D display in Fig. 9, indicating the retention times in either separation mode and the UV intensity, verifies the orthogonal separation power. Only a limited number of redundant peaks appearing in two or more successive fractions are visible.

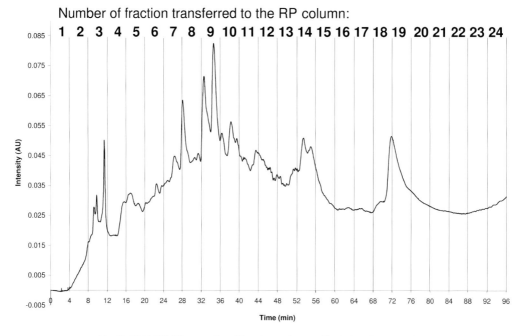

Fig. 7. RAM-CIX-RP separation: The chromatogram illustrates the separation of human hemofiltrate on an analytical cation-exchange column in the first dimension. The fractions (24 in total) were continuously transferred to the second dimension at 4 min intervals.

The reproducibility of the complete system over all three of the separation steps was evaluated with standard proteins and human hemofiltrate. The relative standard deviation (RSD) of the retention times was less than 0.5% on average, while the peak areas showed an RSD of around 10%. The peak capacity in the second dimension could be calculated as approximately 200 per column. Considering the 24 fractions from the first dimension as the first-dimension peak capacity, the total theoretical peak capacity can be calculated to be as high as 5000.

This example clearly demonstrates the resolving power of multidimensional approaches, especially for the on-line technique employing four parallel reversed-phase columns. The system comprises commercially available components and can be established with an acceptable amount of effort. The analyst should be encouraged to risk unconventional approaches.

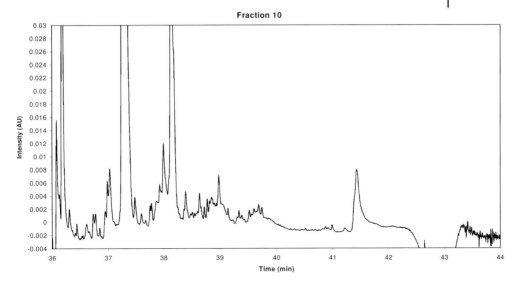

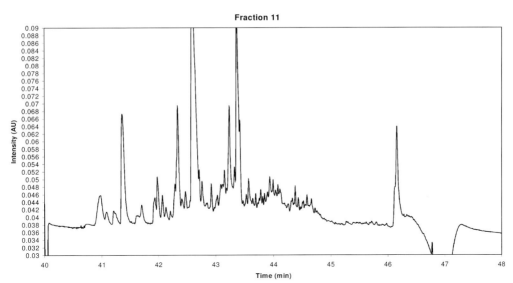

Fig. 8. Two successive RP chromatograms (fractions 10 and 11) as examples of a total of 24 highly resolved separations in the second dimension.

Intensity (AU)

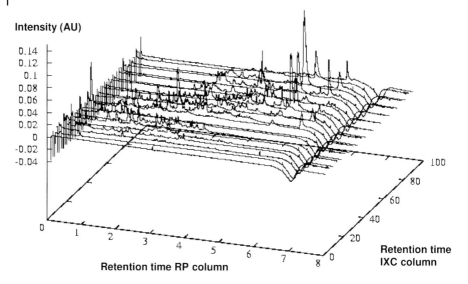

Retention time RP column

Retention time IXC column

Fig. 9. 3D-display of the RAM/CIX/RP separation of peptides from human hemofiltrate.

5.3.6
Critical Parameters of Multidimensional HPLC

It cannot be denied that several critical parameters have to be considered in order to benefit from multidimensional separations.

A fundamental premise is that there are robust methods in the individual dimensions. For the column switching, manifold valves with fast circuit times and a low void volume are essential. During the time of column switch the flow is completely interrupted for a moment, leading to pressure fluctuations. Therefore, steel or PEEK capillaries with a large wall thickness and especially tightened fittings are necessary.

Precise, reproducible gradients with a low delay volume are an important requirement for fast separations. High-pressure gradient systems with a low-volume dynamic mixing chamber are dedicated for this purpose. Fast and high-resolving separations are especially aimed at the second dimension. A compromise between the analysis time and the maximum achievable peak capacity has to be achieved. Optimum fractionation intervals in the first dimension are also critical. A good rule of thumb for an adequate fractionation interval in the first dimension is approximately one peak width. If the sampling rate is too high, the peak capacity is only apparently increased, and subsequent chromatograms in the second dimension display substantially the same information. If the number of fractions is too low, valuable peak capacity is not efficiently used.

The chromatography data system must be capable of controlling internal as well as external switching valves at certain times by means of electrical impulses. Much attention needs to be paid to the correct plumbing of the switching valve ports, columns, pumps, autosampler, and detectors. For an easy comprehension of the eluent flow through the columns and the valves, it is recommended that a schematic overview is drawn for each system state, indicating the valve positions. Pneumatically operated switching valves should be favorable, although fast, electrically-driven valves can be used as well.

References

K. Wagner, T. Miliotis, G. Marko-Varga, R. Bischoff, K. K. Unger, *Anal. Chem.* 74 (2002), 809–820.

J. C. Giddings, in Multidimensional Chromatography (Ed.: H. J. Cortes), Marcel Dekker, New York, 1990.

G. J. Opiteck, J. W. Jorgenson, R. J. Anderegg, *Anal. Chem.* 69 (1997), 2283–2291.

J. Link, J. Eng, D. M. Schielz, E. Charmack, G. J. Mize, D. R. Morris, B. M. Garvik, J. R. Yates, *Nat. Biotechnol.* 17 (1999), 676.

5.4
Evaluation of an Integrated Procedure for the Characterization of Chemical Libraries on the Basis of HPLC-UV/MS/CLND

Mario Arangio, Federico R. Sirtori, Katia Marcucci, Giuseppe Razzano, Maristella Colombo, Roberto Biancardi, and Vincenzo Rizzo

5.4.1
Introduction

In the pharmaceutical industry, various new and improved tools have been introduced to screen at high-throughput pace the compounds synthesized by the medicinal chemists. These new synthetic molecules are stored in huge chemical collections and are routinely screened against validated targets. To obtain reliable results so as to be able to make consistent statements, the quality of the synthesized compounds has to be assessed. In our laboratory, there has been continuous attention not only to the identification but also to the quantification of the new molecules. This is particularly relevant when compounds are synthesized according to combinatorial chemistry protocols, where contamination by residual solvents or salts is often significant. To allow the characterization of large numbers of samples in an extremely fast way, a high-throughput platform has been developed, which is capable of providing, in a single run, information about identity, purity, and quantity. The core part of the platform is the HPLC instrument. The injected crude sample is separated on a reversed-phase column and the obtained peaks are detected first by means of a UV detector for purity assessment, then the flow is partitioned by means of a PEEK cross-piece into three capillaries, one going to a single quadrupole mass spectrometer, where structural identification is confirmed; one to a chemiluminescent nitrogen detector (CLND); and one to waste. The chemiluminescent nitrogen detector is used to obtain information about concentration, as its signal is proportional to the number of moles of nitrogen present. Since the introduction of this new detection method, a series of studies has been published [1, 2]. Some special applications are also described as an alternative approach to classical determination methods [3–5]. The possible role of such a detector in drug discovery can easily be recognized by taking into account that about 90% of all drugs contain nitrogen [5]. In the available literature on this subject, most of the publications deal with the characterization of combinatorial libraries [6–12]. However, a thorough validation with a statistically significant set of compounds is still lacking.

To check the accuracy of the quantification method, a double validation has been performed using caffeine as an external standard. In a first test, a set of 12 commercially available compounds was investigated. These compounds contain nitrogen in different oxidation states in various functional groups. For a cross-validation with a larger set of compounds, the major difficulty was finding another generic quantification method. In our laboratory, a quantification method based on ^{1}H NMR has been developed [13]. In this method, the use of an appropriate internal standard fixes the precision at about 2%. With this technique, a cross-validation was performed on 543 new synthetic molecules from the Medicinal Chemistry Department of Pharmacia, Nerviano.

Pivotal to the whole process of characterization is automated data processing. Using the Discovery corporate database, OpenLynx® software features, and custom-developed Excel VBA macros, we are able to analyze and process in a fully automated procedure about 100 compounds per day.

5.4.2
Materials and Methods

5.4.2.1 Instrumentation

The HPLC system, an Alliance 2790, with thermostatted autosampler and divert valve LabPro, UV detector 2487 and satin interface, and ZQ mass spectrometer with ESI interface, were products of Waters Inc., Milford, Massachusetts. The chemiluminescent nitrogen detector (CLND) model 8060 was a product of ANTEK Instruments Inc., Houston, Texas. The liquid handler Miniprep 75 was a product of Tecan Group Ltd., Maennedorf, Switzerland.

5.4.2.2 Chemicals and Consumables

All chemicals were purchased from Fluka or Aldrich, purity above 97%; methanol, HPLC super gradient grade, was purchased from Riedel-de Haen. HPLC grade water was obtained from a Milli-Q Gradient A 10 autopurification system, Millipore Corp., Billerica, Massachusetts. The samples of new synthetic molecules were synthesized in the Chemistry Department of Pharmacia, Nerviano. Samples were delivered as nominally 10 mM solution in [D$_6$]DMSO. All of these molecules were characterized with regard to identity and purity by NMR and LC-MS. The HPLC column used was a Zorbax SB C8, 50 × 4.6 mm, 5 μm from Agilent Technologies, Palo Alto, California.

A complete list of the 12 selected commercial standards is given in Table 1 and the corresponding structures are shown in Fig. 1. For each of these, a 1.6 mM stock solution was prepared by dissolving an accurately weighed amount in the appropriate volume of DMSO. Four standard solutions at 100 μM, 200 μM, 400 μM, and 800 μM were prepared by diluting each stock solution with DMSO. Each standard was injected at least three times.

A slightly different procedure was used for the new chemical entities from the medicinal chemistry department, which were delivered in semi-deep-well plates. In order to obtain 0.5 mM sample solutions, 15 μL aliquots of 10 mM stock solutions were withdrawn and diluted with DMSO to a volume of 300 μL (1 : 20 dilution factor). The whole process was performed on a Tecan Miniprep 75 liquid handler. In this case, each sample was injected only once.

Table 1. Physico-chemical properties of the commercial standards.

#	Compound name	Molecular formula	MW	t_R (min)
1	Benzyl carbamate	$C_8H_9NO_2$	151.17	5.65
2	4-Nitrophenol	$C_6H_5NO_3$	139.11	5.67
3	4-Amino-methyl benzoate	$C_8H_9NO_2$	151.17	4.82
4	1,5-Dinitronaphthalene	$C_{10}H_6N_2O_4$	218.17	8.38
5	1,3-Dicyanobenzene	$C_8H_4N_2$	128.13	4.71
6	1-Benzylimidazole	$C_{10}H_{10}N_2$	158.20	3.01
7	3,5-Dinitro-*N*-(1-phenylethyl)-benzamide	$C_{15}H_{29}N_3O_5$	315.28	8.62
8	Dibucaine·HCl	$C_{20}H_{29}N_3O_5$	379.90	7.65
9	2,6-Dinitroaniline	$C_6H_5N_3O_4$	183.12	6.36
10	Caffeine	$C_8H_{10}N_4O_2$	194.19	4.14
11	1-(2-Pyrimidyl)piperazine·2 HCl	$C_8H_{12}N_4$	237.13	2.03
12	Theophylline	$C_7H_8N_4O_2$	180.17	3.60

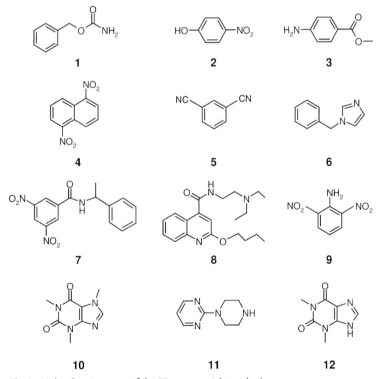

Fig. 1. Molecular structures of the 12 commercial standards.

5.4.2.3 High-Throughput Platform (HTP1) Method Set-up

A multi-detector system was assembled, with the core part consisting of an HPLC separation instrument. The three different detectors were connected on-line. To maintain optimal running of the MS and CLN detectors and to avoid any problems with interference and contamination, a two-way divert valve was used to direct the first minute of elution flow (containing most of the DMSO and salts in the sample) to waste. The overall scheme of the instrument is shown in Fig. 2. During a chromatographic run, the flow rate of the HPLC system was set at 1 mL min^{-1}. After the UV detector, a PEEK cross-piece was inserted to split the flow to the CLND and mass spectrometer. The flow was adjusted with different capillaries in order to obtain 0.1 mL min^{-1} to the CLND and 0.1 mL min^{-1} to the mass spectrometer. The remainder was diverted to waste. Because the hydrodynamic properties of the eluent depend on its composition, the flow to the CLND varies during the gradient. An experimental verification proved that these changes remained within ± 5% of the mean value over the entire range of solvent compositions (from 5% to 95% methanol). This level of variation was considered acceptable for the purpose of our method. The use of a dynamic flow splitter was avoided to keep peaks narrow and to prevent instrument shutdown due to occlusion of the fused silica capillaries.

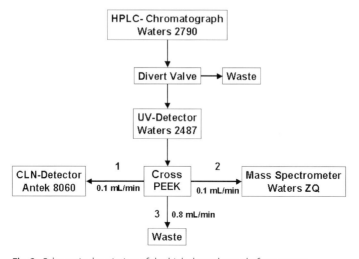

Fig. 2. Schematic description of the high-throughput platform set-up.

5.4.2.4 Chromatographic Conditions

The flow rate was set at 1 mL min^{-1}. Two mobile phases (mobile phase A: 0.1% formic acid in water, mobile phase B: 0.1% formic acid in methanol) were employed to run in 10 min a linear gradient from 5% B to 95% B, which was maintained for 2 min; this was followed by re-equilibration at 5% B for the next 3 min. The run time was 15 min; injection volume 10 μL, autosampler temperature 25 °C, detection wavelength 220 nm.

5.4.2.5 Mass Spectrometer and CLND Conditions

The ESI source and tuning value parameters were as follows: ionization mode: API-ES positive, data acquisition full scan, mass range 120–1000 a.m.u., scan time 0.5 s, desolvation gas flow 615 L h^{-1}, cone gas flow 100 L h^{-1}, source temperature 115 °C, desolvation temperature 250 °C, cone voltage 32 V, capillary voltage 2800 V.

CLND conditions: vacuum 29–30 mbar, make-up flow 25 mL min^{-1}, inlet Ar flow 60 mL min^{-1}, inlet O$_2$ flow 240 mL min^{-1}, ozone flow 25 mL min^{-1}, pressure 22 bar, furnace temperature 1050 °C.

5.4.2.6 Data Processing and Reporting

Data are acquired through MassLynx and processed by OpenLynx Diversity software. The Diversity Report thus obtained shows an overview of the results. The visualization of the microplate with a summary of the most important information is a really useful tool for a qualitative statement (Fig. 3). In the lower part of the screen, the different traces are reported: UV/vis at 220 nm, TIC trace, CLND chromatogram, and mass spectrum of the integrated peaks. The software automatically looks for the peak corresponding to the expected protonated molecular ion of each given target compound (MW in the window in the upper part of the picture) and denotes with a green spot the peaks with the correct mass/charge ratio. The software identifies the main peaks and integrates the areas of both the CLND and UV signals. For the peaks with an incorrect molecular ion, a red spot is shown in the Diversity Report. All the data are exported into an Excel worksheet, combined, processed by custom-developed macros, and transferred to the Discovery Database.

Results are reported in the following database fields:

- **MS ID:** positive (POS) when the mass spectrum corresponding to one of the UV peaks shows a molecular ion consistent with the target one; not determined (ND) when there are no UV peaks corresponding to the expected molecular ion; negative (NEG) if no UV peak corresponds to the target molecular ion but there is a principal UV peak (area > 60%).
- **Purity UV:** UV area % of the target compound peak.
- **Largest impurity:** UV area % of the largest non-target peak ("largest impurity").
- **No UV:** flag denoting absence of significant UV peak at 220 nm.
- **Concentration:** concentration of the sample (µM).

To calculate the concentration, a compound-independent calibration (CIC) is performed using the CLND trace. The relationship between the signal and the concentration is given by the following equation:

$$S = k \cdot c \cdot nN \tag{1}$$

where

k = instrument constant
S = CLND signal

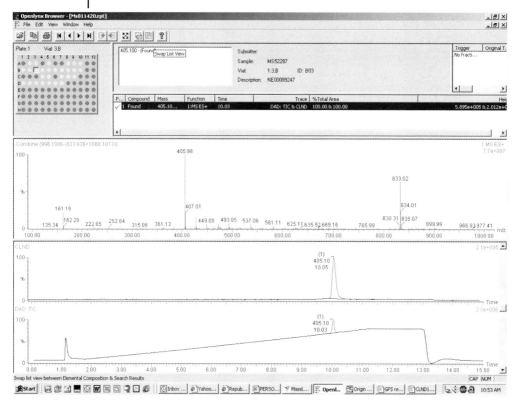

Fig. 3. Example of a MassLynx browser report.

c = concentration

nN = number of nitrogen atoms in the molecule

Therefore, a calibration curve with caffeine at four different concentrations (0.1, 0.2, 0.4 and 0.8 mM) was constructed. The concentration of the target compound was calculated using Eq. (2):

$$c_{CLND,t} = \frac{S_t}{nN_t \cdot s_1} \tag{2}$$

where

$c_{CLND,t}$ = CLND-derived concentration of the target compound
S_t = CLND peak area of the target compound
nN_t = number of nitrogen atoms in the target molecule
s_1 = slope of the caffeine calibration curve

5.4.2.7 Multilinear Regression Analysis for the Derivation of CLND Response Factors

For the derivation of specific CLND response factors of nitrogen-containing chemical groups, Eq. (2) was modified as follows:

$$c_{\text{corr,t}} = \frac{S_t}{n N_{\text{eff,t}} \cdot s_1} \tag{3}$$

where

$c_{\text{corr,t}}$ = true molar concentration of the target compound
$n N_{\text{eff,t}}$ = number of nitrogen atoms in the target molecule, corrected for the respective CLND response factors.

Specifically:

$$n N_{\text{eff,t}} = \sum_i RRF_i \cdot X_{i,t} \tag{4}$$

where RRF_i is the relative response factor for the *i*-th chemical group; $X_{i,t}$ is the number of nitrogen atoms of *i*-th sub-type in the target molecule.

Assuming that the NMR concentration is a much more accurate representation of the true molar concentration than the CLND-derived one, c_{NMR} is used as the best estimate of c_{corr}. The ratio of Eqs. (2) and (3) then provides a useful relationship for the determination of the RRF_i:

$$\frac{c_{\text{CLND,t}}}{c_{\text{corr,t}}} = \frac{n N_{\text{eff,t}}}{n N_t} \dots \Rightarrow n N_t \frac{c_{\text{CLND,t}}}{c_{\text{NMR,t}}} = n N_{\text{eff,t}} \tag{5}$$

$$n N_t \frac{c_{\text{CLND,t}}}{c_{\text{NMR,t}}} = Q_t = \sum_i RRF_i \cdot X_{i,t} \tag{6}$$

In practice, the RRF_i are unknown coefficients that can be estimated from multilinear regression analysis with an adequate set of experimental data. In the investigated set of 545 new chemical compounds, ten different N-containing chemical groups were identified, and for each molecule the number of nitrogen atoms belonging to each given group was assigned by visual inspection.

Multilinear regression analysis was carried out with the Excel data analysis add-in.

5.4.3
Results and Discussion

5.4.3.1 Liquid Chromatography and UV Detection

The chromatographic conditions were chosen in order to obtain a baseline separation of all peaks. Particular care had to be taken in order to meet the needs of the different detectors. The chromatographic column was a Zorbax SB-C$_8$, which is a brush-type reversed phase. A C$_8$ phase allows good selectivity between peaks, but the compounds are not excessively retained (as with a C$_{18}$ column). New generation columns with carbamide groups proved to be unsuitable, because after prolonged usage bleeding was found to interfere with the CLND signal. The eluents used were water and methanol, both containing 0.1% formic acid. Methanol gives a high back-pressure and prevents the possibility of setting up fast-gradient methods; moreover, it has a low absorbance at 220 nm, which produces a slow,

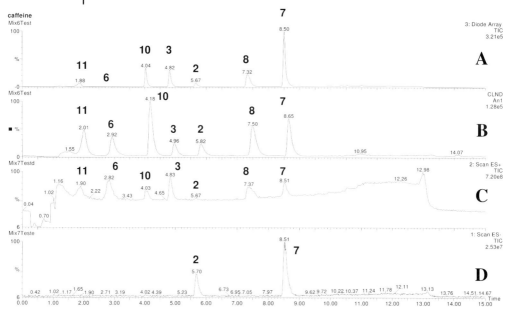

Fig. 4. Multidetector chromatogram for a test mixture of seven commercial compounds. The molecular structures of the seven components are reported in Table 1. The following traces are shown: (a) UV signal at 220 nm; (b) CLND signal; (c) ESI (+) MS signal (TIC); (d) ESI (–) MS signal (TIC).

but still acceptable increase in the baseline. Formic acid represents a good compromise for a modifier: it suppresses silanophilic activity and at the same time favors ionization of the analytes in the ESI(+) interface. Under the experimental conditions described and by injecting a test mixture containing seven out of 12 standard compounds (Table 1), an average peak capacity of around 80 was obtained (Fig. 4), which is sufficient for our purposes.

The UV detector wavelength was set at 220 nm, allowing the detection of most new chemical entities and impurities. The area percent of the target peak was taken as the overall purity of the synthesized compound.

5.4.3.2 Mass Spectrometric Method Development

Mass spectrometric detector selectivity provides unequivocal identification of target compounds. By targeting the MW and by using specific software, fully automated peak recognition is possible. Compounds need to be efficiently ionized for their detection. The single-quadrupole Waters ZQ is equipped with the two commonly available ionization sources: electrospray ionization (ESI) and atmospheric pressure chemical ionization (APCI). As most of the compounds to be examined were basic, electrospray with positive ionization was our method of choice. The selection of ESI was supported by previous experience gained over the last three years in our laboratory: about 96% of the identified compounds, corresponding to more than 9,000 different molecules, were successfully analyzed with the ESI interface.

5.4.3.3 CLND Set-Up

The CLND is the most critical component in the platform. With our current experience, the most crucial factor is the stability at the sprayer exit. This may well be the source of the large deviations (up to 80%) observed for some of the examined compounds. The unpredictable nature of such operational malfunction strongly suggests the need for replicate determinations as the minimal requirement for reliable quantitative estimates. Other factors (injection error of the autosampler, dilution error in the preparation of solutions or standards) may account for only a limited fraction of the total measured variance and could be easily corrected for by the introduction of an appropriate internal standard.

5.4.3.4 Validation with Commercial Standards

During the optimization of mass spectrometry conditions both positive and negative ionization, ESI (+) and ESI (–), were investigated. In Fig. 4, the TIC obtained in positive and negative mode are compared for a test mixture. Even though compound no. 2 (4-nitrophenol) shows a weak ionization in ESI (+), it is still recognizable, whereas compounds 11, 3, 6, 8, and 10 are visible only in the positive mode. As this example illustrates, the positive ionization mode is preferable, and it was chosen as default.

To validate the CLN detector, the same set of 12 commercially available standards was used at different concentrations and the results were compared in order to check precision, reproducibility, and the limit of detection. According to Eq. (1), the signal should depend only on the nitrogen content of the molecule. The compounds are divided into four groups, the members of each group containing

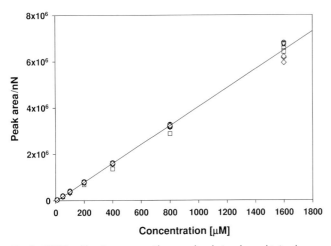

Fig. 5. CLND calibration curve with normalized signal, as obtained for 12 commercial standards with different numbers of nitrogen atoms. The following symbols are used to distinguish compounds with different numbers of nitrogen atoms: (1) nitrogen = circles, (2) nitrogens = triangles, (3) nitrogens = squares, (4) nitrogens = diamonds.

the same number of nitrogen atoms: the calibration curves overlap for compounds in the same group. In Fig. 5, all of the calibration curves have been superimposed. The peak area has been normalized by dividing by the number of nitrogens in each molecule. Furthermore, the retention times are included in Table 1 in order to ascertain whether the different eluent compositions could have an effect on the oxidation of the nitrogen or the signal of the detector.

Reproducibility was determined by repeated injections of caffeine (ninefold) at different concentration levels (Fig. 6). With the exception of data obtained at 10 μM, where a poor signal-to-noise ratio prevents accurate integration, over the whole range the RSD is lower than 4%. This reproducibility value is actually quite similar to that measured for the UV signal (1–3% in the same concentration range). An estimate of the generic method precision is obtained by comparing the normalized signals (CLND signal/number of nitrogens) of the 12 compounds (Fig. 7). In the

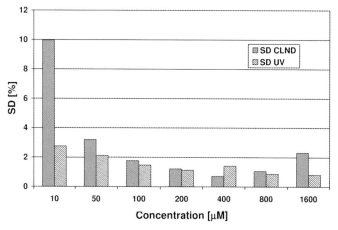

Fig. 6. Reproducibility of the CLND signal at different injected caffeine concentrations.

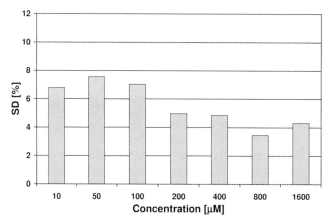

Fig. 7. Precision of the CLND calibration curve using the 12 commercial standards of Table 1.

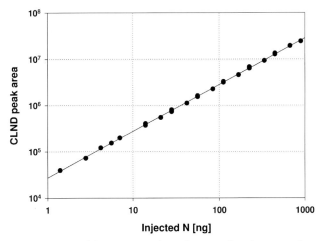

Fig. 8. Linearity of the CLND signal as a function of total amount of injected nitrogen. Data are averages for each subgroup of commercial standards with a given number of nitrogen atoms.

lower concentration range ($< 200\ \mu$M), the RSD increases up to about 7%. This is a consequence of limited integration accuracy for CLND peaks in this range. In the higher concentration range (from 200 to 1600 μM), precision improves to 5% or better.

The final investigation concerned linearity. In Fig. 8, the CLND signal is plotted against the absolute amount of injected nitrogen for all 12 compounds, after averaging so as to give compounds with the same number of nitrogens. The linear range extends over three decades with very good correlation ($R^2 = 0.9995$).

5.4.3.5 Validation with Proprietary Compounds

A set of 2006 new synthetic molecules produced by medicinal chemists was analyzed for purity and identity validation purposes. A lack of a UV signal at the chosen wavelength due to the absence of a suitable chromophore prevented collection of purity data for just five of these compounds. The molecular ion was detected for 97.5% of all samples. To characterize the remaining 2.5%, either APCI or EI was necessary. The favorable response rate of ESI for the identity confirmation is consistent with our former observations and with literature reports. Failures are generally due to a lack of protonation sites in the molecules; the occurrence of such molecules is limited to a few percent in typical medicinal chemistry production. This aspect, together with the limitation of measuring purity by HPLC-UV, have been exhaustively reviewed in the literature and will not be considered further here.

A different set of 543 NCEs has been quantitatively analyzed by both ^{1}H NMR and CLND detection. The samples were diluted with DMSO to a nominal concentration of 500 μM, and 10 μL aliquots of these solutions were injected into the high-throughput platform. The actual concentrations, as determined by quantitative NMR (our reference method), were all above the 200 μM limit, which ensured reproducible and precise results (see above). When the results obtained with the CLND are compared with the ^{1}H NMR data, the correlation appears unsatisfactory (Fig. 9). A possible origin of this discrepancy is that different chemical groups could have different response factors as a result of different NO yields during combustion. A few observations in the literature indicate a reduced CLND response of compounds with N–N bonds [1, 2]. The liberation of molecular nitrogen rather than nitrogen oxide during combustion of these compounds is an obvious explanation for this anomalous behavior. This point has been addressed in our comparative analysis, as detailed in Section 5.4.2. The results are reported in Table 2, in which the structures of the nitrogen-containing chemical groups and the experimentally determined CLND response factors are listed. These data confirm that compounds with an N–N bond have anomalously low responses. There is, however, an important distinction: when the two atoms in the N–N bond are fully engaged in N–C bonds, the response factor (0.76) is considerably higher than that for molecules with N–NH groups (0.34). Acylated N–N bonds, whether or not hydrogen atoms are attached to the two nitrogens, have an intermediate response factor (0.5). The only azide in the set can be placed alongside the molecules with N–NH groups. Compounds containing an NO bond (nitro groups, oximes, hydroxylamines, etc.) also have a low response factor (0.82), but the large observed standard error suggests some heterogeneity in this group. Surprisingly, the three nitrogens involved in guanidinium groups have a measurably higher response (1.13) with respect to the standard (caffeine), and a modest deviation from unity is also observed for amines (1.08). No other relevant effects could be detected in the investigated set of compounds, even though up to 15 different classes of chemical groups were identified and used for a more detailed analysis. The unitary response for such diverse chemical groups as amides, cyano groups, heterocycles, and thioamides without a dependence on the nature or number of substituents is remarkable and confirms the postulated role of CLND as a universal detector for nitrogen-containing compounds [5, 8].

With the factors in Table 2, the corrected concentration was calculated using Eq. (3) for all 543 different compounds. These values are plotted against c_{NMR} in Fig. 9 (lower panel) and an improvement is observed with respect to the uncorrected values (upper panel). As expected, the regression line through all data has a slope of unity; nevertheless, the correlation is still poor as a result of a large variance. The distribution of concentration errors was thus investigated with the help of the quantity $\Delta c/c$, defined as follows:

$$\frac{\Delta c}{c} = \frac{c_{corr} - c_{NMR}}{c_{NMR}} \cdot 100 \tag{7}$$

This quantity was computed for the set of 543 investigated compounds, and the corresponding standard deviation (14.6%) was derived. It appears that a group of positive outliers significantly contributes to the total variance. These may derive from instrument malfunction or from any other failure in the process concerning specific analyses, a potential issue since all data are results of single determinations.

Table 2. CLND relative response factors for ten different chemical groups.

Chemical group	Examples	Total N-atoms in set	Coefficient	Standard error
Amines	R1—N with R3, R2	324	1.08	0.03
Amides	R1—C(=O)—N with R3, R2	664	1.00	0.02
5- and 6-Membered heterocycles	(pyrrole, pyridine structures)	369	1.02	0.03
Guanidinium group	R1—NH—C(=NH)—NH$_2$	146	1.13	0.03
RNHNR^{2}R^3 (cyclic and linear)	HN (pyrazole structure)	305	0.34	0.03
R^{1}R^{2}NNR^{3}R^4 (cyclic and linear)	R1—N (pyrazole structure)	58	0.76	0.05
Hydrazides (cyclic and linear)	R1—C(=O)—N(R2)—N(R3)—R4	96	0.50	0.05
Oximes, hydroxylamines, nitro groups	R1,H—N—OH ; R1—N$^+$(=O)O$^-$	32	0.82	0.11
Cyano	R1—≡N	25	1.01	0.11
Thioamides	R1—C(=S)—N with R3, R2	43	1.01	0.09

Therefore, we investigated the distribution of $\Delta c/c$ with a tool that underestimates these large outliers by imposing a symmetric distribution: populations were derived for the $\Delta c/c$ quantity at 5% intervals with the help of the Excel histogram function and then fitted with a symmetric Gaussian distribution by means of a nonlinear data-fitting package (SigmaPlot, SPSS Inc., Chicago, IL). These populations are shown in the bar chart of Fig. 10, together with the computed, best fitting Gaussian curve. The distribution is centered very close to zero and corresponds to $\sigma = 10.0\%$, a significantly lower standard deviation with respect to that above. This result is close to that obtained for the commercial standards and provides hope that with appropriate correction for chemical group effect and proper control of instrument and process performance, the concentration of generic molecules may be obtained with an accuracy of about 10% from CLND measurements and a single standard.

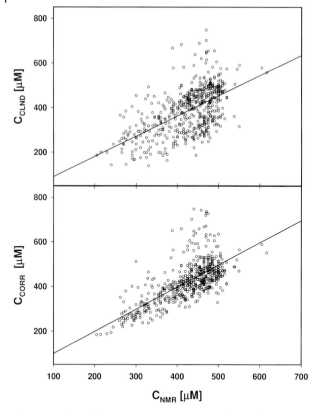

Fig. 9. Correlation between CLND concentration data and NMR-derived concentration before (upper panel) and after (lower panel) correction for chemical group response factors. Regression lines through the data are also shown.

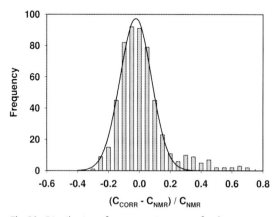

Fig. 10. Distribution of concentration errors for the 543 investigated new synthetic molecules after correction for relative CLND response factors. With these data, grouped at 5% intervals, the best fitting Gaussian distribution (solid line) corresponds to a standard deviation $\sigma = 10\%$ and is centered around 0.

5.4.4
Conclusions

A high-throughput analytical platform has been developed and validated. In one chromatographic run, this platform provides successful evaluation of identity, purity, and quantity for any new chemical entity. By using a systematic approach, all components were evaluated and the corresponding parameters tuned. The present version of the platform has a maximum throughput of 100 compounds per day, which seems sufficient for our current needs. Further decrease in analysis time and therefore increased productivity should be possible by reducing the gradient time.

Comments and recommendations about each component of this analytical platform are collected in Table 3. For the design and development of a similar apparatus, the reader is referred to these suggestions. In particular, the construction of dedicated equipment is highly recommended, since the combined information from the three detectors greatly facilitates data analysis, thus easing one of the bottlenecks of high-throughput generic analysis of small molecules. Moreover, contamination of any component by acetonitrile (or other nitrogen-containing solvents or buffers) should be avoided because of its deleterious consequences on the CLND.

Table 3. Recommendations on high-throughput platform components.

Component	Item	Relevance	Comments
Chromato-graphy	Column	Medium	Do not use stationary phases with amide groups
	Eluent	High	Use methanol/water mixtures
	Modifier	Medium	Formic acid allows positive and negative ionization
HPLC	Autosampler	High	Must be able to accommodate microtiter plates
	UV detector	Low	Normally set at 220 nm
	Pumps/flow	Low	
Mass Spectro-meter	Probe	Low	ESI probe sufficient
	Analyzer	Low	Single quadrupole enough
	Flow	Medium	Flow should not exceed the maximum recommended value
CLND	Operational settings	High	Gas flow, nebulization, and vacuum strongly influence the response
	Flow	High	Do not exceed 0.1 mL min^{-1}: incomplete combustion
Data System	Hardware	High	Must control all detectors
	Software	High	Must provide easy accessibility to data for quick transformation in Excel environment

The evaluation of CLND relative response factors for various chemical groups is a significant achievement of our preliminary tests with this analytical platform. Some 543 well-characterized synthetic molecules were analyzed and the quantitative results were compared with those of quantitative ^{1}H NMR, our reference method. After specific response factors for each chemical group were identified and applied, the average method accuracy in the determination of concentration for a generic molecule was found to be ±15%. Statistical analysis indicates that this may improve to about 10% when outliers (probably due to some kind of failure in the process) are excluded. This strongly enforces the use of multiple determinations for accurate data. The results of our findings compare favorably with literature data on compounds containing an N–N bond [1, 3], but our observation of distinctive classes of responders (the fully carbon-substituted and those with N–H bonds) is new and probably originates from the large set of investigated molecules. Several minor effects have been detected for other nitrogen-containing groups. In the light of the large observed variance, fine-tuning of these response factors will have to wait for a more robust version of the analytical platform.

References

1 M. A. Nussbaum, S. W. Baertschi, P. J. Jansen: Determination of relative UV response factors for HPLC by use of a chemiluminescent nitrogen-specific detector, *J. Pharm. Biomed. Analysis* 27 (2002), 983–993.

2 C. A. Lucy, C. R. Harrison: Chemiluminescence nitrogen detection in ion chromatography for the determination of nitrogen-containing anions, *Can. J. Chromatogr. A* 920 (2001), 135–141.

3 K. Petritis, C. Elfakir, M. Dreux: A comparative study of commercial liquid chromatographic detectors for the analysis of underivatized amino acids, *J. Chromatogr. A* 961 (2002), 9–21.

4 E. W. Taylor, W. Jia, M. Bush, G. D. Dollinger: Accelerating the drug optimization process: identification, structure elucidation, and quantification of in vivo metabolites using stable isotopes with LC/MSn and the chemiluminescent nitrogen detector, *Anal. Chem.* 74 (2002), 3232–3238.

5 W. L. Fitch, A. K. Szardenings, E. M. Fujinari: Chemiluminescent nitrogen detection for HPLC: an important new tool in organic analytical chemistry, *Tetrahedron Lett.* 38 (1997), 1689–1692.

6 X. Cheng, J. Hochlowski: Current application of mass spectrometry to combinatorial chemistry, *Anal. Chem.* 74 (2002), 2679–2690.

7 D. A. Yurek, D. L. Branch, M.-S. Kuo: Development of a system to evaluate compound identity, purity, and concentration in a single experiment and its application in quality assessment of combinatorial libraries and screening hits, *J. Comb. Chem.* 4 (2002), 138–148.

8 W. Li, M. Piznik, K. Bowman, J. Babiak: "Universal" HPLC detector for combinatorial library quantitation in drug discovery? *Abstr. Pap. – Am. Chem. Soc.* (2001) 221st ANYL-008.

9 K. Lewis, D. Phelps, A. Sefler: Automated high-throughput quantification of combinatorial arrays, *Am. Pharm. Rev.* 3 (2000), 63–68.

10 N. Shah, M. Gao, K. Tsutsui, A. Lu, J. Davis, R. Scheuerman, W. L. Fitch, R. L. Wilgus: A novel approach to high-throughput quality control of parallel synthesis libraries, *J. Comb. Chem.* 2 (2000), 453–460.

11 E. W. Taylor, M. G. Qian,
G. D. Dollinger: Simultaneous on-line
characterization of small organic
molecules derived from combinatorial
libraries for identity, quantity, and purity
by reversed-phase HPLC with chemiluminescent nitrogen, UV, and mass
spectrometric detection, *Anal. Chem.* 70
(1998), 3339–3347.

12 B. Yan, L. Fang, M. Irving, S. Zhang
A. M. Boldi, F. Woolard, C. R. Johnson,
T. Kshirsagar, G. M. Figliozzi,

C. A. Krueger, N. Collins: Quality control
in combinatorial chemistry: determination of the quantity, purity and quantitative purity of compounds in combinatorial libraries, *J. Comb. Chem.* 5 (2003),
547–559.

13 V. Pinciroli, R. Biancardi, N. Colombo,
M. Colombo, V. Rizzo: Characterization
of small combinatorial chemistry libraries by ^{1}H NMR. Quantification with a
convenient and novel internal standard,
J. Comb. Chem. 3 (2001), 434–440.

Appendix

Cluster Analysis

Cluster analysis provides a method for discrimination of different classes without *a priori* information about possible class memberships. A group of objects is classified into smaller subgroups through the different realization of the corresponding characteristic features (variables). Objects in the same class should be as "similar" as possible and significantly different from objects in other classes. Similarity can be most easily defined in terms of the distance of objects in the variable space. In cluster analysis, most frequently the Euclidean distance d_{ij} between two objects i and j with p variables is used:

$$d_{ij} = \sqrt{\sum_{k=1}^{p} (x_{ik} - x_{jk})^2} \qquad (1)$$

Based on these multivariate distances or similarities, clusters of objects are generated. The distance between single objects within the clusters is minimized while the distance between clusters is maximized. The objective of the procedure is a clear representation of the objects with fewer dimensions. Cluster analysis is a multivariate explorative method which needs no *a priori* information and yields no statistical evidence about group memberships.

Here, only hierarchical methods based on a symmetrical distance matrix, which contains the distances between the pairs of objects, are utilized. According to their distances, objects are successively clustered and then merged into larger clusters. At the beginning of the procedure every object forms its own cluster, then a stepwise fusion of clusters follows (agglomerative algorithm). Hence, the clustering is determined by way of the distance measure *and* the clustering algorithm. Two algorithms, the *k*-nearest-neighbor or single-linkage algorithm and the *k*-means or centroid-linkage algorithm, were employed in this study. The *k*-nearest-neighbor procedure uses the minimum distance between adjacent objects. Correspondingly, the generated clusters are relatively large and linked in a linear fashion. This allows the easy identification of outliers. The *k*-means algorithm uses instead the mean distances of clusters, which often results in a better representation of the internal structure of the data set compared to the *k*-nearest-neighbor algorithm.

HPLC Made to Measure: A Practical Handbook for Optimization. Edited by Stavros Kromidas
Copyright © 2006 WILEY-VCH Verlag GmbH & Co. KGaA, Weinheim
ISBN: 3-527-31377-X

The fusion process is usually depicted in a dendrogram with a tree-like structure. The dendrogram displays the sequence in which clusters are merged at different distances (similarities). As the clusters are not unique in a statistical sense, the interpretation is focussed on the number of clusters and membership of objects. Due to the hierarchical structure, stable clusters are merged only at the end of the procedure, i.e. at large distances. However, depending on the method of cluster fusion, all hierarchical methods introduce a bias compared to the original n-dimensional structure. An example is the tendency of the k-nearest-neighbor algorithm to generate chains of objects and clusters. Therefore, more than one fusion algorithm is normally employed in an explorative study. Clusters that can be identified with all methods can be safely assumed to be a real feature of the data.

Principal Component Analysis

The objective of a principal component analysis (PCA) is to transform a number of correlated variables into a smaller set of new, uncorrelated variables (factors or latent variables). The first few factors should then explain most of the relevant variation in the data set. To allow this reduction in dimensionality, the variables are characterized by a partial correlation. The new variables can then be generated through a linear combination of the original variables, i.e. the original matrix $\mathbf{X}$ is then the product of a score matrix $\mathbf{P}$ and the transpose ($^\mathrm{T}$) of the loading matrix $\mathbf{A}$:

$$\mathbf{X} = \mathbf{PA}^\mathrm{T} \tag{2}$$

Note that Eq. (2) does not result in a reduction of the original dimensionality. A lower dimensionality is achieved through discarding factors which describe only a small part of the total variance and therefore contain less meaningful information. Typically, several indicator functions are employed to estimate the number of useful factors. Geometrically, PCA is a rotation of the original m-dimensional coordinate system into a new n-dimensional system ($m \gg n$) of the principal components. The first new axis is generated in a least-squares sense so that the first principal component points in the direction of the maximum variance of the data. The other axes are mutually orthogonal and point in the directions of the remaining variance. Due to the fact that only the relevant factors are utilized, the final coordinate system is a sub-space of the original coordinate system.

As the original data matrix can be represented in the variable and object space, the results of the PCA can be displayed in a corresponding score and loading plot. The score plot is the realization of the principal components and represents the position of the objects in the new coordinate system. In this study, score plots were employed to identify similar objects (i.e., columns). Similarly, the loading plots depict the influence of the original variables on a principal component. Usually, it is assumed that principal components which describe a large part of the variance also represent important physical or chemical factors (e.g., in this case, the special characteristics of the different columns). Smaller factors can be interpreted as noise.

For a simultaneous visualization of variables and objects within the new coordinate system, biplots can be used. In these, objects and variables are scaled with the variance described by the corresponding principal component. Variables around the origin have only a small influence on the plotted principal component, while interactions between objects and variables are indicated through their distance and angle.

References

D. L. Massart, L. Kaufman, *The Interpretation of Analytical Chemical Data by the Use of Cluster Analysis*, John Wiley & Sons, New York, 1983.

D. L. Massart, B. G. M. Vandeginste, L. M. C. Buydens, S. De Jong, P. J. Lewi, J. Smeyers-Verbeke, *Handbook of Chemometrics and Qualimetrics: Parts A and B*, Elsevier, Amsterdam, 1997.

K. Danzer, H. Hobert, C. Fischbacher, K. U. Jagemann, *Chemometrik*, Springer, Berlin, 2001.

Additional Remarks

The data analysis was based on the differences between the two solvents used in this study, namely methanol, MeOH, and acetonitrile, MeCN. Additionally, the pH was varied and a global comparison between MeOH and MeCN was made. The explorative investigations were based on two hierarchical cluster procedures and a PCA. The data analysis was performed on various subsets of the data (k values, α values, k and α values together). For the PCA, the number of factors and the described total variance, or the variance per principal component, are given. Results were visualized by means of dendrograms or loading and score plots, including biplots. As many data sets were incomplete, a zero-filling procedure was employed and data sets were only considered if the zero-filling was below 20%. Preliminary tests with and without zero-filling demonstrated that no serious bias was introduced in this way. However, a small bias through zero-filling can introduce a partial correlation in the PCA.

Generally, columns were denoted as objects, while k values and α values represented the corresponding variables. For investigation of the k values and α values, the data were mean-centered, while for a combined analysis the data were autoscaled.

Column name	Abbreviation	Column name	Abbreviation
Aqua	AQA	Prontosil ACE	PAC
Bondapak	BonP	Prontosil AQ	PAQ
Chromolith Performance	CHP	Prontosil C18	PC8
Discovery C16	DC6	Purospher	PUP
Discovery C18	DC8	Purospher Star	PUS
Fluofix IEW	FEW	Repro-Sil AQ	RAQ
Fluofix INW	FNW	Repro-Sil ODS	ROD
Gromsil AB	GAB	Resolve	RES
Gromsil CP	GCP	SilicaRod	SIR
Hypercarb	HCB	SMT	SMT
Hypersil Advance	HAD	Spherisorb ODS 1	SO1
Hypersil BDS	HBD	Spherisorb ODS 2	SO2
Hypersil Elite	HEL	Supelcosil ABZ plus	SUA
Hypersil ODS	HOD	Superspher	SUP
Inertsil ODS 2	IO2	Superspher Select B	SUS
Inertsil ODS 3	IO3	Symmetry	SYM
Jupiter	JUP	Symmetry Shield	SYS
Kromasil	KRO	Synergi Max	SNM
LiChrosorb	LCB	Synergi Polar	SNP
LiChrospher	LCP	TSK	TSK
LiChrospher Select B	LCS	Ultrasep ES	UES
Luna	LUN	Ultrasep	ULT
MP-Gel	MPG	VYDAC	VYD
Nova-Pak	NOP	XTerra	XTA
Nucleosil 100	N10	XTerra MS	XTM
Nucleosil 50	N50	YMC AQ	YMA
Nucleosil AB	NAB	YMC C18	YM8
Nucleosil HD	NHD	Zorbax Bonus	ZOB
Nucleosil Nautilus	NNA	Zorbax Extend	ZOE
Nucleosil Protect 1	NP1	Zorbax ODS	ZOO
Platinum C18	PC8	Zorbax SB C18	Z18
Platinum EPS	PEP	Zorbax SB C8	ZC8
Prodigy	PRD		

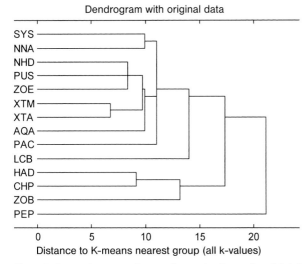

Chem. 1 Comparison according to all retention factors in acidic/alkaline methanol/acetonitrile phosphate buffers, S2.

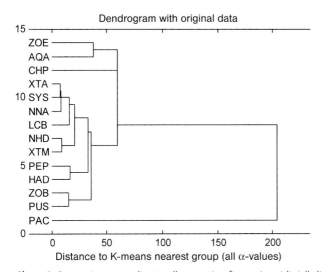

Chem. 2 Comparison according to all separation factors in acidic/alkaline methanol/acetonitrile phosphate buffers, S2.

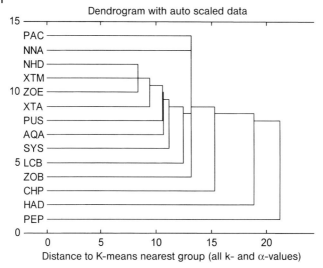

Chem. 3 Comparison according to all retention and separation factors in acidic/alkaline methanol/acetonitrile phosphate buffers, S2 (dendrogram).

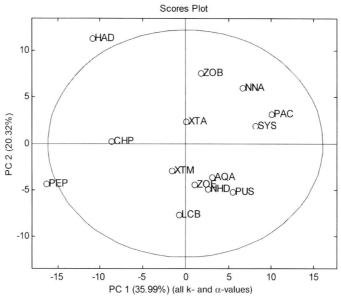

Chem. 4 Comparison according to all retention and separation factors in acidic/alkaline methanol/acetonitrile phosphate buffers, S2 (scores plot).

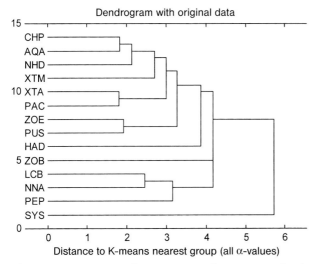

Chem. 5 Comparison according to separation factors in neutral and alkaline methanol/acetonitrile phosphate buffers, S2.

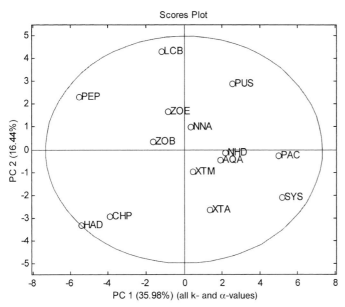

Chem. 6 Comparison according to retention and separation factors in neutral and alkaline methanol/acetonitrile phosphate buffers, S2 (scores plot).

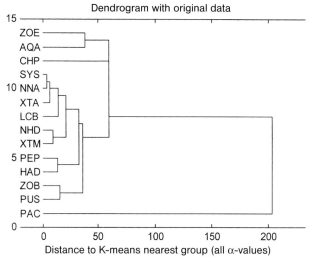

Chem. 7 Comparison according to separation factors in neutral and acidic methanol/acetonitrile phosphate buffers, S2.

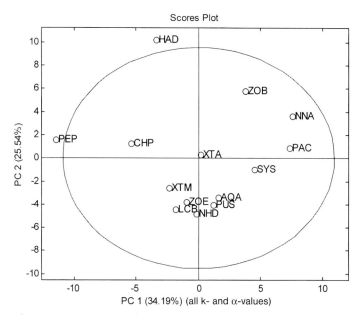

Chem. 8 Comparison according to retention and separation factors in neutral and acidic methanol/acetonitrile phosphate buffers, S2 (scores plot).

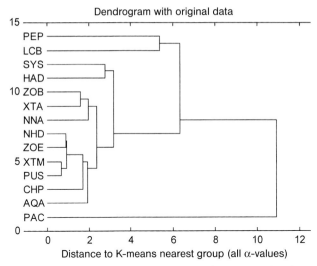

Chem. 9 Comparison according to separation factors in methanol/ acetonitrile eluents, S2 (dendrogram).

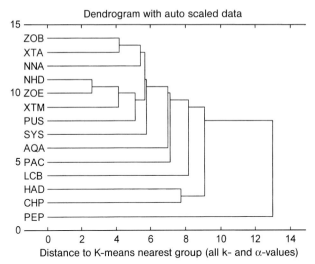

Chem. 10 Comparison according to retention and separation factors in methanol/acetonitrile eluents, S2 (dendrogram).

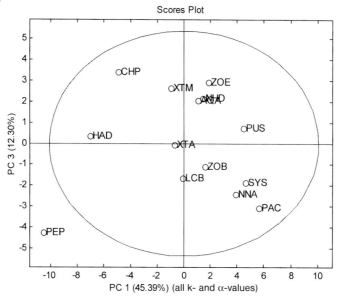

Chem. 11 Comparison according to retention and separation factors in methanol/acetonitrile eluents, S2 (scores plot).

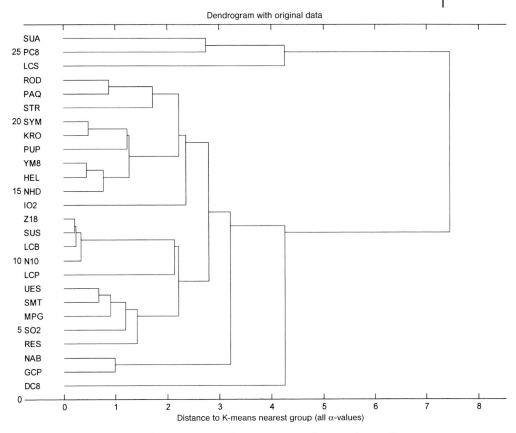

Chem. 12 Comparison according to separation factors in acetonitrile/water eluents, S1.

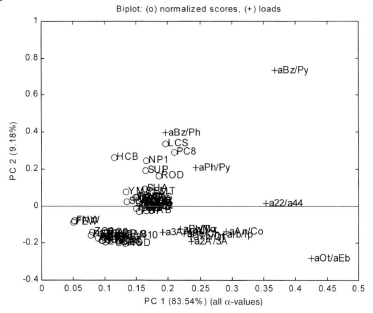

Chem. 13 Biplot of the separation factors and the analytes, S1.

Dendrogram with original data

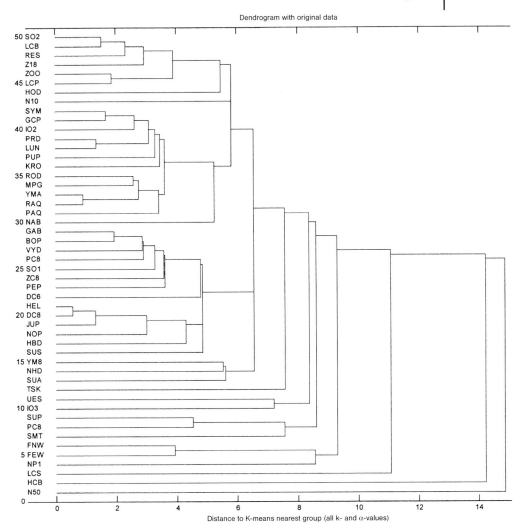

Distance to K-means nearest group (all k- and α-values)

Chem. 14 Comparison according to retention and separation factors
in methanol eluents, S1 (dendrogram).

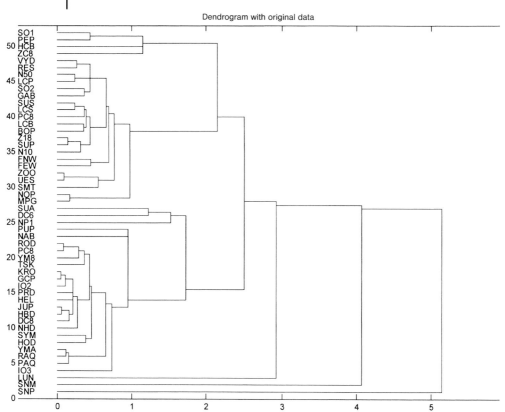

Chem. 15 Comparison according to separation factors in alkaline
methanol/phosphate buffer, S1.

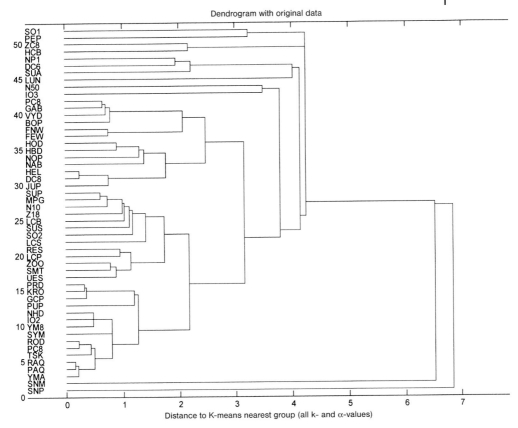

Chem. 16 Comparison according to retention and separation factors in alkaline methanol/phosphate buffer, S1 (dendrogram).

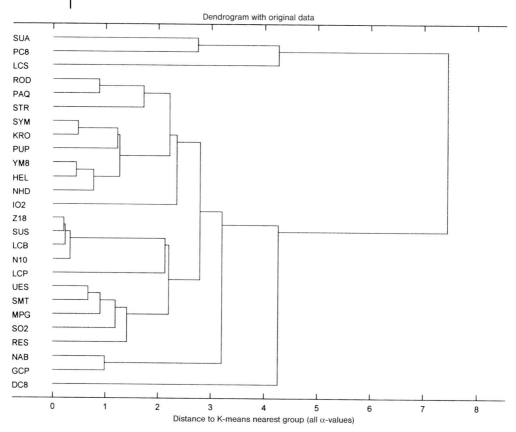

Chem. 17 Comparison according to separation factors in acetonitrile/water, S1.

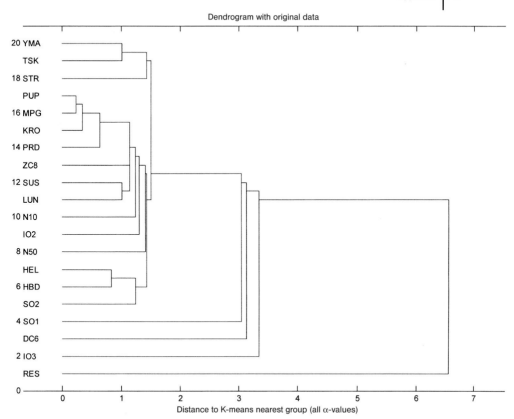

Dendrogram with original data

Distance to K-means nearest group (all α-values)

Chem. 18 Comparison according to separation factors in acidic
acetonitrile/phosphate buffer, S1.

Dendrogram with original data

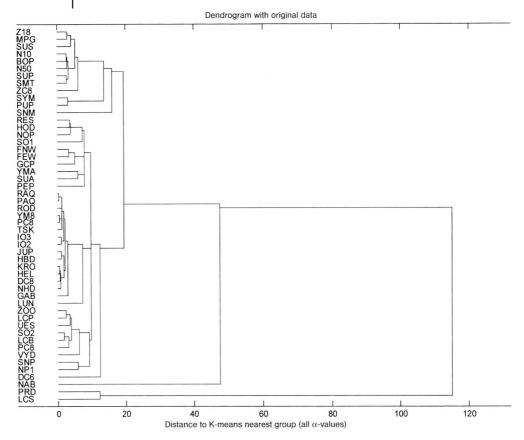

Distance to K-means nearest group (all α-values)

Chem. 19 Comparison according to separation factors in acidic methanol/phosphate buffer, S1.

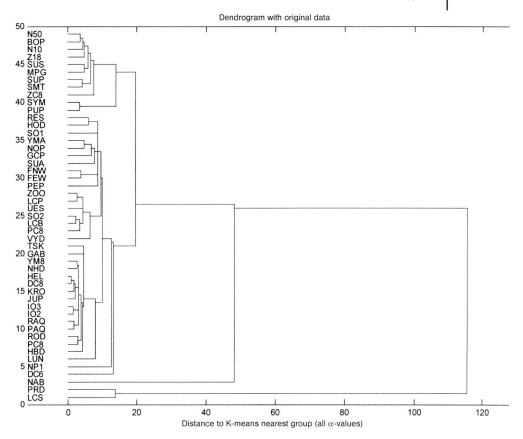

Chem. 20 Comparison according to separation factors in all methanol-containing eluents.

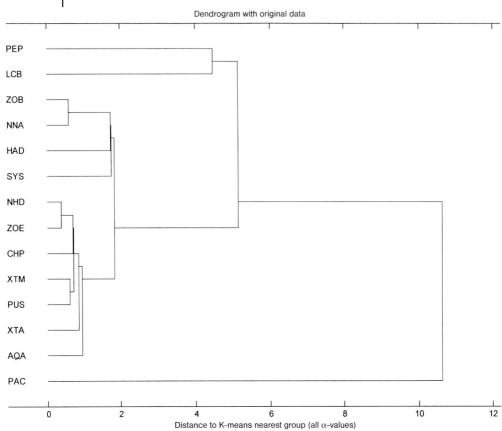

Chem. 21 Comparison according to separation factors in methanol/water, S2.

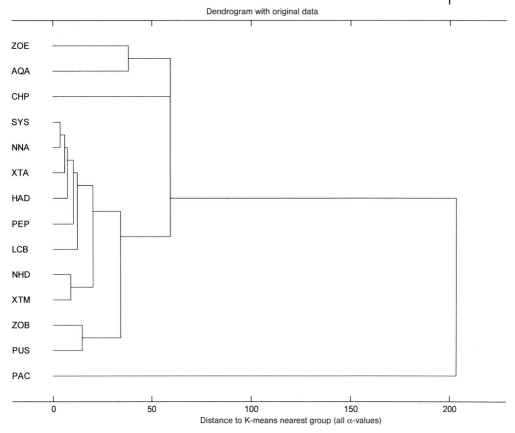

Chem. 22 Comparison according to separation factors in acidic
methanol/phosphate buffer, S2.

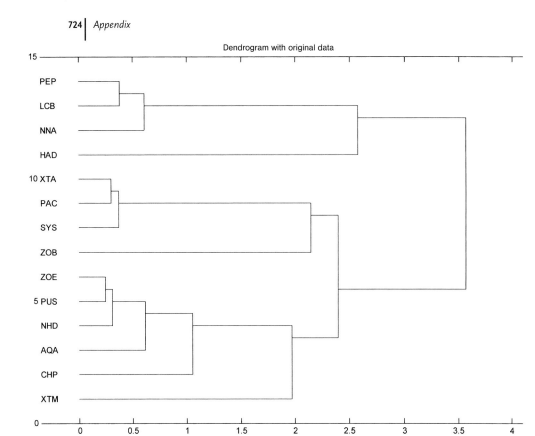

Chem. 23 Comparison according to separation factors in alkaline methanol/phosphate buffer, S2.

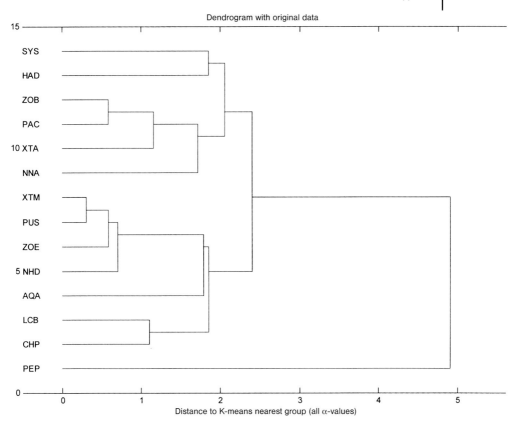

Chem. 24 Comparison according to separation factors in acetonitrile/water.

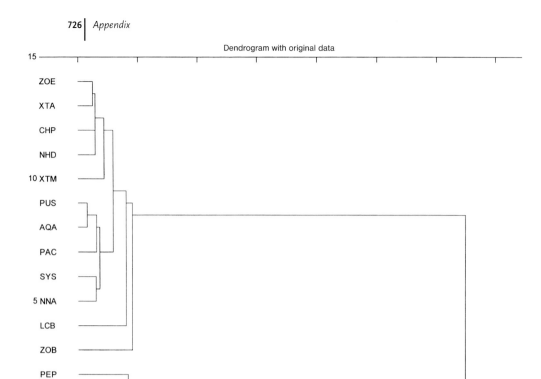

Chem. 25 Comparison according to separation factors in acidic acetonitrile/phosphate buffer.

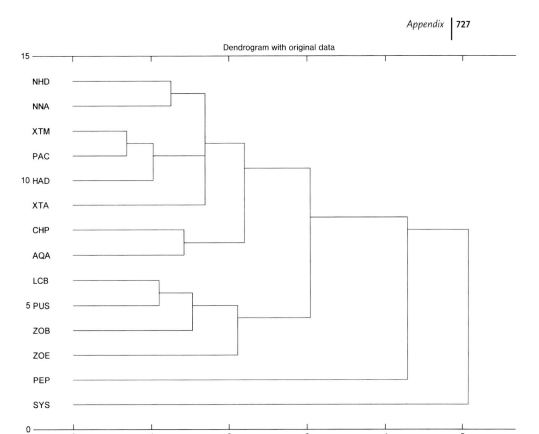

Chem. 26 Comparison according to separation factors in alkaline acetonitrile/phosphate buffer.

Subject Index

HPLC Made to Measure: A Practical Handbook for Optimization. Edited by Stavros Kromidas
Copyright © 2006 WILEY-VCH Verlag GmbH & Co. KGaA, Weinheim
ISBN: 3-527-31377-X